LONDON MATHEMATICAL SOCIETY LECTURE NOTE

Managing Editor: Professor N.J. Hitchin, Mathematical Inst
University of Oxford, 24–29 St Giles, Oxford OX1 3LB, Un

The titles below are available from booksellers, or from Ca

London Mathematical Society Lecture Note Series. 294

Introduction to Operator Space Theory

Gilles Pisier
Texas A&M University & University of Paris 6

CAMBRIDGE
UNIVERSITY PRESS

PUBLISHED BY THE PRESS SYNDICATE OF THE UNIVERSITY OF CAMBRIDGE
The Pitt Building, Trumpington Street, Cambridge, United Kingdom

CAMBRIDGE UNIVERSITY PRESS
The Edinburgh Building, Cambridge, CB2 2RU, UK
40 West 20th Street, New York, NY 10011-4211, USA
477 Williamstown Road, Port Melbourne, VIC 3207, Australia
Ruiz de Alarcón 13, 28014 Madrid, Spain
Dock House, The Waterfront, Cape Town 8001, South Africa

http://www.cambridge.org

First published 2003

Printed in the United Kingdom at the University Press, Cambridge

Typeface CMR 10/12 pt.   *System* LATEX 2$_\varepsilon$ [TB]

A catalog record for this book is available from the British Library.

Library of Congress Cataloging in Publication Data

Pisier, Gilles, 1950–
Introduction to operator space theory / Gilles Pisier.
    p.   cm. – (London Mathematical Society lecture note series ; 294)
Includes bibliographical references and index.
ISBN 0-521-81165-1 (pbk.)
1. Operator spaces.   I. Title.   II. Series.
QA322.2 .P545   2003
515′.732–dc21   2002031358

ISBN 0 521 81165 1 paperback

CONTENTS

Contents

Chapter 0. Introduction

The theory of operator spaces is very recent. It was developed after Ruan's thesis (1988) by Effros and Ruan and Blecher and Paulsen. It can be described as a noncommutative Banach space theory. An *operator space* is simply a Banach space given together with an isometric linear embedding into the space $B(H)$ of all bounded operators on a Hilbert space H. In this new category, the objects remain Banach spaces but the morphisms become the completely bounded maps (instead of the bounded linear ones). The latter appeared in the early 1980s following Stinespring's pioneering work (1955) and Arveson's fundamental results (1969) on completely positive maps. We study completely bounded (in short c.b.) maps in Chapter 1. This notion became important in the early 1980s through the independent work of Wittstock [Wit1-2], Haagerup [H4], and Paulsen [Pa2]. These authors independently discovered, within a short time interval, the fundamental factorization and extension property of c.b. maps (see Theorem 1.6).

For the reader who might wonder why c.b. maps are the "right" morphisms for the category of operator spaces, here are two arguments that come to mind: Consider $E_1 \subset B(H_1)$ and $E_2 \subset B(H_2)$ and let $\pi\colon B(H_1) \to B(H_2)$ be a C^*-morphism (i.e. a *-homomorphism) such that $\pi(E_1) \subset E_2$. Then, quite convincingly, $u = \pi_{|E_1}\colon E_1 \to E_2$ should be an "admissible" morphism in the category of operator spaces. Let us call these morphisms of the "first kind." On the other hand, if a linear map $u\colon E_1 \to E_2$ is of the form $u(x) = VxW$ with $V \in B(H_1, H_2)$ and $W \in B(H_2, H_1)$, then again such an innocent-looking map should be "admissible" and we consider it to be of the "second kind."

But precisely, the factorization theorem of c.b. maps says that any c.b. map $u\colon E_1 \to E_2$ between operator spaces can be written as a composition $E_1 \xrightarrow{u_1} E_3 \xrightarrow{u_2} E_2$ with u_1 of the first kind and u_2 of the second. This is one argument in support of c.b. maps.

Another justification goes via the minimal tensor product: If $E_1 \subset B(H_1)$ and $E_2 \subset B(H_2)$ are operator spaces, their minimal tensor product $E_1 \otimes_{\min} E_2$ is defined as the completion of their algebraic tensor product (denoted by $E_1 \otimes E_2$) with respect to the norm induced on $E_1 \otimes E_2$ by the space $B(H_1 \otimes_2 H_2)$ of all bounded operators on the Hilbertian tensor product $H_1 \otimes_2 H_2$ (this norm coincides with the minimal C^*-norm when E_1 and E_2 are C^*-algebras). Moreover, the isometric embedding

$$E_1 \otimes_{\min} E_2 \subset B(H_1 \otimes_2 H_2)$$

turns $E_1 \otimes_{\min} E_2$ into an operator space. The minimal tensor product is discussed in more detail in §2.1. It is but a natural extension of the "spatial" tensor product of C^*-algebras. In some sense, the minimal tensor product is the most natural operation that is defined using the "operator space structures" of E_1 and E_2 (and not only their norms). This brings us to the second argument supporting the assertion that c.b. maps are the "right" morphisms. Indeed, one can show that a linear map $u\colon E_1 \to E_2$ is c.b. if and only if (iff) for any operator space F the mapping $I_F \otimes u\colon F \otimes_{\min} E_1 \to F \otimes_{\min} E_2$ is bounded in the usual sense. Moreover, the c.b. norm of u could be equivalently defined as

$$\|u\|_{cb} = \sup \|I_F \otimes u\|,$$

where the supremum runs over all possible operator spaces F. (Similarly, u is a complete isometry iff $I_F \otimes u$ is an isometry for all F).

As an immediate consequence, if $v\colon F_1 \to F_2$ is another c.b. map between operator spaces, then $v \otimes u\colon F_1 \otimes_{\min} E_1 \to F_2 \otimes_{\min} E_2$ also is c.b. and

$$\|v \otimes u\|_{CB(F_1 \otimes_{\min} E_1, F_2 \otimes_{\min} E_2)} = \|v\|_{cb}\|u\|_{cb}. \tag{0.1}$$

In conclusion, the c.b. maps are precisely the largest possible class of morphisms for which the minimal tensor product satisfies the "tensorial" property (0.1). So, if one agrees that the minimal tensor product is natural, then one should recognize c.b. maps as the right morphisms.

While the notion of c.b. map (which dates back to the early 1980s, if not sooner) is fundamental to this theory, this new field really took off around 1987 with the thesis of Z. J. Ruan [Ru1], who gave an "abstract characterization" of operator spaces (described in §2.2). Roughly, his result provides a "quantized" counterpart to the norm of a Banach space. When E is an operator space, the norm has to be replaced by the sequence of norms ($\| \ \|_n$) on the spaces $M_n(E) \cong M_n \otimes_{\min} E$ of all $n \times n$ matrices with entries in E. (The usual norm corresponds to the case $n = 1$.)

In this text, we prefer to replace this sequence of norms by a single one, namely, the norm on the space $\mathcal{K} \otimes_{\min} E$ with $\mathcal{K} = K(\ell_2)$ (= compact operators on ℓ_2). Since $\mathcal{K} = \overline{\cup M_n}$, it is of course very easy to pass from one viewpoint to the other.

The main advantage of Ruan's Theorem is that it allows one to manipulate operator spaces independently of the choice of a "concrete" embedding into $B(H)$. In particular, Ruan's Theorem leads to natural definitions for the dual E^* of an operator space E (independently introduced in [ER2, BP1]) and for the quotient E_1/E_2 of an operator space E_1 by a subspace $E_2 \subset E_1$ (introduced in [Ru1]). These notions are explained in §§2.3 and 2.4. It should be emphasized that they respect the underlying Banach spaces: The dual operator space E^* is the dual Banach space equipped with an additional (specific) operator space structure (i.e., for some $\mathcal{H}$ we have an isometric embedding $E^* \subset B(\mathcal{H})$) and similarly for the quotient space. In addition, the general rules of the duality of Banach spaces (for example, the duality between subspaces and quotients) are preserved in this "new" duality.

More operations can be defined following the same basic idea: complex interpolation (see §2.7) and ultraproducts (see §2.8). We will also use some more elementary constructions, such as direct sums (§2.6), complex conjugates (§2.9), and opposites (§2.10).

Although we described Ruan's thesis as the starting point of the theory, there are many important "prenatal" contributions that shaped this new area. Among them, Christensen and Sinclair's factorization of multilinear maps [CS1] stands out (with its extension to operator spaces by Paulsen and Smith [PaS]). Going back further, there is an important paper by Effros and Haagerup [EH], who discovered that operator spaces may fail the local reflexivity principle, a very interesting phenomenon that is in striking contrast with the Banach space case (their results were inspired by Archbold and Batty's results [AB] for C^*-algebras). Closely connected to theirs, Kirchberg's work on "exact" C^*-algebras (see [Ki1, Wa2]) has also been very influential.

Given a normed space E, there are of course many different ways to embed it into $B(H)$. Two embeddings $j_1 \colon E \to B(H_1)$ and $j_2 \colon E \to B(H_2)$ are considered equivalent if the associated norms on $M_n(E)$ (or on $\mathcal{K} \otimes E$) are the same for all $n \geq 1$. By an *operator space structure* on E (compatible with the norm) what is meant usually is the data of an equivalence class of such isometric embeddings $j \colon E \to B(H)$ for this equivalence relation.

Blecher and Paulsen [BP1] observed that the set of all operator space structures admissible on a given normed space E has a minimal and a maximal element, which they denoted by $\min(E)$ and $\max(E)$. We summarize their results and various related open questions in Chapter 3.

The minimal tensor product naturally appears as the analog for operator spaces of the "injective" tensor product of Banach spaces. Thus, both Effros-Ruan [ER6–8] and Blecher and Paulsen [BP1] were led to study the operator space analog of the "projective" tensor product in Grothendieck's sense. Recall that a mapping $u \colon E \to F$ (between Banach spaces) is nuclear iff it admits a factorization of the form

$$E \xrightarrow{\alpha} c_0 \xrightarrow{\Delta} \ell_1 \xrightarrow{\beta} F,$$

where α, β, Δ are bounded mappings and Δ is diagonal with coefficients (Δ_n) in ℓ_1.

Moreover, the nuclear norm $N(u)$ is defined as $N(u) = \inf\{\|\alpha\| \sum |\Delta_n| \|\beta\|\}$, where the infimum runs over all possible factorizations. As is well known, the space $\mathcal{K} = K(\ell_2)$ of all compact operators on ℓ_2 is the noncommutative analog of c_0, while the space S_1 of all trace class operators on ℓ_2 (with the norm $\|x\|_{S_1} = \mathrm{tr}(|x|)$) is the noncommutative analog of ℓ_1. Now, if E, F are operator spaces, a mapping $u \colon E \to F$ is called "nuclear in the o.s. sense" (introduced in [ER6]) if it admits a factorization of the form

$$E \xrightarrow{\alpha} \mathcal{K} \xrightarrow{\Delta} S_1 \xrightarrow{\beta} F,$$

where α, β are c.b. maps and $\Delta \colon \mathcal{K} \to S_1$ is of the form

$$\Delta(x) = axb$$

with a, b Hilbert-Schmidt. Then the os-nuclear norm is defined as

$$N_{os}(u) = \inf\{\|\alpha\|_{cb}\|a\|_2\|b\|_2\|\beta\|_{cb}\}.$$

(Here, of course, $\|a\|_2, \|b\|_2$ denote the Hilbert-Schmidt norms.)

We describe some of the developments of these notions in Chapter 4.

In Banach space theory, Grothendieck's approximation property has played an important role. Recall that Enflo [En] gave the first counterexample in 1972 and Szankowski [Sz] proved around 1980 that the space $B(\ell_2)$ of all bounded operators on ℓ_2 fails the approximation property. Quite naturally, this notion has an operator space counterpart.

In the Banach space case, Grothendieck proved that a space E has the approximation property iff the natural morphism

$$E^* \overset{\wedge}{\otimes} E \to E^* \overset{\vee}{\otimes} E$$

from the projective to the injective tensor product is one to one. We describe in Chapter 4 Effros and Ruan's operator space version of this result.

In Chapter 5, we introduce the Haagerup tensor product $E_1 \otimes_h E_2$ of two operator spaces E_1, E_2. This notion is of paramount importance in this young theory, and we present it from a somewhat new viewpoint. We prove that if E_1, E_2 are subspaces of two unital C^*-algebras A_1, A_2, respectively, then $E_1 \otimes_h E_2$ is naturally embedded (completely isometrically) into the (C^*-algebraic) free product $A_1 * A_2$ ([CES]). We also prove the factorization of completely bounded multilinear maps due to Christensen and Sinclair [CS1] (and to Paulsen and Smith [PaS] for operator spaces) and many more important properties like the self-duality, the shuffle theorem (inspired by [ER10, EKR]), or the embedding of $E_1 \otimes_h E_2$ into the space of maps factoring through the row or column Hilbert space ([ER4]). We also include a brief study of the symmetrized Haagerup tensor product recently introduced in [OP], and we describe the "commutation" between complex interpolation and the Haagerup tensor product ([Ko,P1]).

As an application, we prove in Chapter 6 a characterization of operator algebras due to Blecher-Ruan-Sinclair ([BRS]). The question they answer can be explained as follows: Consider a unital Banach algebra A with a normalized unit and admitting also an operator space structure (i.e., we have $A \subset B(H)$ as a closed linear subspace). When can A be embedded into $B(H)$ as a closed unital subalgebra without changing the operator space structure? They prove that a necessary and sufficient condition is that the product mapping defines a completely contractive map from $A \otimes_h A$ into A. The isomorphic (as opposed to isometric) version of this result was later given by Blecher ([B4]). We include new proofs for these results based on the fact (due to Cole-Lumer-Bernard) that the class of operator algebras (i.e. closed subalgebras of $B(H)$) is stable under quotients by closed ideals. We also give an analogous characterization of operator modules, following [CES].

Curiously, the simplest of all Banach spaces, namely, the Hilbert space ℓ_2, can be realized in many different ways as an operator space. Theoretical physics provides numerous examples of the sort, several of which are described in Chapter 9. Nevertheless, there exists a particular operator space, which we denote by OH, that plays exactly the same central role for operator spaces as the space ℓ_2 among Banach spaces. This space OH is characterized by the property of being canonically completely isometric to its antidual; it also satisfies some remarkable properties with regard to complex interpolation. The space OH is the subject of Chapter 7 (mainly based on [P1]). Since this space gives a nice operator space analog of ℓ_2 or L_2, it is natural to investigate the case of ℓ_p or L_p for $p \neq 2$, as well as the case of a noncommutative vector valued L_p. We do this at the end of Chapter 7, and we return to this in several sections in Chapter 9. However, on that particular topic, we should warn the reader of a certain paradoxical bias: If we do not give to this subject the space it deserves, the sole reason is that we have written an extensive monograph [P2] entirely devoted to it, and we find it easier to refer the reader to the latter for further information.

In Chapter 8 we introduce the group C^*-algebras (full and reduced) and the universal C^*-algebra $C^*\langle E \rangle$ of an operator space E, as well as its universal operator algebra (resp. unital operator algebra) $OA(E)$ (resp. $OA_u(E)$).

Every theory, even one as young as this, displays a collection of "classical" examples that are constantly in the back of the mind of researchers in the field. Our aim in Chapter 9 is to present a preliminary list of such examples for operator spaces. Most of the classical examples of C^*-algebras possess a natural generating subset. Almost always the *linear* span of this subset gives rise to an interesting example of operator space. The discovery that this generating operator space (possibly finite-dimensional) carries a lot of information on the C^*-algebra that it generates has been one of the arguments supporting operator space theory.

In Chapter 9 we make a special effort (directed toward the uninitiated reader) to illustrate the theory with numerous concrete "classical" examples of this type, appearing in various areas of analysis, such as Hankel operators, Fock spaces, and Clifford matrices. Moreover, we describe the linear span of the free unitary generators in the "full" C^*-algebra of the free group (§9.6) as well as in the "reduced" one (§9.7). We emphasize throughout §9 the class of homogeneous Hilbertian operator spaces, and we describe the span of independent Gaussian random variables (or the Rademacher functions) in L_p, in the operator space framework (§9.8). Our treatment underlines the similarity between the latter space and its analog in Voiculescu's free probability theory (see §9.9).

Indeed, it is rather curious that for each $1 \leq p < \infty$ the linear span in L_p of a sequence of independent standard Gaussian variables is completely isomorphic to the span in noncommutative L_p of a free semi-circular (or circular) sequence in Voiculescu's sense (see Theorem 9.9.7). Thus, if we work in L_p with $1 \leq p < \infty$, the operator space structure seems to be roughly the same in the "independent" case and in the "free" one, which is rather surprising.

Our description in §9.8 of the operator space spanned in L_p by Gaussian variables (or the Rademacher functions) is merely a reinterpretation of the noncommutative Khintchine inequalities due to F. Lust-Piquard and the author (see [LuP, LPP]). These inequalities also apply to "free unitaries" (see Theorem 9.8.7) or to "free circular" variables (see Theorem 9.9.7). In view of the usefulness and importance of the classical Khintchine inequalities in commutative harmonic analysis, it is natural to believe that their noncommutative (i.e., operator space theoretic) analog will play an important role in noncommutative L_p-space theory. This is why we have devoted a significant amount of space to this topic in §§9.8 and 9.9. Moreover, in §9.10, we relate these topics to random matrices by showing that the von Neumann algebra of the free group embeds into a (von Neumann sense) ultraproduct of matrix algebras. One can do this by using either the residual finiteness of the free group (as in [Wa1]) or by using one of Voiculescu's matrix models involving independent Gaussian random matrices suitably normalized, and Paul Lévy's concentration of measure phenomenon (see [MS]). Finally, in §9.11, we discuss the possible analogs of Dvoretzky's Theorem for operator spaces (following [P9]).

In Chapter 10 we compare the various examples reviewed in Chapter 9, and we show (by rather elementary arguments) that, except for the few isomorphisms encountered in Chapter 9, these operator spaces are all distinct.

This new theory can already claim some applications to C^*-algebras, many of which are described in the second part of this book. For instance, the existence of an "exotic" C^*-algebra norm on $B(H) \otimes B(H)$ was established in [JP] (see Chapter 22). Moreover, this new ideology allows us to "transfer" into the field of operator algebras several techniques from the "local" (i.e. finite-dimensional) theory of Banach spaces (see Chapter 21). The main applications so far have been to C^*-algebra tensor products. Chapters 11 to 22 are devoted to this topic. We review in Chapters 11 and 12 the basic facts on C^*-norms and nuclear C^*-algebras. Since we are interested in *linear* spaces (rather than cones) of mappings, we strongly emphasize the "decomposable maps" between two C^*-algebras (i.e., those that can be decomposed as a linear combination of (necesssarily at most four) completely positive maps) rather than the completely positive (in short c.p.) ones themselves. Our treatment owes much to Haagerup's landmark paper [H1]. For a more traditional one emphasizing c.p. maps, see [Pa1].

Recall that, if A, B are C^*-algebras, there is a smallest and a largest C^*-norm on $A \otimes B$ and the resulting tensor products are denoted by $A \otimes_{\min} B$ and $A \otimes_{\max} B$. (This notation is coherent with the previous one for the minimal tensor product of operator spaces.) Moreover, a C^*-algebra A is called nuclear if $A \otimes_{\min} B = A \otimes_{\max} B$ for any C^*-algebra B.

In analogy with the Banach space case explored by Grothendieck, the maximal tensor product is projective but not injective, and the minimal one is injective but not projective. Therefore we are naturally led to distinguish two classes: first, the class of C^*-algebras A for which

$$\text{the ``functor'' } B \to A \otimes_{\max} B \text{ is injective}$$

(this means that $B \subset C$ implies $A \otimes_{\max} B \subset A \otimes_{\max} C$), and, second, the class of C^*-algebras A for which

$$\text{the ``functor'' } B \to A \otimes_{\min} B \text{ is projective}$$

(this means that $B = C/I$ implies $A \otimes_{\min} B = (A \otimes_{\min} C)/(A \otimes_{\min} I)$). The first class is that of nuclear C^*-algebras reviewed in Chapter 12 (see Exercise 15.2), and the second one is that of exact C^*-algebras studied in Chapter 17.

In Chapter 12 we first give a somewhat new treatment of the well-known equivalences between nuclearity and several forms of approximation properties. We also apply our approach to multilinear maps into a nuclear C^*-algebra (see Theorem 12.11) in analogy with Sinclair's and Smith's recent work [SS3] on injective von Neumann algebras.

Then, Chapter 13 is devoted to a proof of Kirchberg's Theorem, which says that there is a unique C^*-norm on the tensor product of $B(H)$ with the full C^*-algebra of a free group.

The next chapter, Chapter 14, is devoted to an unpublished result of Kirchberg showing that the decomposable maps (i.e. linear combinations of completely positive maps) are the natural morphisms to use if one replaces the minimal tensor product of C^*-algebras by the maximal one. The same proof actually gives a necessary and sufficient condition for a map defined only on a subspace of a C^*-algebra, with range another C^*-algebra, to admit a decomposable extension.

The next two chapters are closely linked together. In Chapters 15 and 16 we present respectively the "weak expectation property" (WEP) and the "local lifting property" (LLP). We start with the C^*-algebra case in connection with Kirchberg's results from Chapter 13, and then we go on to the generalizations to operator spaces. In particular, we study Ozawa's OLLP from [Oz3]. At the end of Chapter 16 we discuss at length several equivalent reformulations of Kirchberg's fundamental conjecture on the uniqueness of the C^*-norm on $C^*(\mathbb{F}_\infty) \otimes C^*(\mathbb{F}_\infty)$. For instance, it is the same as asking whether LLP implies WEP.

In Chapter 17 we concentrate on the notion of "exactness" for either operator spaces or C^*-algebras.

Assume that A embedded into $B(H)$ as a C^*-subalgebra. Then A is exact iff $A \otimes_{\min} B$ embeds isometrically into $B(H) \otimes_{\max} B$ for any B. Equivalently, this means that the norm induced on $A \otimes B$ by $B(H) \otimes_{\max} B$ coincides with the min-norm.

This is not the traditional definition of exactness, but it is equivalent to it (see Theorem 17.1). The traditional one is in terms of the exactness of the functor $B \to A \otimes_{\min} B$ in the C^*-category (see (17.1)), and for operator spaces there is also a more appealing reformulation in terms of ultraproducts: Exact operator spaces X are those for which the operation $Y \to Y \otimes_{\min} X$ essentially commutes with ultraproducts (see Theorem 17.7).

The concept of "exactness" owes a lot to Kirchberg's fundamental contributions [Ki1–3]. In particular, Kirchberg proved recently the remarkable definitive result that every separable exact C^*-algebra embeds (as a C^*-subalgebra) into a nuclear one. However, in the operator space framework, the situation is not as clear. When X, Y are operator spaces, we will say that X "locally embeds" into Y if there is a constant C such that, for any finite-dimensional subspace $E \subset X$, there is a subspace $\widetilde{E} \subset Y$ and an isomorphism $u \colon E \to \widetilde{E}$ with $\|u\|_{cb}\|u^{-1}\|_{cb} \leq C$.

We denote

$$d_{SY}(X) = \inf\{C\}, \tag{0.2}$$

that is, $d_{SY}(X)$ is the smallest constant C for which this holds.

With this terminology, an operator space X is exact iff it locally embeds into a nuclear C^*-algebra B. Actually, for such a local embedding we can always take simply $B = \mathcal{K}$ (and $\widetilde{E}$ can be a subspace of M_n with n large enough).

It is natural to introduce (see Chapter 17) the "constant of exactness" $ex(E)$ of an operator space E (and we will prove in Chapter 17 that it coincides with the just defined constant $d_{S\mathcal{K}}(E)$). This is of particular interest in the finite-dimensional case, and, while many of Kirchberg's results extend to the operator space case, many interesting questions arise concerning the asymptotic growth of these constants for specific E when the dimension of E tends to infinity. (See, e.g., Theorems 21.3 and 21.4.)

In Chapter 18 we describe the main known facts concerning "local reflexivity." While every Banach space is "locally reflexive" (cf. [LiR]), it is not so in the operator space category, and this raises all sorts of interesting questions.

existing manuscript served as notes for advanced graduate courses given at Texas A&M and Paris VI at several occasions during the period 1996–2001. I am very grateful to all those who made critical comments and pointed out numerous errors and misprints, particularly, A. Arias, G. Aubrun, P. Biane, D. Blecher, M. Junge, C. Le Merdy, A. Nou, N. Ozawa, G. Racher, E. Ricard, and Z.J. Ruan. I am especially grateful to Sophie Grivaux and Xiang Fang for their careful reading of a nearly final version, which allowed many corrections.

The support of NSF and Texas Advanced Research Program 010366-163 is gratefully acknowledged.

Finally, one more time, my warmest thanks go to Robin Campbell, whose incredible efficiency in typing this book has been extremely helpful.

PART I

INTRODUCTION TO OPERATOR SPACES

Chapter 1. Completely Bounded Maps

Let us start by recalling the definition and a few facts on C^*-algebras:

Definition 1.1. *A C^*-algebra is a Banach $*$-algebra satisfying the identity*

$$\|x^*x\| = \|x\|^2$$

for any element x in the algebra.

The simplest example is the space

$$B(H)$$

of all bounded operators on a Hilbert space H, equipped with the operator norm. More generally, any closed subspace

$$A \subset B(H)$$

stable under product and involution is a C^*-algebra.

By classical results (Gelfand and Naimark) we know that every C^*-algebra can be realized as a closed self-adjoint subalgebra of $B(H)$. Moreover, we also know that every commutative unital C^*-algebra can be identified with the space $C(T)$ of all complex-valued continuous functions $f\colon T \to \mathbb{C}$ on some compact space T. If A has no unit, A can be identified with the space $C_0(T)$ of all complex-valued continuous functions, vanishing at infinity, on some locally compact space T.

Of course the object of C^*-algebra theory (as developed in the last 50 years; cf. [KaR, Ta3]) is the classification of C^*-algebras. Similarly, the object of Banach space theory is the classification of Banach spaces.

In the last 25 years, it is their classification *up to isomorphism* (and *NOT* up to isometry) that has largely predominated (cf., e.g., [LT1–3, P4]).

This already indicates one major difference between these two fields since, if A_1 and A_2 are two C^*-algebras,

$$A_1 \text{ isomorphic to } A_2 \Rightarrow A_1 \text{ isometric to } A_2.$$

In particular, a C^*-algebra admits a *unique C^*-norm*. So there is no "isomorphic theory" of C^*-algebras. However, in recent years, operator algebraists have found the need to relax the structure of C^*-algebras and consider more general objects called *operator systems*. These are subspaces of $B(H)$ containing the unit that are stable under the involution but not under the product. The theory of operator systems was developed using the order structure repeatedly, and it is still mostly an isometric theory. The natural morphisms here are the "completely positive" maps (cf. [St, Ar1]). We refer the reader to a survey by Effros [E1] and a series of papers by Choi and Effros (especially [CE3]). Even more recently, operator algebraists have done a radical simplification and considered just "operator spaces":

Definition 1.2. *An operator space is a closed subspace of $B(H)$.*

Equivalently, since we can think of C^*-algebras as closed self-adjoint sub-algebras of $B(H)$, we can think of operator spaces as closed subspaces of C^*-algebras.

Operator space theory can be considered as a merger of C^*-algebra theory and Banach space theory.

It is important to immediately observe that any Banach space can appear as a closed subspace of a C^*-algebra. Indeed, for any Banach space X (with the dual unit ball denoted by B_{X^*}), if we let

$$T = (B_{X^*}, \sigma(X^*, X)),$$

then T is compact and we have an isometric embedding

$$X \subset C(T).$$

Hence, since $C(T)$ is a C^*-algebra (and $C(T) \subset B(H)$ with $H = \ell_2(T)$), X also appears among operator spaces. So operator spaces are just ordinary Banach spaces X but equipped with an extra structure in the form of an embedding

$$X \subset B(H).$$

The main difference between the category of Banach spaces and that of operator spaces lies not in the spaces but in the *morphisms*. We need morphisms that somehow keep track of the extra information contained in the data of the embedding $X \subset B(H)$; the maps that do just this are the *completely bounded maps*.

Definition 1.3. *Let $E \subset B(H)$ and $F \subset B(K)$ be operator spaces and consider a map*

$$
\begin{array}{ccc}
B(H) & & B(K) \\
\cup & & \cup \\
E & \xrightarrow{\ u\ } & F
\end{array}
$$

For any $n \geq 1$, let

$$M_n(E) = \{(x_{ij})_{ij \leq n} \mid x_{ij} \in E\}$$

be the space of $n \times n$ matrices with entries in E. In particular, we have a natural identification

$$M_n(B(H)) \simeq B(\ell_2^n(H)),$$

where $\ell_2^n(H)$ means $\underbrace{H \oplus H \oplus \cdots \oplus H}_{n \text{ times}}$. Thus, we may equip $M_n(B(H))$ and a fortiori its subspace

$$M_n(E) \subset M_n(B(H))$$

(see Proposition 1.5) when E, F are C^*-algebras, the resulting C^*-algebra $E \otimes_{\min} F$ does not depend on the concrete embeddings $E \subset B(H)$ and $F \subset B(K)$, but only on the abstract C^*-algebra structures on E, F.

Remark 2.1.6. Let $E \subset B(H)$ and $F \subset B(K)$ be operator spaces and let A be an auxiliary C^*-algebra. Then the bilinear mapping

$$p_A \colon A \otimes_{\min} E \times A \otimes_{\min} F \to A \otimes_{\min} E \otimes_{\min} F$$

defined by

$$p_A(a_1 \otimes x, a_2 \otimes y) = (a_1 a_2) \otimes x \otimes y$$

has norm ≤ 1.

Indeed, it suffices to prove this when $E = B(H)$ and $F = B(K)$. Then p_A appears as a restriction of the *ordinary* product map of the C^*-algebra $A \otimes_{\min} B(H) \otimes_{\min} B(K)$; more precisely, we can write

$$p_A(a_1 \otimes x, a_2 \otimes y) = (a_1 \otimes x \otimes 1) \cdot (a_2 \otimes 1 \otimes y),$$

whence the announced result. ∎

Exercises

Exercise 2.1.1. Prove (2.1.2). More generally, let $H_\alpha \subset H$ be a directed net of subspaces such that $\cup H_\alpha$ is dense in H. Then let $v_\alpha \colon E \to B(H_\alpha)$ be defined by $v_\alpha(e) = P_{H_\alpha} e|_{H_\alpha}$. We then have

$$\|v(e)\| = \lim_\alpha \uparrow \|v_\alpha(e)\|$$

for any e in E, and for any $x = \sum \alpha_i \otimes b_i \in E \otimes F$ we have

$$\|x\|_{\min} = \lim_\alpha \uparrow \left\| \sum v_\alpha(a_i) \otimes b_i \right\|_{B(H_\alpha) \otimes_{\min} F}.$$

Exercise 2.1.2. With the notation of the preceding exercise, consider a linear map $u \colon F \to E \subset B(H)$. Show that

$$\|u\|_{cb} = \sup_\alpha \|v_\alpha u\|_{cb}.$$

Exercise 2.1.3. Let H_1, H_2 be Hilbert spaces. Consider $T_i \in B(H_i)$ ($i = 1, 2$). Let $T = T_1 \oplus T_2 \in B(H_1 \oplus H_2)$. (This means that $T(h_1, h_2) = (T_1 h_1, T_2 h_2) \ \forall h_i \in H_i$.) Note that $\|T\| = \max\{\|T_1\|, \|T_2\|\}$. Let E be an operator space. Consider $u_i \in CB(E, B(H_i))$ ($i = 1, 2$), and let $u \in CB(E, B(H_1 \oplus H_2))$ be defined by $u(x) = u_1(x) \oplus u_2(x)$. Show that $\|u\|_{cb} = \max\{\|u_1\|_{cb}, \|u_2\|_{cb}\}$.

Exercise 2.1.4. Consider $x_i \in B(H)$, $y_i \in B(K)$ $(1 \leq i \leq n)$. Show that

$$\left\| \sum x_i \otimes y_i \right\|_{\min} \leq \left\| \sum x_i x_i^* \right\|^{1/2} \left\| \sum y_i^* y_i \right\|^{1/2}$$

and also

$$\left\| \sum x_i \otimes y_i \right\|_{\min} \leq \left\| \sum x_i^* x_i \right\|^{1/2} \left\| \sum y_i y_i^* \right\|^{1/2}.$$

Moreover, the first inequality becomes an equality either if $x_i = e_{i1}$ for all i or if $y_i = e_{1i}$ for all i.

2.2. Ruan's Theorem

We will now state Ruan's Fundamental Theorem ([Ru1]). We consider a complex vector space E given together with a sequence of norms on the spaces $M_n(E)$. By this we mean that, for each n, we are given a norm α_n on $M_n(E)$; moreover, we assume them compatible in the following sense: If we view $M_n(E)$ as embedded into $M_{n+1}(E)$ via the embedding that completes a matrix by adding zeroes, then α_n coincides with the restriction of α_{n+1} to $M_n(E)$. Thus, rather than work with a sequence of norms, we may as well work with a single norm on the "union" of all the spaces $M_n(E)$. More precisely, let us denote by $\mathcal{K}_0$ the union of the increasing sequence $M_1 \subset \ldots \subset M_n \subset M_{n+1} \ldots$. We equip $\mathcal{K}_0$ with the norm induced by the spaces M_n or, equivalently, by $B(\ell_2)$, so that the completion of $\mathcal{K}_0$ for this norm can be identified with $\mathcal{K}$. Let

$$\mathcal{K}_0[E] = \cup M_n(E) \simeq \mathcal{K}_0 \otimes E.$$

We define a norm α on $\mathcal{K}_0[E]$ in the obvious way, that is,

$$\forall x \in \mathcal{K}_0[E] \qquad \alpha(x) = \lim_{n \to \infty} \alpha_n(x).$$

Note that the preceding limit is actually stationary: when $x \in M_n(E)$ then $\alpha_m(x) = \alpha_n(x)$ for all $m \geq n$.

We will denote by $\mathcal{K} \otimes_\alpha E$ the completion of $\mathcal{K}_0 \otimes E$ with respect to the norm α. For any x in $\mathcal{K}_0[E]$ and any a, b in $\mathcal{K}_0$, we denote by $a.x.b$ the (left and right, respectively), matricial product of the *scalar* matrices a and b by the matrix $x \in \mathcal{K}_0[E]$.

It is easy to verify that all the sequences of norms that come from an operator space structure on E satisfy the following two axioms (R_1) and (R_2).

AXIOM (R_1): For all n we have

$$\forall x \in M_n(E) \qquad \forall a, b \in M_n \qquad \alpha_n(a.x.b) \leq \|a\|_{M_n} \alpha_n(x) \|b\|_{M_n}.$$

Then (R_1) implies

$$\alpha\left(\sum a_i \cdot x_i \cdot b_i\right) \le \|a\| \alpha(X) \|b\| \qquad (2.2.7)$$

and (R_2) (iterated $(N-1)$-times) implies

$$\alpha(X) = \max\{\alpha(x_1), \ldots, \alpha(x_N)\}.$$

But we know (Remark 1.13) that

$$\|a\| = \left\|\sum a_i a_i^*\right\|^{1/2} \quad \text{and} \quad \|b\| = \left\|\sum b_i^* b_i\right\|^{1/2};$$

hence (2.2.7) implies (R). Thus we have verified (i) $\Rightarrow$ (ii). Since (iv) $\Rightarrow$ (iii) and (iii) $\Rightarrow$ (i) are obvious, this completes the proof. ∎

Exercises

Exercise 2.2.1. Let S be a set and let $\mathcal{F} \subset \ell_\infty(S)$ be a convex cone of real-valued functions on S such that

$$\forall f \in \mathcal{F} \qquad \sup_{s \in S} f(s) \ge 0.$$

Then there is a net (λ_α) of finitely supported probability measures on S such that

$$\forall f \in \mathcal{F} \qquad \lim \int f d\lambda_\alpha \ge 0.$$

Exercise 2.2.2. Let B_1, B_2 be C^*-algebras, let $F_1 \subset B_2$ and $F_2 \subset B_2$ be two linear subspaces, and let $\varphi \colon F_1 \times F_2 \to \mathbb{C}$ be a bilinear form such that, for any finite sets (x_1^j) and (x_2^j) in F_1 and F_2, respectively, we have

$$\left|\sum \varphi(x_1^j, x_2^j)\right| \le \left\|\sum x_1^j x_1^{j*}\right\|^{1/2} \left\|\sum x_2^{j*} x_2^j\right\|^{1/2}.$$

Then there are states f_1 and f_2 on B_1 and B_2, respectively, such that

$$\forall (x_1, x_2) \in F_1 \times F_2 \qquad |\varphi(x_1, x_2)| \le (f_1(x_1 x_1^*) f_2(x_2^* x_2))^{1/2}.$$

Exercise 2.2.3. In the same situation as in Exercise 2.2.2, assume given a vector space G equipped with a norm α. Let $\psi \colon F_1 \times G \times F_2 \to \mathbb{C}$ be a trilinear form such that, for any finite sets (x_1^j) in F_1, (x_2^j) in F_2 and (x_j) in G, we have

$$\left|\sum \psi(x_1^j, x_j, x_2^j)\right| \le \left\|\sum x_1^j x_1^{j*}\right\|^{1/2} \sup_j \alpha(x_j) \left\|\sum x_2^{j*} x_2^j\right\|^{1/2}.$$

Then there are states f_1 and f_2 on B_1 and B_2, respectively, such that

$$\forall (x_1, x, x_2) \in F_1 \times G \times F_2 \qquad |\psi(x_1, x, x_2)| \le (f_1(x_1 x_1^*) f_2(x_2^* x_2))^{1/2} \alpha(x).$$

2.3. Dual space

Let E be an operator space. Let E^* be the Banach space dual of E, and then let α be the norm induced on $\mathcal{K} \otimes E^*$ by the space $CB(E, \mathcal{K})$. Then this satisfies (R) or, equivalently, (R_1) and (R_2). Consequently, E^* can be equipped with an operator space structure for which we have an isometric embedding

$$\mathcal{K} \otimes_{\min} E^* \subset CB(E, \mathcal{K}), \tag{2.3.1}$$

or, equivalently, for which we have isometrically

$$\forall n \ge 1 \qquad M_n(E^*) = CB(E, M_n). \tag{2.3.1}'$$

E^* equipped with this structure will be called the *operator space dual* (in short the o.s. dual) of E.

This notion, which is the key to numerous developments, was introduced independently in [BP1] and [ER2]. It is important to notice that the embedding $E^* \subset B(H)$ is not constructed in any explicit way.

The isometric inclusion (2.3.1) remains valid when $\mathcal{K}$ is replaced by an arbitrary operator space F. More precisely, we have isometric inclusions (note that $F \otimes_{\min} E^* \simeq E^* \otimes_{\min} F$ by transposition)

$$F \otimes_{\min} E^* \subset CB(E, F) \quad \text{and} \quad E^* \otimes_{\min} F \subset CB(E, F). \tag{2.3.2}$$

By Remark 2.1.2 and Exercise 1.4, this can be reduced to the case when $F = M_n$; hence it is a consequence of $(2.3.1)'$. These inclusions illustrate one more time the analogy between, [Minimal tensor product/Completely bounded maps] on one hand, and [Injective tensor product/Bounded maps] on the other. Moreover, it is easy to check that, for a linear operator $u: E \to F$, we have:

$$u \in CB(E, F) \text{ iff } u^* \in CB(F^*, E^*)$$

and

$$\|u^*\|_{cb} = \|u\|_{cb}. \tag{2.3.3}$$

More precisely, for any $n \ge 1$ we have

$$\|I_{M_n} \otimes u^*\|_{M_n(F^*) \to M_n(E^*)} = \|I_{M_n} \otimes u\|_{M_n(E) \to M_n(F)}. \tag{2.3.4}$$

Lemma 2.4.6. *For any x in A, there is $\widetilde{x}$ in A such that $q(\widetilde{x}) = q(x)$ and $\|\widetilde{x}\|_A = \|q(x)\|_{A/I}$.*

Proof. Using Lemma 2.4.5 we can select by induction a sequence $x, x_1, x_2, \ldots$ in A such that $q(x_n) = q(x)$, $\|x_n\| \leq \|q(x)\| + 2^{-n}$ and $\|x_n - x_{n-1}\| \leq \|x_{n-1}\| - \|q(x_{n-1})\| \leq 2^{-n+1}$. Since it is Cauchy, this sequence converges and its limit $\widetilde{x}$ has the announced property.   ∎

The following generalization was kindly pointed out to me by N. Ozawa.

Lemma 2.4.7. *Let $I \subset A$ be an ideal in a C^*-algebra A. Consider an operator space E, let $Q(E) = (E \otimes_{\min} A)/(E \otimes_{\min} I)$, and let $q_E\colon E \otimes_{\min} A \to Q(E)$ denote the quotient map. Then, for any $\widehat{y}$ in $Q(E)$, there is an element y in $E \otimes_{\min} A$ that lifts $\widehat{y}$ (i.e., $q_E(y) = \widehat{y}$) such that $\|y\|_{\min} = \|\widehat{y}\|_{Q(E)}$.*

Proof. Choose y_0 in $E \otimes_{\min} A$ such that $q_E(y_0) = \widehat{y}$. It is easy to check that q_E has the properties appearing in Lemmas 2.4.4 and 2.4.5 suitably modified. Therefore, repeating the argument appearing before Lemma 2.4.6, we obtain a Cauchy sequence $y_0, y_1, \ldots, y_n, \ldots$ in $E \otimes_{\min} A$ such that $q_E(y_n) = \widehat{y}$ for all n and $\|y_n\|_{\min} \to \|\widehat{y}\|_{Q(E)}$ when $n \to \infty$. Thus $y = \lim y_n$ is a lifting with the same norm as $\widehat{y}$.   ∎

Remark. In approximation theory, subspaces for which the infimum is attained in (2.4.5) are called "proximinal." See [HWW, p. 50] for generalizations of the preceding facts to M-ideals.

In the second part of this book, we will need the following useful lemma, for which we need to first introduce a specific notation. Let $\mathcal{I} \subset B$ be a (closed two-sided) ideal in a C^*-algebra B. Let E be an operator space. As before, we denote for simplicity

$$Q(E) = \frac{B \otimes_{\min} E}{\mathcal{I} \otimes_{\min} E}.$$

Then, if F is another operator space and if $u\colon E \to F$ is a c.b. map, we clearly have a c.b. map

$$u_Q\colon Q(E) \to Q(F)$$

naturally associated to $I_B \otimes u$ such that

$$\|u_Q\|_{cb} \leq \|u\|_{cb}.$$

Lemma 2.4.8. *If u is a complete isometry, then u_Q is also one.*

Proof. As above, the proof uses the classical fact that the ideal $\mathcal{I}$ has a two-sided approximate unit formed of elements a_α with $0 \leq a_\alpha$ and $\|a_\alpha\| \leq 1$

(see, e.g., [Ta3, p. 27]). For simplicity we assume that $E \subset F$ and u is the inclusion map. Replacing $E \subset F$ by $M_n(E) \subset M_n(F)$, we see that it clearly suffices to prove that u_Q is isometric.

Let $T_\alpha \colon B \otimes_{\min} F \to \mathcal{I} \otimes_{\min} F$ be the operator defined by $T_\alpha(b \otimes y) = a_\alpha b \otimes y$. Note that $\|T_\alpha\| \leq 1$ and $\|I - T_\alpha\| \leq 1$. Moreover, $T_\alpha(\varphi) \to \varphi$ for any φ in $\mathcal{I} \otimes_{\min} F$. Let us denote by $d(\cdot, \cdot)$ the distance in the norm of $B \otimes_{\min} F$. To show that u_Q is isometric it suffices to show that for any x in $B \otimes E$ we have

$$d(x, \mathcal{I} \otimes_{\min} F) = d(x, \mathcal{I} \otimes_{\min} E).$$

Let $e = d(x, \mathcal{I} \otimes_{\min} E)$ and $f = d(x, \mathcal{I} \otimes_{\min} F)$. Since $E \subset F$, it actually suffices to check that $e \leq f$. Assume $f < 1$. Then there is φ in $\mathcal{I} \otimes_{\min} F$ such that $\|x + \varphi\|_{\min} < 1$. We have then, since $\|1 - T_\alpha\| \leq 1$,

$$\|x - T_\alpha x\|_{\min} \leq \|(I - T_\alpha)(x + \varphi)\|_{\min} + \|(I - T_\alpha)(\varphi)\|_{\min} < 1 + \|(I - T_\alpha)(\varphi)\|_{\min}.$$

Finally, since $x \in B \otimes E$, we have $T_\alpha x \in \mathcal{I} \otimes E$ and $\|(I - T_\alpha)(\varphi)\|_{\min} \to 0$; therefore we obtain

$$e \leq \lim_\alpha \|x - T_\alpha x\|_{\min} \leq 1,$$

and we conclude (by homogeneity) that $e \leq f$. ∎

Exercises

Exercise 2.4.1. Let $u \colon E \to F$ be a c.b. map between operator spaces. Show that the following are equivalent:

(i) For any operator space G, the map $I_G \otimes u \colon G \otimes_{\min} E \to G \otimes_{\min} F$ is surjective.

(ii) There is a constant C such that, for any finite-dimensional subspace $F_1 \subset F$, there is a linear map $v \colon F_1 \to E$ with $\|v\|_{cb} \leq C$ such that $uv(x) = x$ for all x in F_1. Moreover, when $E = B(H)$, the above are also equivalent to the following kind of "local liftability":

(iii) There is a net of of cb maps $v_i \colon F \to E$ with $\sup_i \|v_i\|_{cb} \leq C$ such that $uv_i \to I_F$ pointwise.

Exercise 2.4.2. Let E be an operator space. Let $I \subset A$ be an ideal in a C^*-algebra and let $q \colon A \to A/I$ be the quotient map. Consider a bounded linear map $u \colon E \to A/I$. Assume that there is a net of complete contractions $v_\alpha \colon E \to A$ such that $qv_\alpha \to u$ pointwise on E. Show that, if E is separable, then u admits a completely contractive lifting $v \colon E \to A$, that is, we have $\|v\|_{cb} \leq 1$ and $qv = u$. (This is essentially due to Arveson [Ar4].) **Hint:** Let $\{x_i\}$ be a dense sequence in the unit ball of E. Assume given a complete contraction (c.c. in short) v_n such that

$$\|qv_n x_i - u x_i\| < 2^{-n} \qquad \forall i = 1, \ldots, n.$$

Show that there is a c.c. $v_{n+1}\colon E \to A$ such that

$$\|qv_{n+1}x_i - ux_i\| < 2^{-n-1} \qquad \forall i = 1,\ldots,n+1$$

and

$$\|(v_{n+1} - v_n)x_i\| < 2.2^{-n} \qquad \forall i = 1,\ldots,n.$$

Then $v(x) = \lim v_n(x)$ is the desired lifting.

2.5. Bidual. Von Neumann algebras

For the convenience of the reader, we recall now a few facts concerning von Neumann algebras. A von Neumann algebra on a Hilbert space H is a self-adjoint subalgebra of $B(H)$ that is equal to its bicommutant. For $M \subset B(H)$ we denote by M' (resp. M'') its commutant (resp. bicommutant). By a well-known result due to Sakai (see [Sa]), a C^*-algebra A is C^*-isomorphic to a von Neumann algebra iff it is isometric to a dual Banach space, that is, iff there is a closed subspace $X \subset A^{**}$ such that $X^* = A$ isometrically. If it exists, the predual X is unique and is denoted by A_*. In particular, $B(H)_*$ can be identified with the space of all trace class operators on H, equipped with the trace class norm, and it is easy to check that a von Neumann algebra $M \subset B(H)$ is automatically $\sigma(B(H), B(H)_*)$-closed. Thus, when it is a dual space, A can be realized as a von Neumann algebra on a Hilbert space H and its predual can be identified with the quotient of $B(H)_*$ by the preannihilator of A.

Note that any C^*-isomorphism $u\colon M_1 \to M_2$ from a von Neumann algebra onto another is automatically bicontinuous for the weak-$*$ topologies ($\sigma(M_1, M_{1*})$ and $\sigma(M_2, M_{2*})$). Therefore the preduals are automatically isometric (see, e.g., [Ta3, p. 135]).

A map $u\colon M_1 \to M_2$ between von Neumann algebras is called *normal* if it is continuous for the $\sigma(M_1, M_{1*})$ and $\sigma(M_2, M_{2*})$ topologies or, equivalently, if there is a map $v\colon M_{2*} \to M_{1*}$ of which u is the adjoint.

Let A be a C^*-algebra. The bidual of A can be equipped with a C^*-algebra structure as follows: Let $\pi_u\colon A \to B(H)$ be the universal representation of A (i.e., the direct sum of all cyclic representations of A). Then the bicommutant $\pi_u(A)''$, which is a von Neumann algebra on H, is isometrically isomorphic (as a Banach space) to the bidual A^{**}. Using this isomorphism as an identification, we can now view A^{**} as a von Neumann algebra, so that the canonical inclusion $A \to A^{**}$ is a $*$-homomorphism.

The reader who prefers a more "concrete" description of the product operation on the bidual of A can use the following alternate viewpoint: For any pair a, b in A^{**} let (a_i) and (b_j) be nets in A with $\|a_i\| \leq \|a\|$ and $\|b_j\| \leq \|b\|$, converging respectively to a and b in the Mackey topology $\tau(A^{**}, A^*)$. Then the products $(a_i b_j)$ converge in the same topology to an element of A^{**} denoted by $a \cdot b$. Similarly, (a_i^*) converges, for $\tau(A^{**}, A^*)$, to an element of A^{**} denoted by a^*. It can be shown that A^{**} is equipped with this product (called the Arens product), and this involution becomes a C^*-algebra, which can be identified with the one just defined. We refer the reader to, for example, [BD, p. 213] for more information.

We now return to operator spaces. Of course the bidual of an operator space E can be equipped with an o.s.s. simply by reiteration of the preceding definition in the preceding section, that is, by letting $E^{**} = (E^*)^*$. This of course immediately raises the natural question of the relationship between the Banach bidual and its operator space counterpart. Fortunately, at this point the situation is very nice.

Indeed ([BP1, Theorem 2.1] or [ER2, Theorem 2.2]) the inclusion $E \subset E^{**}$ is a complete isometry (for a proof, see the solution to Exercise 2.3.1). Therefore, if the underlying Banach space is reflexive, E and E^{**} are identical operator spaces. In short, the Banach space reflexivity alone guarantees the operator space.

More generally, we have

Theorem 2.5.1. *([B2, Th. 2.5]) For any operator space E, we have an isometric identity*

$$M_n(E)^{**} = M_n(E^{**}). \tag{2.5.1}$$

Note that the two sides of (2.5.1) can clearly be identified as vector spaces. But we also claim in (2.5.1) that their norms are the same.

For a proof, see Corollary 5.11 below.

A moment of thought shows that, since (2.5.1) holds for any n, actually it must also be a complete isometry.

In particular, let A be a C^*-algebra. We equip A with its natural o.s.s. (coming from its Gelfand embedding into $B(H)$). Then, by the preceding definition, the successive duals A^*, A^{**}, A^{***}, and so on, can now be viewed as operator spaces. We will refer to these operator space structures on A^*, A^{**}, A^{***}, and so on, as the *natural* ones.

Now assume that A is isometric to a dual Banach space, that is, we have $A = (A_*)^*$ and, as explained above, A can be realized as a von Neumann algebra. Then, the inclusion $A_* \subset (A_*)^{**} = A^*$ allows us to equip the predual A_* with the o.s.s. induced by the one just defined on the dual A^*, so we obtain an operator space, denoted by A_*^{os}, having A_* as its underlying Banach space. Here again, a natural question arises: If we now consider the dual operator space to the one just defined, namely, $(A_*^{os})^*$, do we recover the same operator space structure on A? Fortunately, the answer is still affirmative ([B2, Theorem 2.9]): We have $(A_*^{os})^* = A$ completely isometrically.

associated to the parameter

$$0 < \theta < 1.$$

Roughly, the intermediate space E_θ is a mixture of E_0 and E_1; it varies continuously from E_0 to E_1 when θ goes from 0 to 1 (cf. [C, BeL]); when $\theta = 1/2$, its unit ball behaves as some sort of "geometric mean" of the unit balls of E_0 and E_1.

Let us briefly recall the main points of this theory (see [BeL, KPS]). One of the equivalent definitions of E_θ is as follows.

Let $S = \{z \in \mathbb{C} \mid 0 < \operatorname{Re}(z) < 1\}$ be the usual vertical strip. Its boundary ∂S decomposes as $\partial S = \partial_0 \cup \partial_1$ with $\partial_0 = \{z \mid \operatorname{Re}(z) = 0\}$ and $\partial_1 = \{z \mid \operatorname{Re}(z) = 1\}$. We denote by $\mathcal{F}(E_0, E_1)$ the set of all bounded continuous functions $f \colon \overline{S} \to E_0 + E_1$ that are holomorphic on S and such that $f_{|\partial_0}$ and $f_{|\partial_1}$ are bounded continuous functions with values, respectively, in E_0 and E_1. We define

$$\|f\|_{\mathcal{F}(E_0,E_1)} = \max\{\sup_{z\in\partial_0} \|f(z)\|_{E_0}, \sup_{z\in\partial_1} \|f(z)\|_{E_1}\}.$$

Then the space E_θ is defined as the subspace of $E_0 + E_1$ formed of all the values $f(\theta)$ when f runs over $\mathcal{F}(E_0, E_1)$. Equipped with the norm

$$\|x\|_{E_\theta} = \inf\{\|f\|_{\mathcal{F}(E_0,E_1)} \mid f \in \mathcal{F}(E_0, E_1), f(\theta) = x\}, \tag{2.7.1}$$

it becomes a Banach space. Actually, it is often convenient to work with a smaller class of functions than $\mathcal{F}(E_0, E_1)$, namely, the functions f for which there is a finite-dimensional subspace $M \subset E_0 \cap E_1$ such that

$$\forall z \in \overline{S} \qquad f(z) \in M.$$

Note that, viewed as an M-valued function, f is then (automatically) bounded continuous on $\overline{S}$ and analytic on S. We will denote by $\mathcal{F}_0(E_0, E_1)$ this subspace, which is useful because of the following (cf. e.g. [Sta]).

Lemma 2.7.1. *The space $E_0 \cap E_1$ is dense in E_θ for any $0 < \theta < 1$. Moreover, for any x in $E_0 \cap E_1$ we have*

$$\|x\|_{E_\theta} = \inf\{\|f\|_{\mathcal{F}(E_0,E_1)} \mid f \in \mathcal{F}_0(E_0, E_1), f(\theta) = x\}. \tag{2.7.2}$$

Lemma 2.7.2. *Let $\widehat{E}_0$ and $\widehat{E}_1$ denote, respectively, the closure of $E_0 \cap E_1$ in E_0 and E_1. Then we have an isometric identity*

$$(\widehat{E}_0, \widehat{E}_1)_\theta = (E_0, E_1)_\theta.$$

Proof. This is an immediate consequence of the previous lemma. ∎

We will not define the space E^θ precisely; it is a larger space (containing E_θ isometrically; see [Be]) obtained as the values $f(\theta)$ when f is allowed to admit noncontinuous and even non-Bochner measurable boundary values, which can be described in terms of vector measures. This space is essential if one wants a precise description of the dual of E_θ; however, it is possible, for our purposes here, to mostly avoid using E^θ, which explains why we choose not to give more details.

Now if E_0, E_1 are equipped with operator space structures, we define the norm α by the identity

$$\mathcal{K} \otimes_\alpha E_\theta \overset{\text{def}}{=} (\mathcal{K} \otimes_{\min} E_0, \mathcal{K} \otimes_{\min} E_1)_\theta, \qquad (2.7.3)$$

and similarly for E^θ.

More generally, for any $n \geq 1$ we have

$$M_n(E_\theta) = (M_n(E_0), M_n(E_1))_\theta, \qquad (2.7.4)$$

and also for rectangular matrices for any $n, m \geq 1$

$$M_{n,m}(E_\theta) = (M_{n,m}(E_0), M_{n,m}(E_1))_\theta. \qquad (2.7.5)$$

See Corollary 2.7.7 for an extension of (2.7.5).

Again, it is then easy to check the axioms (R_1, R_2) so that we now have extended the complex interpolation functor to the category of operator spaces.

Remark 2.7.3. For instance, given two arbitrary operator spaces E, F, we may consider

$$E_0 = E \oplus F \quad \text{and} \quad E_1 = E \oplus_1 F$$

in the sense of §2.6. Since these have the same underlying vector space, they clearly form a compatible couple. By §2.6, both E_0 and E_1 have natural o.s. structures. Thus we can define the o.s.s. on the space $E \oplus_p F$ ($1 < p < \infty$) as the one obtained by interpolation, that is, we set

$$E \oplus_p F = (E_0, E_1)_\theta \quad \text{with} \quad \theta = 1/p.$$

A similar argument applies for $\ell_p(\{E_i \mid i \in I\})$. The resulting operator space has many of the nice properties one would expect, for instance, we have a completely isometric identity

$$(E_0 \oplus_p E_1)^* = E_0^* \oplus_{p'} E_1^* \quad \left(\frac{1}{p} + \frac{1}{p'} = 1\right).$$

In interpolation theory, the intersection $E_0 \cap E_1$ and the sum $E_0 + E_1$ play an important role. We will first review the analogous definitions for operator spaces. Recall that our assumption that (E_0, E_1) form a *compatible* couple means that there are continuous injections

$$j_0 \colon E_0 \to X \quad \text{and} \quad j_1 \colon E_1 \to X$$

of E_0 and E_1 into a topological vector space X. For simplicity, one usually identifies E_0 and E_1, respectively, with $j_0(E_0) \subset X$ and $j_1(E_1) \subset X$. Then $E_0 \cap E_1$ is the setwise intersection, equipped with the norm

$$\|x\|_{E_0 \cap E_1} = \max\{\|x\|_{E_0}, \|x\|_{E_1}\}.$$

Similarly, $E_0 + E_1 = \{x_0 + x_1 \mid x_0 \in E_0, x_1 \in E_1\}$ with

$$\|x\|_{E_0 + E_1} = \inf\{\|x_0\|_{E_0} + \|x_1\|_{E_1} \mid x = x_0 + x_1\}.$$

With this norm, $E_0 + E_1$ becomes a Banach space, which we may substitute, without loss of generality, to the space X. Thus X can always be assumed to be a Banach space.

Now suppose that E_0 and E_1 are equipped with operator space structures. We define an operator space structure on $E_0 \cap E_1$ via the isometric embedding

$$j \colon E_0 \cap E_1 \to E_0 \oplus E_1$$

defined by $j(x) = (x, x)$, where, of course, $E_0 \oplus E_1$ is equipped with the o.s.s. defined in §2.6. Similarly, we can equip the sum $E_0 + E_1$ with the operator space structure induced by the isometric identity

$$E_0 + E_1 = \frac{E_0 \oplus_1 E_1}{\Delta},$$

where $\Delta = \{(x, -x) \mid x \in E_0 \cap E_1\}$. Indeed, we have already defined the natural o.s.s. of $E_0 \oplus_1 E_1$ (§2.6) and of quotient spaces (§2.4), so this identity endows $E_0 + E_1$ with a natural o.s.s.

Equivalently, if we let I' be the set of all maps $u \colon E_0 + E_1 \to B(H_u)$ such that $\|u_{|E_0}\|_{cb} \le 1$ and $\|u_{|E_1}\|_{cb} \le 1$, then the mapping $j \colon E_0 + E_1 \to \oplus_{u \in I'} B(H_u)$ defined by $j(x) = \oplus_{u \in I'} u(x)$ is a complete isometry. Therefore, the resulting o.s. $E_0 + E_1$ is characterized by the property that a linear map $u \colon E_0 + E_1 \to B(H)$ is a complete contraction iff $\|u_{|E_0}\|_{cb} \le 1$ and $\|u_{|E_1}\|_{cb} \le 1$.

It is easy to check that the Banach space $\mathcal{K} \otimes_{\min} (E_0 \cap E_1)$ can be identified isometrically with $(\mathcal{K} \otimes_{\min} E_0) \cap (\mathcal{K} \otimes_{\min} E_1)$. The analogous property for the sum $E_0 + E_1$ is valid but only with an extra numerical factor equal to 2; namely, we have a natural isomorphism

$$\psi \colon \mathcal{K} \otimes_{\min} (E_0 + E_1) \to \mathcal{K} \otimes_{\min} E_0 + \mathcal{K} \otimes_{\min} E_1$$

that satisfies $\|\psi\|_{cb} \leq 2$ and $\|\psi^{-1}\|_{cb} \leq 1$. (Note however that we have isometrically

$$\mathcal{K}^* \otimes^\wedge (E_0 + E_1) \simeq \mathcal{K}^* \otimes^\wedge E_0 + \mathcal{K}^* \otimes^\wedge E_1.)$$

Now assume that $E_0 \cap E_1$ is dense both in E_0 and in E_1 (for their respective norms). Consequently, we have natural continuous injections

$$E_0^* \to (E_0 \cap E_1)^* \quad \text{and} \quad E_1^* \to (E_0 \cap E_1)^*$$

that allow us to consider (E_0^*, E_1^*) as a compatible couple. It is then easy to verify that

$$(E_0 \cap E_1)^* \simeq E_0^* + E_1^* \text{ and } (E_0 + E_1)^* \simeq E_0^* \cap E_1^*$$

completely isometrically.

Then, by Calderón's fundamental duality theorem (cf. [C, BeL]), we have an isometric identity

$$[(E_0, E_1)_\theta]^* = (E_0^*, E_1^*)^\theta.$$

More explicitly, the duality can be described as follows. There is a natural inclusion

$$(E_0^*, E_1^*)_\theta \to (E_\theta)^*. \tag{2.7.6}$$

Indeed, consider ξ in $(E_0^*, E_1^*)_\theta$ and x in $(E_0, E_1)_\theta$. By density, we may assume in addition that $x \in E_0 \cap E_1$. We will show that

$$|\langle \xi, x \rangle| \leq \|x\|_{E_\theta} \|\xi\|_{(E_0^*, E_1^*)_\theta}.$$

(Note that this makes sense since $\xi \in E_0^* + E_1^* = (E_0 \cap E_1)^*$.) Indeed, for any $\varepsilon > 0$, there are $f \in \mathcal{F}_0(E_0, E_1)$ and $g \in \mathcal{F}(E_0^*, E_1^*)$ such that $f(\theta) = x$, $g(\theta) = \xi$,

$$\|f\|_{\mathcal{F}(E_0, E_1)} < \|x\| + \varepsilon, \quad \text{and} \quad \|g\|_{\mathcal{F}(E_0^*, E_1^*)} < \|\xi\| + \varepsilon.$$

Note that since the range $f(\overline{S})$ is bounded in $E_0 \cap E_1$, the function

$$F \colon z \to \langle g(z), f(z) \rangle$$

is bounded and continuous on $\overline{S}$ and analytic on S. By the maximum principle we have

$$|\langle \xi, x \rangle| = |F(\theta)| \leq \sup_{z \in \partial S} |F(z)|$$

$$\leq (\|x\| + \varepsilon)(\|\xi\| + \varepsilon),$$

which shows, since $\varepsilon > 0$ is arbitrary, that (2.7.6) makes sense and is contractive.

Proof. We recall (2.7.3) and apply Proposition 2.7.6 with $(\mathcal{S}_0, \mathcal{S}_1)$ replaced by the couple $(X \otimes_{\min} E_0, X \otimes_{\min} E_1)$. ∎

Remark 2.7.8. In general $(E_0, E_1)_\theta$ depends very much on the choice of "compatibility" underlying the couple (E_0, E_1). Nevertheless, there is a simple "invariance principle" as follows. Let (E_0, E_1) and (F_0, F_1) be two compatible pairs of Banach spaces. We assume the "compatibility" given by continuous injections

$$E_0 \xrightarrow{\ i_0\ } \mathcal{X}, \quad E_1 \xrightarrow{\ i_1\ } \mathcal{X} \quad \text{and} \quad F_0 \xrightarrow{\ j_0\ } \mathcal{X}, \quad F_1 \xrightarrow{\ j_1\ } \mathcal{X}$$

into (say) a Banach space $\mathcal{X}$. Moreover, we assume that there are isometric isomorphisms $u_0 \colon E_0 \to F_0$ and $u_1 \colon E_1 \to F_1$ such that $i_k = j_k u_k$ $(k = 0, 1)$. This implies that (u_0, u_1) unambiguously defines a map $u \colon E_0 + E_1 \to F_0 + F_1$ (and also one from $E_0 \cap E_1$ to $F_0 \cap F_1$). Then, this mapping u induces an isometric isomorphism from $(E_0, E_1)_\theta$ to $(F_0, F_1)_\theta$ for any $0 < \theta < 1$. Clearly, in the operator space case, if we assume furthermore that u_0 and u_1 are actually *completely* isometric, then we obtain

$$(E_0, E_1)_\theta \simeq (F_0, F_1)_\theta \quad \text{(completely isometrically)}.$$

Remark. In analogy with the complex case presented in [P1], the "real interpolation theory" (à la Lions-Peetre; cf. [BeL]) has been developed for operator spaces recently by Quanhua Xu (cf. [Xu]).

2.8. Ultraproduct

We first recall several definitions. Let $\mathcal{U}$ be a nontrivial ultrafilter on a set I. Let $(E_i)_{i \in I}$ be a family of Banach spaces. We denote by ℓ the space of all families $x = (x_i)_{i \in I}$ with $x_i \in X_i$ such that $\sup_{i \in I} \|x_i\| < \infty$. We equip this space with the norm $\|x\| = \sup_{i \in I} \|x_i\|$. Let $n_\mathcal{U} \subset \ell$ be the subspace formed of all families such that $\lim_\mathcal{U} \|x_i\| = 0$.

Definition. *The quotient $\ell/n_\mathcal{U}$ is called the ultraproduct of the family $(E_i)_{i \in I}$ with respect to $\mathcal{U}$. We denote it by $\Pi_{i \in I} E_i / \mathcal{U}$.*

It is important to observe that, for every element $\dot{x}$ in $\ell/n_\mathcal{U}$ admitting $x \in \ell$ as its representative modulo $n_\mathcal{U}$, we have

$$\|\dot{x}\| = \lim_\mathcal{U} \|x_i\|.$$

Hence, the ultraproduct $\Pi_{i \in I} E_i / \mathcal{U}$ appears as "the limit" of the spaces $(E_i)_{i \in I}$ along $\mathcal{U}$.

Remark. It is easy to check that, if $(S_i)_{i\in I}$ is a family of closed subspaces (i.e., $S_i \subset E_i$ for each i), then the ultraproduct of the subspaces $\Pi_{i\in I}S_i/\mathcal{U}$ can be identified isometrically with a closed subspace of the original ultraproduct $\Pi_{i\in I}E_i/\mathcal{U}$. Similarly, the quotient of the ultraproducts can be identified with the ultraproduct of the quotients. Thus Banach space ultraproducts are both "injective" and "projective."

Now assume that each space E_i is given equipped with an operator space structure. It is very easy to extend the notion of ultraproduct to the operator space setting. We simply define

$$M_n(\Pi_{i\in I}X_i/\mathcal{U}) = \Pi_{i\in I}M_n(X_i)/\mathcal{U}. \tag{2.8.1}$$

This identity endows $M_n \otimes (\Pi_{i\in I}X_i/\mathcal{U})$ with a norm satisfying Ruan's axioms, whence after completion a norm on $\mathcal{K} \otimes (\Pi_{i\in I}X_i/\mathcal{U})$.

Alternatively, we can view an operator space as a subspace of a C^*-algebra, observe that C^*-algebras are stable by ultraproducts, and apply the preceding remark to realize any ultraproduct of operator spaces as a subspace of an ultraproduct of C^*-algebras. This alternate route leads to the same operator space structure as in (2.8.1).

Yet another equivalent description of $\prod_{i\in I} E_i$ as an operator space is the following: Consider the direct sum $\ell = \bigoplus_{i\in I} E_i$ as defined in the previous section. Then $\prod_{i\in I} E_i/\mathcal{U}$ can be identified with the quotient operator space $\ell/n_{\mathcal{U}}$. It is easy to verify that the injectivity and projectivity of ultraproducts is preserved in the operator space setting: Let $S_i \subset E_i$ be a family of subspaces of operator spaces E_i. Then we have a completely isometric embedding

$$\prod_{i\in I} S_i/\mathcal{U} \subset \prod_{i\in I} E_i/\mathcal{U}; \tag{2.8.2}$$

moreover, the quotient $\left(\prod_{i\in I} E_i/\mathcal{U}\right) \Big/ \left(\prod_{i\in I} S_i/\mathcal{U}\right)$ can be identified completely isometrically with the ultraproduct $\prod_{i\in I}(E_i/S_i)/\mathcal{U}$. These properties follow easily from the definition (2.8.1) and the corresponding isometric identities mentioned in the above remark. Let us denote $\widehat{E}_{\mathcal{U}} = \prod_{i\in I} E_i/\mathcal{U}$. Now let $(F_i)_{i\in I}$ be another family of operator spaces and, again, let $\widehat{F}_{\mathcal{U}} = \prod_{i\in I} F_i/\mathcal{U}$. For any i in I, consider a map u_i in $CB(E_i, F_i)$ with $\sup_{i\in I}\|u_i\|_{cb} < \infty$. Then, $(u_i)_{i\in I}$ defines a linear map from $\widehat{E}_{\mathcal{U}}$ to $\widehat{F}_{\mathcal{U}}$ taking the equivalence class of $(x_i)_{i\in I}$ to that of $(u_i(x_i))_{i\in I}$. Let

$$\widehat{u}\colon \widehat{E}_{\mathcal{U}} \to \widehat{F}_{\mathcal{U}}$$

be the resulting mapping. In the Banach space setting, it is well known that

$$\|\widehat{u}\| = \lim_{\mathcal{U}} \|u_i\|. \tag{2.8.3}$$

We claim that $\widehat{u}$ is c.b. and

$$\|\widehat{u}\|_{cb} \leq \lim_{\mathcal{U}} \|u_i\|_{cb}. \tag{2.8.4}$$

More precisely, for any integer N and any map $u\colon E \to F$ between operator spaces, let

$$\|u\|_N = \|I_{M_N} \otimes u\|_{M_N(E) \to M_N(F)}.$$

As an immediate consequence of (2.8.1) and (2.8.3), we have

$$\|\widehat{u}\|_N = \lim_{\mathcal{U}} \|u_i\|_N. \tag{2.8.5}$$

Whence

$$\|\widehat{u}\|_{cb} = \sup_N \|\widehat{u}\|_N = \sup_N \lim_{\mathcal{U}} \|u_i\|_N \leq \lim_{\mathcal{U}} \|u_i\|_{cb},$$

so we obtain (2.8.4) as announced.

Note that if F is a finite-dimensional operator space and if $F_i = F$ for all i in I, then $\widehat{F}_{\mathcal{U}} = F$ canonically and completely isometrically (the identification $\widehat{F}_{\mathcal{U}} \to F$ is the map that takes $(x_i)_{i \in I}$ to $\lim_{\mathcal{U}} x_i$). We now fix $N \geq 1$ and consider the particular case when $F_i = M_N$ for all i in I. Then, by Proposition 1.12, we have

$$\|u_i\|_{cb} = \|u_i\|_N.$$

Therefore, if we identify $\widehat{F}_{\mathcal{U}}$ with M_N, we can rewrite (2.8.5) as follows:

$$\|\widehat{u}\|_{CB(\widehat{E}_{\mathcal{U}}, M_N)} = \lim_{\mathcal{U}} \|u_i\|_{CB(E_i, M_N)}. \tag{2.8.6}$$

We claim that (2.8.6) implies that we have a completely isometric embedding

$$\prod_{i \in I} E_i^* / \mathcal{U} \hookrightarrow \left(\prod_{i \in I} E_i / \mathcal{U} \right)^*. \tag{2.8.7}$$

Indeed, the case $N = 1$ in (2.8.6) yields that the correspondence $(u_i)_{i \in I} \to \widehat{u}$ defines an isometric embedding and the general case of (2.8.6) (together with $(2.3.2)'$) shows that it is completely isometric. In particular:

Lemma 2.8.1. *Fix $d \geq 1$. Let $(E_i)_{i \in I}$ be a sequence of d-dimensional operator spaces. Then we have a completely isometric identification*

$$\left(\prod_{i \in I} E_i / \mathcal{U} \right)^* \simeq \prod_{i \in I} E_i^* / \mathcal{U}. \tag{2.8.8}$$

In other words, the ultraproducts "commute" with the duality.

Proof. Elementary (linear algebraic) considerations show that $\prod_{i \in I} E_i / \mathcal{U}$ is d-dimensional and the same for $\prod_{i \in I} E_i^* / \mathcal{U}$. Hence (2.8.8) follows from (2.8.7). ∎

Analogous results can be formulated for the minimal tensor product.

Proposition 2.8.2. *Let $(E_i)_{i \in I}$ be an arbitrary family of operator spaces with I and $\mathcal{U}$ as above. Then, for any finite-dimensional operator space E, we have a completely contractive inclusion*

$$\prod_{i \in I}(E_i \otimes_{\min} E)/\mathcal{U} \longrightarrow \left(\prod_{i \in I} E_i/\mathcal{U}\right) \otimes_{\min} E. \qquad (2.8.9)$$

Proof. Since E is finite-dimensional, the two spaces appearing in (2.8.9) are clearly identifiable as vector spaces. Observe that, by (2.6.2), we have an isometric identity

$$\left(\bigoplus_{i \in I} E_i\right) \otimes_{\min} E = \bigoplus_{i \in I}(E_i \otimes_{\min} E).$$

Changing E to $M_N(E)$ $(N = 1, 2, \ldots)$ shows that this is even a completely isometric equality. The quotient mapping $q \colon \ell \to \ell/n_{\mathcal{U}}$ is completely contractive, and hence it defines a complete contraction

$$q \otimes I_E \colon \ell \otimes_{\min} E \to (\ell/n_{\mathcal{U}}) \otimes_{\min} E$$

that clearly vanishes on $n_{\mathcal{U}} \otimes_{\min} E$. Passing to the quotient modulo $\ker(q \otimes I_E)$, we obtain a completely contractive map

$$(\ell \otimes_{\min} E)/(n_{\mathcal{U}} \otimes_{\min} E) \longrightarrow \left(\prod_{i \in I} E_i/\mathcal{U}\right) \otimes_{\min} E.$$

By the preceding observation, $\ell \otimes_{\min} E \simeq \bigoplus_{i \in I}(E_i \otimes_{\min} E)$ and $n_{\mathcal{U}} \otimes_{\min} E$ can clearly be identified with the subspace of $\bigoplus_{i \in I}(E_i \otimes_{\min} E)$ formed of all families that tend to zero along $\mathcal{U}$. In other words, $(\ell \otimes_{\min} E)/(n_{\mathcal{U}} \otimes_{\min} E)$ can be identified with $\prod_{i \in I}(E_i \otimes_{\min} E)/\mathcal{U}$. Thus we obtain the proposition. ∎

Remark 2.8.3. Note that by inverting the mapping (2.8.9) we have, for any operator space X, a natural inclusion of the *algebraic* tensor product:

$$\left(\prod_{i \in I} E_i/\mathcal{U}\right) \otimes X \subset \prod_{i \in I}(E_i \otimes_{\min} X)/\mathcal{U}.$$

We will study in Chapter 17 the conditions on X that ensure that this inclusion extends to an embedding of the minimal tensor product $\left(\prod_{i \in I} E_i/\mathcal{U}\right) \otimes_{\min} X$.

Convention. Whenever we are discussing an ultraproduct $\prod_{i \in I} E_i/\mathcal{U}$ of a family of Banach spaces or operator spaces, it will be convenient to identify abusively a bounded family $(x_i)_{i \in I}$ with $x_i \in E_i$ for all i in I with the corresponding equivalence class modulo $\mathcal{U}$, which it determines in $\prod_{i \in I} E_i / \mathcal{U}$. Thus, when we speak of $(x_i)_{i \in I}$ as an element of $\prod_{i \in I} E_i / \mathcal{U}$, we really are referring to the equivalence class that it determines. This abuse is consistent with the one routinely done in standard measure theory.

Remark 2.8.4. Let $(E_\alpha)_{\alpha \in I}$ and $(F_\alpha)_{\alpha \in I}$ be families with $E_\alpha = E$ and $F_\alpha = F$ for all α in the index set I equipped with an ultrafilter $\mathcal{U}$. Let $\widehat{E}$ and $\widehat{F}$ be the associated ultraproducts (or ultrapowers). Let $\varphi \colon E \to \widehat{E}$ be the canonical (completely isometric) inclusion and let $\psi \colon \widehat{F} \to F^{**}$ be the canonical (completely contractive) map obtained by compactness of $(B_{F^{**}}, \sigma(F^{**}, F^*))$. Then, let $T \colon E \to F$ be a bounded linear map with the associated ultraproduct map $\widehat{T} \colon \widehat{E} \to \widehat{F}$. It is an easy exercise to check that, if $i_F \colon F \to F^{**}$ denotes the canonical inclusion, we have

$$\psi \widehat{T} \varphi = i_F T \colon E \to F^{**}.$$

Note in passing that $i_F T = T^{**} i_E$.

Let $(E_\alpha)_{\alpha \in I}$ be a net of subspaces of E directed by inclusion and such that $\overline{\bigcup_{\alpha \in I} E_\alpha} = E$. We can then still define $\varphi \colon \bigcup_\alpha E_\alpha \to \prod E_\alpha / \mathcal{U}$ by setting $\varphi(x) = (\varphi_\alpha(x))_\alpha$, where we set $\varphi_\alpha(x) = x$ if $x \in E_\alpha$ and $= 0$ (say) otherwise. By density, φ extends to a (completely isometric) map $\varphi \colon E \to \prod_{\alpha \in I} E_\alpha / \mathcal{U}$. Moreover, we have a canonical complete contraction $\psi \colon \prod_{\alpha \in I} E_\alpha / \mathcal{U} \to E^{**}$ defined by $\psi((x_\alpha)) = \lim_{\mathcal{U}} x_\alpha$ (the limit is relative to $\sigma(E^{**}, E^*)$). Thus, we have proved:

Theorem 2.8.4. *Any operator space E can be embedded completely isometrically into a suitable ultraproduct $\mathcal{X}$ of finite-dimensional subspaces of E. More precisely, the inclusion map $E \subset E^{**}$ factors completely contractively through $\mathcal{X}$.*

The Banach space analog of this is well known.

2.9. Complex conjugate

Let E be a Banach space. We will denote by $\overline{E}$ the complex conjugate of E, that is, the vector space E with the same norm but with the conjugate multiplication by a complex scalar. We will denote by $x \to \overline{x}$ the identity map from E to $\overline{E}$. Thus, the space $\overline{E}$ is anti-isometric to E.

To be more "concrete," if an operator space E is given as a collection of bi-infinite matrices $\{(a_{ij})\}$ (representing operators acting on ℓ_2), then $\overline{E}$ can be thought of as the space of all operators with complex conjugate matrices $\{(\overline{a_{ij}})\}$.

It is easy to see that $\overline{B(H)}$ can be canonically identified with $B(\overline{H})$, so that the embedding

$$\overline{E} \subset \overline{B(H)} = B(\overline{H})$$

allows us to equip $\overline{E}$ with an operator space structure. Moreover, if $E \subset B(H)$ is a C^* (resp. von Neumann) subalgebra, then so is $\overline{E} \subset B(\overline{H})$.

It is easy to see that the corresponding norm α on $\mathcal{K} \otimes \overline{E}$ is characterized by the following identity:

$$\forall a_i \in \mathcal{K} \quad \forall x_i \in E \quad \alpha(\sum_1^n a_i \otimes \overline{x_i}) = \|\sum_1^n \overline{a_i} \otimes x_i\|_{\overline{\mathcal{K}} \otimes_{\min} E}. \qquad (2.9.1)$$

In other words, the operator space structure of $\overline{E}$ is precisely defined so that $\mathcal{K} \otimes_{\min} \overline{E}$ is naturally anti-isometric to $\overline{\mathcal{K}} \otimes_{\min} E$. For each matrix (a_{ij}) in $M_n(E)$, we simply have

$$\|(\overline{a_{ij}})\|_{M_n(\overline{E})} = \|(a_{ij})\|_{M_n(E)}.$$

Note, however, that there are examples (see [Bou2]) showing that E and $\overline{E}$ may fail to be isomorphic Banach spaces. Moreover ([Co3]), there are examples of von Neumann algebras E that fail to be C^*-isomorphic to $\overline{E}$ (actually, the example in [Co3] has trivial center, i.e., is a factor).

Proposition 2.9.1. *Let H, K be Hilbert spaces. We denote by $S_2(K, H)$ the space of all Hilbert-Schmidt operators $x \colon K \to H$ and we denote its Hilbert-Schmidt norm simply by $\|x\|_2$. Consider finite sequences (a_i) in $B(H)$ and (b_i) in $B(K)$. Then we have*

$$\left\|\sum a_i \otimes \overline{b_i}\right\|_{B(H) \otimes_{\min} \overline{B(K)}} = \sup\left\{\left\|\sum a_i x b_i^*\right\|_2 \ \Big| \ x \in S_2(K, H), \|x\|_2 \le 1\right\}.$$

Proof. Note that $B(H) \otimes_{\min} \overline{B(K)}$ has the norm induced by $B(H \otimes_2 \overline{K})$, and if we identify $H \otimes_2 \overline{K}$ with $S_2(K, H)$, then the operator $\sum a_i \otimes \overline{b_i}$ can be identified with $x \to \sum a_i x b_i^*$. ∎

2.10. Opposite

Let $u \colon H_1 \to H_2$ be a bounded linear mapping between Hilbert spaces. In this section, we will denote by $^t u \colon H_2^* \to H_1^*$ the *transposed* mapping. Note that the dual spaces H_1^* and H_2^* are Hilbert spaces that can be canonically (i.e., in a basis free manner) identified respectively with the conjugates $\overline{H}_1$ and $\overline{H}_2$. Let $E \subset B(H)$ be an operator space. The *opposite* is the operator space E^{op} associated to the same space E but equipped with the following norms on $M_n(E)$. For any (a_{ij}) in $M_n(E)$ we define

$$\|(a_{ij})\|_{M_n(E^{op})} \overset{\text{def}}{=} \|(a_{ji})\|_{M_n(E)}.$$

Equivalently, for any $a_i \in \mathcal{K}$, $x_i \in E$ we have

$$\left\| \sum a_i \otimes x_i \right\|_{\mathcal{K} \otimes_{\min} E^{op}} = \left\| \sum {}^t a_i \otimes x_i \right\|_{\mathcal{K} \otimes_{\min} E}.$$

In other words, E^{op} can be defined as the operator space structure on E for which the transposition: $x \to {}^t x \in B(H^*)$ defines a completely isometric embedding of E^{op} into $B(H^*)$.

Let A be a C^*-algebra. In that case, note that both $\overline{A} \subset B(\overline{H})$ and $A^{op} \subset B(H^*)$ are C^*-subalgebras and, moreover, A^{op} and $\overline{A}$ are C^*-isomorphic; hence they are completely isometric. Indeed, the involution $x \to x^*$ is an antilinear isomorphism between A and A^{op} and hence a linear one between $\overline{A}$ and A^{op}.

Note however that this is special to C^*-algebras: When E is a general operator space, we have $E^{op} \not\simeq \overline{E}$, for instance, $\overline{R}$ and R are completely isometric but R^{op} is completely isomorphic to C, and hence not to R.

2.11. Ruan's Theorem and quantization

Ruan's Theorem illustrates very well the idea of *quantization* (see [E2]). This usually refers to the basic idea of quantum mechanics to replace (real or complex) numbers by operators (hermitian or not) on a Hilbert space.

Here our field of scalars is replaced by $\mathcal{K}$ (or by $B(\ell_2)$) and the product is replaced by the tensor product. Seen through this light, the formulas become quite familiar. While a Banach space structure can be summarized by a norm on a vector space

$$E \simeq \mathbb{C} \otimes E,$$

an operator space can be summarized by a norm (verifying R_1 and R_2) on

$$\mathcal{K} \otimes E.$$

Let I be any set. Given a vector space V, we denote by $V^{(I)}$ the space of all finitely supported families $(v_i)_{i \in I}$ in V^I.

Let $(e_i)_{i \in I}$ be an algebraic basis for E. If we identify E to $\mathbb{C}^{(I)}$, then

$$\mathcal{K} \otimes E \simeq \mathcal{K}^{(I)},$$

which shows that the elements of $\mathcal{K}$ play the role of the "scalar coordinates" of a vector: The norm of E is defined on the linear combinations

$$\sum_{i \in I} \lambda_i e_i \qquad (\lambda_i) \in \mathbb{C}^{(I)}$$

while the norm α is defined on the tensors

$$\sum_{i \in I} a_i \otimes e_i \qquad (a_i) \in \mathcal{K}^{(I)}.$$

Seen with this interpretation, the various identities concerning operator spaces become much more transparent: For all the formulas defining the dual, the quotient, the interpolated space, and so on, the same observation is relevant.

Let us illustrate this principle for the dual: Let $E \subset B(H)$ be an operator space. Suppose E is finite-dimensional to simplify. Let $e_1, \ldots, e_n$ be a basis for E. For all coefficients $a_1, \ldots, a_n$ in $\mathcal{K}$ let

$$\|(a_1, \ldots, a_n)\|^* = \sup \left\{ \left\| \sum a_i \otimes b_i \right\|_{\mathcal{K} \otimes_{\min} \mathcal{K}} \mid b_i \in \mathcal{K} \right.$$
$$\left. \left\| \sum_1^n e_i \otimes b_i \right\|_{E \otimes_{\min} \mathcal{K}} \leq 1 \right\}. \tag{2.11.1}$$

Then, by Ruan's Theorem, there are operators $T_1, \ldots, T_n$ in $B(H)$ (for $H = \ell_2$) such that, for all a_i in $\mathcal{K}$,

$$\|(a_1, \ldots, a_n)\|^* = \left\| \sum_1^n T_i \otimes a_i \right\|_{B(H) \otimes_{\min} \mathcal{K}},$$

and the dual operator space E^* can be identified to $\overline{\text{span}}[T_1, \ldots, T_n] \subset B(H)$ ([ER2, BP1]).

Actually, (2.11.1) remains valid in arbitrary dimension: Let $(e_i)_{i \in I}$ be a linear (i.e. algebraic) basis for an operator space E and let $(e_i^*)_{i \in I}$ be the biorthogonal system in E^*. Then, we can write for all "coefficients" $(a_i)_{i \in I}$, $a_i \in \mathcal{K}$ with only finitely many nonzero terms,

$$\| \sum a_i \otimes e_i^* \|_{\mathcal{K} \otimes_{\min} E^*} = \sup \{ \| \sum a_i \otimes b_i \|_{\mathcal{K} \otimes_{\min} \mathcal{K}} \mid b_i \in \mathcal{K},$$
$$\| \sum b_i \otimes e_i \|_{\mathcal{K} \otimes_{\min} E} \leq 1 \}. \tag{2.11.2}$$

Moreover, if we wish, we can replace $\mathcal{K}$ by $B(H)$ in (2.11.1) and (2.11.2). For example, the o.s. structure of R and C is described by the following formulas (which, by Exercise 2.1.4, are dual to each other in the sense of (2.11.2)) established in Remark 1.13:

$$\forall x_i \in \mathcal{K} \qquad \left\| \sum x_i \otimes e_{1i} \right\|_{\mathcal{K} \otimes_{\min} R} = \left\| \sum x_i x_i^* \right\|^{1/2}$$
$$\left\| \sum x_i \otimes e_{i1} \right\|_{\mathcal{K} \otimes_{\min} C} = \left\| \sum x_i^* x_i \right\|^{1/2}.$$

Note that these two expressions are not equivalent, which reflects the fact that R and C are not completely isomorphic (compare with (1.3)).

In particular, note for the record the following inequality (see Exercise 2.1.4)

$$\forall x_i, \ y_i \in B(H) \ (1 \leq i \leq n) \quad \|\sum x_i \otimes y_i\|_{\min} \leq \|\sum x_i x_i^*\|^{1/2} \|\sum y_i^* y_i\|^{1/2}. \tag{2.11.3}$$

This follows directly from (1.11) using $a_i = x_i \otimes I$, $b_i = I \otimes y_i$. We note in passing that (2.11.3) implies

$$\forall x_i \in B(H) \ (1 \leq i \leq n) \quad \|\sum x_i \otimes \overline{x_i}\|_{\min}^{1/2} \leq \max\{\|\sum x_i x_i^*\|^{1/2},$$

$$\|\sum x_i^* x_i\|^{1/2}\}. \tag{2.11.4}$$

2.12. Universal objects

Every Banach space E embeds into a space of the form $\ell_\infty(I)$ for some I, for instance, with $I = B_{E^*}$. When E is separable we can take $I = \mathbb{N}$. The operator space counterpart is obvious: Every o.s. E embeds into $B(H)$, and if E is separable, we can take H separable. More precisely, we have

Proposition 2.12.1. *Let $\mathcal{M} = \bigoplus_{n \geq 1} M_n$. Then any separable operator space embeds completely isometrically into $\mathcal{M}$. More generally, for any operator space E, there is a family of integers $(n_i)_{i \in I}$ (possibly with repetitions) such that E embeds completely isometrically into $\bigoplus_{i \in I} M_{n_i}$.*

Proof. Let I be the collection of all maps $v \colon E \to M_{n_v}$ with $\|v\|_{cb} \leq 1$. We define $J \colon E \to \bigoplus_{v \in I} M_{n_v}$ by $J(x) = \bigoplus_{v \in I} v(x)$. Then, by (2.1.6) and (2.6.2), J is a complete isometry. Now, when E is separable, each space $CB(E, M_n) = M_n(E^*)$ is weak-$*$ separable, and hence the supremum can be restricted a countable collection I, so we can assume $I = \mathbb{N}$. Finally, adding zero entries whenever necessary, we can assume $n_{i+1} > n_i$ for all $i \in \mathbb{N}$. Thus we obtain an embedding into $\mathcal{M}$. ∎

Remark. In the Banach space case, every separable space embeds into $C[0, 1]$, which itself is separable. Moreover, any n-dimensional space can be $1 + \varepsilon$-embedded into ℓ_∞^N for some $N = N(\varepsilon, n)$ suitably large (see Exercise 2.13.2). The analogous results for operator spaces fail dramatically: No single separable o.s. contains all of them, and the possibility of $(1 + \varepsilon)$-embedding a space E into M_N for some N is *very restrictive*. This is related to the notion of exactness that will be studied in Chapter 17.

The situation for quotients is a bit simpler: At least we do have a separable universal object. In the Banach space case, every Banach space E is isometric to a quotient of $\ell_1(I)$ for some set I, with $I = \mathbb{N}$ in the separable case. The o.s. analog is as follows.

Proposition 2.12.2. Let $\mathcal{M}_* = \ell_1(\{M_n^* \mid n \geq 1\})$. Then every separable operator space is completely isometric to a quotient of $\mathcal{M}_*$. More generally, for every o.s. E, there is a set I and a family of integers $(n_i)_{i \in I}$ so that E is a quotient of $\ell_1(\{M_{n_i}^* \mid i \in I\})$.

Proof. Let I be the collection of all possible maps $v \colon M_{n_v}^* \to E$ with $\|v\|_{cb} \leq 1$. Let $X_I = \ell_1(\{M_{n_i}^* \mid i \in I\})$. We define $q \colon X_I \to E$ by $q((\xi_v)_{v \in I}) = \sum_{v \in I} v(\xi_v)$. Then $\|q\|_{cb} \leq 1$. By construction, for any n, any $x \in M_n(E) = CB(M_n^*, E)$ admits a lifting with the same norm in $CB(M_n^*, X_I)$. Hence q is a complete metric surjection. (Alternatively, a simple argument shows that the adjoint $J = q^*$ is a complete isometry.) Hence q is a complete surjection onto E, so that $E \simeq X_I / \ker(q)$. Again, when E is separable we can restrict to I countable and $n_{i+1} > n_i$, and hence E is a quotient of $\mathcal{M}_*$. ∎

The next corollary explains why S_1 is viewed as the operator space analog of ℓ_1.

Corollary 2.12.3. Every separable operator space is completely isometric to a quotient of $S_1 = (\mathcal{K})^*$.

Proof. Since $c_0(\{M_n \mid n \geq 1\})$ is a subspace of $\mathcal{K}$, by Exercise 2.6.2 and by (2.4.3), $\mathcal{M}_* = \ell_1(\{M_n^* \mid n \geq 1\})$ is a quotient of $\mathcal{K}^* = S_1$, so this follows from the preceding statement. ∎

2.13. Perturbation lemmas

We end this chapter with several simple facts from the Banach space folklore that have been easily transferred to the operator space setting. We start by a well-known fact (the proof is the same as for ordinary norms of operators).

Lemma 2.13.1. Let $v \colon X \to Y$ be a complete isomorphism between operator spaces. Then clearly any map $w \colon X \to Y$ with $\|v - w\|_{cb} < \|v^{-1}\|_{cb}^{-1}$ is again a complete isomorphism, and if we let $\Delta = \|v - w\|_{cb}\|v^{-1}\|_{cb}$, we have

$$\|w^{-1}\|_{cb} \leq \|v^{-1}\|_{cb}(1 - \Delta)^{-1} \quad \text{and} \quad \|w^{-1} - v^{-1}\|_{cb} \leq \|v^{-1}\|_{cb}^2(1 - \Delta)^{-1}.$$

Recall that the c.b. distance between two n-dimensional operator spaces E_1, E_2 is defined as follows:

$$d_{cb}(E_1, E_2) = \inf\{\|w\|_{cb}\|w^{-1}\|_{cb}\},$$

where the infimum runs over all possible complete isomorphisms $w \colon E_1 \to E_2$.

Lemma 2.13.2. *Fix $0 < \varepsilon < 1$. Let X be an operator space. Consider a biorthogonal system (x_i, x_i^*) $(i = 1, 2, \ldots, n)$ with $x_i \in X$, $x_i^* \in X^*$ and let $y_1, \ldots, y_n \in X$ be such that*

$$\sum \|x_i^*\| \, \|x_i - y_i\| < \varepsilon.$$

Then there is a complete isomorphism $w \colon X \to X$ such that $w(x_i) = y_i$ and

$$\|w\|_{cb} \le 1 + \varepsilon \qquad \|w^{-1}\|_{cb} \le (1 - \varepsilon)^{-1}.$$

In particular, if $E_1 = \mathrm{span}(x_1, \ldots, x_n)$ and $E_2 = \mathrm{span}(y_1, \ldots, y_n)$, we have

$$d_{cb}(E_1, E_2) \le (1 + \varepsilon)(1 - \varepsilon)^{-1}.$$

Proof. Recall (Proposition 1.10 (ii)) that any rank one linear map $v \colon X \to X$ satisfies $\|v\| = \|v\|_{cb}$. Let $\delta \colon X \to X$ be the map defined by setting $\delta(x) = \sum x_i^*(x)(y_i - x_i)$ for all x in X. Then $\|\delta\|_{cb} \le \sum \|x_i^*\| \, \|y_i - x_i\| < \varepsilon$. Let $w = I + \delta$. Note that $w(x_i) = y_i$ for all $i = 1, 2, \ldots, n$, $\|w\|_{cb} \le 1 + \|\delta\|_{cb} \le 1 + \varepsilon$, and by the preceding lemma we have $\|w^{-1}\|_{cb} \le (1 - \varepsilon)^{-1}$. ∎

Corollary 2.13.3. *Let X be any separable operator space. Then, for any n, the set of all the n-dimensional subspaces of X is separable for the distance associated to d_{cb}.*

Proof. Let $(x_1(m), \ldots, x_n(m))$ be a dense sequence in the set of all linearly independent n-tuples of elements of X. Let $E_m = \mathrm{span}(x_1(m), \ldots, x_n(m))$. Then, by the preceding lemma, for any $\varepsilon > 0$ and any n-dimensional subspace $E \subset X$, there is an m such that $d_{cb}(E, E_m) \le 1 + \varepsilon$. ∎

Lemma 2.13.4. *Consider an operator space E and a family of subspaces $E_\alpha \subset E$ directed by inclusion and such that $\overline{\cup E_\alpha} = E$. Then, for any $\varepsilon > 0$ and any finite-dimensional subspace $S \subset E$, there exists α and $\widetilde{S} \subset E_\alpha$ such that $d_{cb}(S, \widetilde{S}) < 1 + \varepsilon$. Let $u \colon F_1 \to F_2$ be a linear map between two operator spaces. Assume that u admits the factorization $F_1 \xrightarrow{a} E \xrightarrow{b} F_2$ with c.b. maps a, b such that a is of finite rank. Then for each $\varepsilon > 0$ there exists α and a factorization $F_1 \xrightarrow{\widetilde{a}} E_\alpha \xrightarrow{\widetilde{b}} F_2$ with $\|\widetilde{a}\|_{cb} \|\widetilde{b}\|_{cb} < (1 + \varepsilon)\|a\|_{cb}\|b\|_{cb}$ and $\widetilde{a}$ of finite rank.*

Proof. For the first part, let $x_1, \ldots, x_n$ be a linear basis of S and let x_i^* be the dual basis extended (by Hahn-Banach) to elements of E^*. Fix $\varepsilon' > 0$. Choose α large enough and $y_1, \ldots, y_n \in E_\alpha$ such that $\sum \|x_i^*\| \, \|x_i - y_i\| < \varepsilon'$. Let $\widetilde{S} = \mathrm{span}(y_1, \ldots, y_n)$. Then, by the preceding lemma, there is a complete isomorphism $w\colon E \to E$ with $\|w\|_{cb}\|w^{-1}\|_{cb} < (1 + \varepsilon')(1 - \varepsilon')^{-1}$ such that $w(S) = \widetilde{S} \subset E_\alpha$. In particular, $d_{cb}(S, \widetilde{S}) \le (1 + \varepsilon')(1 - \varepsilon')^{-1}$, so it suffices to adjust ε' to obtain the first assertion.

Now consider a factorization $F_1 \overset{a}{\longrightarrow} E \overset{b}{\longrightarrow} F_2$ and let $S = a(F_1)$. Note that S is finite-dimensional by assumption. Applying the preceding to this S, we find α and a complete isomorphism $w\colon E \to E$ with $\|w\|_{cb}\|w^{-1}\|_{cb} < 1 + \varepsilon$ such that $w(S) \subset E_\alpha$. Thus, if we take $\widetilde{a} = wa\colon F_1 \to E_\alpha$ and $\widetilde{b} = bw^{-1}_{|E_\alpha}$, we obtain the announced factorization.   ∎

Exercises

Exercise 2.13.1. Let $0 < \varepsilon < 1$. Let $\{x_1, \ldots, x_N\}$ be an ε-net in the unit sphere of a finite-dimensional Banach space E. Let $u\colon E \to F$ be a linear map into another Banach space with $\|u\| \le 1$. Assume that $\varepsilon' \ge 0$ satisfies $\varepsilon + \varepsilon' < 1$ and

$$\|u(x_i)\| \ge 1 - \varepsilon' \qquad \forall i = 1, 2, \ldots, N.$$

Then

$$\|u(x)\| \ge (1 - \varepsilon - \varepsilon')\|x\| \qquad \forall x \in E,$$

and u defines an isomorphism from E to $u(E)$ such that $\|u^{-1}\colon u(E) \to E\| \le (1 - \varepsilon - \varepsilon')^{-1}$. Consequently, $d(E, u(E)) \le (1 - \varepsilon - \varepsilon')^{-1}$.

Exercise 2.13.2. Let E be a finite-dimensional Banach space. Show that, for any $\delta > 0$, there is an integer N and a subspace $\widetilde{E} \subset \ell^N_\infty$ such that $d(E, \widetilde{E}) \le 1 + \delta$.

Exercise 2.13.3. Let A be the (commutative) C^*-algebra of all continuous functions on a compact set K. Show that there is a net of maps $u_\alpha\colon A \to A$ admitting factorizations of the form $A \overset{v_\alpha}{\longrightarrow} \ell^{n_\alpha}_\infty \overset{w_\alpha}{\longrightarrow} A$ with $\|v_\alpha\|\|w_\alpha\| \le 1$ that tend pointwise to the identity on A.

Chapter 3. Minimal and Maximal Operator Space Structures

Let E be a normed space. Then each linear embedding of E into $B(H)$ with H Hilbert defines an operator space structure on E. If the embedding is isometric (i.e., preserves the original norm of E), then we will say (in this section only) that the o.s.s. is admissible. Note that the associated norm α on $\mathcal{K} \otimes E$ then satisfies, for any $a \in \mathcal{K}$ and any $e \in E$, $\alpha(a \otimes e) = \|a\| \|e\|$. Blecher and Paulsen observed that the set of all admissible o.s.s. on a given normed space E admits both a minimal and a maximal element. The minimal one is easy to describe: Simply embed E isometrically into a commutative C^*-algebra C (for instance, we can take $C = C(T)$ the algebra of continuous functions on the compact set $T = (B_{E^*}, \sigma(E^*, E))$). Let us denote by $\min(E)$ the resulting operator space. By Proposition 1.10(ii), $\min(E)$ does not depend on the choice of C. Moreover, by (1.9), we have $\forall (a_{ij}) \in M_n(\min(E))$

$$\|(a_{ij})\|_{M_n(\min(E))} = \sup_{\xi \in B_{E^*}} \|(\xi(a_{ij}))\|_{M_n} = \Big\| \sum e_{ij} \otimes a_{ij} \Big\|_{M_n \overset{\vee}{\otimes} E}. \tag{3.1}$$

Equivalently, the norm $\alpha_{\min}$ on $\mathcal{K} \otimes E$ associated to $\min(E)$ coincides with the injective tensor norm (see Remark 1.11) on $\mathcal{K} \otimes E$.

Clearly (see Proposition 1.10(ii)), for any operator space F, any linear map $u \colon F \to E$ satisfies

$$\|u \colon F \to \min(E)\|_{cb} = \|u\|. \tag{3.2}$$

The maximal tensor product can be described as follows: Let I be the collection of all maps $u \colon E \to B(H_u)$ with $\|u\| \le 1$.

(Warning: A "collection" is not necessarily a set. Throughout these notes, we deliberately ignore this set theoretic difficulty wherever it appears since it is obvious how to fix it.)

Consider the embedding

$$j \colon E \to \bigoplus_{u \in I} B(H_u) \subset B\left(\bigoplus_{u \in I} H_u \right)$$

defined by

$$\forall x \in E \qquad j(x) = \bigoplus_{u \in I} u(x).$$

Then (as was explained in §2.6) this embedding (which clearly is isometric) defines an admissible o.s.s. on E. We denote by $\max(E)$ the resulting operator space.

We have, by (2.6.2), $\forall (a_{ij}) \in M_n(\max(E))$

$$\|(a_{ij})\|_{M_n(\max(E))} = \sup\{\|(u(a_{ij}))\|_{M_n(B(H_u))} \mid u \colon E \to B(H_u), \|u\| \le 1\}. \tag{3.3}$$

Clearly, for any operator space F, any linear map $u: E \to F$ satisfies

$$\|u: \max(E) \to F\|_{cb} = \|u\|. \tag{3.4}$$

Let $\alpha_{\max}$ be the norm on $\mathcal{K} \otimes E$, corresponding to the operator space $\max(E)$. (Warning: It is *not* the projective norm in Grothendieck's sense!) For any admissible o.s.s. on E, the identity on E defines completely contractive maps

$$\max(E) \to E \to \min(E).$$

In other words, the set of norms α on $\mathcal{K} \otimes E$ satisfying Axioms (R_1) and (R_2) and (say) $\alpha(e_{11} \otimes e) = \|e\|$ $\forall e \in E$ admits a minimal element $\alpha_{\min}$ and a maximal one $\alpha_{\max}$ and

$$\alpha_{\min} \leq \alpha \leq \alpha_{\max}.$$

We have completely isometrically (see [BP1, B2])

$$\min(E)^* = \max(E^*) \quad \text{and} \quad \max(E)^* = \min(E^*). \tag{3.5}$$

For the proof, see Exercise 3.2.

In particular, we have for any set I and any measure space (Ω, μ)

$$[\max(\ell_1(I))]^* = \ell_\infty(I) \quad \text{and} \quad [\max(L_1(\Omega, \mu))]^* = L_\infty(\Omega, \mu),$$

where $\ell_\infty(I)$ or $L_\infty(\Omega, \mu)$ are equipped with their o.s. structures as commutative C^*-algebras.

In [Pa2], Paulsen observed that the norm of the space $M_n(\max(E))$ can be described as follows:

Theorem 3.1. *Let* $x \in M_n(E)$. *Then* $\|x\|_{M_n(\max(E))} < 1$ *iff* x *admits for some* N *a factorization of the form* $x = \alpha_0 D \alpha_1$, *where* α_0, α_1 *are (rectangular) scalar matrices of size* $n \times N$ *and* $N \times n$, *respectively, and where* $D \in M_N(E)$ *is a diagonal matrix such that* $\|\alpha_0\| \, \|D\| \, \|\alpha_1\| < 1$.

Proof. Let $\|\|x\|\|_n = \inf\{\|\alpha_0\| \, \|D\| \, \|\alpha_1\|\}$, where the infimum runs over all possible factorizations as above. It is not hard to check that Axioms (R_1) and (R_2) hold in this case, so that, by Ruan's Theorem (see §2.2), these norms come from an operator space structure on E. Let $\widetilde{E}$ be the resulting operator space so that $\|x\|_{M_n(\widetilde{E})} = \|\|x\|\|_n$. By examining the case $n = 1$, we see that $\widetilde{E}$ is isometric to E. Moreover, by the very definition of $\|\| \ \|\|_n$, it is easy to see that the identity defines a complete contraction from $\widetilde{E}$ to $\max(E)$, but then by the maximality of $\max(E)$ we conclude $\widetilde{E} = \max(E)$ completely isometrically. ∎

A variant of the preceding argument yields the following:

Theorem 3.2. *([Pa2]) Let Γ be an arbitrary set, let $\ell_1(\Gamma)$ be the classical ℓ_1-space over the index set Γ, and let $E = \max(\ell_1(\Gamma))$. We denote by $E_0 \subset E$ the linear span of the canonical basis vectors $(e_\gamma)_{\gamma \in \Gamma}$ in E. Consider x in $M_n(E_0)$. Then $\|x\|_{M_n(E)} < 1$ iff x admits (for some integer N) a factorization of the form $x = \alpha_0 D \alpha_1$, where α_0, α_1 are scalar matrices of size $n \times N$ and $N \times n$, respectively, and where D is a diagonal matrix with entries of the form $(e_{\gamma_1}, \ldots, e_{\gamma_N})$ for some N-tuple $\gamma_1, \ldots, \gamma_N$ in Γ.*

Naturally, we will say that an operator space is "minimal" (resp. "maximal") if $\min(E) = E$ (resp. $\max(E) = E$) completely isometrically.

A minimal (resp. maximal) operator space E is characterized by the following property:

$$\forall F \ \forall u : F \to E \text{ (resp. } \forall u : E \to F) \quad \|u\|_{cb} = \|u\|.$$

Indeed, this property clearly holds if E is minimal (resp. maximal), and the converse is easy to show: Take $F = \min(E)$ (resp. $F = \max(E)$) and let $u = I_E$.

Obviously, minimality passes to subspaces, but it need not pass to quotients (see, e.g., §9.1). On the other hand, maximality does not pass in general to subspaces, but it does to quotients, as follows.

Proposition 3.3. *Any quotient of a maximal operator space is itself maximal. More precisely, let $E = \max(E)$ be a maximal operator space and let $S \subset E$ be a closed subspace. Then the following completely isometric identity holds:*

$$\max(E/S) = \max(E)/S.$$

In particular, any maximal operator space $E = \max(E)$ is completely isometric to a quotient of $\max(\ell_1(I))$ for some set I, and if E is separable, we can take $I = \mathbb{N}$.

Proof. Let F be an arbitrary operator space and let $u : E/S \to F$ be an arbitrary map. We denote by $q : E \to E/S$ the canonical surjection. Clearly, it suffices to show that

$$\|u : \max(E/S) \to F\|_{cb} = \|u : \max(E)/S \to F\|_{cb}. \tag{3.6}$$

But the left side of (3.6) is equal to $\|u\|$ and, by (2.4.2), the right side is $= \|uq\|_{cb} = \|uq\| = \|u\|$. Thus we indeed have (3.6). Since any Banach space is isometric to a quotient of $\ell_1(I)$, with $I = \mathbb{N}$ in the separable case, the last assertion is clear. ∎

Let E be a Banach space. Recall we denote by I_E the identity mapping on E. Paulsen [Pa2] introduced the constant

$$\alpha(E) = \|I_E : \min(E) \to \max(E)\|_{cb}.$$

It is easy to show (exercise) that $\alpha(E)$ is equal to the supremum of the ratio $\|u\|_{cb}/\|u\|$ when u runs over all possible maps $u: \widetilde{E} \to F$ with $\widetilde{E}$ isometric to E and F arbitrary.

The finite-dimensional case is quite interesting:

Proposition 3.4. *([Pa2]) Let $E = (\mathbb{C}^n, \| \ \|_E)$ be any n-dimensional normed space. Then*

$$\alpha(E) = \sup\left\{\left\|\sum_1^n a_k \otimes b_k\right\|_{\min}\right\}, \tag{3.7}$$

where the supremum runs over all sets $a_1, \ldots, a_n \in B(H)$, $b_1, \ldots, b_n \in B(K)$ $(H, K$ arbitrary Hilbert spaces) such that

$$\left\|\sum t_k a_k\right\|_{B(H)} \leq \|t\|_E \qquad\qquad \forall t \in \mathbb{C}^n$$

and

$$\left\|\sum t_k b_k\right\|_{B(K)} \leq \|t\|_{E^*}. \qquad\qquad \forall t \in \mathbb{C}^n$$

Proof. By the definition of $\alpha(E)$, we have

$$\alpha(E) = \sup_m \sup\{\|x\|_{M_m(\max(E))} \mid \|x\|_{M_m(\min(E))} \leq 1\}. \tag{3.8}$$

Consider $x = (x_{ij}) \in M_m(E)$ with $\|x\|_{M_m(\min(E))} \leq 1$. Let $\{e_k\}$ be the canonical basis of $E = \mathbb{C}^n$ and let $\xi_k \in E^*$ be the dual basis. Let $b_k = \sum_{i,j} e_{ij} \otimes \xi_k(x_{ij}) \in M_m$. By (3.1), we have

$$\sup\left\{\left\|\sum t_k b_k\right\|_{M_m} \;\middle|\; (t_k) \in \mathbb{C}^n, \left\|\sum t_k \xi_k\right\|_{E^*} \leq 1\right\} = \|x\|_{M_m(\min(E))} \leq 1. \tag{3.9}$$

On the other hand, by (2.6.2) and the definition of $\max(E)$, we have

$$\|x\|_{M_m(\max(E))} = \sup_{\|v\| \leq 1} \left\|\sum e_{ij} \otimes v(x_{ij})\right\|_{M_m(B(H_v))}. \tag{3.10}$$

Now fix $v: E \to B(H)$ with $\|v\| \leq 1$ and let $a_k = v(e_k)$. Note that, for any t in $\mathbb{C}^n$, we have $\|\sum t_k a_k\| \leq \|v\| \|t\|_E \leq \|t\|_E$. Moreover,

$$\left\|\sum e_{ij} \otimes v(x_{ij})\right\| = \left\|\sum_{i,j,k} e_{ij} \xi_k(x_{ij}) \otimes v(e_k)\right\|$$

$$= \left\|\sum b_k \otimes a_k\right\|_{M_m(B(H))} = \left\|\sum a_k \otimes b_k\right\|_{B(H) \otimes_{\min} M_m}.$$

Combining this with (3.8), (3.9), and (3.10), we obtain

$$\alpha(E) \leq \sup\left\{\left\|\sum a_k \otimes b_k\right\|_{\min}\right\}$$

with the supremum as in (3.7). Moreover, we have equality when the supremum is restricted to K finite-dimensional. Then, invoking (2.1.7), we obtain the equality in (3.7), as announced. ∎

Remark. By (2.1.7), the supremum is the same in (3.7) if we restrict H and K both to be finite-dimensional.

Corollary 3.5. *([Pa2]) For any n-dimensional normed space E, we have*

$$\alpha(E) = \alpha(E^*). \tag{3.11}$$

Note that this also can be deduced from (3.5) and (2.3.3).

We now return to the number $\alpha(n)$ considered at the end of Chapter 1.

Proposition 3.6. *([Pa2]) Let*

$$\alpha(n) = \sup\left\{ \frac{\|u\|_{cb}}{\|u\|} \;\middle|\; u\colon E \to F,\ \mathrm{rk}(u) \le n \right\},$$

where the supremum runs over all possible operator spaces E, F. Then

$$\alpha(n) = \sup\{\alpha(G) \mid \dim G \le n\}. \tag{3.12}$$

The proof is left as an exercise.

To majorize $\alpha(n)$, we will need the following classical lemma.

Lemma 3.7. *(Auerbach's Lemma) Let E be an arbitrary n-dimensional normed space. There is a biorthogonal system $x_i \in E, \xi_i \in E^*$ $(i = 1, 2, \ldots, n)$ such that $\|x_i\| = \|\xi_i\| = 1$ for all $i = 1, \ldots, n$.*

Proof. Choose $x_1, \ldots, x_n$ in the unit sphere of E on which the function $x \to |\det(x_1, \ldots, x_n)|$ attains its maximum, supposed equal to $C > 0$. Then let $\xi_i(y) = C^{-1} \det(x_1, \ldots, x_{i-1}, y, x_{i+1}, \ldots, x_n)$. The desired properties are easy to check. ∎

Theorem 3.8. *([Pa2]) For any $n \ge 1$, we have*

$$n/2 \le \alpha(\ell_2^n) \le \alpha(n) \le n$$

and

$$(n/2)^{1/2} \le \alpha(\ell_\infty^n) \le \sqrt{n}.$$

Proof. We first show that $\alpha(n) \le n$. Consider E with $\dim(E) = n$. Let (x_i, ξ_i) be as in the preceding lemma. We have $I_E(x) = \sum_1^n u_i(x)$ with

$u_i(x) = \xi_i(x)x_i$. Hence $\alpha(E) = \|I_E: \min(E) \to \max(E)\|_{cb} \leq \sum_1^n \|u_i:$
$\min(E) \to \max(E)\|_{cb}$. But since u_i is of rank ≤ 1, we have (see Proposition 1.10.(ii)) $\|u_i\|_{cb} = \|u_i\| = \|x_i\| \|\xi_i\| = 1$, whence $\alpha(E) \leq n$, and by (3.12) $\alpha(n) \leq n$. This bound is improved later in Theorem 7.15.

We now turn to the converse estimate. Let $(U_1, \ldots, U_n)$ be a "spin system," that is, an n-tuple of unitary self-adjoint operators such that

$$\forall i \neq j \qquad U_i U_j + U_j U_i = 0.$$

A simple calculation shows that, for any x in $\mathbb{C}^n$, we have

$$\left\| \sum x_i U_i \right\| \leq \sqrt{2} \left(\sum |x_i|^2 \right)^{1/2}.$$

Indeed, let $T = \sum x_i U_i$; we then have $T^*T + TT^* = 2\sum |x_i|^2 \cdot I$, and hence $\|T\|^2 \leq 2\sum |\xi_i|^2$.

Moreover, it is easy to see that such a system generates a finite-dimensional C^*-algebra (indeed the linear span of all the 2^n products $U_{i_1} U_{i_2} \ldots U_{i_k}$ with $i_1 < i_2 < \ldots < i_k$ is a C^*-algebra). Therefore, such a system can be realized in $B(H)$ with $\dim(H) < \infty$ (actually with $\dim(H) = 2^n$). From this, we will deduce the following claim:

$$\left\| \sum_1^n U_k \otimes {}^t U_k \right\|_{\min} = n. \tag{3.13}$$

Indeed, let (e_p) be an orthonormal basis of H. Let $t = \sum e_p \otimes e_p$ be the tensor in $H \otimes_2 H$ corresponding to the identity of H. Since U_k is self-adjoint and unitary, a simple calculation shows that $U_k \otimes {}^t U_k(t) = t$, and hence $\sum_1^n U_k \otimes {}^t U_k(t) = nt$, whence $\|\sum_1^n U_k \otimes {}^t U_k\| \geq n$. The converse is obvious, whence the claim (3.13). Applying (3.7) with $a_k = {}^t b_k = 2^{-1/2} U_k$, we obtain

$$n/2 \leq \alpha(\ell_2^n) \tag{3.14}$$

and a fortiori $n/2 \leq \alpha(n)$.

We now turn to the second line. The lower bound follows again from (3.7) with $a_k = U_k/(2n)^{1/2}$ and $b_k = {}^t U_k$. The remaining upper bound is left as an exercise (Hint: Use, e.g., the following factorization of the identity $\min(\ell_1^n) \to C_n \to \max(\ell_1^n)$, where the first arrow has c.b. norm $= 1$ and the second one $= \sqrt{n}$). ∎

The next result comes from [Pa2] for $\dim(E) > 4$ (we could include dimensions 2 and 3 thanks to exercise 3.7).

Corollary 3.9. *Any infinite-dimensional Banach space E satisfies $\alpha(E) = \infty$. Moreover, we have $\alpha(E) > 1$ as soon as $\dim(E) > 2$.*

Proof. Observe that, if F is another Banach space, we have obviously

$$\alpha(F) \leq d(E, F)\alpha(E). \qquad (3.15)$$

In particular, $\alpha(\ell_2^n) \leq d(E, \ell_2^n)\alpha(E)$. By a well-known result due to F. John (see, e.g., [P8] or [TJ1]), if $\dim(E) = n$, then $d(E, \ell_2^n) \leq \sqrt{n}$. Thus, together with Theorem 3.8, this gives us that any n-dimensional space E satisfies $\sqrt{n}/2 \leq \alpha(E)$. In particular, $\alpha(E) > 1$ whenever $n > 4$. Using Exercise 3.7, we can extend this $n > 2$.

When E is infinite-dimensional, we use a variant of F. John's Theorem: For any n, there is a subspace $E_n \subset E$ with $\dim(E_n) = n$ and an isomorphism $u\colon E_n \to \ell_2^n$ with $\|u^{-1}\| \leq 1$, $\|u\| \leq \sqrt{n}$ but, moreover, such that u admits an extension $\widetilde{u}\colon E \to \ell_2^n$ with $\|\widetilde{u}\| \leq \sqrt{n}$. Then again we claim that $\alpha(E) \geq \alpha(\ell_2^n)/\sqrt{n}$. Indeed, the identity $\min(\ell_n^2) \to \max(\ell_2^n)$ factors as

$$\min(\ell_2^n) \xrightarrow{u^{-1}} \min(E) \xrightarrow{I_E} \max(E) \xrightarrow{\widetilde{u}} \max(\ell_2^n),$$

which implies that (recalling (3.2) and (3.4))

$$\alpha(\ell_2^n) = \|I\colon \min(\ell_2^n) \to \max(\ell_2^n)\|_{cb} \leq \|u^{-1}\|\alpha(E)\|\widetilde{u}\| \leq \sqrt{n}\alpha(E).$$

Letting n go to ∞, we obtain $\alpha(E) = \infty$ by Theorem 3.8. ∎

Remark. By Corollary 3.9, any infinite-dimensional Banach space E can be embedded isometrically into some $B(H)$ so that there is a bounded map (and actually an isometric one) $u\colon E \to B(H)$ that is not c.b.

There are some interesting questions left open in the dimensions $n = 2, 3$, or 4 that are aimed at measuring how small (or how large) the set of all admissible o.s.s. on E can be. For instance, what is exactly its diameter? Which Banach spaces E admit a unique admissible operator space structure? Equivalently, for which E do we have $\alpha(E) = 1$? By Corollary 3.9, the dimension of such an E must be ≤ 2, but the only known examples of dimension > 1 are those of the following result: the spaces ℓ_∞^2 and ℓ_1^2 (i.e. the two-dimensional versions of ℓ_∞ and ℓ_1). Are these the only examples?

Proposition 3.10. *[Pa2]. There is a unique operator space structure (respecting the norm) on the spaces ℓ_1^2 and ℓ_∞^2. Equivalently, we have $\alpha(\ell_1^2) = \alpha(\ell_\infty^2) = 1$.*

Proof. Let e_1, e_2 be the canonical basis of ℓ_1^2. The space $\min(\ell_1^2)$ can be realized in $C(\partial D)$ (here $D = \{z \in \mathbb{C} \mid |z| < 1\}$) as the linear span of the pair $[1, z]$, so that we have for any pair (x_1, x_2) in M_n:

$$\|x_1 \otimes e_1 + x_2 \otimes e_2\|_{M_n(\min(\ell_1^2))} = \sup_{z \in \partial D} \|x_1 + z x_2\|_{M_n}. \qquad (3.16)$$

On the other hand, note that the data of a contraction u: $\ell_1^2 \to B(H)$ boil down to that of a pair of contractions T_1, T_2 in $B(H)$ with $T_i = u(e_i)$. Therefore we have

$$\|x_1 \otimes e_1 + x_2 \otimes e_2\|_{M_n(\max(\ell_1^2))} = \sup\{\|x_1 \otimes T_1 + x_2 \otimes T_2\|_{\min}\}, \quad (3.17)$$

where the supremum runs over all H and all pairs (T_1, T_2) of contractions in $B(H)$. Since any contraction is in the closed convex hull of the unitaries (by the Russo-Dye Theorem [Ped, p.4]), the last supremum remains the same if we restrict ourselves to pairs of unitary operators (T_1, T_2). But then we may multiply by T_1^{-1} (say on the left), and we find that we can restrict ourselves further to pairs of the form $(1, T)$ with T unitary. Then, the pair $(1, T)$ lies in a commutative C^*-algebra that can be identified with $C(\sigma(T))$, $\sigma(T)$ being the spectrum of T, and $\sigma(T) \subset \partial D$. Thus we have clearly

$$\|x_1 \otimes 1 + x_2 \otimes T\| = \sup_{z \in \sigma(T)} \|x_1 + zx_2\|_{M_n} \leq \sup_{z \in \partial D} \|x_1 + zx_2\|_{M_n}.$$

Recalling (3.16) and (3.17), the previous discussion shows that $\|\min(\ell_1^2) \to \max(\ell_1^2)\|_{cb} = 1$, whence $\alpha(\ell_1^2) = 1$. By (3.11), we have $\alpha(\ell_\infty^2) = 1$. ∎

Remark. In [Pa2], Paulsen actually shows that

$$\max\{\sqrt{n}, n/2\} \leq \alpha(\ell_2^n) \leq n/\sqrt{2}. \quad (3.18)$$

and

$$(n/2)^{1/2} \leq \alpha(\ell_\infty^n) = \alpha(\ell_1^n) \leq \min\{\sqrt{n}, n/2\}, \quad (3.19)$$

Note that, by (3.15), $\alpha(\ell_2^n) \leq \sqrt{n}\alpha(\ell_\infty^n)$.

That $\sqrt{n} \leq \alpha(\ell_2^n)$ follows easily for instance from (1.5). The upper bound $\alpha(\ell_2^n) \leq n/\sqrt{2}$ follows from Proposition 3.10. Indeed, let (a_k) and (b_k) be as in Proposition 3.4 with $E = \ell_2^n$. By (3.15), we have $\alpha(\ell_2^n) \leq \sqrt{2}$. Hence for any $k \neq \ell$ we have

$$\|a_k \otimes b_k + a_\ell \otimes b_\ell\| \leq \sqrt{2},$$

and, on the other hand,

$$2(n-1) \sum a_k \otimes b_k = \sum_{k \neq \ell} a_k \otimes b_k + a_\ell \otimes b_\ell;$$

hence by the triangle inequality

$$2(n-1) \left\| \sum a_k \otimes b_k \right\| \leq \sum_{k \neq \ell} \|a_k \otimes b_k + a_\ell \otimes a_\ell\|$$

$$\leq n(n-1)\sqrt{2},$$

and thus we obtain $\|\sum a_k \otimes b_k\| \leq n/\sqrt{2}$. Hence, by Proposition 3.4, $\alpha(\ell_2^n) \leq n/\sqrt{2}$. A similar argument shows that $\alpha(\ell_1^n) \leq n/2$.

Finally, the bound $\alpha(\ell_1^n) \leq \sqrt{n}$ follows from (say) the factorization $\min(\ell_1^n) \xrightarrow{v} C_n \xrightarrow{w} \max(\ell_1^n)$ defined by $v(e_i) = e_{i1}$ and $w(e_{i1}) = e_i$, since it is easy to check that $\|v\|_{cb} = 1$ and $\|w\|_{cb} = \sqrt{n}$.

Paulsen [Pa2] conjectures that $\alpha(\ell_2^n) = n/\sqrt{2}$ and $\alpha(\ell_\infty^n) = \sqrt{n/2}$, but (contrary to this) we believe that the lower bound appearing in Exercise 3.7 is sharp.

We refer the reader to Paulsen's paper [Pa2] and to the work by Arias, Figiel, Johnson, and Schechtman [AFJS] for more on this theme.

Open problems

1. Are the spaces $\mathbb{C}, \ell_1^2$, and ℓ_∞^2 the only operator spaces $E \neq \{0\}$ admitting a unique operator space structure, that is, for which $\alpha(E) = 1$?

2. What is the value of $\alpha(\ell_2^n)$ for $n > 2$? What about $\alpha(\ell_\infty^n)$ for $n > 2$? Note that, by (3.18) and (3.19), we have $\alpha(\ell_2^2) = \sqrt{2}$ and $\alpha(\ell_\infty^2) = 1$.

More importantly:

3. What are the values of:

$$\limsup_{n\to\infty} \alpha(\ell_\infty^n)/n^{1/2}? \qquad \limsup_{n\to\infty} \alpha(\ell_2^n)/n? \qquad \limsup_{n\to\infty} \alpha(n)/n?$$

4. What are the n-dimensional normed spaces E such that $\alpha(E) = \alpha(n)$?

We discuss more open problems concerning maximal operator spaces and their subspaces in Chapter 18.

Exercises

Exercise 3.1. (Principle of local reflexivity [LiR].) Let E, G be Banach spaces with $\dim(G) < \infty$. We then have $(E \overset{\vee}{\otimes} G)^{**} = E^{**} \overset{\vee}{\otimes} G$ (isometrically). As a consequence, for any $v \colon E^* \to G$ there is a net of weak-$*$ continuous maps $v_\alpha \colon E^* \to G$ with $\|v_\alpha\| \leq \|v\|$ that tend to v in the topology of simple convergence.

Exercise 3.2. Let E be an arbitrary Banach space. Prove that $\min(E^*) = \max(E)^*$ and $\max(E^*) = \min(E)^*$ ([BP1]).

Exercise 3.3. Show that E is minimal (resp. maximal) iff its bidual E^{**} is also ([BP1]).

Exercise 3.4. Show that the direct sum of a family of minimal spaces is again minimal.

Exercise 3.5. Let $\{E_i \mid i \in I\}$ be a family of maximal operator spaces (for instance, we could have $E_i = \mathbb{C}$ for all i in I). Show that $\ell_1(\{E_i \mid i \in I\})$ is again a maximal operator space.

Exercise 3.6. Let $U_1, \ldots, U_n$ be a spin system. Show that the spaces

$$E = \text{span}\{U_i \otimes U_i \mid 1 \leq i \leq n\} \quad \text{and} \quad F = \text{span}\{U_i \otimes \overline{U_i} \mid 1 \leq i \leq n\}$$

are minimal operator spaces.

Exercise 3.7. Let C_i be the creation operators $(i = 1, \ldots, n)$ on the anti-symmetric Fock space (see §9.3) associated to ℓ_2^n. Show that $\|\sum_1^n x_k C_k\| = \left(\sum |x_k|^2\right)^{1/2}$ for any x in $\mathbb{C}^n$ and deduce that

$$\alpha(\ell_2^n) \geq \left\|\sum_1^n C_k \otimes C_k\right\|_{\min}.$$

Show that

$$\left\|\sum_1^n C_k \otimes C_k\right\| \geq (n+1)/2 \quad \text{if } n \text{ is odd}$$

and that

$$\left\|\sum_1^n C_k \otimes C_k\right\| \geq \left(\frac{n}{2} + 1\right)^{1/2} \left(\frac{n}{2}\right)^{1/2} \quad \text{if } n \text{ is even.}$$

Exercise 3.8. Let X be a maximal o.s. Let $X_1 \subset X$ be a separable subspace. Show that there is a subspace X_2 with $X_1 \subset X_2 \subset X$ that is separable and maximal.

Chapter 4. Projective Tensor Product

Since the minimal tensor product is analogous to the (Banach space) injective tensor product, it is natural to search for an analog in the operator space setting of the (Banach space) projective tensor product. This question is treated in [BP1] and [ER2] independently. Effros and Ruan went further and considered a version for operator spaces (in short, o.s.) of Grothendieck's approximation property. They proved the o.s. analog of many of Grothendieck's Banach space results, introduced integral operators and absolutely summing operators, and proved a version for o.s. of the Dvoretzky-Rogers Lemma. This program meets several interesting difficulties and leaves open several problems, mostly related to the absence of local reflexivity for operator spaces. We will limit ourselves here to a brief description of the o.s. version of the projective tensor product of two operator spaces E, F, which we will denote by $E \otimes^\wedge F$. The latter space is defined in [BP1] as the natural "predual" of the space $CB(E, F^*)$. The equivalent but more explicit definition of [ER2] is as follows.

Let t be an element of the algebraic tensor product $E \otimes F$. Of course, t admits a (non-unique) representation of the form

$$t = \sum_{i,j \leq \ell \ p,q \leq m} \alpha_{ip} \, x_{ij} \otimes y_{pq} \, \beta_{jq},$$

where ℓ, m are integers and where $x \in M_\ell(E)$, $y \in M_m(F)$, and α, β are rectangular scalar matrices.

Then the "projective" tensor norm (in the sense of o.s.) $\|t\|_{E \otimes^\wedge F}$ is defined as:

$$\|t\|_{E \otimes^\wedge F} = \inf\{\|\alpha\|_{HS} \|x\|_{M_\ell(E)} \|y\|_{M_m(F)} \|\beta\|_{HS}\},$$

where $\| \ \|_{HS}$ is the Hilbert-Schmidt norm and where the infimum runs over all possible representations (actually, by adding zeroes we may restrict attention to the case $\ell = m$ if we wish).

We denote by $E \otimes^\wedge F$ the completion of $E \otimes F$ with respect to this norm. More generally, this space can be equipped with an operator space structure corresponding to the norm defined (for each n) on $M_n(E \otimes^\wedge F)$ as follows: Let $t = (t_{rs}) \in M_n(E \otimes F)$ and assume

$$t = \alpha \cdot (x \otimes y) \cdot \beta, \tag{4.1}$$

where the dot denotes the matricial product and where $x \in M_\ell(E)$, $y \in M_m(F)$, and α (resp. β) is a matrix of size $n \times (\ell m)$ (resp. $(\ell m) \times n$). Note that $x \otimes y$ is viewed here as an element of the space $M_{\ell m}(E \otimes F)$. More explicitly, we may index $[1, \ldots, m\ell]$ either by a pair (ip) $(1 \leq i \leq \ell, 1 \leq p \leq m)$ or by a pair (jq) $(1 \leq j \leq \ell, 1 \leq q \leq m)$, so that by dropping the parentheses we may write

$$\alpha = [\alpha_{r,ip}]_{r,ip} \quad \text{and} \quad \beta = [\beta_{jq,s}]_{jq,s}.$$

Then (4.1) can be rewritten more explicitly as

$$t_{rs} = \sum_{i,p,j,q} \alpha_{r,ip}(x_{ij} \otimes y_{pq})\beta_{jq,s} \qquad (1 \leq r \leq n, 1 \leq s \leq n). \tag{4.2}$$

Then (following [ER2]) we can define

$$\|t\|_{M_n(E\otimes^\wedge F)} = \inf\left\{\|\alpha\|_{M_{n,\ell m}} \|x\|_{M_\ell(E)} \|y\|_{M_m(F)} \|\beta\|_{M_{\ell m,n}}\right\}. \tag{4.3}$$

We then obtain an operator space structure on $E \otimes^\wedge F$. Moreover (see [ER2, BP1]):

Theorem 4.1. *We have completely isometrically*

$$(E \otimes^\wedge F)^* = CB(E, F^*). \tag{4.4}$$

More generally, for any operator space G, we have completely isometrically

$$CB(E \otimes^\wedge F, G) = CB(E, CB(F, G)). \tag{4.5}$$

Proof. Note that (4.4) is a particular case of (4.5) (take $G = \mathbb{C}$). Therefore we concentrate on (4.5). We will show that (4.5) is an isometric identity. It follows "automatically" (replacing G by $M_n(G)$) that it is completely isometric.

Let $\varphi\colon E \otimes F \to G$ be a linear map with associated mapping $u\colon E \to CB(F, G)$ defined by $u(x)(y) = \varphi(x \otimes y)$. To show that (4.5) is isometric, it clearly suffices to show that $I = II$, where

$$I = \|\varphi\|_{CB(E\otimes^\wedge F,G)} \quad \text{and} \quad II = \|u\|_{CB(E,CB(F,G))}.$$

By (4.2) and (4.3), we have

$$I = \sup\left\{\left\|\left[\sum_{ijpq}\alpha_{r,ip}\varphi(x_{ij} \otimes y_{pq})\beta_{jq,s}\right]_{r,s}\right\|_{M_n(G)}\right\},$$

where the supremum runs over all n and all m, ℓ, α, β, x, y with $\|\alpha\|_{M_{n,m\ell}} \leq 1$, $\|\beta\|_{M_{m\ell,n}} \leq 1$, $\|x\|_{M_\ell(E)} \leq 1$, $\|y\|_{M_m(F)} \leq 1$. Let $a \in M_{\ell m}(G)$ be the $\ell m \times \ell m$ matrix with entries as follows:

$$a(i, p; j, q) = \varphi(x_{ij} \otimes y_{pq}).$$

Clearly, the matrix product $\alpha.a.\beta$ satisfies

$$\sup\|\alpha.a.\beta\|_{M_n(G)} \leq \|a\|_{M_{\ell m}(G)}, \tag{4.6}$$

where the supremum runs over all α, β in the unit ball respectively of $M_{n,m\ell}$ and $M_{m\ell,n}$. In addition, when $n = \ell m$ the choice of α, β equal to the identity shows that (4.6) is an equality. A fortiori, we have equality whenever $n \leq \ell m$. Hence we conclude from this that

$$I = \sup\{\|[\varphi(x_{ij} \otimes y_{pq})]_{(ip),(jq)}\|_{M_{\ell m}(G)}\}, \tag{4.7}$$

where the supremum runs over all ℓ, m and all x, y as before.

On the other hand, we have by definition

$$II = \sup\{\|(I_{M_\ell} \otimes u)(x)\|_{M_\ell(CB(F,G))} \mid \ell \geq 1, \|x\|_{M_\ell(E)} \leq 1\}.$$

Then, using $M_\ell(CB(F,G)) = CB(F, M_\ell(G))$, we find

$$\|(I_{M_\ell} \otimes u)(x)\|_{M_\ell(CB(F,G))} =$$
$$\sup\{\|I_{M_m} \otimes [(I_{M_\ell} \otimes u)(x)](y)\|_{M_m(CB(F,M_\ell(G)))} \mid m \geq 1, \|y\|_{M_m}(F) \leq 1\}.$$

Now again $M_m(CB(F, M_\ell(G))) \simeq CB(F, M_m(M_\ell(G)) \simeq CB(F, M_{\ell m}(G))$. Moreover, in this correspondence

$$I_{M_m} \otimes [(I_{M_\ell} \otimes u)(x)](y)$$

is nothing but the $\ell m \times \ell m$ matrix with entries

$$a(i, p; j, q) = u(x_{ij})(y_{pq}) = \varphi(x_{ij} \otimes y_{pq}).$$

Thus by (4.7) we obtain $I = II$ as announced. ∎

Remark. In particular, taking $F = M_n^*$ and recalling that $M_n(E^*) \simeq CB(E, M_n)$, (4.4) implies

$$(M_n^* \otimes^\wedge E)^* \simeq (E \otimes^\wedge M_n^*)^* \simeq M_n(E^*). \tag{4.8}$$

The projective tensor product is a "good" one in the sense that, for any operator spaces E_i, F_i, if $u_1 \colon E_1 \to F_1$ and $u_2 \colon E_2 \to F_2$ are c.b., then $u_1 \otimes u_2 \colon E_1 \otimes^\wedge E_2 \longrightarrow F_1 \otimes^\wedge F_2$ is c.b. and we have

$$\|u_1 \otimes u_2\|_{cb} \leq \|u_1\|_{cb}\|u_2\|_{cb}.$$

The *projective* nature of this tensor product can be seen through the following property: If $S \subset F$ is a closed subspace, then we have a completely isometric identification

$$E \otimes^\wedge (F/S) = (E \otimes^\wedge F)/N, \tag{4.9}$$

where N is the kernel of the natural (surjective) mapping from $E \otimes^\wedge F$ onto $E \otimes^\wedge (F/S)$.

More generally, if in the preceding situation u_1 and u_2 are complete surjections, then $u_1 \otimes u_2 \colon E_1 \otimes^\wedge E_2 \longrightarrow F_1 \otimes^\wedge F_2$ is also one. This is fairly easy to check with the definition of $E \otimes^\wedge F$.

However, of course, just like in the Banach space case, *injectivity* fails in general: $u_1 \otimes u_2 \colon E_1 \otimes^\wedge E_2 \longrightarrow F_1 \otimes^\wedge F_2$ may even fail to be injective when u_1 and u_2 are complete isometries.

In the Banach space case, the projective tensor norm is the largest reasonable tensor norm. An analogous property also holds in the operator space case (see [BP1]).

Recall that, following Sakai ([Sa]), one can define a von Neumann algebra as a C^*-algebra M that is the dual of a Banach space, which we call its predual (see §2.5). This predual is unique up to isometry, and we denote it by M_*. It is what is usually called a "noncommutative L_1-space." Actually this terminology is quite abusive, since the commutative case (which corresponds to the standard L_1-spaces) is not really excluded! Let M, N be two von Neumann algebras equipped with their natural o.s. structure, and let M^* and N^* be their duals equipped with the dual structure, as defined in (2.3.1). We can equip M_* and N_* with the o.s. structure induced by M^* and N^*. One can then show ([ER8]) that the projective tensor product $M_* \otimes^\wedge N_*$ is completely isometric to the predual of the von Neumann algebra $M \overline{\otimes} N$ generated by the algebraic tensor product $M \otimes N$. In other words (see [ER8]):

Theorem 4.2. *We have a completely isometric identification*

$$M_* \otimes^\wedge N_* \simeq (M\overline{\otimes}N)_*. \tag{4.10}$$

Proof. By Theorem 2.5.2, we already know that $M\overline{\otimes}N \simeq CB(M_*, N)$. Thus, the fact that $M_* \otimes^\wedge N_* \simeq (M\overline{\otimes}N)_*$ isometrically can be deduced from (4.4) and the unicity of the predual of $M\overline{\otimes}N$. (Actually, since the unicity of the predual is also valid in the completely isometric sense, this argument yields a proof that $M_* \otimes^\wedge N_* \simeq CB(M_*, N)$ completely isometrically.)   ∎

Therefore, if E, F are "noncommutative L_1-spaces," then $E \otimes^\wedge F$ is also one. This is analogous to Grothendieck's classical result ([Gr]) that the (Banach space) projective tensor product of $L_1(\mu)$ and $L_1(\nu)$ is isometric to an L_1-space, namely, to the space $L_1(\mu \times \nu)$.

It is possible to check that the natural morphism

$$E \otimes^\wedge F \to E \otimes_{\min} F$$

is a complete contraction, but in general it is *not injective*. Its injectivity is related to the operator space version of the approximation property for E or F, as follows.

Following [ER8], an o.s. E is said to have the OAP if there is a net of finite rank (c.b.) maps $u_i \colon E \to E$ such that the net $I \otimes u_i$ converges pointwise to the identity on $\mathcal{K} \otimes_{\min} E$. This is the o.s. analog of Grothendieck's approximation property (AP) for Banach spaces. When the net (u_i) is bounded in $CB(E, E)$, we say that E has the CBAP (this is analogous to the BAP for Banach spaces). To quote a sample result from [ER8]: E has the OAP iff the natural map $E^* \otimes^\wedge E \to E^* \otimes_{\min} E$ is injective. The class of groups G for which the reduced C^*-algebra of G has the OAP is studied in [HK] (see also §9 in [Ki8]).

Remark. It should be emphasized that the AP for the underlying Banach space is totally irrelevant for the OAP: Indeed, Alvaro Arias [A2] recently constructed an operator space isometric to ℓ_2 but failing the OAP! Building on previous unpublished work by T. Oikhberg, Oikhberg and Ricard [ORi] obtained more dramatic examples of the same nature. In particular, they constructed a Hilbertian operator space X such that a linear map T on X is c.b. iff it is the sum of a multiple of the identity and a Hilbert-Schmidt map, or iff it is the sum of a multiple of the identity and a nuclear map in the o.s. sense (they can even produce finite-dimensional versions of the space X). In particular, every $T \in CB(X)$ has a nontrivial invariant subspace.

The ideas revolving around the OAP or the CBAP are likely to lead to a simpler and more conceptual proof of the main result of [Sz] (that $B(H)$ fails the AP), but unfortunately this challenge has resisted all attempts so far. We will return to these topics (the OAP and the CBAP) when we discuss "exactness" in Chapter 17. We refer the reader to [BP1, ER2,5,6,8] for more information on all of this.

Chapter 5. The Haagerup Tensor Product

Curiously, the category of operator spaces admits a special kind of tensor product (called the Haagerup tensor product) that does not really have any counterpart (with similar properties) in the Banach category. This tensor product leads to a very rich multilinear theory (initiated in [CS2]), the equivalent of which does not exist for Banach spaces.

The Haagerup tensor norm was introduced by Effros and Kishimoto [EK], who, in view of its previous use by Haagerup in [H3], called it this way. They only considered the resulting Banach spaces; but actually, it is the *operator space* structure of the Haagerup tensor product that has proved most fruitful, and the latter was introduced in [PaS], extending the fundamental work of Christensen and Sinclair [CS2] in the C^*-algebra case.

Basic properties. Let E_1, E_2 be operator spaces. Let $x_1 \in \mathcal{K} \otimes E_1$, $x_2 \in \mathcal{K} \otimes E_2$. We will denote by $(x_1, x_2) \to x_1 \odot x_2$ the bilinear mapping from $(\mathcal{K} \otimes E_1) \times (\mathcal{K} \otimes E_2)$ to $\mathcal{K} \otimes (E_1 \otimes E_2)$ that is defined on rank 1 tensors by

$$(k_1 \otimes e_1) \odot (k_2 \otimes e_2) = (k_1 k_2) \otimes (e_1 \otimes e_2).$$

Let us denote for any x_i in $\mathcal{K} \otimes E_i$

$$\alpha_i(x_i) = \|x_i\|_{\mathcal{K} \otimes_{\min} E_i} \quad (i = 1, 2).$$

Recall that $\mathcal{K}_0$ denotes the *linear* span of the system $\{e_{ij}\}$ in $\mathcal{K}$. Then, for any x in $\mathcal{K}_0 \otimes (E_1 \otimes E_2)$, we define

$$\alpha_h(x) = \inf \left\{ \sum_{j=1}^{n} \alpha_1(x_1^j) \alpha_2(x_2^j) \right\}, \tag{5.1}$$

where the infimum runs over all possible decompositions of x as a finite sum

$$x = \sum_{j=1}^{n} x_1^j \odot x_2^j \tag{5.2}$$

with

$$x_1^j \in \mathcal{K}_0 \otimes E_1, \quad x_2^j \in \mathcal{K}_0 \otimes E_2.$$

In the particular case when $x \in E_1 \otimes E_2 \simeq M_1(E_1 \otimes E_2) \simeq M_1 \otimes (E_1 \otimes E_2)$, this definition means that

$$\|x\|_h = \inf \left\{ \left\| \sum e_{1i} \otimes a_i \right\|_{\min} \left\| \sum e_{i1} \otimes b_i \right\|_{\min} \right\},$$

where the infimum runs over all possible ways to write x as a finite sum $x = \sum_1^n a_i \otimes b_i$ with $a_i \in E_1$, $b_i \in E_2$. Hence, if $E_i \subset B(H_i)$, we have (cf. Remark 1.13)

$$\|x\|_h = \inf \left\{ \left\| \sum a_i a_i^* \right\|^{1/2} \left\| \sum b_i^* b_i \right\|^{1/2} \mid x = \sum a_i \otimes b_i \right\}.$$

Notation. Let E be an operator space and let $x = \sum a_k \otimes e_k$ be in $\mathcal{K} \otimes E$ ($a_k \in \mathcal{K}, e_k \in E$). Then we will denote for any a, b in $B(\ell_2)$

$$a \cdot x = \sum a a_k \otimes e_k, \quad x \cdot b = \sum a_k b \otimes e_k$$

and also of course

$$a \cdot x \cdot b = \sum a a_k b \otimes e_k.$$

One surprisingly nice property of (5.1) is that it suffices to take $n = 1$ in (5.1), that is, we have

$$\alpha_h(x) = \inf\{\alpha_1(x_1)\alpha_2(x_2) \mid x = x_1 \odot x_2\}. \tag{5.3}$$

Indeed, let x_1^j, x_2^j be as in (5.2). Let $s_j \colon \ell_2 \to \ell_2$ be a sequence of isometries with orthogonal ranges, that is, such that $s_j^* s_j = I$ and $s_i^* s_j = 0$ for all $i \neq j$. Then let

$$x_1 = \sum_{j=1}^n x_1^j \cdot s_j^* \quad \text{and} \quad x_2 = \sum_{k=1}^n s_k \cdot x_2^k.$$

Clearly, we have $x = x_1 \odot x_2$ and, moreover, if we embed E_i into a C^*-algebra A (completely isometrically), then we have in the C^*-algebra $\mathcal{K} \otimes_{\min} A_i$,

$$\alpha_1(x_1)^2 = \|x_1 x_1^*\|_{\mathcal{K} \otimes_{\min} A_1} = \left\| \sum_{j=1}^n x_1^j x_1^{j*} \right\|_{\mathcal{K} \otimes_{\min} A_1} \tag{5.4}$$

and similarly

$$\alpha_2(x_2)^2 = \|x_2^* x_2\|_{\mathcal{K} \otimes_{\min} A_2} = \left\| \sum_{j=1}^n x_2^{j*} x_2^j \right\|_{\mathcal{K} \otimes_{\min} A_2}, \tag{5.5}$$

which in particular yields

$$\alpha_1(x_1)\alpha_2(x_2) \leq \left(\sum_{j=1}^n \alpha_1(x_1^j)^2 \right)^{1/2} \left(\sum_{j=1}^n \alpha_2(x_2^j)^2 \right)^{1/2}.$$

Now by a homogeneity trick this clearly yields the identity of the right sides of (5.1) and (5.3), respectively.

It is very easy to check that the norm α_h satisfies the axioms of Ruan's Theorem. Hence, after completion, we obtain an operator space denoted by $E_1 \otimes_h E_2$ and called the Haagerup tensor product. But actually, we do not need Ruan's Theorem here; the fact that $E_1 \otimes_h E_2$ is an operator space follows from Theorem 5.1.

Note that, by Remark 2.1.6, we have for any x in $\mathcal{K}_0 \otimes E_1 \otimes E_2$,

$$\|x\|_{\mathcal{K} \otimes_{\min} E_1 \otimes_{\min} E_2} \leq \alpha_h(x).$$

Therefore, we have a completely contractive mapping

$$E_1 \otimes_h E_2 \to E_1 \otimes_{\min} E_2.$$

By an entirely similar process we can define the Haagerup tensor product of an N-tuple $E_1, \ldots, E_N$ of operator spaces. Once again, for any x in $\mathcal{K}_0 \otimes (E_1 \otimes E_2 \ldots \otimes E_N)$, we can define

$$\alpha(x) = \inf\{\alpha_1(x_1)\alpha_2(x_2)\ldots\alpha_N(x_N) \mid x = x_1 \odot x_2 \ldots \odot x_N, \ x_i \in \mathcal{K}_0 \otimes E_i\}. \tag{5.6}$$

This coincides with the extension of (5.1). Again this satisfies Ruan's axioms, so that we obtain an operator space denoted by $E_1 \otimes_h E_2 \ldots \otimes_h E_N$. The very definition of the norm (5.6) clearly shows that this tensor product is associative, that is, for instance, we have

$$(E_1 \otimes_h E_2) \otimes_h E_3 = E_1 \otimes_h (E_2 \otimes_h E_3) = E_1 \otimes_h E_2 \otimes_h E_3.$$

However, it is important to underline that it is *not* commutative (i.e., $E_1 \otimes_h E_2$ can be very different from $E_2 \otimes_h E_1$).

It is immediate from the definition that $E_1 \otimes_h E_2$ enjoys the classical "tensorial" properties required of a decent tensor product; that is, for any operator spaces F_1, F_2 and any c.b. maps $u_i \colon E_i \to F_i$ ($i = 1, 2$), the mapping $u_1 \otimes u_2$ extends to a c.b. map from $E_1 \otimes_h E_2$ into $F_1 \otimes_h F_2$ with

$$\|u_1 \otimes u_2\|_{cb} \leq \|u_1\|_{cb}\|u_2\|_{cb}.$$

Moreover, this remains valid with N factors instead of 2.

The great power of the Haagerup tensor product stems from the fact that it admits two distinct descriptions, a "projective" one, as above, and an "injective" one, as follows.

Although we will soon abandon this notation, for any x in $\mathcal{K}_0 \otimes E_1 \otimes E_2$ of the form say $x = \sum a_k \otimes e_k^1 \otimes e_k^2$, we define

$$\alpha_f(x) = \sup\left\{\left\|\sum a_k \otimes \sigma_1(e_k^1)\sigma_2(e_k^2)\right\|_{\min}\right\}, \tag{5.7}$$

where the supremum runs over all Hilbert spaces $\mathcal{H}$ and possible maps $\sigma_i \colon E_i \to B(\mathcal{H})$ with $\|\sigma_i\|_{cb} \leq 1$.

This norm clearly corresponds to an operator space structure on $E_1 \otimes E_2$ associated to the mapping

$$\sum e_k^1 \otimes e_k^2 \to \bigoplus_{\sigma_1, \sigma_2} \sum \sigma_1(e_k^1)\sigma_2(e_k^2),$$

where the direct sum runs over the collection of all possible pairs of complete contractions

$$\sigma_1 \colon E_1 \to B(\mathcal{H}) \quad \text{and} \quad \sigma_2 \colon E_2 \to B(\mathcal{H})$$

on the *same* Hilbert space. We will denote by $E_1 \otimes_f E_2$ the resulting operator space (after completion). We will denote by $\sigma_1 \cdot \sigma_2 \colon E_1 \otimes E_2 \to B(\mathcal{H})$ the linear mapping taking $e_1 \otimes e_2$ to $\sigma_1(e_1)\sigma_2(e_2)$. Thus, we have

$$[I_{\mathcal{K}} \otimes (\sigma_1 \cdot \sigma_2)](x) = \sum a_k \otimes \sigma_1(e_k^1)\sigma_2(e_k^2).$$

In other words, if we denote

$$\Phi_1(e_1) = \bigoplus_{\sigma_1,\sigma_2} \sigma_1(e_1)$$

and

$$\Phi_2(e_2) = \bigoplus_{\sigma_1,\sigma_2} \sigma_2(e_2),$$

where the direct sum runs over all pairs as above, and if we denote by $\mathcal{H}$ the Hilbert space on which these are acting, then the mapping

$$\Phi_1 \cdot \Phi_2 \colon E_1 \otimes_f E_2 \to B(\mathcal{H}) \tag{5.7$'$}$$

is a complete isometry. Moreover, this definition clearly extends to an arbitrary number of factors $E_1, ..., E_N$ and leads to an operator space denoted by $E_1 \otimes_f \cdots \otimes_f E_N$.

We will show that, if we have $E_i \subset A_i$ (completely isometrically) for some C^*-algebra A_i as above, then $E_1 \otimes_f E_2$ naturally embeds into the *free* product $A_1 * A_2$ of the two C^*-algebras, whence our notation. But first we will show that the tensor products $E_1 \otimes_h E_2$ and $E_1 \otimes_f E_2$ are actually identical, as follows.

Theorem 5.1. *For any operator spaces* E_1, E_2 *we have* $\alpha_h = \alpha_f$.

Proof. Let $x = x_1 \odot x_2$ with $x_i \in \mathcal{K}_0 \otimes E_i$, $(i = 1, 2)$ and $x \in \mathcal{K}_0 \otimes (E_1 \otimes E_2)$. Then, for any pair of complete contractions

$$\sigma_1 \colon E_1 \to B(\mathcal{H}), \quad \sigma_2 \colon E_2 \to B(\mathcal{H}),$$

we clearly have

$$[I_{\mathcal{K}} \otimes (\sigma_1 \cdot \sigma_2)](x) = (I_{\mathcal{K}} \otimes \sigma_1)(x_1) \cdot (I_{\mathcal{K}} \otimes \sigma_2)(x_2),$$

where the product on the right is the product in the C^*-algebra $\mathcal{K} \otimes_{\min} B(\mathcal{H})$. Hence

$$\|[I_{\mathcal{K}} \otimes (\sigma_1 \cdot \sigma_2)](x)\| \leq \|(I_{\mathcal{K}} \otimes \sigma_1)(x_1)\|\|(I_{\mathcal{K}} \otimes \sigma_2)(x_2)\| \leq \alpha_1(x_1)\alpha_2(x_2).$$

This proves that
$$\alpha_f(x) \le \alpha_h(x). \tag{5.8}$$

To prove the converse, we use the Hahn-Banach Theorem. Let $\xi \in (\mathcal{K} \otimes E_1 \otimes E_2)^*$. We will show that
$$\alpha_f^*(\xi) \le \alpha_h^*(\xi).$$

This clearly yields the converse to (5.8). Assume $\alpha_h^*(\xi) = 1$. Then, by (5.4) and (5.5), for all finite sequences $(x_i^j)_{1 \le j \le n}$ in $\mathcal{K}_0 \otimes E_i$ $(i = 1, 2)$, we have

$$\left| \sum \xi(x_1^j \odot x_2^j) \right| \le \left\| \sum x_1^j x_1^{j*} \right\|_{\mathcal{K} \otimes_{\min} A_1}^{1/2} \left\| \sum x_2^{j*} x_2^j \right\|_{\mathcal{K} \otimes_{\min} A_2}^{1/2}.$$

For simplicity, we denote $B = B(\ell_2)$ in the rest of this proof.

By a standard application of the Hahn-Banach Theorem (see Exercise 2.2.2) this implies the existence of states f_1 and f_2 on $B \otimes_{\min} A_1$ and $B \otimes_{\min} A_2$, respectively, such that, for all x_i in $\mathcal{K}_0 \otimes E_i$,

$$|\xi(x_1 \odot x_2)| \le (f_1(x_1 x_1^*) f_2(x_2^* x_2))^{1/2}.$$

If we now consider the GNS (unital) representations $\pi_i \colon B \otimes_{\min} A_i \to B(H_i)$ associated respectively to f_1 and f_2, we find elements $\xi_1 \in H_1$, $\xi_2 \in H_2$ of norm 1 such that

$$|\xi(x_1 \odot x_2)| \le \|\pi_1(x_1^*)\xi_1\| \, \|\pi_2(x_2)\xi_2\|. \tag{5.9}$$

Clearly, (5.9) implies the existence of an operator $T \colon H_2 \to H_1$ with $\|T\| \le 1$ such that
$$\xi(x_1 \odot x_2) = \langle T\pi_2(x_2)\xi_2, \pi_1(x_1^*)\xi_1 \rangle,$$

and hence we find

$$\xi(x_1 \odot x_2) = \langle \pi_1(x_1) T \pi_2(x_2)\xi_2, \xi_1 \rangle. \tag{5.10}$$

Now let $r_i \colon B \to B(\mathcal{H})$ and $\rho_i \colon A_i \to B(\mathcal{H})$ $(i = 1, 2)$ be the representations determined by the identities

$$\pi_1 = r_1 \cdot \rho_1 \quad \text{and} \quad \pi_2 = r_2 \cdot \rho_2.$$

Note that r_i and ρ_i have commuting ranges. From (5.10) we deduce that, for any i, j and any $a_1 \in E_1$, $a_2 \in E_2$, we have (using $e_{ij} = e_{i1} e_{1j}$, hence $e_{ij} \otimes a_1 \otimes a_2 = (e_{i1} \otimes a_1) \odot (e_{1j} \otimes a_2)$)

$$\xi(e_{ij} \otimes a_1 \otimes a_2) = \langle \rho_1(a_1) T \rho_2(a_2) k_j, \ell_i \rangle \tag{5.11}$$

with k_j, $\ell_i \in \mathcal{H}$ defined by

$$k_j = r_2(e_{1j})\xi_2 \quad \text{and} \quad \ell_i = r_1(e_{i1})^* \xi_1.$$

Note that

$$\sum_j \|k_j\|^2 = \sum \langle r_2(e_{1j})^* r_2(e_{1j})\xi_2, \xi_2 \rangle \le \|\xi_2\| \le 1$$

and similarly $\sum \|\ell_i\|^2 \le 1$.

Let us define $\sigma_2 \colon E_2 \to B(H_2, H_1)$ by setting $\sigma_2(a_2) = T\rho_2(a_2)\ \forall a_2 \in E_2$. Now consider an element x in $\mathcal{K}_0 \otimes (E_1 \otimes E_2)$ of the form

$$x = \sum_{i,j=1}^m e_{ij} \otimes t_{ij} \qquad (t_{ij} \in E_1 \otimes E_2).$$

By (5.11) we have

$$\xi(x) = \sum_{i,j} \langle (\rho_1 \cdot \sigma_2)(t_{ij})k_j, \ell_i \rangle.$$

Hence (note that we may as well assume $H_1 = H_2$ if we wish)

$$|\xi(x)| \le \left\| \sum e_{ij} \otimes (\rho_1 \cdot \sigma_2)(t_{ij}) \right\|_{\mathcal{K} \otimes_{\min} B(H_2, H_1)}$$
$$\le \alpha_f(x).$$

This proves that $\alpha_f^*(\xi) \le 1$; hence, by homogeneity, we have $\alpha_f^*(\xi) \le \alpha_h^*(\xi)$, and the proof is complete. ∎

Remark 5.2. Using the factorization of c.b. maps (Theorem 1.6) it is easy to check that the tensor product $E_1 \otimes_f E_2 \otimes_f \ldots \otimes_f E_N$ is associative. Indeed, using the complete isometry (5.7)', Theorem 1.6 implies that any c.b. map $u \colon E_1 \otimes_f E_2 \to B(H)$ can be written as $u(x_1 \otimes x_2) = \sigma_1(x_1)\sigma_2(x_2)$ with $\|\sigma_1\|_{cb}\|\sigma_2\|_{cb} = \|u\|_{cb}$. From this it is then very easy to verify, for example, that $(E_1 \otimes_f E_2) \otimes_f E_3 = E_1 \otimes_f E_2 \otimes_f E_3$.

Now, using the associativity of both $\otimes_f$ and $\otimes_h$, we can extend Theorem 5.1 (by induction on N) to the case of an arbitrary number N of factors. Whence the following statement.

Corollary 5.3

(i) Let $E_1, \ldots, E_N$ be arbitrary operator spaces. Then $E_1 \otimes_f \cdots \otimes_f E_N = E_1 \otimes_h \cdots \otimes_h E_N$ (as operator spaces).

(ii) There are complete isometries $\Phi_i \colon E_i \to B(H)$ on some Hilbert space H such that

$$\Phi_1 \cdot \Phi_2 \ldots \Phi_N \colon E_1 \otimes_h \cdots \otimes_h E_N \to B(H)$$

is a complete isometry.

(iii) Consider an element x in $\mathcal{K} \otimes E_1 \otimes \cdots \otimes E_N$ of the form

$$x = \sum_i \lambda_i \otimes x_i^1 \otimes \cdots \otimes x_i^N$$

with $\lambda_i \in \mathcal{K}$ and $x_i^1 \in E_1, \ldots, x_i^N \in E_N$. Then we have

$$\|x\|_{\mathcal{K} \otimes_{\min}[E_1 \otimes_h E_2 \cdots \otimes_h E_N]}$$
$$= \sup \left\{ \left\| \sum_i \lambda_i \otimes \sigma^1(x_i^1) \sigma^2(x_i^2) \ldots \sigma^N(x_i^N) \right\|_{\min} \right\},$$

where the supremum runs over all possible choices of H and of complete contractions $\sigma^1 \colon E_1 \to B(H), \ldots, \sigma^N \colon E_N \to B(H)$.

Multilinear factorization. The next corollary is the fundamental factorization of multilinear maps due to Christensen and Sinclair [CS1]. It was extended to operator spaces in [PaS]. See [LeM1] for an extension with H and H_i replaced by Banach spaces.

Corollary 5.4. Let $E_1, \ldots, E_N$ be operator spaces and let $u \colon E_1 \otimes \cdots \otimes E_N \to B(H)$ be a linear mapping. The following are equivalent.

(i) The map u extends to a complete contraction from $E_1 \otimes_h \cdots \otimes_h E_N$ into $B(H)$.

(ii) There are Hilbert spaces H_i and complete contractions $\sigma_i \colon E_i \to B(H_{i+1}, H_i)$ with $H_{N+1} = H$ and $H_1 = H$ such that

$$u(x_1 \otimes \cdots \otimes x_N) = \sigma_1(x_1)\sigma_2(x_2) \ldots \sigma_N(x_N). \qquad (5.12) \qquad \forall x_i \in E_i$$

Moreover, if $E_1, \ldots, E_N$ are all separable (say) and if $\dim(H) = \infty$, then we can take $H_i = H$ for all i.

Remark 5.5. Recall that, by the factorization Theorem 1.6, if E_i is completely isometrically embedded in a C^*-algebra A_i, each σ_i itself admits a factorization of the form

$$\sigma_i(x_i) = V_i \pi_i(x_i) W_i$$

with $\|V_i\| \|W_i\| \le 1$ and some representation $\pi_i \colon A_i \to B(\widehat{H}_i)$. Thus (5.12) implies a decomposition

$$u(x_1 \otimes \cdots \otimes x_N) = T_0 \pi_1(x_1) T_1 \pi_2(x_2) \ldots \pi_N(x_N) T_N \qquad (5.13)$$

with contractive "bridging maps" $T_i \colon \widehat{H}_{i+1} \to \widehat{H}_i$ with $\widehat{H}_{N+1} = H$, $\widehat{H}_0 = H$.

Remark 5.6. If we assume H and all the spaces $E_1, \ldots, E_N$ finite-dimensional, it is easy to see that in Corollary 5.4(ii) we can assume (by suitably restricting) that all the spaces H_i also are finite dimensional.

Proof of Corollary 5.4. (i) $\Rightarrow$ (ii) follows from the factorization Theorem 1.6 using Corollary 5.3 (ii), as in Remark 5.2.

Conversely, assume (ii). Then clearly by definition of $\otimes_f$, u extends to a complete contraction on $E_1 \otimes_f \cdots \otimes_f E_N$, but by Corollary 5.3(i) this coincides with $E_1 \otimes_h \cdots \otimes_h E_N$. The last assertion is immediate since, if $\dim(H) = \infty$, any countable direct sum of copies of H can be identified with H. ∎

Injectivity/projectivity. Quite strikingly, the Haagerup tensor product turns out to be *both* injective and projective, as explicited in the next corollary. Concerning the projective case, we recall that a linear mapping $u\colon E \to F$ between Banach spaces is called a metric surjection if u is surjective and the associated map $\hat{u}\colon E/\ker(u) \to F$ is an isometry. When E, F are operator spaces, recall (see §2.4) that u is a complete metric surjection iff $I_{\mathcal{K}} \otimes u\colon \mathcal{K} \otimes_{\min} E \to \mathcal{K} \otimes_{\min} F$ is a metric surjection.

Corollary 5.7. *Let E_1, E_2, F_1, F_2 be operator spaces.*

(i) *("Injectivity") If $E_i \subset F_i$ completely isometrically, then $E_1 \otimes_h E_2 \subset F_1 \otimes_h F_2$ completely isometrically.*

(ii) *("Projectivity") If $q_i\colon E_i \to F_i$ is a complete metric surjection, then $q_1 \otimes q_2\colon E_1 \otimes_h E_2 \to F_1 \otimes_h F_2$ is also one.*

Proof. (i) Observe that, by the Arveson-Wittstock extension theorem, any pair $\sigma_i\colon E_i \to B(\mathcal{H})$ $(i = 1, 2)$ of complete contractions admits completely contractive extensions $\tilde{\sigma}_i\colon F_i \to B(\mathcal{H})$. Hence, by definition of $\otimes_f$, we have $E_1 \otimes_f E_2 \hookrightarrow F_1 \otimes_f F_2$ completely isometrically; whence the result by Theorem 5.1.

(ii) This is immediate from the very definition of $E_1 \otimes_h E_2$. ∎

Self-duality. In some sense, Corollary 5.4 describes precisely the dual operator space $(E_1 \otimes_h \cdots \otimes_h E_n)^*$. However, this information can be inscribed into some very nice formulas that reflect the self-duality of the Haagerup tensor product, as follows. For simplicity, we state this only for two factors, but it trivially extends to any number of factors by the associativity of $\otimes_h$. This fact is apparently due to Blecher, Effros, and Ruan (see [ER4]).

Corollary 5.8. *Let E_1, E_2 be two finite-dimensional operator spaces. Then*

$$(E_1 \otimes_h E_2)^* = E_1^* \otimes_h E_2^* \text{ completely isometrically.} \tag{5.14}$$

Actually, it suffices for this to have one of the spaces E_1, E_2 finite-dimensional. Moreover, in the infinite-dimensional case, we have a natural completely isometric embedding

$$E_1^* \otimes_h E_2^* \subset (E_1 \otimes_h E_2)^*. \tag{5.15}$$

Proof. We need a preliminary observation. Consider maps $\sigma_i\colon E_i \to \mathcal{K}_0$ ($i = 1, 2$) and let $x_i \in \mathcal{K}_0 \otimes E_i^*$ be the corresponding tensors. Let $u = \sigma_1 \cdot \sigma_2\colon E_1 \otimes E_2 \to \mathcal{K}_0$ be defined as before by $u(x_1 \otimes x_2) = \sigma_1(x_1)\sigma_2(x_2)$, and let $x \in \mathcal{K}_0 \otimes E_1^* \otimes E_2^*$ be the tensor corresponding to u. Then a simple verification shows that

$$x = x_1 \odot x_2.$$

Assume E_1, E_2 finite-dimensional. We will identify for simplicity $(E_1 \otimes E_2)^*$ with $E_1^* \otimes E_2^*$. Recall the notation $\mathcal{K}_0 = \bigcup_n M_n \subset \mathcal{K}$. Recall that, by definition, we have

$$M_n \otimes_{\min} (E_1 \otimes_h E_2)^* = CB(E_1 \otimes_h E_2, M_n).$$

Hence, by Remark 5.6 and our preliminary observation, for any x in $\mathcal{K}_0 \otimes (E_1 \otimes_h E_2)^*$ (say $x \in M_n \otimes (E_1 \otimes_h E_2)^*$ for some n) with $\|x\|_{\min} \leq 1$, there are $x_i \in \mathcal{K}_0 \otimes E_i^*$ with $\|x_i\|_{\min} \leq 1$ such that $x_1 \odot x_2 = x$. Hence, we have $\alpha_h(x) = \|x\|_{\mathcal{K} \otimes_{\min}(E_1^* \otimes_h E_2^*)} \leq 1$.

Conversely, if $\|x\|_{\mathcal{K} \otimes_{\min}(E_1^* \otimes_h E_2^*)} \leq 1$, then we have $\alpha_h(x) \leq 1$ (with E_i^* instead of E_i in the definition of α_h), and the (easy) implication (ii) $\Rightarrow$ (i) in Corollary 5.4 then shows that $\|x\|_{CB(E_1 \otimes_h E_2, \mathcal{K})} \leq 1$. This establishes (5.14).

We now easily verify (5.15). Consider $y \in M_n(E_1^* \otimes E_2^*)$. Then there are finite-dimensional subspaces $G_i \subset E_i^*$ with $y \in M_n(G_1 \otimes G_2)$. We have $G_i = (E_i/F_i)^*$ with $F_i \subset E_i$ equal to the preannihilator of G_i. Now, by Corollary 5.7, $G_1 \otimes_h G_2 \to E_1^* \otimes_h E_2^*$ is a complete isometry and $q\colon E_1 \otimes_h E_2 \to E_1/F_1 \otimes_h E_2/F_2$ is a complete metric surjection. Hence (see §2.4) q^* is a complete isometry from $(E_1/F_1 \otimes_h E_2/F_2)^*$ into $(E_1 \otimes_h E_2)^*$. But, by the first part of the proof, $(E_1/F_1 \otimes_h E_2/F_2)^* = G_1 \otimes_h G_2$. Hence we have

$$\|y\|_{M_n((E_1 \otimes_h E_2)^*)} = \|y\|_{M_n(G_1 \otimes_h G_2)} = \|y\|_{M_n(E_1^* \otimes_h E_2^*)},$$

whence (5.15). Note in particular that if either E_1 or E_2 is finite-dimensional, we have equality in (5.15). ∎

Remark. More generally, using a classical result of Bessaga and Pełczyński (cf. [LT1, p.22]) it is not too hard to show that, if the space $E_1^* \otimes_h E_2^*$ does not contain c_0 as a Banach subspace, then (5.14) holds. See Exercise 5.1 for more details.

In the next statement (and from now on) we use an obvious extension of our previous notation: If A is any operator algebra (for instance if $A = B(H)$), consider elements y_1, y_2 of the form $y_1 = \sum_i a_i(1) \otimes e_i(1) \in A \otimes E_1$, $y_2 = \sum_j a_j(2) \otimes e_j(2) \in A \otimes E_2$ (finite sums). Then we denote

$$y_1 \odot y_2 = \sum_{i,j} a_i(1)a_j(2) \otimes e_i(1) \otimes e_j(2) \in A \otimes E_1 \otimes E_2.$$

Corollary 5.9. Let E_1, E_2 be arbitrary operator spaces. Consider y in $B(H) \otimes E_1 \otimes E_2$. If $\dim(H) = \infty$, then the following are equivalent:

(i) $\|y\|_{B(H) \otimes_{\min}(E_1 \otimes_h E_2)} \leq 1$.

(ii) There are y_i in $B(H) \otimes E_i$ with $\|y_i\|_{\min} \leq 1$ $(i = 1, 2)$ such that $y = y_1 \odot y_2$.

Proof. By Remark 2.1.6, (ii) $\Rightarrow$ (i) is valid in full generality (for any H), so we concentrate on the converse. Assume (i). By the injectivity of $\otimes_h$ we can assume that E_1, E_2 are both finite-dimensional. Then, let $u \colon E_1^* \otimes E_2^* \to B(H)$ be the mapping associated to y. By (5.14), we have $\|u\|_{cb} = \|y\|_{B(H) \otimes_{\min}(E_1 \otimes_h E_2)} \leq 1$. Hence, by the last assertion in Corollary 5.4, we can write $u(\xi_1 \otimes \xi_2) = \sigma_1(\xi_1)\sigma_2(\xi_2)$ $(\xi_i \in E_i^*)$ with $\|\sigma_i\|_{CB(E_i^*, B(H))} \leq 1$. Let $y_i \in B(H) \otimes E_i$ be associated to σ_i. We then have $\|y_i\|_{\min} = \|\sigma_i\|_{cb} \leq 1$ and $y = y_1 \odot y_2$. ∎

At this stage, it is not hard to check the following completely isometric identities (dual to each other):

$$C_n \otimes_h C_k \simeq C_{nk} \quad \text{and} \quad R_n \otimes_h R_k \simeq R_{nk} \tag{5.16}$$

which are valid for any integers n, k. The infinite-dimensional analogs are also valid. More generally, for any Hilbert spaces H, K, we have the following completely isometric identities:

$$H_c \otimes_h K_c = (H \otimes_2 K)_c \quad \text{and} \quad H_r \otimes_h K_r = (H \otimes_2 K)_r. \tag{5.16}'$$

For the proof see Exercise 5.2.

The following identity is a much more significant and useful result (cf. [BP1, Proposition 5.5]. See also [ER4]).

Corollary 5.10. Let E be an arbitrary operator space. Then, for any integer n, we have completely isometric isomorphisms

$$C_n \otimes_h E \otimes_h R_n \simeq M_n(E) \tag{5.17}$$

and

$$C \otimes_h E \otimes_h R \simeq \mathcal{K} \otimes_{\min} E \tag{5.18}$$

via the mapping

$$\sum e_{i1} \otimes x_{ij} \otimes e_{1j} \to \sum e_{ij} \otimes x_{ij}.$$

More generally, for any pair E, F of operator spaces, we have a complete isometry

$$(C \otimes_h E) \otimes_h (F \otimes_h R) \simeq \mathcal{K} \otimes_{\min} (E \otimes_h F). \tag{5.19}$$

Proof. That (5.17) or (5.18) is completely isometric is an easy exercise left to the reader (see Exercise 5.3). Then (5.19) follows by associativity. ∎

By duality we obtain

Corollary 5.11. We have $R_n \otimes_h E^* \otimes_h C_n \simeq M_n(E)^*$ and hence $M_n(E)^{**} \simeq$ $M_n(E^{**})$ completely isometrically. Moreover, we have completely isometric isomorphisms

$$R_n \otimes_h E \otimes_h C_n \simeq M_n^* \otimes^\wedge E \tag{5.20}$$

$$R \otimes_h E \otimes_h C \simeq \mathcal{K}^* \otimes^\wedge E \tag{5.21}$$

via the mapping

$$\sum e_{1i} \otimes x_{ij} \otimes e_{j1} \to \sum \xi_{ij} \otimes x_{ij},$$

where $\xi_{ij} \in \mathcal{K}^*$ is the functional defined by $\xi_{ij}(a) = a_{ij}$.

Proof. By (5.14), (5.17) implies $C_n^* \otimes_h E^* \otimes_h R_n^* \simeq M_n(E)^*$, and hence (since $R^* \simeq C$ and $C^* \simeq R$ by Exercise 2.3.5) $R_n \otimes_h E^* \otimes_h C_n \simeq M_n(E)^*$. Iterating this, we find $M_n(E)^{**} \simeq C_n \otimes_h E^{**} \otimes_h R_n$ and hence by (5.17) again $\simeq M_n(E^{**})$.

We now turn to (5.21). Note that $M_n^* \otimes^\wedge E$ (resp. $R_n \otimes_h E \otimes_h C_n$) is completely isometrically embedded in $\mathcal{K}^* \otimes^\wedge E$ (resp. $R \otimes_h E \otimes_h C$), and the union over n of these spaces is dense in $\mathcal{K}^* \otimes^\wedge E$ (resp. $R \otimes_h E \otimes_h C$). Hence it actually suffices to prove (5.20). Then by the first part of the proof we have $R_n \otimes_h E^{**} \otimes_h C_n \simeq M_n(E^*)^*$, and hence by (4.8) $R_n \otimes_h E^{**} \otimes_h C_n \simeq (M_n^* \otimes^\wedge E)^{**}$. Since the inclusion $X \subset X^{**}$ is completely isometric (see Exercise 2.3.1 and (2.5.1)) both for $X = R_n \otimes_h E \otimes_h C_n$ and for $M_n^* \otimes^\wedge E$, we obtain (5.20). ∎

In particular, when E is one-dimensional, the two preceding corollaries yield

$$R_n \otimes_h C_n \simeq M_n^*, \qquad C_n \otimes_h R_n \simeq M_n,$$

and

$$R \otimes_h C \simeq \mathcal{K}^*, \qquad C \otimes_h R \simeq \mathcal{K}.$$

More generally, for any Hilbert spaces H, K, the product $H_c \otimes_h K_r$ (resp. $H_r \otimes_h K_c$) can be canonically identified, completely isometrically, with the space of all compact (resp. trace class) operators from K^* to H. Here, the space of trace class operators is equipped with its operator space structure as the predual of $B(H, K^*)$, as explained in §2.5.

This illustrates the noncommutativity of $\otimes_h$. This shows (since $\mathcal{K}$ and $\mathcal{K}^*$ are not isomorphic) that the spaces $E_1 \otimes_h E_2$ and $E_2 \otimes_h E_1$ may fail to be even isomorphic as Banach spaces.

However, it is easy to check (using the "opposite" spaces as defined in §2.10) that, if E_1, E_2 are arbitrary operator spaces, the space $E_2 \otimes_h E_1$ is completely isometric to the space $(E_1^{op} \otimes_h E_2^{op})^{op}$, via the linear mapping that takes $x_2 \otimes x_1$ to $x_1 \otimes x_2$. Equivalently, we have $(E_1 \otimes_h E_2)^{op} \simeq E_2^{op} \otimes_h E_1^{op}$, completely isometrically. (This was observed in [BP1].)

For the complex conjugates (see §2.9), the analogous question has a simpler answer: We have

$$\overline{E_1 \otimes_h E_2} \simeq \overline{E}_1 \otimes_h \overline{E}_2$$

completely isometrically. We leave the proofs of these last two identities as (easy) exercises for the reader.

The next result originates (essentially) in Haagerup's unpublished work [H3] (see [EK, BS] for more on the same theme).

Theorem 5.12. *Let $A \subset B(H)$ and $B \subset B(K)$ be C^*-algebras. We have a natural completely isometric embedding*

$$J \colon A \otimes_h B \to CB(B(K,H), B(K,H))$$

defined by

$$J(a \otimes b) \colon T \to aTb.$$

In particular, when $A = B = M_n$ $(n \geq 1)$, this map is a completely isometric isomorphism

$$M_n \otimes_h M_n \simeq CB(M_n, M_n).$$

More generally, for all integers, n, m, r, s, the same map defines a completely isometric isomorphism

$$M_{n,m} \otimes_h M_{r,s} \simeq CB(M_{m,r}, M_{n,s}).$$

Proof. We only prove the last assertion and leave the rest as an exercise. We have, by Corollary 5.10 (and associativity),

$$
\begin{aligned}
M_{n,m} \otimes_h M_{r,s} &= C_h \otimes_h R_m \otimes_h C_r \otimes_h R_s \\
&= C_n \otimes_h (R_m \otimes_h C_r) \otimes_h R_s \\
&= M_{n,s}(M_{m,r}^*) \\
&\simeq CB(M_{m,r}, M_{n,s}). \qquad \blacksquare
\end{aligned}
$$

Remark. Let $\mathcal{S}_n \subset CB(M_n, M_n)$ be the subspace of all the Schur multipliers of M_n (i.e., maps taking each e_{ij} to a multiple of itself) and let $D_n \subset M_n$ be the subalgebra of all diagonal matrices. It is easy to check that $J(D_n \otimes D_n) = \mathcal{S}_n$ and, since $D_n \simeq \ell_\infty^n$ (completely isometrically), we obtain completely isometrically

$$\ell_\infty^n \otimes_h \ell_\infty^n \simeq \mathcal{S}_n \subset CB(M_n, M_n).$$

Actually, Haagerup proved that, for any Schur multiplier $T \colon M_n \to M_n$, we must have $\|T\|_{cb} = \|T\|$ (cf., e.g., [P10, p. 100]); hence we also find an isometric embedding of $\ell_\infty^n \otimes_h \ell_\infty^n$ into $B(M_n, M_n)$. See [ChS] and references there for more on this theme.

Free products. Let A_1, A_2 be two C^*-algebras (resp. unital C^*-algebras). We will denote by $A_1 \dot{*} A_2$ (resp. $A_1 * A_2$) their free product (resp. free product as unital C^*-algebras). This is defined as follows (cf., e.g., [VDN]): We consider the involutive algebra (resp. unital algebra) $\dot{\mathcal{A}}$ (resp. $\mathcal{A}$) that is the free product in the algebraic sense. By the universal property of free products, for any pair $\pi_i \colon A_i \to B$ $(i = 1, 2)$ of homomorphisms (resp. unital ones) into a single algebra (resp. unital algebra) B, there is a unique homomorphism (resp. unital) $\pi_1 \dot{*} \pi_2 \colon \dot{\mathcal{A}} \to B$ (resp. $\pi_1 * \pi_2 \colon \mathcal{A} \to B$) that extends π_1 and π_2 when A_1 and A_2 are embedded into $A_1 \dot{*} A_2$ (resp. $A_1 * A_2$) via their natural embedding. Now assume $B = B(H)$ and π_1, π_2 representations. Then $\pi_1 * \pi_2$ and $\pi_1 \dot{*} \pi_2$ are $*$-homomorphisms, so that we can introduce a C^*-norm on $\dot{\mathcal{A}}$ (resp. $\mathcal{A}$) by defining for all x in $\dot{\mathcal{A}}$ (resp. $\mathcal{A}$)

$$\|x\| = \sup \|\pi_1 * \pi_2(x)\|_{B(H)},$$

where the supremum runs over all pairs (resp. unital pairs) (π_1, π_2) of representations of A_1, A_2 on an arbitrary Hilbert space H. The completions of $\dot{\mathcal{A}}$ (resp. $\mathcal{A}$) for these norms are C^*-algebras denoted, respectively, by $A_1 \dot{*} A_2$ and $A_1 * A_2$. Clearly, a similar definition leads to the free products $\dot{\underset{i \in I}{*}} A_i$ and $\underset{i \in I}{*} A_i$ for an *arbitrary* family $(A_i)_{i \in I}$ of C^*-algebras (resp. unital ones).

Note that we have a canonical bilinear map $(a_1, a_2) \to a_1 \cdot a_2$ from $A_1 \times A_2$ into $A_1 \dot{*} A_2$ (resp. $A_1 * A_2$), and hence a canonical linear map from $A_1 \otimes A_2$ into $A_1 \dot{*} A_2$ (resp. $A_1 * A_2$). More generally, for any N-tuple $A_1, \ldots, A_N$ of C^*-algebras, we have a canonical linear map $A_1 \otimes \cdots \otimes A_N \to A_1 \dot{*} \cdots \dot{*} A_N$ (resp. $A_1 * \cdots * A_N$) that takes $a_1 \otimes \cdots \otimes a_N$ to $a_1 a_2 \ldots a_N$ (where the product on the right is the product in the free product viewed as containing canonically each algebra $A_1, \ldots, A_N$).

The next result shows that the Haagerup tensor product is intimately connected to free products.

Theorem 5.13. *Let $E_1, \ldots, E_N$ be a family of operator spaces given with completely isometric embeddings $E_i \subset A_i$ $(i = 1, \ldots, N)$ into C^*-algebras (resp. unital ones) $A_1, \ldots, A_N$. Then the canonical map, restricted to $E_1 \otimes \cdots \otimes E_N$, defines a completely isometric embedding of $E_1 \otimes_h \cdots \otimes_h E_N$ into $A_1 \dot{*} \cdots \dot{*} A_N$ (resp. into $A_1 * \cdots * A_N$).*

This is proved in [CES] in the nonunital case. The unital case is worked out in detail in [P11]. The proof requires a dilation trick in order to replace N-tuples of complete contractions (as in the definition of $\otimes_f$ above) by N-tuples of C^*-representations.

Lemma 5.14. *Consider C^*algebras $A_1, \ldots, A_N$ and operator spaces $E_i \subset A_i$ (completely isometrically). The equivalent assertions considered in Corollary 5.4 are equivalent to:*

(iii) *There are a Hilbert space $\mathcal{H}$, representations $\widehat{\pi}_i \colon A_i \to B(\mathcal{H})$, and operators $\xi \colon \mathcal{H} \to H$ and $\eta \colon H \to \mathcal{H}$ with $\|\xi\| \leq 1$ and $\|\eta\| \leq 1$ such that*

$$\forall x_i \in E_i \qquad u(x_1 \otimes \cdots \otimes x_N) = \xi\widehat{\pi}_1(x_1)\widehat{\pi}_2(x_2)\ldots\widehat{\pi}_N(x_N)\eta. \qquad (5.22)$$

Moreover, if the C^*-algebras are all unital, the representations $\widehat{\pi}_i$ can also be assumed all unital.

Proof. (iii) $\Rightarrow$ (ii) is obvious. Conversely, assume (ii). By Remark 5.5, we can assume (5.13). Replacing the spaces $\widehat{H}_i$ by suitable enlargements, we can assume (5.13) with $\widehat{H}_i = K$ for all $i = 2, \ldots, N$. Since $\|T_i\| \leq 1$ in (5.13), it is easy to show that there are unitary elements U_i in $M_2(B(K))$ such that $T_i = (U_i)_{11}$. $\Big($Indeed, one just considers for each contraction T the matrix

$$U = \begin{pmatrix} T & (1 - TT^*)^{1/2} \\ -(1 - T^*T)^{1/2} & T^* \end{pmatrix}.\Big)$$

Let $p \colon K \oplus K \to K$ (resp. $j \colon K \to K \oplus K$) be the canonical projection (resp. injection) for the first coordinate. Let

$$r_i(x) = \begin{pmatrix} \pi_i(x) & 0 \\ 0 & 0 \end{pmatrix}.$$

Then it is very easy to verify that

$$T_0\pi_1(x_1)T_1 \ldots \pi_N(x_N)T_N = T_0p[r_1(x_1)U_1r_2(x_2)U_2 \ldots U_{N-1}r_N(x_N)]jT_N.$$

Without loss of generality, we may assume that there are unitaries V_i such that $U_1 = V_1V_2^*$, $U_2 = V_2V_3^*, \ldots, U_{N-1} = V_{N-1}V_N^*$ (indeed, we just choose V_1 arbitrarily, say, $V_1 = 1$; then the relations determine $V_2, V_3, \ldots$ successively). Then, if we replace $r_i(\cdot)$ by $\widehat{\pi}_i = V_i^*r_i(\cdot)V_i$, we obtain (5.22) with $\mathcal{H} = K \oplus K$, $\xi = T_0pV_1$, and $\eta = V_N^*jT_N$.

To check the unital case, we need to go one step further. So we assume that the A_i are unital and that (5.22) holds but with a priori nonunital representations $\widehat{\pi}_i$. Let $p_i = \widehat{\pi}_i(1)$. Note that p_i is a projection on $\mathcal{H}$ commuting with the range of $\widehat{\pi}_i$.

By associativity (see below), we may and do assume that $N = 2$.

Note that (by considering, e.g. , $A_1 \otimes_{\min} A_2$) we know that there exists a pair $\rho_i \colon A_i \to B(L)$ $(i = 1, 2)$ of unital (and faithful) representations on the same Hilbert space L.

Note also that, by suitably augmenting $\mathcal{H}$, we can assume that $(1-p_1)(\mathcal{H})$ and $(1-p_2)(\mathcal{H})$ are of the same Hilbertian dimension, and that they are both isometric to some direct sum of copies of L. This allows us to define (using ρ_1 and ρ_2) $*$-homomorphisms $\widehat{\rho}_i \colon A_i \to B(\mathcal{H})$ $(i=1,2)$ such that

$$\widehat{\rho}_1(1) = 1-p_1 \quad \text{and} \quad \widehat{\rho}_2(1) = 1-p_2.$$

Then we define for $a_i \in A_i$

$$\sigma_1(a_1) = p_1 \widehat{\pi}_1(a_1) p_1 + (1-p_1)\widehat{\rho}_1(a_1)(1-p_1)$$
$$\sigma_2(a_2) = p_2 \widehat{\pi}_2(a_2) p_2 + (1-p_2)\widehat{\rho}_2(a_2)(1-p_2).$$

We have (note that $\widehat{\pi}_1(1) = p_1$ implies $\widehat{\pi}_1(a_1)p_1 = p_1\widehat{\pi}_1(a_1) = \widehat{\pi}_1(a_1),\dots$)

$$\sum \widehat{\pi}_1(a_i^1)\widehat{\pi}_2(a_i^2) = p_1 \sum \sigma_1(a_i^1)\sigma_2(a_i^2)p_2,$$

but now σ_1, σ_2 are unital representations (i.e. $*$-homomorphisms); hence this yields finally, for $x_i \in E_i$,

$$u(x_1 \otimes x_2) = \xi p_1 \sigma_1(x_1)\sigma_2(x_2)p_2 \eta,$$

which completes the unital case of (5.22) when $N = 2$. It follows from (5.22) that $E_1 \otimes_h E_2$ is naturally embedded into the unital free product $A_1 * A_2$ (see below). Then, if $N = 3$, we may apply what we just proved to $(E_1 \otimes_h E_2) \otimes_h E_3$ and this yields (5.22) for $N = 3$ and so on for larger N.   ∎

Remark. Lemma 5.14 is proved in [CES] (see also [Y]) for the nonunital free product, but nothing is said there about the unital case, which is verified in [P11].

Proof of Theorem 5.13. By the factorization Theorem 1.6, the equivalence of (i) and (iii) in Lemma 5.14 means, in the unital case, that a mapping $u \colon E_1 \otimes_h \dots \otimes_h E_N \to B(H)$ is a complete contraction iff it extends to a complete contraction on the free product $A_1 * \dots * A_N$. Equivalently, this means that we have a completely isometric embedding

$$E_1 \otimes_h \dots \otimes_h E_N \subset A_1 * \dots * A_N.$$

In the nonunital case, the proof is identical.   ∎

The next result (inspired by [ER10, EKR]) on the "tensor shuffling" is one of the many applications of the free product connection.

Theorem 5.15. *Let E_i ($i = 1, 2, 3, 4$) be operator spaces. We have a natural completely contractive map*

$$(E_1 \otimes_{\min} E_2) \otimes_h (E_3 \otimes_{\min} E_4) \to (E_1 \otimes_h E_3) \otimes_{\min} (E_2 \otimes_h E_4)$$

that reduces to a permutation on the algebraic tensor product.

Proof. Assume $E_i \subset A_i$ for some C^*-algebra A_i. Let $B = A_1 \ast A_3$ and $C = A_2 \ast A_4$. Consider the C^*-algebra $A = B \otimes_{\min} C$. Clearly the product map $p \colon A \times A \to A$ defines a complete contraction from $A \otimes_h A$ to A. Now, if we view A_1 and A_3 (resp. A_2 and A_4) as subalgebras of B (resp. C), we have

$$E_1 \otimes_{\min} E_2 \subset A \quad \text{and} \quad E_3 \otimes_{\min} E_4 \subset A,$$

which yields that p restricted to $(E_1 \otimes_{\min} E_2) \otimes_h (E_3 \otimes_{\min} E_4)$ is completely contractive. But the latter restriction takes value in

$$\overline{\operatorname{span}}(E_1 \cdot E_3) \otimes_{\min} \overline{\operatorname{span}}(E_2 \cdot E_4),$$

where the products are meant in $A_1 \ast A_3$ and $A_2 \ast A_4$, respectively. By Theorem 5.13, $\overline{\operatorname{span}}(E_1 \cdot E_3) \simeq E_1 \otimes_h E_3$ and $\overline{\operatorname{span}}(E_2 \cdot E_4) \simeq E_2 \otimes_h E_4$, so we obtain the announced result. ∎

Factorization through R or C. Let X be a fixed operator space. A linear map $u \colon E \to F$ (between operator spaces) is said to factor through X if there are c.b. maps $v \colon X \to F$ and $w \colon E \to X$ such that $u = vw$. We will denote by $\Gamma_X(E, F)$ the set of all such maps, and we define

$$\gamma_X(u) = \inf\{\|v\|_{cb}\|w\|_{cb}\},$$

where the infimum runs over all possible such factorizations. Note that in general $\gamma_X(\cdot)$ is not a norm and $\Gamma_X(E, F)$ may fail to be stable under addition, but when X is "nice" (as below), this "pathology" disappears. In the following section, we will study the cases $X = R$ and $X = C$.

Proposition 5.16. *([ER4]) Let E, F be operator spaces. Then the natural mappings $E^* \otimes F \to CB(E, F)$ and $E \otimes F^* \to CB(F, E)$ extend to isometric embeddings*

$$E^* \otimes_h F \subset \Gamma_R(E, F) \quad \text{and} \quad E \otimes_h F^* \subset \Gamma_C(F, E).$$

Moreover, if either E or F is finite-dimensional, these embeddings are actually isometric isomorphisms. Similarly, we have isometric embeddings $E \otimes_h F \subset \Gamma_R(E^*, F)$ and $E \otimes_h F \subset \Gamma_C(F^*, E)$.

Proof. Consider $x \in E^* \otimes F$ and let $u \colon E \to F$ be the associated finite rank linear map. Let $E \xrightarrow{w} R \xrightarrow{v} F$ be a factorization of u through R. Since

u has finite rank, say, equal to n, and R is "homogeneous" (which means that (1.7) holds), we can replace R by a suitable subspace (and project onto it) to find a modified factorization $E \xrightarrow{w_1} R_n \xrightarrow{v_1} F$ with $\|w_1\|_{cb} = \|w\|_{cb}$ and $\|v_1\|_{cb} = \|v\|_{cb}$. Identifying w_1 with $a = \sum e_{1i} \otimes a_i \in R_n \otimes E^*$ and v_1 with $b = \sum e_{i1} \otimes b_i \in C_n \otimes F$, we find $x = a \odot b$ with

$$\|w_1\|_{cb} = \|a\|_{M_n(E^*)}, \quad \|v_1\|_{cb} = \|b\|_{M_n(F)}.$$

Conversely, to any a, b as above corresponds a factorization of u so that we have

$$\gamma_R(u) = \inf\{\|a\|_{\min}\|b\|_{\min} \mid a \in R_n \otimes E^*, b \in C_n \otimes F, x = a \odot b\},$$

which means precisely that

$$\gamma_R(u) = \|x\|_h.$$

The other assertions are proved similarly. ∎

Remark. If we equip $\Gamma_R(E, F)$ and $\Gamma_C(F, E)$ with appropriate o.s.s., then the preceding isometries become completely isometric. The appropriate structure on $\Gamma_R(E, F)$ (resp. $\Gamma_C(F, E)$) is the one that gives to the space $M_n(\Gamma_R(E, F))$ (resp. $M_n(\Gamma_C(F, E))$) the norm of the space $\Gamma_R(R_n(E), R_n(F))$ (resp. $\Gamma_C(C_n(F), C_n(E))$), where we have denoted

$$R_n(E) = R_n \otimes_{\min} E \quad \text{and} \quad C_n(F) = C_n \otimes_{\min} F.$$

Symmetrized Haagerup tensor product. It is natural to investigate what happens to $\otimes_f$ if one restricts the maps (σ_1, σ_2) to have commuting ranges. This produces a symmetrized version of the Haagerup tensor product that is studied extensively in [OP]. We will describe only one basic result from [OP] and its consequences. Let (E_1, E_2) be operator spaces. We denote by $\mathcal{C}$ the collection of all pairs $\sigma = (\sigma_1, \sigma_2)$ of complete contractions $\sigma_i \colon E_i \to B(H_\sigma)$ (into the same Hilbert space) with commuting ranges. Let $\Phi \colon E_1 \otimes E_2 \to \bigoplus_{\sigma \in \mathcal{C}} B(H_\sigma)$ be the embedding defined by $\Phi = \oplus_{\sigma \in \mathcal{C}} \sigma_1 \cdot \sigma_2$ or, more explicitly:

$$\Phi(x_1 \otimes x_2) = \bigoplus_{\sigma \in \mathcal{C}} \sigma_1(x_1)\sigma_2(x_2).$$

This linear embedding induces an operator space structure on $E_1 \otimes E_2$; we denote by $E_1 \otimes_\mu E_2$ the resulting operator space after completion.

A similar construction makes sense for N-tuples of operators $(E_1, \ldots, E_N)$ and produces an operator space denoted by $(E_1 \otimes \cdots \otimes E_N)_\mu$ (so that $(E_1 \otimes$

$E_2)_\mu$ is the same as $E_1 \otimes_\mu E_2$). This new tensor product is projective, and, by its very construction, it is commutative, but it fails to be either associative (see [OP] for Le Merdy's argument for this) or injective.

Theorem 5.17. *([OP]) Let E_1, E_2 be two operator spaces. Consider the mapping*

$$Q\colon (E_1 \otimes_h E_2) \oplus_1 (E_2 \otimes_h E_1) \to E_1 \otimes_\mu E_2$$

defined on the direct sum of the linear tensor products by $Q(u \oplus v) = u + {}^t v$. Then Q extends to a complete metric surjection from $(E_1 \otimes_h E_2) \oplus_1 (E_2 \otimes_h E_1)$ onto $E_1 \otimes_\mu E_2$. In particular, for any u in $E_1 \otimes E_2$, we have: $\|u\|_\mu < 1$ iff there are v, w in $E_1 \otimes E_2$ such that $u = v + w$ and $\|v\|_{E_1 \otimes_h E_2} + \|{}^t w\|_{E_2 \otimes_h E_1} < 1$.

The preceding statement means that $E_1 \otimes_\mu E_2$ is completely isometric to the "sum" $E_1 \otimes_h E_2 + E_2 \otimes_h E_1$ (in the style of interpolation theory; see §2.7) in analogy with the space $R + C$ described later in §9.8.

The case of N-tuples is also considered in [OP], but only the completely isomorphic analog of the preceding result is proved. Moreover, [OP] also contains a different result (completely isometric, this time) for N-tuples of completely contractive maps $\sigma_i\colon E_i \to B(H_\sigma)$ with "cyclically commuting" ranges (see [OP, Theorem 19]).

By Corollary 5.4, one can see that the following statement is a dual reformulation of Theorem 5.17.

Theorem 5.18. *([OP]) Let E_1, E_2 be two operator spaces, and let $\varphi\colon E_1 \otimes E_2 \to B(\mathcal{H})$ be a linear mapping. The following are equivalent:*

(i) $\|\varphi\|_{CB(E_1 \otimes_\mu E_2, B(\mathcal{H}))} \le 1$.

(ii) *For some Hilbert space H, there are complete contractions $\alpha_1\colon E_1 \to B(H, \mathcal{H})$, $\alpha_2\colon E_2 \to B(\mathcal{H}, H)$, and $\beta_1\colon E_1 \to B(\mathcal{H}, H)$, $\beta_2\colon E_2 \to B(H, \mathcal{H})$ such that*

$$\forall (x_1, x_2) \in E_1 \times E_2 \qquad \varphi(x_1 \otimes x_2) = \alpha_1(x_1)\alpha_2(x_2) = \beta_2(x_2)\beta_1(x_1).$$

(iii) *For some Hilbert space H, there are complete contractions $\sigma_i\colon E_i \to B(H)$ $(i = 1, 2)$, with commuting ranges, and contractions $V\colon \mathcal{H} \to H$ and $W\colon H \to \mathcal{H}$ such that*

$$\forall (x_1, x_2) \in E_1 \times E_2 \qquad \varphi(x_1 \otimes x_2) = W\sigma_1(x_1)\sigma_2(x_2)V.$$

Proof. Assume (i). Then φ defines a complete contraction into $B(\mathcal{H})$ both from $E_1 \otimes_h E_2$ and from $E_2 \otimes_h E_1$. Then (ii) follows from Corollary 5.4. Now assume (ii). Let $H_1 = \mathcal{H}$, $H_2 = H$, and $H_3 = \mathcal{H}$. We define maps

$\sigma_1 \colon E_1 \longmapsto B(H_1 \oplus H_2 \oplus H_3)$ and $\sigma_2 \colon E_2 \longrightarrow B(H_1 \oplus H_2 \oplus H_3)$ using matrix notation, as follows:

$$\sigma_1(x_1) = \begin{pmatrix} 0 & \alpha_1(x_1) & 0 \\ 0 & 0 & \beta_1(x_1) \\ 0 & 0 & 0 \end{pmatrix}$$

$$\sigma_2(x_2) = \begin{pmatrix} 0 & \beta_2(x_2) & 0 \\ 0 & 0 & \alpha_2(x_2) \\ 0 & 0 & 0 \end{pmatrix}.$$

Then, by (ii), we have

$$\sigma_1(x_1)\sigma_2(x_2) = \sigma_2(x_2)\sigma_1(x_1) = \begin{pmatrix} 0 & 0 & \varphi(x_1 \otimes x_2) \\ 0 & 0 & 0 \\ 0 & 0 & 0 \end{pmatrix};$$

hence σ_1, σ_2 have commuting ranges and are complete contractions. Therefore, if we let $W \colon H_1 \oplus H_2 \oplus H_3 \to \mathcal{H}$ be the projection onto the first coordinate and $V \colon \mathcal{H} \to H_1 \oplus H_2 \oplus H_3$ be the isometric inclusion into the third coordinate, then we obtain (iii). Finally, the implication (iii) implies (i) is obvious by the very definition of $E_1 \otimes_\mu E_2$. ∎

Proof of Theorem 5.17. By duality, it clearly suffices to show that, for any linear map $\varphi \colon E_1 \otimes E_2 \to B(\mathcal{H})$, the norms $\|\varphi\|_{CB(E_1 \otimes_\mu E_2 \to B(\mathcal{H}))}$ and $\|\varphi Q\|_{cb}$ are equal. But this is precisely the meaning of the equivalence between (i) and (ii) in Theorem 5.18. Thus we conclude that Theorem 5.18 implies Theorem 5.17. ∎

Remark. Let $X = R \oplus C$. Let $\tilde{u} \colon E_1 \to E_2$ be a finite rank map with associated tensor $u \in E_1^* \otimes E_2$. Using Proposition 5.15, it is easy to check that Theorem 5.17 implies

$$\tfrac{1}{2}\|u\|_\mu \le \gamma_X(\tilde{u}) \le \|u\|_\mu. \tag{5.23}$$

Moreover, it is easy to see that the identity of X factors completely contractively through $\mathcal{K}$, and hence

$$\gamma_\mathcal{K}(\tilde{u}) \le \gamma_X(\tilde{u}). \tag{5.24}$$

Corollary 5.19. ([OP]) *Let E be an n-dimensional operator space. Let $i_E \in E^* \otimes E$ be associated to the identity of E and let*

$$\mu(E) = \|i_E\|_\mu.$$

Then

$$\max\{\gamma_{\mathcal{K}}(I_E), \gamma_{\mathcal{K}}(I_{E^*})\} \leq \mu(E). \tag{5.25}$$

Moreover, $\mu(E) = 1$ iff either $E = R_n$ or $E = C_n$ (completely isometrically).

Proof. Note that (5.25) clearly follows from (5.23). Assume that $\mu(E) = 1$. Then, by Theorem 5.17 and Proposition 5.16 (using also an obvious compactness argument), we have a decomposition $I_E = u_1 + u_2$ with

$$\gamma_R(u_1) + \gamma_C(u_2) = 1. \tag{5.26}$$

In particular, this implies that $\gamma_2(I_E) = 1$ (where $\gamma_2(.)$ denotes the norm of factorization through Hilbert space; see, e.g., [P4, chapter 2] for more background), whence that E is isometric to ℓ_2^n ($n = \dim E$). Moreover, for any e in the unit sphere of E we have

$$1 = \|e\| \leq \|u_1(e)\| + \|u_2(e)\| \leq \|u_1\| + \|u_2\| \leq \gamma_R(u_1) + \gamma_C(u_2) \leq 1.$$

Therefore we must have

$$\|u_1(e)\| = \|u_1\| = \gamma_R(u_1) \text{ and } \|u_2(e)\| = \|u_2\| = \gamma_C(u_2). \tag{5.27}$$

Let $\alpha_i = \|u_i\|$, so that (by (5.26)) $\alpha_1 + \alpha_2 = 1$. Assume that both $\alpha_1 > 0$ and $\alpha_2 > 0$. We will show that this is impossible if $n > 1$. Indeed, then $U_i = (\alpha_i)^{-1} u_i$ ($i = 1, 2$) is an isometry on ℓ_2^n, such that, for any e in the unit sphere of E, we have $e = \alpha_1 U_1(e) + \alpha_2 U_2(e)$. By the strict convexity of ℓ_2^n, this implies that $U_1(e) = U_2(e) = e$ for all e. Moreover, by (5.27) we have $\gamma_R(U_1) = 1$ and $\gamma_C(U_2) = 1$. This implies that $E = R_n$ and $E = C_n$ completely isometrically, which is absurd when $n > 1$. Hence, if $n > 1$, we conclude that either $\alpha_1 = 0$ or $\alpha_2 = 0$, which implies either $\gamma_C(I_E) = 1$ or $\gamma_R(I_E) = 1$, or, equivalently, either $E = C_n$ or $E = R_n$ completely isometrically. The remaining case $n = 1$ is trivial. ∎

Remark. Here is an alternate argument: If $\mu(E) = 1$, then, using (5.23), (5.24) and the reflexivity of E, we see that, for both E and E^*, the identity factors through $\mathcal{K}^{**}$; therefore E is an injective operator space as well as its dual. Now, in [Ru2], Ruan gives the complete list of the injective operator subspaces of finite-dimensional C^*-algebras (see also [EOR] for more on this theme). Running down this list, and using an unpublished result of R. Smith which says that a finite-dimensional injective operator space is completely contractively complemented in a finite-dimensional C^*-algebra (see [B2]), we find that R_n and C_n are the only possibilities.

For the isomorphic version of the last statement, we will use the following result from [O1]:

Theorem 5.20. *([O1]) Let E be an operator space such that I_E can be factorized completely boundedly through the direct sum $X = H_c \oplus_1 K_r$ of a column space and a row space (i.e., there are c.b. maps $u\colon E \to X$ and $v\colon X \to E$ such that $I_E = vu$). Then there are subspaces $E_1 \subset H_c$ and $E_2 \subset K_r$ such that E is completely isomorphic to $E_1 \oplus_1 E_2$. More precisely, if we have $\|u\|_{cb}\|v\|_{cb} \le c$ for some number c, then we can find a complete isomorphism $T\colon E \to E_1 \oplus_1 E_2$ such that $\|T\|_{cb}\|T^{-1}\|_{cb} \le f(c)$, where $f\colon \mathbb{R}_+ \to \mathbb{R}_+$ is a certain function.*

Theorem 5.21. *([OP]) The following properties of an an operator space E are equivalent:*

  (i) *For any operator space F, we have $F \otimes_{\min} E = F \otimes_\mu E$ isomorphically.*

  (ii) *E is completely isomorphic to the direct sum of a row space and a column space.*

Proof. The implication (ii) $\Rightarrow$ (i) is easy and left to the reader. Conversely, assume (i). Then, a routine argument shows that there is a constant K such that for all F and all u in $F \otimes E$ we have $\|u\|_\mu \le K\|u\|_{\min}$. Let $S \subset E$ be an arbitrary finite-dimensional subspace, let $j_S\colon S \to E$ be the inclusion map, and let $t_S \in S^* \otimes E$ be the associated tensor. Then we have by (5.23) $\sup_S \gamma_{R\oplus_1 C}(j_S) = \sup_S \|t_S\|_\mu \le K$. By a routine ultraproduct argument, this implies that the identity of E can be written as in Theorem 5.20 with $c = K$; thus we conclude that $E \simeq E_1 \oplus_1 E_2$, where E_1 is a row space and E_2 is a column space. Note that we obtain an isomorphism $T\colon E \to E_1 \oplus_1 E_2$ such that $\|T\|_{cb}\|T^{-1}\|_{cb} \le f(K)$. In particular, if E is finite-dimensional, we find T such that

$$\|T\|_{cb}\|T^{-1}\|_{cb} \le f(\|i_E\|_\mu). \qquad \blacksquare$$

Complex interpolation. The Haagerup tensor product also has a nice "commutation property" with respect to complex interpolation (defined in §2.7) that can be briefly described as follows. Let (E_0, E_1) and (F_0, F_1) be two compatible pairs of operator spaces. Then the couple $(E_0 \otimes_h F_0, E_1 \otimes_h F_1)$ can be viewed as compatible. Indeed, assume that (E_0, E_1) (resp. (F_0, F_1)) are continuously injected into a Banach space $\mathcal{E}$ (resp. $\mathcal{F}$), which allows us to view the couple as compatible. Then, it is easy to check that the tensor product of the suitable injections defines a continuous injection from $E_0 \otimes_h F_0$ into the injective tensor product $\mathcal{E} \overset{\vee}{\otimes} \mathcal{F}$. Similarly, $E_1 \otimes_h F_1$ is continuously injected into the same space $\mathcal{E} \overset{\vee}{\otimes} \mathcal{F}$. (Note: The injectivity of the tensor product of two injective mappings can sometimes fail, but not for tensor norms as nice as the injective one or the Haagerup one. See [P1, Theorem 6.6] for a general result related to this technical point.) Thus we may view the couple $(E_0 \otimes_h F_0, E_1 \otimes_h F_1)$ as compatible. Then we have (cf. [P1, Theorem 2.3]):

Theorem 5.22. *Let (E_0, E_1) and (F_0, F_1) be two compatible couples of operator spaces. Then, viewing the couple $(E_0 \otimes_h F_0, E_1 \otimes_h F_1)$ as compatible as explained above, we have a complete isometry*

$$(E_0 \otimes_h F_0, E_1 \otimes_h F_1)_\theta = (E_0, E_1)_\theta \otimes_h (F_0, F_1)_\theta.$$

Exercises

Exercise 5.1. Let E_1, E_2 be operator spaces such that $E_1^* \otimes_h E_2^*$ does not contain a subspace isomorphic (as a Banach space) to c_0. Then

$$(E_1 \otimes_h E_2)^* = E_1^* \otimes_h E_2^*.$$

Exercise 5.2. Prove (5.16) and (5.16)'.

Exercise 5.3. Prove (5.17) and (5.18). Deduce from (5.18) that $C \otimes_h E = C \otimes_{\min} E$ and $E \otimes_h R = E \otimes_{\min} R$ for any operator space E. Use this to give an alternate solution to Exercise 5.2.

Exercise 5.4. Let $E_i \subset B(H_i)$ $(i = 1, 2)$ be operator spaces. Then any x in $E_1 \otimes_h E_2$ can be written as a series of the form $\sum_1^\infty a_n \otimes b_n$, where $a_n \in E_1$ $b_n \in E_2$ with the series $\sum_1^\infty a_n a_n^*$ and $\sum_1^\infty b_n^* b_n$ converging in norm in $B(H_1)$ and $B(H_2)$, respectively. Moreover, in that case, the series $\sum_1^\infty a_n \otimes b_n$ converges in $E_1 \otimes_h E_2$, and, if $\|x\|_h < 1$, we can choose (a_n) and (b_n) so that

$$\left\| \sum_1^\infty a_n a_n^* \right\| < 1 \quad \text{and} \quad \left\| \sum_1^\infty b_n^* b_n \right\| < 1.$$

Exercise 5.5. Show that for any operator space E we have completely isometrically

$$(\mathcal{K} \otimes_{\min} E)^* \simeq (C \otimes_h E \otimes_h R)^* \simeq R \otimes_h E^* \otimes_h C \simeq \mathcal{K}^* \otimes^\wedge E^*.$$

Exercise 5.6. Show that if E is finite-dimensional, we have completely isometrically

$$(\mathcal{K} \otimes_{\min} E)^{**} \simeq \mathcal{K}^{**} \otimes_{\min} E.$$

Moreover, if A is any C^*-algebra, we have

$$(\mathcal{K} \otimes_{\min} A)^{**} \simeq \mathcal{K}^{**} \overline{\otimes} A^{**}.$$

Exercise 5.7. Let E_1, E_2 be maximal operator spaces. Then for any x in $E_1 \otimes E_2$ we have

$$\|x\|_h = \|{}^t x\|_h = \|x\|_\mu = \inf \left\{ \|a\|_{M_n} \cdot \left(\sum \|x_i\|^2 \right)^{1/2} \left(\sum \|y_j\|^2 \right)^{1/2} \right\},$$

where the infimum runs over all n and all possible ways to decompose x as

$$x = \sum_{i,j=1}^{n} a_{ij} x_i \otimes y_j.$$

Exercise 5.8. Let A, B be arbitrary C^*-algebras and let $A \ast B$ be their (nonunital) free product. For any $k \geq 1$, let $T_{2k} = A \otimes_h B \otimes_h \cdots \otimes_h A \otimes_h B$ (where A and B each appear k times) and $T_{2k+1} = T_{2k} \otimes_h A$. For any $d > 1$, let

$$\psi_d \colon T_d \to A \ast B$$

be the completely contractive map induced by the product map in $A \ast B$. Show that ψ_d is a completely isomorphic embedding of T_d in $A \ast B$. More precisely, the range $\psi_d(T_d)$ is the closed subspace $\mathcal{W}_d \subset A \ast B$ spanned by all the "words" of length d that begin in A, and we have

$$\|\psi_{d|\mathcal{W}_d}^{-1}\|_{cb} \leq (d-1)^{d-1}.$$

Exercise 5.9. Fix an integer n. The aim of this exercise is to produce explicit completely isometric embeddings of C_n and R_n into $\ell_\infty^n \otimes_h \ell_\infty^n$. We will denote by (e_i) the canonical basis of ℓ_∞^n. Let $w = (w_{ik})$ be an $n \times n$ matrix with unimodular entries (i.e., $|w_{ik}| = 1$ for all i, k) such that

$$\|n^{-1/2} w\|_{M_n} = 1.$$

In $\ell_\infty^n \otimes_h \ell_\infty^n$ we consider the vectors

$$x_i = e_i \otimes \sum_k w_{ik} e_k \qquad y_i = \sum_k w_{ik} e_k \otimes e_i.$$

(Note that $y_i = {}^t x_i$.) Let $E_x \subset \ell_\infty^n \otimes_h \ell_\infty^n$ (resp. $E_y \subset \ell_\infty^n \otimes_h \ell_\infty^n$) be the operator subspace spanned by $\{x_i \mid i = 1, 2, \ldots, n\}$ (resp. $\{y_i \mid i = 1, \ldots, n\}$). Prove that $E_x \simeq C_n$ (resp. $E_y \simeq R_n$) completely isometrically.

Exercise 5.10. The aim of this exercise is to produce an explicit completely isometric embedding of M_n into $\ell_\infty^n \otimes_h \ell_\infty^n \otimes_h \ell_\infty^n$. Let w, w' be two $n \times n$ matrices with the same properties as in the preceding exercise. We introduce the following elements of the operator space $\ell_\infty^n \otimes_h \ell_\infty^n \otimes_h \ell_\infty^n$:

$$z_{ij} = e_i \otimes \left(\sum_k w_{ik} w'_{kj} e_k \right) \otimes e_j.$$

Then the span of $\{z_{ij} \mid i, j = 1, \ldots, n\}$ in $\ell_\infty^n \otimes_h \ell_\infty^n \otimes_h \ell_\infty^n$ is completely isometric to M_n. Let H be an infinite-dimensional Hilbert space. Show that a matrix (a_{ij}) in $M_n(B(H))$ is in the unit ball iff it can be factorized as a matrix product $D_1(n^{-1/2}w) D_2 (n^{-1/2}w') D_3$, where D_1, D_2, D_3 are three diagonal matrices in the unit ball of $B(H)$.

Chapter 6. Characterizations of Operator Algebras

In the Banach algebra literature, an operator algebra is just a closed subalgebra of $B(H)$. A uniform algebra is a closed unital subalgebra of the space $C(T)$ of all continuous functions on a compact set T, which is (usually) assumed to separate the points of T. In the 1970s, the theory of uniform algebras gave birth to the notion of Q-algebras (quotients of a uniform algebra by a closed ideal) and ultimately uncovered some surprising stability properties of the class of operator algebras.

Specifically, in the late 1960s, B. Cole (see [We]) discovered that Q-algebras are necessarily operator algebras, and soon after that, G. Lumer and A. Bernard proved that the class of operator algebras is stable under quotients (see Theorem 6.3). One of the natural problems considered during that period was to characterize the Banach algebras that are isomorphic to a uniform algebra, or a Q-algebra, or an operator algebra. Among the many contributions from that time, those that stand out are Craw's Lemma characterizing uniform algebras and Varopoulos's work on operator algebras ([V1–2]).

Varopoulos discovered that, if the product mapping p_A: $x \otimes y \to xy$ of a Banach algebra A is continuous on the tensor product $A \otimes A$ equipped with the γ_2-norm, then A is an operator algebra. (From this he deduced, using Grothendieck's inequality, that any Banach algebra structure on a commutative C^*-algebra is necessarily an operator algebra.) Several authors then tried to characterize operator algebras by a property of this type but with a different tensor norm than the γ_2-norm. Variations on this theme were given by P. Charpentier and A. Tonge (see [DJT, Chapter 18]) until K. Carne [Ca] somewhat closed that chapter by showing that operator algebras cannot be characterized by the continuity of the product map on a suitable tensor product. More precisely, he showed that there is no reasonable tensor norm γ such that the continuity of p_A: $A \otimes_\gamma A \to A$ characterizes operator algebras. In sharp contrast with Carne's result, it turns out that in the category of operator spaces the situation is much nicer. We have

Theorem 6.1. *([BRS]) Let A be a unital Banach algebra with unit 1_A given with an operator space structure such that $\|1_A\| = 1$. The following are equivalent.*

(i) *The product map*

$$p_A: A \otimes_h A \to A$$

is completely contractive.

(ii) *There is, for some Hilbert space H, a completely isometric unital homomorphism j: $A \to B(H)$.*

Remark. Let us denote here simply $x \cdot y$ for $p_A(x, y)$. Note that, by the definition of $\otimes_h$, (i) is clearly the same as:

(i)′ The natural matrix product

$$ab = \left(\sum_k a_{ik} \cdot b_{kj} \right)_{ij}$$

defines a unital Banach algebra structure on $M_n(A)$ for all $n \geq 1$.
In other words, (i) holds iff $\|e\| = 1$ and

$$\forall n \geq 1 \; \forall a, b \in M_n(A) \qquad \|ab\|_{M_n(A)} \leq \|a\|_{M_n(A)} \|b\|_{M_n(A)}.$$

The implication (i)′ ($\Leftrightarrow$ (i)) $\Rightarrow$ (ii) shows that this implies that, for any unital operator algebra B, $B \otimes_{\min} A$ is equipped with a natural unital operator algebra structure corresponding to the tensor product of p_B and p_A.

Curiously, the isomorphic version of this theorem resisted until Blecher [B4] recently proved the following statement.

Theorem 6.2. *([B4]) Let A be a Banach algebra given with an operator space structure. The following are equivalent.*

(i) *The product map p_A: $A \otimes_h A \to A$ is completely bounded.*

(ii) *There is, for some Hilbert space H, a homomorphism j: $A \to B(H)$, which is a complete isomorphism from A to $j(A)$.*

Remark. Note that (i) in Theorem 6.2 holds iff $\mathcal{K}[A] = \mathcal{K} \otimes_{\min} A$ is a Banach algebra up to a constant for its natural matrix product, in other words, iff there is a constant C such that $\|ab\| \leq C\|a\| \, \|b\|$ for all a, b in $\mathcal{K}[A]$.

The original proofs of these theorems do not use the results from the 1970s, and can be used to give new proofs of the latter. However, we have recently found a new approach: We will deduce the preceding two theorems from the stability under quotients of operator algebras. Note, however, that, both for Theorems 6.1 and 6.2, some trick evocative of the original proofs of [BRS, B4] still remains.

Our approach is based on the following statement (due to A. Bernard, to G. Lumer independently in some form, and originally due to Cole when A is a uniform algebra).

Theorem 6.3. *Let $A \subset B(H)$ be a closed subalgebra of $B(H)$ and let $I \subset A$ be a closed ideal. Then there is, for some Hilbert space $\mathcal{H}$, an isometric (and completely contractive) homomorphism*

$$\varphi \colon A/I \longrightarrow B(\mathcal{H}).$$

In other words, the class of operator algebras is stable under quotients.

Proof. We essentially follow Dixon's exposition in [Dix1]. Consider $x \in A/I$ with $\|x\| = 1$. We claim that there is, for some $\widehat{H}$, a homomorphism

φ_x: $A/I \to B(\widehat{H})$ with $\|\varphi_x\|_{cb} \leq 1$ and $\|\varphi_x(x)\| = \|x\| = 1$. From this claim the conclusion of Theorem 6.3 follows immediately: We simply let $\varphi = \bigoplus_x \varphi_x$ (with x running over the unit sphere of A). We now briefly justify this claim. Consider ξ in the unit sphere of $(A/I)^*$ such that $\xi(x) = 1$. Let us denote by q: $A \to A/I$ the quotient map and let $\widehat{\xi} \in A^*$ be defined by $\widehat{\xi}(a) = \xi(q(a))$, so that $\|\widehat{\xi}\|_{A^*} = 1$. Then there is a representation π: $B(H) \to B(\widehat{H})$ and elements s, t in the unit ball of $\widehat{H}$ such that

$$\widehat{\xi}(a) = \langle \pi(a)s, t \rangle. \qquad (6.1) \qquad \forall a \in A$$

(This is entirely elementary and well known, but we can view it as a very special case of Theorem 1.6 since $\|\widehat{\xi}\|_{A^*} = \|\widehat{\xi}\|_{CB(A,\mathbb{C})}$.) Now let $E_1 \subset \widehat{H}$ and $E_2 \subset E_1$ be defined by

$$E_1 = \overline{\mathrm{span}}[s, \pi(a)s \mid a \in A]$$
$$E_2 = \overline{\mathrm{span}}[\pi(i)s \mid i \in I].$$

Then E_1 and E_2 are clearly invariant under $\pi(A)$; therefore the subspace

$$E = E_1 \ominus E_2$$

is semi-invariant with respect to $\pi(A)$, so that (by Sarason's well-known ideas, see, e.g., [P10, Theorem 1.7]) the compression $\widehat{\pi}$: $A \to B(E)$ defined by

$$\widehat{\pi}(a) = P_E \pi(a)_{|E}$$

is a homomorphism. Indeed, for $a, b \in A$, we have $P_E \pi(a) P_{E_2} = 0$, and hence $\widehat{\pi}(a)\widehat{\pi}(b) = \widehat{\pi}(ab)$. Moreover, $\widehat{\pi}$ is clearly contractive (actually completely contractive). Observe that since $\widehat{\xi}$ vanishes on I, we have $t \in E_2^{\perp}$. Thus, it is now easy to check that (6.1) yields

$$\widehat{\xi}(a) = \langle \widehat{\pi}(a) P_E s, t \rangle, \qquad (6.2)$$

and also, since $\pi(I)E_1 \subset E_2$, that $\widehat{\pi}$: $A \to B(E)$ vanishes on I. Hence, if we let φ_x: $A/I \to B(E)$ be defined by $\widehat{\pi} = \varphi_x q$, we have (take a with $q(a) = x$ in (6.2))

$$1 = \|x\| = \xi(x) = \langle \varphi_x(x) P_E s, t \rangle,$$

whence $1 \leq \|\varphi_x(x)\|$. Since (by (2.4.2)) $\|\varphi_x\|_{cb} = \|\widehat{\pi}\|_{cb} \leq 1$, this proves our claim. ∎

Corollary 6.4. *In the situation of Theorem 6.3, there is, for some Hilbert space $\mathcal{H}$, a completely isometric homomorphism*

$$\psi: A/I \to B(\mathcal{H}).$$

Proof. Two simple proofs of this can be given. A first proof consists of applying Theorem 6.3 for each integer n to the algebra $M_n(A)/M_n(I) \simeq M_n(A/I)$. Assuming A and φ: $M_n(A)/M_n(I) \to B(\mathcal{H})$ unital, it is easy to check (by elementary algebra) that φ is necessarily of the form $I_{M_n} \otimes \psi_n$ for some homomorphism ψ_n: $A/I \to B(\mathcal{H})$. A direct sum argument (considering $\psi = \oplus_n \psi_n$) then completes the proof.

A second (and better) proof consists of repeating the proof of Theorem 6.3, but this time with ξ in the unit sphere of $CB(A/I, B(L))$ for some Hilbert space L. The same proof (using the fundamental Theorem 1.6 to factorize $\hat{\xi}$: $A \to B(L)$ instead of (6.1)) then yields Corollary 6.4.   ∎

Remark. The analogs of Theorems 6.1 and 6.2 for dual operator algebras are proved in [LeM6] (see also [B5]).

Remark. See [LeM2] for an extension of the preceding two results to the case when the class of Hilbert spaces is replaced by that of subspaces of quotients of L_p, with $p \neq 2$.

The idea of the proofs of Theorems 6.1 and 6.2 is very natural: We will represent A as a quotient of the "free operator algebra" generated by A. The latter is defined as follows. We will distinguish between the unital and nonunital cases. Let E be an operator space. Let $T(E)$ (resp. $T_u(E)$) be the tensor algebra (resp. the unital tensor algebra) over E, that is,

$$T(E) = E \oplus (E \otimes E) \oplus \cdots$$

and

$$T_u(E) = \mathbb{C} \oplus E \oplus (E \otimes E) \oplus \cdots .$$

We view $T(E)$ as a subspace of $T_u(E)$. Then $T(E)$ (resp. $T_u(E)$) is an algebra (resp. a unital algebra) containing E as a linear subspace and characterized by the property that any linear map v: $E \to B$ from E into an algebra (resp. a unital algebra) B uniquely extends to a homomorphism (resp. unital homomorphism) $T(v)$: $T(E) \to B$ (resp. $T_u(v)$: $T_u(E) \to B$). We will denote the free operator algebra associated to E by $OA(E)$, and we define it as follows: Let

$$C = \{v: E \to B(H_v) \mid \|v\|_{cb} \leq 1\},$$

where H_v is an arbitrary Hilbert space with dimension at most, say, the density character of E. We define a linear embedding

$$J: T_u(E) \to \bigoplus_{v \in C} B(H_v) \subset B\left(\bigoplus_{v \in C} H_v\right)$$

by setting

$$J(x) = \bigoplus_{v \in C} T_u(v)(x).$$

Note that J is a unital homomorphism. We define $OA_u(E)$ (resp. $OA(E)$) as the closure of $J(T_u(E))$ (resp. $J(T(E))$) in $B\left(\bigoplus_{v \in C} H_v\right)$. Thus $OA_u(E)$ is a unital operator algebra and $OA(E) \subset OA_u(E)$ is a closed subalgebra. It is easy to check that $OA_u(E)$ is characterized by the following universal property: For any map $v \colon E \to B(H)$ with $\|v\|_{cb} \le 1$ there is a unique unital homomorphism $\widehat{v} \colon OA_u(E) \to B(H)$, extending v, with $\|\widehat{v}\|_{cb} \le 1$.

The elements of $T_u(E)$ can be described as the vector space of formal sums:

$$P = \lambda_0 1 + P_1 + \cdots + P_N,$$

where each P_j $(1 \le j \le N)$ is "homogeneous" of degree j, that is, P_j is of the form

$$P_j = \sum_\alpha \lambda_\alpha(j) e_1^\alpha(j) \otimes \cdots \otimes e_j^\alpha(j)$$

with $\lambda_\alpha(j) \in \mathbb{C}$ and $e_i^\alpha(j) \in E$ $(1 \le i \le j)$. Similarly, any element Q of $T(E)$ can be written as $Q = P_1 + \cdots + P_N$. For any $v \in C$, let

$$v(P_j) = \sum_\alpha \lambda_\alpha(j) v(e_1^\alpha(j)) \ldots v(e_j^\alpha(j))$$

and let $v(P) = \lambda_0 I + v(P_1) + \cdots + v(P_N)$ and $v(Q) = v(P_1) + \cdots + v(P_N)$.

Clearly $v(P)$ and $v(Q)$ are nothing but $T_u(v)(P)$ and $T(v)(Q)$. Then we have $\|P\|_{OA_u(E)} = \sup_{v \in C} \|v(P)\|$ and $\|Q\|_{OA(E)} = \sup_{v \in C} \|v(Q)\|$. Explicitly, this means that

$$\|P\|_{OA_u(E)} = \sup_{v \in C} \left\| \lambda_0 I + \sum_{j=1}^N \sum_\alpha \lambda_\alpha(j) v(e_1^\alpha(j)) v(e_2^\alpha(j)) \ldots v(e_j^\alpha(j)) \right\|_{B(H_v)}.$$

More generally, let us now assume that the coefficients $\lambda_0, \lambda_\alpha(j)$ are all in $\mathcal{K}$ (or even in $B(\ell_2)$) and let P be an element of $\mathcal{K} \otimes T_u(E)$ of the form

$$P = \lambda_0 \otimes 1 + \sum_{j=1}^N \sum_\alpha \lambda_\alpha(j) \otimes e_1^\alpha(j) \otimes \cdots \otimes e_j^\alpha(j).$$

Then the above definition of $OA_u(E)$ means that

$$\|P\|_{\mathcal{K} \otimes_{\min} OA_u(E)}$$
$$= \sup_{v \in C} \left\| \lambda_0 \otimes I + \sum_{j=1}^N \lambda_\alpha(j) \otimes v(e_1^\alpha(j)) \ldots v(e_j^\alpha(j)) \right\|_{\mathcal{K} \otimes_{\min} B(H_v)}. \tag{6.3}$$

Of course a similar formula holds for $OA(E)$ with the constant term omitted.

Remark 6.5. Let E be any operator space. Consider the iterated Haagerup tensor product $X = E \otimes_h E \dots \otimes_h E$ (N times). Let x be arbitrary in $\mathcal{K} \otimes E \otimes \dots \otimes E$. Then x can be written as a finite sum

$$x = \sum_i \lambda_i \otimes x_i^1 \otimes \dots \otimes x_i^N$$

with $\lambda_i \in \mathcal{K}$ and $x_i^1, \dots, x_i^N \in E$. By Corollary 5.3(iii), we have

$$\|x\|_{\mathcal{K} \otimes_{\min} X} = \sup \left\{ \left\| \sum_i \lambda_i \otimes \sigma^1(x_i^1) \sigma^2(x_i^2) \dots \sigma^N(x_i^N) \right\|_{\min} \right\},$$

where the supremum runs over all possible choices of H and of complete contractions $\sigma^1 \colon E \to B(H), \dots, \sigma^N \colon E \to B(H)$. We claim that this supremum is actually attained when $\sigma^1, \dots, \sigma^N$ are all the same; more precisely, we have

$$\|x\|_{\mathcal{K} \otimes_{\min} X} = \sup \left\{ \left\| \sum_i \lambda_i \otimes v(x_i^1) v(x_i^2) \dots v(x_i^N) \right\|_{\min} \right\}, \qquad (6.4)$$

where the supremum runs over all possible H and all complete contractions $v \colon E \to B(H)$. Indeed, this follows from a trick already used by Blecher in [B1] and which seems to have originated in Varopoulos's paper [V2]. The trick consists of replacing $\sigma^1, \dots, \sigma^N$ by the single map $v \colon E \to B(\underbrace{H \oplus \dots \oplus H}_{N+1 \text{ times}})$

of the form

$$v(e) = \begin{pmatrix} 0 & \sigma_1(e) & & & \\ & \ddots & & \bigcirc & \\ & & \ddots & & \\ & & & & \sigma_N(e) \\ & \bigcirc & & & 0 \end{pmatrix}.$$

(More precisely, $v(e)$ is the $(N+1) \times (N+1)$ matrix having $(\sigma^1(e), \dots, \sigma^N(e))$ above the main diagonal and zero elsewhere.)

Then it is easy to check that $\|v\|_{cb} = \sup_j \|\sigma^j\|_{cb}$, $\forall x^1, \dots, x^N \in E$,

$$[v(x^1) \dots v(x^N)]_{1,N+1} = \sigma^1(x^1) \sigma^2(x^2) \dots \sigma^N(x^N).$$

From this our claim immediately follows.

Proposition 6.6. *Let E be any operator space. Fix $N \geq 1$. Let $E_N = E \otimes \dots \otimes E$ (N times). Consider E_N as embedded into $OA(E)$. Then the identity mapping on E_N extends to a completely isometric embedding of*

$E \otimes_h \cdots \otimes_h E$ *(N times) into* $OA(E)$. *Moreover, there is a completely contractive projection from* $OA_u(E)$ *onto the range of this embedding.*

Proof. For simplicity let us again denote $X = E \otimes_h \cdots \otimes_h E$ (*N* times). Since the algebraic tensor product $E \otimes \cdots \otimes E$ is dense in X, to show the completely isometric part it suffices to prove that, for any element G in $\mathcal{K} \otimes E \otimes \cdots \otimes E$, we have

$$\|G\|_{\mathcal{K} \otimes_{\min} X} = \|G\|_{\mathcal{K} \otimes_{\min} OA(E)}.$$

But this is immediate by (6.3) and (6.4).

We now show that we have a "nice" projection Q_N from $OA(E)$ onto the closure of E_N in $OA(E)$. Let x be a typical element of $T(E)$, of the form $x = \sum_{n=0}^{m} x_n$ with $x_n \in E_n$ viewed as included in $OA_u(E)$ (so J is viewed as an inclusion) and (say) $N \leq m$. Note that by (6.3) we have

$$\|x\|_{OA(E)} = \left\| \sum e^{int} x_n \right\|_{OA(E)} \tag{6.5}$$

for any real t and $x_N = \int e^{-iNt} (\sum_k e^{ikt} x_k) dt / 2\pi$, which implies

$$\|x_N\| \leq \|x\|_{OA(E)}. \tag{6.6}$$

This proves that the projection defined by $Q_N(x) = x_N$ is contractive. But by an obvious modification (using operator coefficients instead of scalars) one easily verifies that it actually is completely contractive. ∎

Let E be an operator space equipped with a Banach algebra structure. Let $p\colon E \otimes E \to E$ be the linear map associated to the product map. Then, by associativity, p uniquely extends to a mapping

$$\widehat{p}\colon T(E) \to E,$$

which is the identity on E and satisfies for any $N \geq 2$ and any x_i in E

$$\widehat{p}(x_1 \otimes \cdots \otimes x_N) = p(\widehat{p}(x_1 \otimes \cdots \otimes x_{N-1}) \otimes x_N).$$

To prove Theorem 6.2, we will need the following simple lemma (already used in [B4]).

Lemma 6.7. *For any z in $\mathbb{C}$, let us denote by $\sigma_z\colon T(E) \to T(E)$ the homomorphism that, when restricted to E, is equal to $z\, I_E$. Then, if the product map extends to a mapping on $E \otimes_h E$ with $\|p\|_{CB(E \otimes_h E, E)} = c$, we have, for any z with $|z| < 1/c$,*

$$\|\widehat{p}\sigma_z\|_{cb} \leq \frac{|z|}{1 - |z|c}. \tag{6.7}$$

Proof. By an obvious iteration p defines a product mapping from $E \otimes_h \cdots \otimes_h E$ (n times) into E with c.b. norm $\leq c^{n-1}$. By Proposition 6.6, the natural embedding of $E \otimes_h \cdots \otimes_h E$ (n times) into $OA(E)$ is a complete isometry; hence the restriction of $\widehat{p}$ to $E \otimes \cdots \otimes E \subset T(E) \subset OA(E)$ has c.b. norm $\leq c^{n-1}$.

We have

$$\widehat{p}\sigma_z(x) = \sum_{1}^{N} z^n \widehat{p}(x_n),$$

and hence by (6.5)

$$\|\widehat{p}\sigma_z(x)\| \leq \sum_{1}^{N} |z|^n c^{n-1} \|x\|_{OA(E)}$$

or, equivalently,

$$\|\widehat{p}\sigma_z\| \leq \sum_{1}^{\infty} |z|^n c^{n-1} = \frac{|z|}{1 - |z|c}.$$

By a simple modification (left to the reader) we obtain the same bound for the c.b. norm of $\widehat{p}\sigma_z$, whence (6.7). ∎

Proof of Theorem 6.2. The implication (ii) ⇒ (i) is immediate by Corollary 5.4. We now prove the converse. Assume (i). We denote by E the operator space underlying A. Fix z with $|z| < 1/c$. Clearly $\widehat{p}\sigma_z$ is a surjective homomorphism from $OA(E)$ onto E. Let I be its kernel and let w: $OA(E)/I \to E$ be the associated isomorphic homomorphism. By (6.7) and the definition of the quotient operator space structure (cf. §2.4), we have $\|w\|_{cb} \leq \frac{|z|}{1-|z|c}$. Let i: $E \to OA(E)$ be the canonical inclusion and let q: $OA(E) \to OA(E)/I$ be the quotient map. Clearly we have $\sigma_z i(x) = z i(x)$; hence $\widehat{p}\sigma_z i(x) = zx$ or $(wq)i(x) = zx$, so that $(1/z)qi = w^{-1}$. Whence

$$\|w^{-1}\|_{cb} = \|(1/z)qi\|_{cb} \leq |z|^{-1}.$$

Thus if we choose for instance the value $z = 1/2c$, we find a homomorphism w: $OA(E)/I \to E$ with $\|w^{-1}\|_{cb} \leq 2c$ and $\|w\|_{cb} \leq 1/c$. By Corollary 6.4, this completes the proof. ∎

Paradoxically, in our approach the proof of Theorem 6.1 is a bit more complicated. The crucial point is the following.

Lemma 6.8. Let E be an operator space that is also a Banach algebra with a unit e. Let p: $E \otimes E \to E$ be associated to the product map. Then, if $\|e\| = 1$ and if p defines a complete contraction from $E \otimes_h E$ into E, the mapping $\widehat{p}$ introduced before the preceding lemma extends to a completely contractive unital homomorphism $\widehat{p}_u$ from $OA_u(E)$ onto E.

Proof. Let $E_0 = \mathbb{C}$ and $E_j = E \otimes \cdots \otimes E$ (j times). Let x be a typical element in $\mathcal{K} \otimes T_u(E)$. Then x can be written as $x = \sum_{j=0}^{n} x_j$ with $x_0 = t \otimes 1 \in \mathcal{K} \otimes \mathbb{C}$ and with each x_j expanded as a finite sum

$$x_j = \sum_{\alpha} t_j(\alpha) \otimes e_1^{\alpha} \otimes \cdots \otimes e_j^{\alpha}$$

with $t_j(\alpha) \in \mathcal{K}$, $e_j^{\alpha} \in E$, $1 \le j \le n$. For $e_1, \ldots, e_j \in E$, we denote simply below by $e_1 e_2 \ldots e_j$ their product in E. Note that $I_{\mathcal{K}} \otimes \widehat{p}_u(x) = \sum_j \sum_{\alpha} t_j(\alpha) \otimes e_1^{\alpha} e_2^{\alpha} \ldots e_j^{\alpha}$, so that the content of Lemma 6.8 is the inequality

$$\left\| \sum_j \sum_{\alpha} t_j(\alpha) \otimes (e_1^{\alpha} e_2^{\alpha} \ldots e_j^{\alpha}) \right\|_{\mathcal{K} \otimes_{\min} E} \le \|x\|_{\mathcal{K} \otimes_{\min} OA_u(E)}. \tag{6.8}$$

This is somewhat reminiscent of von Neumann's inequality for polynomials in a contraction.

Observe that (as in the proof of Theorem 6.2) we clearly have (6.8) if x is reduced to a "homogeneous" term of some degree, that is, if $x = x_j$ for some j. Indeed, by iteration the product map p defines a complete contraction $e_1 \otimes \cdots \otimes e_j \to e_1 e_2 \ldots e_j$ from $E \otimes_h \cdots \otimes_h E$ into E, and $E \otimes_h \cdots \otimes_h E = E \otimes_f \cdots \otimes_f E$ clearly naturally embeds completely isometrically into $OA_u(E)$. Thus we immediately get (6.8) in the homogeneous case.

For the general case, the idea will be to replace a general element $x = \sum x_j$ as above by another one $y(N)$ that is homogeneous of degree N and satisfies moreover $\|y(N)\|_{\min} \le \|x\|_{\min}$ and $\lim_{N \to \infty} \|(I_{\mathcal{K}} \otimes \widehat{p}_u)(y(N))\|_{\min} = \|(I_{\mathcal{K}} \otimes \widehat{p}_u)(x)\|_{\min}$. The conclusion will then be easy to reach. Note that we will denote simply by $\| \ \|_{\min}$ the norms appearing on both sides of (6.8). This should not bring any confusion.

Note that $\mathcal{K} \otimes_{\min} OA_u(E)$ is an operator algebra embedded in the unital operator algebra $B = B(\ell_2) \otimes_{\min} OA_u(E)$. For convenience we introduce the following elements of B: We denote

$$\xi_k = 1_{B(\ell_2)} \otimes \underbrace{e \otimes \cdots \otimes e}_{k-\text{times}}.$$

Then we define

$$\Delta_j(N) = \frac{1}{N - j + 1} \sum_{k=0}^{N-j} \xi_k \cdot x_j \cdot \xi_{N-k-j},$$

where the dot denotes the product in B. Observe that

$$(I_{\mathcal{K}} \otimes \widehat{p}_u)(\Delta_j(N)) = I_{\mathcal{K}} \otimes \widehat{p}_u(x_j). \tag{6.9}$$

We claim that we have, for all $N \geq n$,

$$\left\| \sum_{j=0}^{n} (I_{\mathcal{K}} \otimes \widehat{p}_u)(\Delta_j(N)) \left(\frac{N-j+1}{N+1} \right) \right\|_{\min} \leq \|x\|. \tag{6.10}$$

To verify this let

$$\varphi(t) = (N+1)^{-1/2} \sum_{k=0}^{N} e^{ikt} \xi_k \quad \text{and} \quad \psi(t) = (N+1)^{-1/2} \sum_{\ell=0}^{N} e^{-i\ell t} \xi_{N-\ell}.$$

Note that since $\|e\| = 1$ we have $\|\xi_k\| = 1$ for all k; hence, if we view $B \subset B(H)$, we have

$$\left\| \int \varphi(t)\varphi(t)^* \frac{dt}{2\pi} \right\|_{B(H)} \leq 1 \quad \text{and} \quad \left\| \int \psi(t)^*\psi(t) \frac{dt}{2\pi} \right\|_{B(H)} \leq 1. \tag{6.11}$$

Let us compute

$$y(N) = \int \varphi(t) \cdot \left(\sum_{j=0}^{n} e^{ijt} x_j \right) \cdot \psi(t) \, \frac{dt}{2\pi}. \tag{6.12}$$

A simple verification shows that

$$y(N) = (N+1)^{-1} \sum_{\substack{k+j=\ell \\ 0\leq j\leq n, \ 0\leq k,\ell\leq N}} \xi_k \cdot x_j \cdot \xi_{N-\ell}$$

$$(6.13) \qquad\qquad = \sum_{j=0}^{n} \Delta_j(N) \cdot \left(\frac{N-j+1}{N+1} \right).$$

The preceding formula shows that $y(N)$ is "homogeneous of degree N," that is, we have $y(N) \in \mathcal{K} \otimes E_N$. As mentioned above after (6.8), $\widehat{p}_u$ is clearly completely contractive when restricted to $\mathcal{K} \otimes E_N$. Therefore we have

$$\|(I_{\mathcal{K}} \otimes \widehat{p}_u)(y(N))\|_{\min} \leq \|y(N)\|_{\min}, \tag{6.14}$$

and by (6.12) we have (by a variant of (1.12))

$$\|y(N)\|_{\min} \leq \left\| \int \varphi(t)\varphi(t)^* \frac{dt}{2\pi} \right\|^{1/2} \sup_t \left\| \sum e^{ijt} x_j \right\| \left\| \int \psi(t)^*\psi(t) \frac{dt}{2\pi} \right\|^{1/2},$$

whence by (6.11) and (6.5)

$$\|y(N)\|_{\min} \leq \|x\|_{\min}. \tag{6.15}$$

Finally, we have by (6.9) and (6.13)

$$(I_{\mathcal{K}} \otimes \widehat{p}_u)(y(N)) = \sum_{j=0}^{n} \left(\frac{N - j + 1}{N + 1} \right) (I_{\mathcal{K}} \otimes \widehat{p}_u)(x_j);$$

hence by (6.14) and (6.15) we obtain our claim (6.10). Then, from (6.9) and (6.10), we have

$$\left\| (I_{\mathcal{K}} \otimes \widehat{p}_u) \left(\sum_{j=0}^{n} x_j \right) \right\|_{\min} \leq \limsup_{N \to \infty} \| (I_{\mathcal{K}} \otimes \widehat{p}_u)(y(N)) \|_{\min} \leq \|x\|_{\min}.$$

This concludes the proof of (6.8) and hence of Lemma 6.8. ∎

Proof of Theorem 6.1. Again (ii) ⇒ (i) is immediate by Corollary 5.4. We turn to the converse. We denote by E the operator space underlying A. The proof is analogous to that of Theorem 6.2, but instead we need to know that, if (i) holds, the product map $e_1 \otimes \cdots \otimes e_n \to e_1 \cdot e_2 \ldots e_n$ extends to a completely contractive unital map from $OA_u(E)$ to E. This is provided by Lemma 6.8. Given this fact, we conclude as before: Assume (i). The map $\widehat{p}_u \colon OA_u(E) \to E$ yields after passing to the quotient by its kernel I a unital bijective homomorphism $w \colon OA_u(E)/I \to E$ with $\|w\|_{cb} \leq 1$. As before, we have $w^{-1} = qi$, where $i \colon E \to OA_u(E)$ is the natural injection and $q \colon OA_u(E) \to OA_u(E)/I$ is the natural quotient map.

Hence we conclude that $\|w^{-1}\|_{cb} \leq \|q\|_{cb}\|i\|_{cb} = 1$, that is, w is a completely isometric isomorphism from $OA_u(E)/I$ onto E. By Corollary 6.4, E is unitally and completely isometrically isomorphic to a unital operator algebra. ∎

Remark. We refer the reader to [BLM] for a detailed study of various possible operator algebra structures on "classical" examples such as ℓ_p or S_p (the Schatten p-class).

Operator spaces that are also modules over an operator algebra (in other words, "operator modules") can be characterized in a similar way (see [CES, ER9], and see also [Ma1–2] for dual modules). Let $X \subset B(H)$ be an operator space and let A, B be two subalgebras of $B(H)$. If X is stable under left (resp. right) multiplication by elements of A (resp. B), then X can be viewed as a concrete submodule of $B(H)$ with respect to the actions of A and B by left and right multiplication. The next statement characterizes the "abstract" bimodules that admit such a concrete realization in $B(H)$.

Theorem 6.9. *([CES]) Let A, B be C^*-algebras and let X be an (A, B)-bimodule, that is to say, X is both a left A-module (with action denoted*

$(a, x) \rightarrow a.x)$ and a right B-module (with action denoted $(x, b) \rightarrow x.b$), so that we have a "module multiplication" map $m: A \times X \times B \rightarrow X$ defined by $m(a, x, b) = (a.x).b = a.(x.b)$. For simplicity we write $m(a, x, b) = a \cdot x \cdot b$. We assume that the sets $\{a.x \mid a \in A, x \in X\}$ and $\{x.b \mid x \in X, b \in B\}$ are dense in X. Then, given an operator space structure on X, the following are equivalent:

(i) m defines a complete contraction from $A \otimes_h X \otimes_h B$ to X.

(ii) There exists a completely isometric embedding $j: X \rightarrow B(\mathcal{H})$ and representations $\rho_1: A \rightarrow B(\mathcal{H})$ and $\rho_2: B \rightarrow B(\mathcal{H})$ such that, for all x in X, a in A, and b in B, we have

$$j[m(a, x, b)] = \rho_1(a)j(x)\rho_2(b).$$

Proof. Assume $X \subset B(H)$. Applying Lemma 5.14 to the trilinear map $m: A \times X \times B \rightarrow B(H)$, we obtain a complete contraction $\sigma: X \rightarrow B(\mathcal{H})$; representations $\pi_1: A \rightarrow B(\mathcal{H})$, $\pi_2: B \rightarrow B(\mathcal{H})$; and contractive maps $V, W: H \rightarrow \mathcal{H}$ such that $(\forall x \in X, a \in A, b \in B)$

$$a \cdot x \cdot b = V^*\pi_1(a)\sigma(x)\pi_2(b)W. \tag{6.16}$$

Let p_1 (resp. p_2) be the orthogonal projection onto $\overline{\mathrm{span}}[\pi_1(A)VH]$ (resp. $\overline{\mathrm{span}}[\pi_2(B)WH]$). Let $s = p_1\sigma(a \cdot x \cdot b)p_2$ and $t = p_1\pi_1(a)\sigma(x)\pi_2(b)p_2$. We claim that $s = t$. Indeed, it suffices to check that

$$\langle s\pi_2(\beta)W(h), \pi_1(\alpha)^*V(k)\rangle = \langle t\pi_2(\beta)W(h), \pi_1(\alpha)^*V(k)\rangle$$

for all α in A, β in B, and h, k in H. But this equality is the same as

$$\langle V^*\pi_1(\alpha)\sigma(a \cdot x \cdot b)\pi_2(\beta)W(h), k\rangle = \langle V^*\pi_1(\alpha a)\sigma(x)\pi_2(b\beta)W(h), k\rangle,$$

and indeed by (6.16) both sides are equal to $\langle (\alpha a \cdot x \cdot b\beta)(h), k\rangle$, which proves our claim that

$$p_1\sigma(a \cdot x \cdot b)p_2 = p_1\pi_1(a)\sigma(x)\pi_2(b)p_2. \tag{6.17}$$

Note that p_1 (resp. p_2) commutes with π_1 (resp. π_2) so that $\rho_1(\cdot) = p_1\pi_1(\cdot)$ and $\rho_2(\cdot) = \pi_2(\cdot)p_2$ are representations. Setting $j(x) = p_1\sigma(x)p_2$, we have $\|j\|_{cb} \leq 1$ and we obtain by (6.17)

$$j(a \cdot x \cdot b) = \rho_1(a)j(x)\rho_2(b).$$

It remains to show that j is completely isometric. In the unital case this is immediate because (6.16) then allows us to write $x = V^*j(x)W$, and since V, W are contractive (and $\|j\|_{cb} \leq 1$), j must be completely isometric. In general, our density assumption guarantees that if (a_i) (resp. (b_k)) is an approximate unit in the unit ball of A (resp. B), then $a_i \cdot x \cdot b_k \to x$. Hence, by (6.16) we have

$$V^* \pi_1(a_i) j(x) \pi_2(b_k) W = a_i \cdot x \cdot b_k \to x,$$

and hence we have $\|x\| \leq \|j(x)\|$ and similarly for the matrix norms, so again we conclude that j is completely isometric. ∎

Suitably modified versions of the Haagerup tensor product are available for operator modules (see [BMP, ChS, Ma2, AP, Pop]).

Operator modules play a central role in [BMP], where the foundations of a Morita theory for non-self-adjoint operator algebras are laid. See [BMP], [B4], Blecher's survey in [Kat], and references contained therein for more on all of this.

Exercises

Exercise 6.1. Let A be an operator algebra. Let $E_1, \ldots, E_N$ be operator spaces. Then the product map $A \times \cdots \times A \to A$ induces a complete contraction

$$(E_1 \otimes_{\min} A) \otimes_h (E_2 \otimes_{\min} A) \otimes_h \cdots \otimes_h (E_N \otimes_{\min} A) \to (E_1 \otimes_h \cdots \otimes_h E_N) \otimes_{\min} A.$$

A fortiori, for $N = 3$, the product map $p_A \colon A \otimes_h A \to A$ defines a complete contraction

$$(E_1 \otimes_{\min} A) \otimes_h E_2 \otimes_h (E_3 \otimes_{\min} A) \to (E_1 \otimes_h E_2 \otimes_h E_3) \otimes_{\min} A.$$

Chapter 7. The Operator Hilbert Space

Hilbertian operator spaces. In operator algebra theory and in quantum physics, numerous Hilbertian operator spaces have appeared. Let us say that a subspace $E \subset B(H)$ is Hilbertian if it is isometric (as a Banach space) to a Hilbert space. We have already met the spaces R and C, for instance. A more sophisticated example is the linear span of the Clifford matrices or, equivalently, of the generators of the Fermion algebra (this is sometimes called a spin system). Essentially the same example appears with the linear span of the creation (or annihilation) operators on the antisymmetric Fock space (see §9.3). A different example is the linear span of the generators of the Cuntz algebra O_∞ (see §9.4) and also the generators of the reduced C^*-algebra $C_\lambda^*(F_\infty)$ on the free group with infinitely many generators (this one is only isomorphic to ℓ_2; see §9.7) or the "free" analog of Gaussian variables in Voiculescu's "free" probability theory (see §9.9).

Actually it is possible to show that there is a continuum of distinct (i.e. pairwise not completely isomorphic) isometrically Hilbertian operator spaces. Furthermore (we will come back to this in Chapter 21), if $n > 2$, it can be shown that the set of all Hilbertian operator spaces of a fixed dimension n equipped with the "complete" analog of the Banach-Mazur distance is not compact, and *not even separable*.

As mentioned in Chapter 3, Blecher and Paulsen [BP1] observed that any separable infinite-dimensional Hilbertian operator space $\mathcal{H}$ sits "in between" the extreme cases $\min(\ell_2)$ and $\max(\ell_2)$, that is, we have isometric and completely contractive inclusions

$$\max(\ell_2) \subset \mathcal{H} \subset \min(\ell_2).$$

Moreover, the interval between the two extremes is in some sense very "broad."

Existence and unicity of OH. Basic properties. Despite the multiplicity of (different) examples of Hilbertian operator spaces, it turns out that there is a *central object* in this class, that is, a space that plays the same central role in the category of operator spaces as Hilbert space does in the category of Banach spaces. To motivate the next result, we recall that, if H is a Hilbert space, we have a canonical (i.e. basis free) identification

$$H = \overline{H^*}.$$

Moreover, this characterizes Hilbert spaces in the sense that, if a Banach space E is such that there is an isometry $i\colon E \to \overline{E^*}$ that is positive (i.e., $i(x)(x) \geq 0 \ \forall x \in E$), then E is isometric to a Hilbert space.

In the category of o.s. this becomes (see [P1])

Theorem 7.1. *Let I be any set. Then there is an operator space*

$$OH(I)$$

that is isometric (as a Banach space) to $\ell_2(I)$ and which is such that the canonical identification $\ell_2(I) = \overline{\ell_2(I)}^*$ induces a complete isometry from $OH(I)$ to $\overline{OH(I)}^*$. Moreover, $OH(I)$ is the unique operator space up to a complete isometry with this property. Let K be any Hilbert space and let $(T_i)_{i \in I}$ be any orthonormal basis of $OH(I)$. Then for any finitely supported family $(x_i)_{i \in I}$ in $B(K)$ we have

$$\left\| \sum x_i \otimes T_i \right\|_{\min} = \left\| \sum x_i \otimes \overline{x_i} \right\|_{\min}^{1/2} . \tag{7.1}$$

Notation. If $I = \mathbb{N}$, we denote the space $OH(I)$ simply by OH (we call it "*the* operator Hilbert space"). If $I = \{1, 2, \ldots, n\}$, we denote the space $OH(I)$ simply by OH_n. Note that any n-dimensional subspace of OH is completely isometric to OH_n.

To prove Theorem 7.1, it suffices to produce a norm α on $K \otimes \ell_2(I)$ with a suitable self-dual property. The formula is extremally simple: $\forall x = \sum_{i \in I} x_i \otimes e_i \in K \otimes \ell_2(I)$ (finite sum) with $x_i \in K$, we set

$$\alpha(x) = \left\| \sum_{i \in I} x_i \otimes \overline{x_i} \right\|_{\min}^{1/2} .$$

The proof that this is indeed an operator space structure on $\ell_2(I)$ with the required property is based on a version of the Cauchy-Schwarz inequality due to Haagerup ([H1, Lemma 2.4]), as follows:

$$\left\| \sum_{i \in I} x_i \otimes \overline{y_i} \right\|_{\min} \leq \left\| \sum x_i \otimes \overline{x_i} \right\|_{\min}^{1/2} \left\| \sum y_i \otimes \overline{y_i} \right\|_{\min}^{1/2} \tag{7.2}$$

valid for all $x_i, y_i \in K$ (or, more generally, for all $x_i, y_i \in B(\ell_2)$). These formulas are one more illustration of the idea of "quantization" (see the discussion in §2.11).

Proof of (7.2). By Proposition 2.9.1 we have

$$\left\| \sum x_i \otimes \overline{y_i} \right\|_{\min} = \sup \left\{ \left| \sum \langle x_i a y_i^*, b \rangle \right| \right\}, \tag{7.3}$$

where the supremum runs over all a, b in the unit ball of $S_2 = S_2(\ell_2, \ell_2)$ and where $\langle \, , \, \rangle$ denotes the scalar product in the Hilbert space S_2, that is, $\langle x_i a y_i^*, b \rangle = \operatorname{tr}(x_i a y_i^* b^*)$. Note that $\|a\|_2 \leq 1$ iff a can be written as $a = a_1 a_2$ with $\operatorname{tr}|a_1|^4 \leq 1$ and $\operatorname{tr}|a_2|^4 \leq 1$, and similarly for b. Using this we can write

$$\langle x_i a y_i^*, b \rangle = \operatorname{tr}(x_i a_1 a_2 y_i^* b_2^* b_1^*) = \langle b_1^* x_i a_1, b_2 y_i a_2^* \rangle .$$

Hence, by Cauchy-Schwarz

$$\left|\sum \langle x_i a y_i^*, b\rangle\right| \le \left(\sum \|b_1^* x_i a_1\|_2^2\right)^{1/2} \left(\sum \|b_2 y_i a_2^*\|_2^2\right)^{1/2},$$

but by Proposition 2.9.1 again

$$\sum \|b_1^* x_i a_1\|_2^2 = \sum \mathrm{tr}(b_1^* x_i a_1 a_1^* x_i^* b_1)$$
$$= \sum \mathrm{tr}(x_i a_1 a_1^* x_i^* b_1 b_1^*) \le \left\|\sum x_i \otimes \overline{x_i}\right\|_{\min},$$

and similarly for the other term. Hence we finally derive (7.2) from (7.3). ■

Remark. The preceding proof also establishes the following two additional formulas:

$$(7.3)' \quad \left\|\sum x_i \otimes \overline{x_i}\right\|_{\min} = \sup\left\{\sum \mathrm{tr}(x_i a x_i^* b) \mid a \ge 0, b \ge 0,\ \mathrm{tr}\,a^2 \le 1,\right.$$
$$\left. \mathrm{tr}\,b^2 \le 1\right\}$$

$$(7.3)'' \quad \left\|\sum x_i \otimes \overline{x_i}\right\|_{\min} = \sup\left\{\left(\sum \|\beta x_i \alpha\|_2^2\right)^{1/2} \mid \ \mathrm{tr}|\alpha|^4 \le 1,\ \mathrm{tr}|\beta|^4 \le 1\right\}.$$

Proof of Theorem 7.1. We first show the "existence" of the space $OH(I)$. Let K be a fixed *infinite-dimensional* Hilbert space. Let $B = B(K)$ (actually, the proof works equally well if we set $B = \mathcal{K}$). Let us denote by C the set of all finitely supported families $y = (y_i)_{i \in I}$ in B such that $\|\sum y_i \otimes \overline{y_i}\|_{\min} \le 1$. Let $H_y = \overline{K}$ for all y in C. For any fixed i in I, we define

$$T_i = \bigoplus_{y \in C} \overline{y_i} \in B\left(\bigoplus_{y \in C} H_y\right).$$

Then, for any finitely supported family $(x_i)_{i \in I}$ in B, we have

$$\left\|\sum x_i \otimes T_i\right\|_{\min} = \sup_{y \in C} \left\|\sum x_i \otimes \overline{y_i}\right\|.$$

But by (7.2) we have

$$\sup_{y \in C} \left\|\sum x_i \otimes \overline{y_i}\right\| = \left\|\sum x_i \otimes \overline{x_i}\right\|_{\min}^{1/2}, \tag{7.4}$$

since the supremum is attained for

$$y_i = x_i \left[\left\|\sum x_i \otimes \overline{x_i}\right\|_{\min}^{-1/2}\right].$$

Thus we obtain

$$\left\|\sum x_i \otimes T_i\right\| = \left\|\sum x_i \otimes \overline{x_i}\right\|_{\min}^{1/2}. \tag{7.5}$$

In particular, if $e \in B$ is any element with $\|e\| = 1$, we have for any finitely supported familly of scalars $(\alpha_i)_{i \in I}$

$$\left\|\sum \alpha_i T_i\right\| = \left\|\sum \alpha_i e \otimes T_i\right\|_{\min} = \left\|\sum (\alpha_i e) \otimes \overline{(\alpha_i e)}\right\|^{1/2} = \left(\sum |\alpha_i|^2\right)^{1/2}.$$

Hence if we let

$$E = \overline{\mathrm{span}}[T_i \mid i \in I],$$

then $E \simeq \ell_2(I)$ isometrically. We claim that $E \simeq \overline{E^*}$ completely isometrically. To verify this it suffices to check that, if $(\xi_i)_{i \in I}$ is the basis of E^* biorthogonal to $(T_i)_{i \in I}$, we have for any $x = (x_i)_{i \in I}$ as before

$$\left\|\sum x_i \otimes T_i\right\| = \left\|\sum x_i \otimes \overline{\xi_i}\right\|_{B \otimes_{\min} \overline{E^*}}. \tag{7.6}$$

But by the definition of the complex conjugate (see §2.9) we have

$$\left\|\sum x_i \otimes \overline{\xi_i}\right\|_{B \otimes_{\min} \overline{E^*}} = \left\|\sum \overline{x_i} \otimes \xi_i\right\|_{\overline{B} \otimes_{\min} E^*} = \|u\|_{cb}, \tag{7.7}$$

where $u \colon E \to \overline{B}$ is the mapping defined by $u(x) = \sum \xi_i(x)\overline{x_i}$ for all x in E (or, equivalently, $u(T_i) = \overline{x_i}$ for all i).

By (2.1.9) (and by a density argument) we have $\|u\|_{cb} = \sup\{\|\sum y_i \otimes u(T_i)\|\}$, where the sup runs over all finitely supported families $(y_i)_{i \in I}$ in B with $\|\sum y_i \otimes T_i\|_{\min} \leq 1$. Equivalently,

$$\|u\|_{cb} = \sup\left\{\left\|\sum y_i \otimes \overline{x_i}\right\| \mid y \in C\right\}.$$

Hence, by (7.4),

$$= \left\|\sum x_i \otimes \overline{x_i}\right\|^{1/2}.$$

Thus (7.7) and (7.5) give us (7.6). This proves the existence of a space with the properties in Theorem 7.1. We now address the unicity. Let F be an operator space isometric to $\ell_2(I)$ and such that $F \simeq \overline{F^*}$ completely isometrically. Let $(\theta_i)_{i \in I}$ be any orthonormal basis in F. We will show that, for any $(x_i)_{i \in I}$ as before, we have necessarily

$$\left\|\sum x_i \otimes \theta_i\right\|_{\min} = \left\|\sum x_i \otimes \overline{x_i}\right\|_{\min}^{1/2}.$$

This will show that the correspondence $T_i \to \theta_i$ is a completely isometric isomorphism, thus establishing the announced unicity. Let $(\eta_i)_{i \in I}$ be the basis of F^* biorthogonal to $(\theta_i)_{i \in I}$. Reasoning as above, we see that $F \simeq \overline{F^*}$ implies that, for all $(x_i)_{i \in I}$ as before, we have

$$\left\|\sum x_i \otimes \theta_i\right\|_{B \otimes_{\min} F} = \left\|\sum x_i \otimes \overline{\eta_i}\right\|_{B \otimes_{\min} \overline{F^*}} = \sup\left\{\left\|\sum \overline{x_i} \otimes y_i\right\|\right\}, \quad (7.8)$$

where the supremum runs over all finitely supported families $(y_i)_{i \in I}$ in B such that $\|\sum y_i \otimes \theta_i\|_{\min} \leq 1$. In particular, we must have by homogeneity

$$\left\|\sum \overline{x_i} \otimes y_i\right\|_{\min} \leq \left\|\sum x_i \otimes \theta_i\right\|_{\min} \left\|\sum y_i \otimes \theta_i\right\|_{\min}.$$

Taking $x_i = y_i$, this implies

$$\left\|\sum x_i \otimes \overline{x_i}\right\|_{\min}^{1/2} \leq \left\|\sum x_i \otimes \theta_i\right\|_{\min}.$$

In particular, $\|\sum y_i \otimes \theta_i\|_{\min} \leq 1$ implies $(y_i)_{i \in I} \in C$, and hence (7.8) and (7.4) imply the converse inequality

$$\left\|\sum x_i \otimes \theta_i\right\|_{\min} \leq \left\|\sum x_i \otimes \overline{x_i}\right\|_{\min}^{1/2}.$$

Thus, we conclude that for any finitely supported family $(x_i)_{i \in I}$ in $B = B(K)$ we have

$$\left\|\sum x_i \otimes \theta_i\right\|_{\min} = \left\|\sum x_i \otimes \overline{x_i}\right\|_{\min}^{1/2} = \left\|\sum x_i \otimes T_i\right\|_{\min}.$$

Of course, this holds a fortiori when K is finite-dimensional. Thus if we set $OH(I) = E$, we have checked all the properties in Theorem 7.1. ∎

Definition. Let E be a (complex) vector space. We will say that a linear mapping $J: E \to \overline{E^*}$ (resp. $i: \overline{E^*} \to E$) is "positive definite" if it satisfies

$$\forall\, x \in E \qquad \overline{J(x)}(x) \geq 0 \quad (\text{resp.} \quad \forall\, \xi \in E^* \qquad i(\overline{\xi})(\xi) \geq 0).$$

Remark. It is easy to see that a bounded linear operator $J: E \to \overline{E^*}$ (resp. $i: \overline{E^*} \to E$) is "positive definite" iff there is a Hilbert space H and a bounded operator $v: E \to H$ (resp. $v: H \to E$) such that $J = {}^t v\, v: E \to H \simeq \overline{H^*} \to \overline{E^*}$ (resp. $i = v\, {}^t v: \overline{E^*} \to \overline{H^*} \simeq H \to E$). In the category of Banach spaces, the analog of Theorem 7.1 is the following obvious fact: If a Banach space E admits a positive definite isometry $J: E \to \overline{E^*}$ (resp. $i: \overline{E^*} \to E$), then E is isometric to a Hilbert space, and J (resp. i) is then surjective and coincides with the *canonical* (i.e. basis free) correspondence between $\overline{E^*}$ and E.

Thus Theorem 7.1 says that the operator space $E = OH(I)$ is characterized by the property that there is a positive definite map $J: E \to \overline{E^*}$ (resp. $i: \overline{E^*} \to E$) that is a complete isometry.

The next statement is mainly a useful reformulation of (7.1).

Proposition 7.2. (i) Let E be any operator space, I any set, and let v: $OH(I) \to E$ be a c.b. map. Let ^tv: $E^* \to OH(I)^*$ be its adjoint. Consider the composition

$$v\overline{^tv}: \overline{E^*} \to \overline{OH(I)^*} \simeq OH(I) \to E.$$

We have then

$$\|v\overline{^tv}\|_{cb} = \|v\|_{cb}^2. \tag{7.9}$$

(ii) Moreover, if (T_i) is an orthonormal basis of $OH(I)$, we have

$$\|v\|_{cb}^2 = \sup\left\{ \left\| \sum_{i \in J} v(T_i) \otimes \overline{v(T_i)} \right\|_{\min} \mid J \subset I, |J| < \infty \right\}. \tag{7.10}$$

(iii) For any $v: OH(I) \to OH(I')$ (I, I' arbitrary) we have $\|v\|_{cb} = \|v\|$.

Proof. (i) (Finite Case) Assume first that I is finite. Let $\xi_i \in OH(I)^*$ be the biorthogonal basis to (T_i), so that the mapping v corresponds to $\sum \xi_i \otimes v(T_i)$. Then, by (2.3.2) and (7.5), we have

$$\|v\|_{cb} = \left\| \sum \xi_i \otimes v(T_i) \right\|_{\min} = \left\| \sum \overline{\xi_i} \otimes \overline{v(T_i)} \right\|_{\min}$$

$$= \left\| \sum T_i \otimes \overline{v(T_i)} \right\|_{\min} = \left\| \sum v(T_i) \otimes \overline{v(T_i)} \right\|_{\min}^{1/2}.$$

On the other hand, for any ξ in E^* we have $^tv(\xi) = \sum \langle \xi, v(T_i) \rangle \xi_i$. Hence

$$\overline{^tv}(\overline{\xi}) = \overline{^tv(\xi)} = \sum \overline{\langle \xi, v(T_i) \rangle}\, \overline{\xi_i} = \sum \langle \overline{\xi}, \overline{v(T_i)} \rangle \overline{\xi_i} \simeq \sum \langle \overline{\xi}, \overline{v(T_i)} \rangle T_i$$

and hence

$$v(\overline{^tv}(\overline{\xi})) = \sum \langle \overline{\xi}, \overline{v(T_i)} \rangle v(T_i),$$

which means that the map $v\overline{^tv}: \overline{E^*} \to E$ corresponds to the tensor $\sum \overline{v(T_i)} \otimes v(T_i)$. Thus, by (2.3.2) again we obtain

$$\|v\overline{^tv}\|_{cb} = \left\| \sum \overline{v(T_i)} \otimes v(T_i) \right\|_{\min},$$

and we conclude by the first part of the proof

$$= \|v\|_{cb}^2.$$

(ii) Now let I be an arbitrary (a priori infinite) set. Let $J \subset I$ be a finite subset and let $E(J) = \text{span}[T_i \mid i \in J]$. By a simple density argument, we have

$$\|v\|_{cb} = \sup_J \|v_{|E(J)}\|_{cb}.$$

Hence, by the first part of the proof we obtain (7.10).

(i) (General Case) Assume now that I is infinite but that E is finite-dimensional. Then the series $\sum_{i \in I} \bar{\xi}_i \otimes v(T_i)$ and $\sum_{i \in I} \overline{v(T_i)} \otimes v(T_i)$ are convergent, respectively, in $\overline{OH(I)}^* \otimes E$ and in $\bar{E} \otimes E$, so that the preceding proof of (7.9) is easily extended to this case.

Finally, in the general case, we can use Exercise 1.4 to show that (7.9) reduces to the case when E is finite-dimensional, which we just treated.

(iii) By density, it suffices to check this when I is finite. Then, applying (i) with $E = OH(I')$ we are reduced to showing $\|v^t \bar{v}\|_{cb} = \|v^t \bar{v}\|$. But now, since $w = v^t \bar{v}$ is a non-negative Hermitian finite rank operator, it is diagonalizable; hence, for some orthonormal basis $(T_i)_{i \in I'}$ in $OH(I'), w$ is associated to a tensor of the form

$$\sum_{i \in I'} |\lambda_i|^2 \, \bar{T}_i \otimes T_i$$

with $(\lambda_i)_{i \in I'}$ finitely supported. The inequality $\|w\|_{cb} \le \|w\|$ is thus reduced to the following claim:

$$\left\| \sum_{i \in J} |\lambda_i|^2 \bar{T}_i \otimes T_i \right\| \le \sup_{i \in J} |\lambda_i|^2$$

for any finite subset $J \subset I'$. But then, by (7.3)″, we have

$$\left\| \sum_{i \in J} |\lambda_i|^2 \bar{T}_i \otimes T_i \right\| \le \sup_{i \in J} |\lambda_i|^2 \left\| \sum_{i \in J} \bar{T}_i \otimes T_i \right\|,$$

and by (7.10) applied to the identity map on $OH(I')$ we know that

$$\left\| \sum_{i \in J} \bar{T}_i \otimes T_i \right\| \le 1,$$

so we obtain the announced claim. ∎

Corollary 7.3. *Let I and J be arbitrary sets and let $OH(I)$ and $OH(J)$ be the corresponding operator Hilbert spaces. Then*

$$OH(I) \otimes_h OH(J) \simeq OH(I \times J)$$

completely isometrically.

Proof. For simplicity, we may and will assume that I and J are finite sets. It is easy to reduce the general case (by density) to that case. Let $E = OH(I)$ and $F = OH(J)$. Then, by Corollary 5.8, the following complete isometries hold:

$$\overline{(E \otimes_h F)^*} \simeq \overline{E^* \otimes_h F^*}$$
$$\simeq \overline{E^*} \otimes_h \overline{F^*}$$
$$\simeq E \otimes_h F.$$

Clearly, the corresponding mapping $E \otimes F \to \overline{(E \otimes F)^*}$ is positive definite (by the same argument as the one showing that the tensor product of two scalar products is also one). Thus, by the preceding remark, we conclude that $E \otimes_h F$ is completely isometric to $OH(I')$ with the cardinality of I' equal to $\dim(E \otimes_h F)$, so that we can set $I' = I \times J$. $\blacksquare$

Remarks

(i) By (7.1), the o.s.s. just defined on $OH(I)$ is clearly independent of the choice of an orthonormal basis.

(ii) Thus, given a Hilbert space H, we may equip H with the o.s.s. corresponding to $OH(I)$ with the cardinality of I determined by the Hilbertian dimension of H. Thus, Theorem 7.1 admits the following basis free ("canonical") reformulation: For any Hilbert space H, there is a unique o.s.s. on H for which the canonical isometric identification $H \simeq \overline{H}^*$ becomes a *complete* isometry. We will sometimes denote by H_{oh} the resulting operator space.

(iii But actually, we will rarely use the notation H_{oh} in the sequel. We prefer to make the following convention: We consider that this o.s.s. is the "natural" one on a Hilbert space, so we always assume (unless explicitly stated otherwise) that H is equipped with the o.s.s. of H_{oh}.

(iv) In particular, if $H \simeq \ell_2(I)$, the Hilbert space $S_2(H)_{oh}$ can be identified with $OH(I \times I)$, allowing us to view $S_2(H)$ as an operator space from now on.

(v) It will be convenient to record here the following formula describing this o.s.s. on $S_2(H)$.
For any $x = \sum e_{ij} \otimes x_{ij}$ in $M_n(S_2(H))$ $(x_{ij} \in S_2(H))$, we have

$$\|x\|_{M_n(S_2(H))} = \sup\{\|\alpha \cdot x \cdot \beta\|_{S_2(\ell_2^n \otimes_2 H)}\}, \qquad (7.11)$$

where the supremum runs over all α, β in M_n with $\operatorname{tr}|\alpha|^4 \leq 1$ and $\operatorname{tr}|\beta|^4 \leq 1$.

Indeed we have

$$x = \sum_{i,j} e_{ij} \otimes \sum_{p,q} e_{pq} x_{ij}(p,q) = \sum_{p,q} y_{pq} \otimes e_{pq}$$

with $y_{pq} = \sum_{i,j} e_{ij} x_{ij}(p,q)$. Hence, since $\{e_{pq} \mid p, q \in I\}$ is an orthonormal basis of $S_2(H)$, we have by (7.3)''

$$\|x\|_{\min} = \sup \left(\sum_{p,q} \|\alpha y_{pq} \beta\|_2^2 \right)^{1/2}$$

$$= \sup \|\alpha \cdot x \cdot \beta\|_{S_2(\ell_2^n \otimes_2 H)},$$

where the supremum is as above.   ∎

Lemma 7.4. *Consider a, b in M_N. We denote $\|a\|_4 = (\mathrm{tr}|a|^4)^{1/4}$ and $\|b\|_4 = (\mathrm{tr}|b|^4)^{1/4}$. Let $M_{a,b} \colon M_N \to S_2^N$ be the operator defined by $M_{a,b}(x) = axb$, and let $v \colon OH_n \to M_N$ be any map. We have then (note: here by S_2^N we mean $(S_2^N)_{oh}$)*

$$(7.12) \qquad\qquad \|M_{a,b}v\|_{HS} \le \|v\|_{cb}\|a\|_4\|b\|_4$$

$$(7.13) \qquad\qquad \|M_{a,b}\|_{cb} \le \|a\|_4\|b\|_4.$$

More generally, the same estimates hold on any infinite-dimensional Hilbert space H, with $M_{a,b} \colon B(H) \to S_2(H)$ defined in the same way with a, b in $B(H)$ such that $\mathrm{tr}|a|^4 < \infty$, $\mathrm{tr}|b|^4 < \infty$, and for any c.b. map $v \colon OH \to B(H)$.

Proof. Let (T_i) be an orthonormal basis in OH_n. We then have

$$\|M_{a,b}v\|_{HS} = \left(\sum \|M_{a,b}v(T_i)\|^2\right)^{1/2} = \left(\sum \|av(T_i)b\|_2^2\right)^{1/2}.$$

Hence, by $(7.3)''$ and (7.10),

$$\le \|a\|_4\|b\|_4 \left\|\sum v(T_i) \otimes \overline{v(T_i)}\right\|_{\min}^{1/2} \le \|a\|_4\|b\|_4\|v\|_{cb}.$$

This establishes (7.12). To verify (7.13), consider $x = \sum e_{ij} \otimes x_{ij}$ in $M_n(M_N)$. We have $(I \otimes M_{a,b})(x) = \sum e_{ij} \otimes ax_{ij}b$. Hence, by (7.11),

$$\|(I \otimes M_{a,b})(x)\|_{M_n(S_2^N)} = \sup \left\{ \left\|(\alpha \otimes a) \cdot \sum e_{ij} \otimes x_{ij} \cdot (\beta \otimes b)\right\|_{S_2(\ell_2^n \otimes \ell_2^N)} \right.$$

$$\left. \times\ \|\alpha\|_4 \le 1, \|\beta\|_4 \le 1 \right\},$$

which yields

$$\le \|\alpha \otimes a\|_4 \left\|\sum e_{ij} \otimes x_{ij}\right\|_{M_n \otimes_{\min} M_N} \|\beta \otimes b\|_4 \le \|x\|_{M_n(M_N)}\|a\|_4\|b\|_4.$$

Thus we obtain (7.13).   ∎

Finite-dimensional estimates. By a classical result due to Fritz John (1948), for any n-dimensional Banach space E, there is an isomorphism $u \colon \ell_2^n \to E$ such that $\|u\| = 1$ and $\|u^{-1}\| \le \sqrt{n}$. Equivalently, $d(E, \ell_2^n) \le \sqrt{n}$. In John's original proof, the mapping u is selected so that $u(B_{\ell_2^n})$ is the maximal volume ellipsoid included in B_E. One can also majorize the projection constants: If E is an n-dimensional subspace of a Banach space X, it is known that there is a linear projection $P \colon X \to E$ with norm $\|P\| \le n^{1/2}$ (Kadec-Snobar, 1971; see [KTJ1–2] for more references and more recent information).

In modern Banach space theory (see [P8, TJ1]), the following lemma due to Dan Lewis has turned out to be a very useful generalization of John's idea.

Lemma 7.5. *Let E be an n-dimensional Banach space and let α be any norm on the space $L(\ell_2^n, E)$ of all linear maps from ℓ_2^n to E. For any $v\colon E \to \ell_2^n$, we define*

$$\alpha^*(v) = \sup\{|\mathrm{tr}(vw)| \mid w\colon \ell_2^n \to E,\ \alpha(w) \leq 1\}.$$

Then there is an isomorphism $u\colon \ell_2^n \to E$ such that $\alpha(u) = 1$ and $\alpha^*(u^{-1}) = n$.

Proof. Choose $u\colon \ell_2^n \to E$ with $\alpha(u) = 1$ such that

$$|\det(u)| = \sup\{|\det(w)| \mid w\colon \ell_2^n \to E,\ \alpha(w) \leq 1\}.$$

Clearly $\det(u) \neq 0$. By homogeneity, for any z in $\mathbb{C}$ and for any $w\colon \ell_2^n \to E$ with $\alpha(w) \leq 1$, we have

$$|\det(u + zw)| \leq |\det(u)|(1 + |z|)^n.$$

Hence, letting $|z| \to 0$, we find

$$|\det(I + zu^{-1}w)| \leq (1 + |z|)^n = 1 + n|z| + o(|z|),$$

and since $\det(I + zu^{-1}w) = 1 + z\,\mathrm{tr}(u^{-1}w) + o(|z|)$, we obtain $|\mathrm{tr}(u^{-1}w)| \leq n$. ∎

Remark. If α is invariant under the unitary group $U(n)$ (i.e., if we have $\alpha(u\omega) = \alpha(u)$ for any $n \times n$ unitary matrix ω), then the isomorphism u appearing in Lemma 7.5 is unique modulo $U(n)$; that is, if $u_1\colon \ell_2^n \to E$ is another isomorphism with the same property, we must have $u^{-1}u_1 \in U(n)$. (In geometric language, the corresponding ellipsoid is unique.) Indeed, let $u^{-1}u_1 = b$. Let $b = \omega|b|$ be the polar decomposition. By the choice of u ("maximal volume") we know that $|\det(u_1)| \leq |\det(u)|$ and hence $|\det(b)| = |\det(|b|)| \leq 1$. Therefore the eigenvalues $\lambda_1, \ldots, \lambda_n$ of $|b|$ satisfy $\Pi\lambda_i \leq 1$. But, on the other hand, $b^{-1} = u_1^{-1}u$, and hence $\sum \lambda_i^{-1} = \mathrm{tr}(b^{-1}) \leq \alpha(u)\alpha^*(u_1^{-1}) \leq n$. Hence we must have (since the geometric mean of λ_i^{-1} is equal to its arithmetic mean) $\lambda_i = 1$ for all i, which means that $u^{-1}u_1$ is unitary. ∎

Theorem 7.6. *For any n-dimensional operator space E and any $\varepsilon > 0$, there is an isomorphism $u\colon OH_n \to E$ with $\|u\|_{cb} = 1$ such that $u^{-1}\colon E \to OH_n$ admits for some integer N a factorization of the form $E \xrightarrow{v_1} M_N \xrightarrow{v_2} OH_n$ with $\|v_1\|_{cb} = \sqrt{n}$ and $\|v_2\|_{cb} < 1 + \varepsilon$.*

Proof. By the preceding lemma, applied with $\alpha(u) = \|u\|_{CB(OH_n, E)}$, there is an isomorphism $u\colon OH_n \to E$ with $\|u\|_{cb} = 1$ and $\alpha^*(u^{-1}) = n$. In other words (see Chapter 4), the o.s. nuclear norm of u^{-1} is equal to n. Therefore, for any $\varepsilon > 0$ there is an N and a factorization of u^{-1} of the form $u^{-1} = wTv_1$ as follows:

$$u^{-1}: \; E \xrightarrow{v_1} M_N \xrightarrow{T} S_1^N \xrightarrow{w} OH_n$$

with $\|v_1\|_{cb} = n^{1/2}$, $\|w\|_{cb} < n^{1/2}(1 + \varepsilon)$, and with $T: M_N \to S_1^N$ defined by $T(x) = axb$ with a, b in the unit ball of S_2^N. If we wish, we may as well assume (by the polar decomposition) that $a \geq 0$ and $b \geq 0$. Then, we define $T_1: M_N \to S_2^N$ and $T_2: S_2^N \to S_1^N$ by $T_1(x) = a^{1/2}xb^{1/2}$ and $T_2(y) = a^{1/2}yb^{1/2}$, so that $T = T_2T_1$.

Consider the composition $T_1v_1: E \to S_2^N$. By Lemma 7.4, we have

$$\|T_1\|_{cb} \leq 1 \quad \text{and} \quad \|T_1v_1u\|_{HS} \leq n^{1/2}. \tag{7.14}$$

Let $E_2 = T_1v_1(E) \subset S_2^N$. Clearly, $\dim E_2 = n$ (since u bijective implies T_1v_1 injective). Thus we may (and will) identify E_2 with OH_n. Let $P: S_2^N \to E_2$ be the orthogonal projection. Clearly (see Proposition 7.2(iii)), $\|P\|_{cb} = \|P\| = 1$.

Recapitulating, we have

$$u^{-1} = wTv_1 = wT_2T_1v_1 = wT_2PT_1v_1.$$

Now let $W: E_2 \to OH_n$ be the restriction of wT_2 to E_2 and let $V = PT_1v_1: E \to E_2$. We may then write

$$u^{-1} = WV,$$

and V, W are both invertible. By Lemma 7.4 (note that, with the proper identification, $T_2^* = T_1$) we have

$$\|W^*\|_{HS} \leq \|(wT_2)^*\|_{HS} < n^{1/2}(1 + \varepsilon),$$

and by (7.14) we also have

$$\|Vu\|_{HS} \leq n^{1/2}.$$

We claim that this implies

$$\|W^* - Vu\|_{HS} < f_n(\varepsilon) \tag{7.15}$$

with $f_n(\varepsilon) \to 0$ when $\varepsilon \to 0$. Indeed, using the scalar product of the Hilbert-Schmidt norm, namely, $\langle x, y \rangle = \text{tr}(xy^*)$, the identity $\text{tr}(WVu) = \text{tr}(u^{-1}u) = n$ means that $\langle W^*, Vu \rangle = n$. Therefore we have

$$\|W^* - Vu\|_{HS}^2 = \|Vu\|_{HS}^2 + \|W^*\|_{HS}^2 - 2\,\text{Re}(\langle W^*, Vu \rangle) < n + n(1+\varepsilon)^2 - 2n,$$

from which the above claim becomes obvious. Since $W(Vu) = I$, (7.15) implies

$$\|WW^* - I\|_{HS} < f_n(\varepsilon)\|W\| < f_n(\varepsilon)n^{1/2}(1 + \varepsilon);$$

hence, a fortiori $\|WW^* - I\| < f_n'(\varepsilon)$ with $f_n'(\varepsilon) \to 0$ when $\varepsilon \to 0$. Therefore, a look at the polar decomposition of W shows that there is a unitary operator $U: E_2 \to OH_n$ such that

$$\|W - U\| < f_n''(\varepsilon)$$

with $f_n''(\varepsilon) \to 0$ when $\varepsilon \to 0$.

We can now finally conclude: We define $v_2: M_N \to OH_n$ by $v_2 = WPT_1$. Note $u^{-1} = v_2 v_1$. By Lemma 7.4, we have $\|T_1\|_{cb} \le 1$; on the other hand, by Proposition 7.2(iii) and the preceding bound,

$$\|W\|_{cb} = \|W\| < 1 + f_n''(\varepsilon).$$

Thus we finally obtain

$$\|v_2\|_{cb} \le \|W\|_{cb}\|T_1\|_{cb} < 1 + f_n''(\varepsilon). \qquad \blacksquare$$

Corollary 7.7. *Let $E \subset B(H)$ be any n-dimensional operator space. There are an isomorphism $u: OH_n \to E$ with $\|u\|_{cb} = 1$ and $\|u^{-1}\|_{cb} \le \sqrt{n}$ and a projection $P: B(H) \to E$ with $\|P\|_{cb} \le \sqrt{n}$. In particular,*

$$d_{cb}(E, OH_n) \le \sqrt{n}.$$

Moreover, for any $\varepsilon > 0$, there is an integer N and a subspace $\widehat{E} \subset M_N$ such that $d_{cb}(E, \widehat{E}) \le (1 + \varepsilon)\sqrt{n}$.

Proof. By the preceding statement, for any $\varepsilon > 0$, there is a map $u_\varepsilon: OH_n \to E$ with $\|u_\varepsilon\|_{cb} = 1$ and $\|(u_\varepsilon)^{-1}\|_{cb} < \sqrt{n}(1+\varepsilon)$. By norm-compactness, the net (u_ε) admits a cluster point u when $\varepsilon \to 0$. Clearly, u has the desired property. Note that, by the extension property of M_N, the mapping v_1 appearing in Theorem 7.6 admits an extension $\widetilde{v}_1: B(H) \to M_N$ with $\|\widetilde{v}_1\|_{cb} = \|v_1\|_{cb}$. Therefore, we may assume that $(u_\varepsilon)^{-1}$ admits an extension $v_\varepsilon: B(H) \to OH_n$ with $\|v_\varepsilon\|_{cb} < \sqrt{n}(1 + \varepsilon)$. The mapping $P_\varepsilon = u_\varepsilon v_\varepsilon: B(H) \to E$ is then a projection with $\|P_\varepsilon\|_{cb} \le \sqrt{n}(1 + \varepsilon)$. Taking a cluster point of (P_ε) for the pointwise convergence on $B(H)$, we obtain a projection P with the desired property. Finally, with the notation in Theorem 7.6, let $\widehat{E} = v_1(E) \subset M_N$. Then the restriction of uv_2 to $\widehat{E}$ is the inverse of $v_1: E \to \widehat{E}$, and hence $d_{cb}(E, \widehat{E}) \le \|v_1\|_{cb}\|uv_2\|_{cb} \le \|v_1\|_{cb}\|v_2\|_{cb} \le (1 + \varepsilon)\sqrt{n}$. $\qquad \blacksquare$

Remark. Fritz John's result can be recovered from Corollary 7.7 as a special case: Indeed, for any n-dimensional normed space E we have $d(E, \ell_2^n) \le d_{cb}(\min(E), OH_n) \le \sqrt{n}$.

Remark. For any operator space $X \subset B(H)$, we define the c.b. projection constant of X as follows:

$$\lambda_{cb}(X) = \inf\{\|P\|_{cb} \mid P : B(H) \to X, \text{ projection onto } X\}.$$

By Corollary 1.7, this is invariant under a completely isometric isomorphism. With this notation, Corollary 7.7 implies that any n-dimensional operator space E satisfies $\lambda_{cb}(E) \leq \sqrt{n}$. In particular, $\lambda_{cb}(OH_n) \leq \sqrt{n}$. Surprisingly, however, this estimate is far from optimal. Indeed, Marius Junge [J2] proved quite recently that $\lambda_{cb}(OH_n)$ was $O(\sqrt{n/\text{Log } n})$ when $n \to \infty$, and the paper [PiS] already showed that $\lambda_{cb}(OH_n)$ was not asymptotically smaller. Thus, there is a constant $K > 0$ such that, for any n, we have

$$K^{-1}\sqrt{n/\text{Log } n} \leq \lambda_{cb}(OH_n) \leq K\sqrt{n/\text{Log } n}.$$

See [KTJ1–2] for recent results on the Banach space analog of n-dimensional projection constants.

Theorem 7.8. *Let $T_n \colon R_n \to C_n$ be the linear mapping taking e_{1i} to e_{i1}. For any n-dimensional operator space E, there is a factorization*

$$T_n \colon R_n \xrightarrow{w} E \xrightarrow{v} C_n$$

of T_n through E with $\|w\|_{cb} = 1$ and $\|v\|_{cb} = \sqrt{n}$. Moreover, this is unique in the sense that if v_1, w_1 is another such factorization, there is a unitary U on $\ell_2^n \simeq R_n$ such that $w_1 = wU$.

Proof. We will apply Lemma 7.5 with $\alpha(w) = \|\sum e_{i1} \otimes w(e_i)\|_{C_n \otimes_h E}$. Given $\xi_1, \ldots, \xi_n$ in E^*, let $v \colon E \to \ell_2^n$ be the mapping defined by $v(e) = \sum_1^n \xi_i(e)e_i$. Then, by (5.14), since $C_n^* \simeq R_n$, we have $\alpha^*(v) = \|\sum e_{i1} \otimes \xi_i\|_{R_n \otimes_h E^*}$. By definition of $\otimes_h$ (and by homogeneity), $\alpha^*(v) \leq n$ iff we can write

$$e_{11} \otimes \sum e_{1i} \otimes \xi_i = \left(\sum_k e_{1k} \otimes \sum_k a_{ik}e_{1i}\right) \odot \left(\sum_k e_{k1} \otimes \eta_k\right)$$

with $a_{ik} \in \mathbb{C}, \eta_k \in E^*$ such that $(\sum_{i,k} |a_{ik}|^2)^{1/2} \leq \sqrt{n}$ and $\|\sum e_{k1} \otimes \eta_k\|_{\min} = \sqrt{n}$. Now let $u \colon \ell_2^n \to E$ and $u^{-1} \colon \ell_2^n \to E$ be as in Lemma 7.5. Let $w \colon R_n \to E$ be defined by $w(e_{1i}) = u(e_i)$. By (2.3.2) and Exercise 5.3, we have $\|w\|_{cb} = \alpha(u) = 1$. Let $\xi_i \in E^*$ be such that $u^{-1}(e) = \sum \xi_i(e) \otimes e_i$. Let $a = (a_{ik})$ and (η_k) be as above. We then have

$$\forall i \qquad \xi_i = \sum_k a_{ik}\eta_k.$$

By polar decomposition, we may assume without loss of generality that a is hermitian ≥ 0. Let $v \colon E \to C_n$ be defined by $v(e) = \sum \eta_k(e)e_{k1}$ ($e \in$

E). Then $\|v\|_{cb} = \|\sum e_{k1} \otimes \eta_k\|_{\min} = \sqrt{n}$. Hence, by (1.5), $\|vw\|_{HS} = \|vw\|_{CB(R_n,C_n)} \leq \|v\|_{cb}\|w\|_{cb} \leq \sqrt{n}$. Note that $u^{-1}u = I$ implies $\xi_i(x_j) = \delta_{ij}$. Hence $\sum_k a_{ik}\eta_k(x_j) = \delta_{ij}$. Let $b_{kj} = \eta_k(x_j)$. We then have $b = a^{-1}$ but also $n = \text{tr}(ab) = \langle a^*, b \rangle$ and $\|u\|_2 \leq \sqrt{n}$, $\|b\|_2 \leq \sqrt{n}$. Thus $n^{-1/2}b$ norms $n^{-1/2}a^*$ in S_2^n. Hence we must have $b = a^*$. Since $b = a^{-1}$, a must be unitary, and since $a \geq 0$, we must have $a = I$. Thus we conclude that $\eta_k(x_j) = \delta_{kj}$, so that $vw = T_n$. The unicity follows from the remark appearing after Lemma 7.5. ∎

Notation. In the direct sum $R_n \oplus C_n$, we consider the vectors $\delta_i = e_{1i} \oplus e_{i1}$. In the style of §2.7, we will denote by $R_n \cap C_n$ the linear span of $\{\delta_i \mid 1 \leq i \leq n\}$. More generally, let E be any operator space. We will denote by $E \cap E^{op}$ the operator subspace of $E \oplus E^{op}$ formed of all vectors of the form $e \oplus e$ ($e \in E$). Similarly, we may consider the mapping $Q \colon E \oplus_1 E^{op} \to E$ taking (x, y) to $x + y$. We will denote by $E + E^{op}$ the operator space $(E \oplus_1 E^{op})/\ker(Q)$.

Corollary 7.9. ([J1]) For any n-dimensional operator space E, $d_{cb}(E, R_n \cap C_n) \leq 2\sqrt{n}$.

Proof. We apply Theorem 7.8 to the operator space $E + E^{op}$: We have a factorization

$$T_n \colon R_n \xrightarrow{w} E + E^{op} \xrightarrow{v} C_n$$

with $\|w\|_{cb} = 1$, $\|v\|_{cb} = \sqrt{n}$. Since $E \to E + E^{op}$ and $E^{op} \to E + E^{op}$ are completely contractive, we have $\|v \colon E \to C_n\|_{cb} \leq \sqrt{n}$ and $\|v \colon E \to R_n\|_{cb} = \|v \colon E^{op} \to C_n\|_{cb} \leq \sqrt{n}$ and hence $\|v \colon E \to R_n \cap C_n\|_{cb} \leq \sqrt{n}$. In the preceding, we of course think of R_n, C_n, and $R_n \cap C_n$ as having ℓ_2^n as common underlying vector space, via the obvious identifications. On the other hand, let $w(e_{1i}) = x_i \in E$. Since $\|w \colon R_n \to E + E^{op}\|_{cb} \leq 1$, we can write $x_i = a_i + b_i$ ($a_i, b_i \in E$) with $\|\sum e_{i1} \otimes a_i\|_{C_n \otimes_{\min} E} \leq 1$, $\|\sum e_{i1} \otimes b_i\|_{C_n \otimes_{\min} E^{op}} \leq 1$. Hence, if we assume $E \subset B(H)$, we have $\|\sum a_i^* a_i\|^{1/2} \leq 1$ and $\|\sum b_i b_i^*\|^{1/2} \leq 1$. Therefore, for any c_i in $\mathcal{K}$ we can write

$$\left\|\sum x_i \otimes c_i\right\|_{\min} \leq \left\|\sum a_i \otimes c_i\right\|_{\min} + \left\|\sum b_i \otimes c_i\right\|_{\min};$$

hence, by (1.11),

$$\leq \left\|\sum a_i^* a_i\right\|^{1/2} \left\|\sum c_i c_i^*\right\|^{1/2} + \left\|\sum b_i b_i^*\right\|^{1/2} \left\|\sum a_i^* a_i\right\|^{1/2} \leq 2\left\|\sum \delta_i \otimes c_i\right\|,$$

and we conclude $\|w \colon R_n \cap C_n \to E\|_{cb} \leq 2$. Thus if we define $u(\delta_i) = w(e_{1i})$, we find $\|u \colon R_n \cap C_n \to E\|_{cb} \leq 2$ and we have $\|u^{-1} \colon E \to R_n \cap C_n\|_{cb} = \|v \colon E \to R_n \cap C_n\|_{cb} \leq \sqrt{n}$, so that finally $\|u\|_{cb}\|u^{-1}\|_{cb} \leq 2\sqrt{n}$. ∎

Complex interpolation. After we observed the existence of $OH(I)$, we developed in [P1] the theory of these spaces in analogy with what is known in Banach space theory. In these developments, complex interpolation has played a crucial role. For instance, we have

Theorem 7.10. *Let E be an arbitrary operator space equipped with a continuous injection*

$$i: \overline{E^*} \to E$$

with which we may view the couple $(\overline{E^*}, E)$ as compatible. Assume that i is "positive definite," that is to say, such that $i(\overline{\xi})(\xi) \geq 0$ for all ξ in E^*. Then $(\overline{E^*}, E)_{\frac{1}{2}}$ is completely isometric to $OH(I)$ for some index set I.

Proof. In the first part of the proof we treat E as a Banach space. We denote $E_0 = \overline{E^*}$ and $E_1 = E$. Note that, by Lemma 2.7.2, we have an isometric identity

$$(E_0, E_1)_{1/2} = (E_0, E^{**})_{1/2}. \tag{7.16}$$

Let $F_0 = E$ and $F_1 = \overline{E^*}$. We apply Lemma 2.7.5 to the sesquilinear mapping

$$u: E_0 \cap E_1 \times F_0 \cap F_1 \to \mathbb{C}$$

defined by $u(\overline{\xi}, i(\overline{\eta})) = i(\overline{\xi})(\eta)$ or, equivalently (since $\overline{\xi}$ and $i(\overline{\xi})$ are identified by the "compatibility"), $u(i(\overline{\xi}), \overline{\eta}) = i(\overline{\xi})(\eta)$. Thus,

$$|u(i(\overline{\xi}), \overline{\eta})| \leq \|i(\overline{\xi})\|_E \|\eta\|_{E^*}.$$

Moreover, by the polarization identity, positive definiteness implies self-adjointness, that is, we have $i(\overline{\xi})(\eta) = \overline{i(\overline{\eta})(\xi)}$. It follows that, if x, y belong to $E_0 \cap E_1 = F_0 \cap F_1$, we have simultaneously

$$|u(x, y)| \leq \|x\|_{E_0} \|y\|_{F_0} \quad \text{and} \quad |u(x, y)| \leq \|x\|_{E_1} \|y\|_{F_1}.$$

(Note that *we do not assume* that i is contractive.) Let $X = (E_0, E_1)_{1/2} = (\overline{E^*}, E)_{1/2}$. Note that by symmetry $X = (F_0, F_1)_{1/2}$. By Lemma 2.7.5, u extends to a contractive sesquilinear form (still denoted by u) on $X \times X$. Hence we have

$$\forall\, x, y \in X \qquad |u(x, y)| \leq \|x\|_X \|y\|_X.$$

But for all $x = i(\overline{\xi})$ in $\overline{E^*}$ we have $u(x, x) = i(\overline{\xi})(\xi) \geq 0$. Hence, since $i(\overline{E^*})$ is dense in X, we have $u(x, x) \geq 0$ for all x in X, so that u is a scalar product on X. Taking $x = y$, we obtain

$$\forall\, x \in X \qquad u(x, x)^{1/2} \leq \|x\|_X. \tag{7.17}$$

We now claim that we have a "natural" isometric embedding $X \to \overline{X^*}$. Indeed, by (2.7.6), we have an isometric embedding of $(E_0^*, E_1^*)_{1/2}$ into X^* or, equivalently, of $(\overline{E_0^*}, \overline{E_1^*})_{1/2}$ into $\overline{X^*}$. But $\overline{E_0^*} = E^{**}$ and $\overline{E_1^*} = E_0$; hence (7.16) ensures that $(\overline{E_0^*}, \overline{E_1^*})_{1/2} = X$ and we indeed obtain $X \subset \overline{X^*}$ isometrically. In particular,

$$\forall\, x = i(\overline{\xi}) \in i(\overline{E^*}) \quad \|x\|_X = \sup\{|\langle \xi, y \rangle| \mid y \in i(\overline{E^*})\ \|y\|_X \leq 1\}.$$

Hence (7.17) implies by Cauchy-Schwarz

$$\forall\, x \in i(\overline{E^*}) \quad \|x\|_X \le \sup\{|u(x,y)| \mid u(y,y) \le 1\} \le u(x,x)^{1/2}.$$

Thus we conclude finally that

$$\|x\|_X = u(x,x)^{1/2},$$

which shows that X is isometric to a Hilbert space.

We now turn to the operator space case. By Theorem 2.7.4 (and the above observation (7.16) that takes care of the bidual E^{**} and replaces it by E) the inclusion $X \to \overline{X^*}$ is completely isometric, and since X is Hilbertian by what precedes, it is actually surjective. Thus, by the unicity part of Theorem 7.1, we conclude that X must be completely isometric to $OH(I)$ for a suitable set I. ∎

Remark. See [Wat1-2] for a more general discussion of the complex interpolation space (with $\theta = 1/2$) between a Banach space and its antidual.

Corollary 7.11. *We have completely isometrically*

$$(R, C)_{\frac{1}{2}} = OH$$

$$(\min(\ell_2), \max(\ell_2))_{\frac{1}{2}} = OH.$$

Remark. Let H be a Hilbert space. Recall (see (1.7)) that we have denoted by H_c (resp. H_r) the space H equipped with the column (resp. row) o.s.s. In the same vein, we sometimes denote by H_{oh} the space H equipped with the o.s.s. corresponding to the OH-space of the same (Hilbertian) dimension as H. With this more canonical notation, the first part of Corollary 7.11 says that we have a complete isometry

$$(H_c, H_r)_{1/2} \simeq H_{oh}.$$

More explicitly, we have

Corollary 7.12. *Fix $n \ge 1$. Let $A_0 = B(H)^n$ (resp. $A_1 = B(H)^n$) equipped with the norm*

$$\|(x_1, \ldots, x_n)\|_0 = \left\|\sum x_i x_i^*\right\|^{1/2}$$

$$\left(\text{resp. } \|(x_1, \ldots, x_n)\|_1 = \left\|\sum x_i^* x_i\right\|^{1/2}\right).$$

Then $(A_0, A_1)_{\frac{1}{2}}$ coincides with $B(H)^n$ equipped with the norm

$$\|(x_1,\ldots,x_n)\|_{1/2} = \left\|\sum x_i \otimes \overline{x_i}\right\|_{\min}^{1/2}.$$

Proof. It is easy to reduce to the case $H = \ell_2$. Then, by (2.7.3), Corollary 7.11 implies

$$(\mathcal{K} \otimes_{\min} R_n, \mathcal{K} \otimes_{\min} C_n)_{1/2} \simeq \mathcal{K} \otimes_{\min} OH_n.$$

On the other hand, for any finite-dimensional operator space E, by Exercise 5.6 we have (isometrically)

$$(\mathcal{K} \otimes_{\min} E)^{**} \simeq B(\ell_2) \otimes_{\min} E.$$

Hence, by Theorem 2.7.4, we have isometrically

$$(B(\ell_2) \otimes_{\min} R_n, B(\ell_2) \otimes_{\min} C_n)_{1/2} \simeq B(\ell_2) \otimes_{\min} OH_n,$$

which is the content of Corollary 7.12. ∎

This has been generalized to the case when $B(H)$ is replaced by a general von Neumann algebra ([P7]). These ideas have applications to the study of completely bounded projections $P\colon M \to N$ from a von Neumann algebra M onto a subalgebra $N \subset M$. See Chapters 15 and 23.

Vector-valued L_p-spaces, either commutative or noncommutative. Let (Ω, μ) be a finite measure space. The natural inclusion map $L_\infty(\mu) \to L_1(\mu)$ allows us to view the pair $(L_\infty(\mu), L_1(\mu))$ as compatible in the most classical sense. We feel it is worthwhile to first explain why the classical isometric identity $(L_\infty(\mu), L_1(\mu))_{\frac{1}{2}} \simeq L_2(\mu)$ follows from Theorem 7.10. For any y in $L_\infty(\mu)$, let $\varphi_y \in L_1(\mu)^*$ be the functional defined by $\varphi_y(x) = \int xy\, d\mu$. Let $i\colon \overline{L_1(\mu)^*} \longrightarrow L_1(\mu)$ be the (linear) mapping defined by $i(\overline{\varphi_y}) = \overline{y}$. Clearly i is "positive definite"; in fact, if $v\colon L_2(\mu) \to L_1(\mu)$ is the natural inclusion, we have $i = v^t v$. Thus Theorem 7.10 tells us that $(\overline{L_1(\mu)^*}, L_1(\mu))_{1/2} \simeq L_2(\mu)$ (isometrically).

Now, obviously, we have an isometric map $u_0\colon \overline{L_1(\mu)^*} \to L_\infty(\mu)$ defined by $u_0(\overline{\varphi_y}) = \overline{y}$ such that, if $u_1\colon L_1(\mu) \to L_1(\mu)$ is the identity on $L_1(\mu)$, the assumptions of Remark 2.7.8 hold, and therefore we obtain (isometrically)

$$(L_\infty(\mu), L_1(\mu))_{1/2} \simeq (\overline{L_1(\mu)^*}, L_1(\mu))_{1/2} \simeq L_2(\mu).$$

In the operator space case, exactly the same argument gives us (using Exercise 3.2)

$$(\min(L_\infty(\mu)), \max(L_1(\mu)))_{1/2} \simeq L_2(\mu)_{oh}.$$

We now turn to the noncommutative case. Let M be a von Neumann algebra, equipped with a faithful, normal, finite trace τ. The space $L_p(\tau)$ $(1 \le p < \infty)$ is defined as the completion of M with respect to the norm

$$\|x\|_{L_p(\tau)} = (\tau(|x|^p))^{1/p}.$$

In the case $p = 1$, it is well known that $L_1(\tau)^* \simeq M$ (isometrically) with respect to the duality $\langle y, x \rangle = \tau(xy)$ $(x \in L_1(\tau), y \in M)$. For any y in M, let $\varphi_y \in L_1(\tau)^*$ be the functional now defined by $\varphi_y(x) = \tau(xy)$. Let $i \colon \overline{L_1(\tau)^*} \longrightarrow L_1(\tau)$ be the linear mapping defined by

$$i(\overline{\varphi_y}) = y^*.$$

Again i is clearly positive definite, so that, if the "compatibility" is defined using i, we have

$$(\overline{L_1(\tau)^*}, L_1(\tau))_{1/2} = L_2(\tau) \quad \text{(isometrically)}.$$

Moreover, since the (linear) mapping $u_0 \colon \overline{L_1(\tau)^*} \to M$ defined by $u_0(\overline{\varphi_y}) = y^*$ is clearly isometric, we may invoke Remark 2.7.8 again and we obtain

$$(M, L_1(\tau))_{1/2} \simeq L_2(\tau) \quad \text{(isometrically)},$$

where the compatibility of the pair $(M, L_1(\tau))$ is now defined using the natural inclusion of M into $L_1(\tau)$. More generally, we have

$$(M, L_1(\tau))_{1/p} \simeq L_p(\tau) \quad \text{(isometrically)}.$$

Clearly, we can extend all this to operator spaces as before. There is however an important point that must be emphasized. We must equip $L_1(\tau)$ with the appropriate o.s.s., namely, the one for which u_0 is completely isometric. Unfortunately, this point is insufficiently explained in [P2] (actually page 37 in [P2] is definitely misleading) and I am very grateful to M. Junge and Z. J. Ruan for pointing this out to me, thus allowing me to correct this "loose end" here. So we want $u_0 \colon \overline{L^1(\tau)^*} \to M$ to be completely isometric. Equivalently, we want $\overline{u_0} \colon L_1(\tau)^* \to \overline{M}$ (defined by $\overline{u_0}(\varphi_y) = \overline{y^*}$) to be completely isometric, and since $\overline{M} \simeq M^{op}$ completely isometrically (via the map $\overline{y} \to y^*$), this means that we want $L_1(\tau)^* \simeq M^{op}$, completely isometrically, via the mapping taking φ_y to y. In other words, we must equip $L_1(\tau)$ with the o.s.s. of $(M_*)^{op}$. Provided we make this convention (and we will stick to it throughout) everything ticks just as before and Remark 2.7.8 yields

$$(M, L_1(\tau))_{1/2} \simeq L_2(\tau)_{oh} \quad \text{(completely isometrically)}.$$

Although we assumed τ finite (for simplicity) in the preceding discussion, the situation is the same if τ is only assumed semi-finite, which means that the subspace $V = \{x \in M \mid \tau(|x|) < \infty\}$ is weak-$*$ dense in M. The space $L_1(\tau)$ can then be defined as the completion of V with respect to the norm $x \to \tau(|x|)$. Let $\mathcal{P}$ be the collection of all projections in M with finite trace. We can still view the couple $(M, L_1(\tau))$ as compatible using the natural inclusions

$$M \subset \bigoplus_{p \in \mathcal{P}} pMp \quad \text{and} \quad L_1(\tau) \subset \bigoplus_{p \in \mathcal{P}} L_1(\tau_{|pMp}).$$

Then the preceding discussion remains valid (see [Wat1–2] for more information).

Consider in particular the case when $M = B(\ell_2)$ equipped with the usual (semi-finite) trace denoted simply "tr". We prefer to denote the resulting noncommutative L_1-space by S_1. This is the "trace class," formed of all the compact operators x on ℓ_2 such that $\|x\|_{S_1} = \mathrm{tr}(|x|) < \infty$. More generally, if $1 \le p < \infty$, when $\tau = \mathrm{tr}$ the space $L_p(\tau)$ is the Schatten class S_p of all compact operators x such that $\mathrm{tr}|x|^p < \infty$, equipped with the norm

$$\|x\|_{S_p} = (\mathrm{tr}(|x|^p))^{1/2}. \tag{7.18}$$

When $p = 2$, S_2 is the classical "Hilbert–Schmidt class." The previous discussion leads to

$$(S_1, B(\ell_2))_{1/2} \simeq (S_2)_{oh} \quad \text{(completely isometrically)},$$

or equivalently, by Lemma 2.7.2,

$$(S_1, \mathcal{K})_{1/2} \simeq (S_2)_{oh} \quad \text{(completely isometrically)},$$

where the compatibility is meant with respect to the natural inclusion $S_1 \subset \mathcal{K}$.

Remark 7.13. As is well known, we have $\mathcal{K}^* \simeq S_1$ isometrically. In this isometric identity, the duality can be defined by setting either

$$\langle x, y \rangle_I = \mathrm{tr}(xy) \quad \text{or} \quad \langle x, y \rangle_{II} = \mathrm{tr}(x^t y).$$

In other words, this gives us two distinct isometric identifications between $\mathcal{K}^*$ and S_1. Note that the resulting two o.s. structures on S_1 (for which $\mathcal{K}^* \simeq S_1$ becomes completely isometric) depend on the choice of this duality: Each one is the opposite of the other. But since $\mathcal{K} \simeq \mathcal{K}^{op}$ (completely isometrically, via $x \to {}^t x$), the resulting operator spaces are completely isometric. In the preceding discussion, we chose to work with $\langle \cdot, \cdot \rangle_I$, but actually, since we equip S_1 with the o.s.s. for which $(\mathcal{K}^*)^{op} \simeq S_1$, the resulting o.s.s. on S_1 is exactly the same as the one that we would obtain on S_1 by requiring that the isomorphism $\mathcal{K}^* \simeq S_1$ associated to $\langle \cdot, \cdot \rangle_{II}$ be completely isometric (which is the convention adopted in [P2, §1]).

The preceding identities allow us to equip $L_p(\tau)$ with an o.s.s. More generally, the author has developed ([P2]) a theory of *noncommutative vector-valued L_p-spaces*, which extends the so-called Lebesgue-Bochner theory of the spaces

$$L_p(\Omega, \mu; E),$$

where $1 \le p \le \infty$, (Ω, μ) is a measure space and E is a Banach space. Recall that, in the discrete case, that is, when (Ω, μ) is just the integers (resp.

$\Omega = \{1, ..., n\}$) equipped with the counting measure, $L_p(\Omega, \mu; E)$ is usually denoted by $\ell_p(E)$ (resp. $\ell_p^n(E)$).

In this new theory, E has to be equipped with an operator space structure. Then, using interpolation, the spaces

$$L_p(\Omega, \mu; E)$$

can also be given a natural operator space structure. When $p = 2$ and $E = \mathbb{C}$ this reduces to the theory of the already mentioned space OH.

In the noncommutative case we replace $L_\infty(\Omega, \mu)$ by an injective von Neumann algebra $M \subset B(\mathcal{H})$. We assume that M admits a faithful normal semi-finite trace τ so that, as before, the predual M_* can be identified with $L_1(\tau)$ in the usual way (cf., e.g., [Di3, Se, Ne, Ta3]).

Then let $E \subset B(H)$ be an operator space, and we set

$$L_\infty^0(\tau; E) \stackrel{\text{def}}{=} M \otimes_{\min} E.$$

Note that, in the commutative case, this space is in general smaller than the classical space $L_\infty(\Omega, \mu; E)$ in Bochner's sense but coincides with it if E or $M = L_\infty(\Omega, \mu)$ is finite-dimensional.

The case $p = 1$ has already been treated by Effros and Ruan, who introduced more generally the analog of the projective tensor product, denoted by $\otimes^\wedge$. We simply set

$$L_1(\tau; E) \stackrel{\text{def}}{=} L_1(\tau) \otimes^\wedge E.$$

Then we can define (following [P2])

$$L_p(\tau; E) = (L_\infty^0(\tau; E), L_1(\tau; E))_{1/p}. \tag{7.19}$$

Of course, when $E = \mathbb{C}$, the space $L_p(\tau; E)$ reduces to the previously defined space $L_p(\tau)$.

For example, consider $M = B(\ell_2)$ equipped with the usual trace (somewhat discrete) "tr". Then, if $p < \infty$, $L_p(\tau)$ is the Schatten class S_p of all compact operators such that $\text{tr}|T|^p < \infty$ equipped with the norm defined in (7.18). In this particular case, the space $L_p(\tau; E)$ is denoted by $S_p[E]$.

The space S_p (resp. $S_p[E]$) is analogous to a discrete or atomic L_p-space. It is analogous to ℓ_p or ℓ_p^n (resp. $\ell_p(E)$ or $\ell_p^n(E)$) in the commutative case. These spaces satisfy all the basic properties of the classical Lebesgue-Bochner spaces $L_p(\Omega, \mu; E)$, such as duality, interpolation, injectivity, and projectivity relative to the space of "values" E.

The reader should observe that our theory yields an isometric embedding

$$S_p \subset B(H)$$

that is highly nonstandard! Just as in the case of OH (when $p = 2$), we have no explicit description of this embedding.

We will only give here a brief introduction to that subject and refer the reader to the extensive text [P2] devoted entirely to this topic. To avoid any technicality, we will discuss here only the spaces $S_p^n[E]$. The spaces $S_p[E]$ behave very much the same, and in fact the embeddings $\ldots \subset S_p^n[E] \subset S_p^{n+1}[E] \subset \ldots$ are completely isometric, so that $S_p[E]$ can be viewed as the completion of the union of the spaces $S_p^n[E]$. Throughout this discussion, E will be an arbitrary operator space. The normed space $S_p^n[E]$ can be described as follows. It is the same vector space as $M_n(E)$ but equipped with the following norm ($1 \leq p < \infty$): For any x in $M_n(E)$ we set

$$\|x\|_{S_p^n[E]} = \inf\{\|a\|_{S_{2p}^n}\|\widehat{x}\|_{M_n(E)}\|b\|_{S_{2p}^n}\}, \tag{7.20}$$

where the infimum runs over all possible factorizations of x of the form $x = a \cdot \widehat{x} \cdot b$ with $a, b \in M_n$ and $\widehat{x} \in M_n(E)$.

Actually, the same definition (7.20) makes sense for $S_p[E]$, but $S_p[E]$ is now defined as the subspace of $\mathcal{K} \otimes_{\min} E$ formed of all the x that can be written as $x = a \cdot \widehat{x} \cdot b$ with $a, b \in S_{2p}$ and $\widehat{x}$ in $\mathcal{K} \otimes_{\min} E$. It is natural to denote $S_\infty = \mathcal{K}$ and $S_\infty[E] = \mathcal{K} \otimes_{\min} F$; clearly the elements of $S_\infty[E]$ can be viewed as bi-infinite matrices with entries in E.

This definition is reminiscent of the obvious fact that

$$\left(\sum \|y_n\|_E^p\right)^{1/p} = \inf\{\|c\|_{\ell_p}\|\widehat{y}\|_{\ell_\infty(E)}\}$$

over all factorizations of the form $y_n = c_n\widehat{y}_n$. But here the noncommutativity forces us to consider separately both left and right multiplications, and thus the space S_{2p}^n makes its appearance (note that $\|c\|_{S_p^n} \leq 1$ iff c can be written $c = ab$ with $\|a\|_{S_{2p}^n} \leq 1$ and $\|b\|_{S_{2p}^n} \leq 1$). If $x \in S_p^n[E]$ is a *diagonal* matrix, then we find simply $\|x\|_{S_p^n[E]} = (\sum \|x_{ii}\|_E^p)^{1/p}$, so that $\ell_p^n(E) \subset S_p^n[E]$ isometrically.

In the case $p = 1$, the norm appearing in (7.20) coincides with the norm of $S_1^n \otimes^\wedge E$ as defined in Chapter 4. Therefore we have $S_1^n[E]^* = M_n(E^*)$ isometrically. More generally (see [P2]) we have, for any $1 < p < \infty$, $S_p^n[E]^* = S_{p'}^n[E^*]$ isometrically. Equivalently, we have another description of the norm of $S_p^n[E]$ as follows:

$$\|x\|_{S_p^n[E]} = \sup\{\|a \cdot x \cdot b\|_{S_1^n[E]}\}, \tag{7.21}$$

where the supremum runs over all a, b in the unit ball of S_{2p}^n.

Thus, with (7.20) and (7.21), we have two a priori different ways (dual to each other) to look at the spaces $S_p^n[E]$ (or $S_p[E]$), and this is (to a large extent) the reason why their theory works so nicely. Note that (7.20) is projective (if E is a quotient of F, then $S_p^n[E]$ appears as a quotient of $S_p^n[F]$) while (7.21) is injective (if $E \subset F$, then $S_p^n[E] \subset S_p^n[F]$). Having defined the norm in the spaces $S_p^N[E]$, we can now define their o.s.s. as follows: For any x in $M_n(S_p^N[E])$ we define

$$\|x\|_n = \sup\{\|a \cdot x \cdot b\|_{S_p^{nN}[E]}\},$$

where the supremum runs over all a, b in the unit ball of S_{2p}^n and where we have identified $M_n(M_N(E))$ with $M_{nN}(E)$. Then it can be checked that this sequence of norms satisfies Ruan's axioms (R_1) and (R_2); hence it defines an o.s.s. on $S_p^N[E]$ (and also on $S_p[E]$). With this structure, if $1 < p < \infty$, we have (see [P2]) completely isometrically:

$$S_p[E]^* = S_{p'}[E^*].$$

Since $\ell_p^n(E) \subset S_p^n[E]$ (and $\ell_p(E) \subset S_p(E)$), the preceding structure induces an o.s.s. on $\ell_p^n(E)$ and $\ell_p(E)$. More generally, if (Ω, μ) is any measure space, we can equip $L_p(\Omega, \mu; E)$ with an o.s.s. by introducing the following norm on $M_n(L_p(\Omega, \mu; E))$: Viewing an element x in $M_n(L_p(\Omega, \mu; E))$ as one in $L_p(\Omega, \mu; M_n(E))$, we let

$$\|x\|_n = \sup\{\|a \cdot x \cdot b\|_{L_p(\Omega, \mu; S_p^n[E])}\}, \tag{7.22}$$

where the supremum runs over all a, b in the unit ball of S_{2p}^n.

The resulting o.s.s. on $L_p(\Omega, \mu; E)$ is called the *natural* one. It coincides with the one described above using complex interpolation in (7.19).

The introduction of these vector-valued (actually operator space-valued) noncommutative L_p spaces opens the way for noncommutative analogs of a number of natural questions originating in either Banach space theory, classical analysis, or probability. To name a few of those: We can study uniform convexity and the Radon-Nikodym property, maximal inequalities for martingales or the unconditionality of martingale differences (UMD) for operator spaces. One can also ask which Fourier (or Schur) multipliers are *completely bounded* on L_p, and this is of interest already for the most classical groups such as $\mathbb{Z}$ or $\mathbb{T}$.

We will return to this subject in §9.5, and we illustrate it with "concrete" examples in the subsequent §§9.6–9.9. See [P2] for a full treatment.

To give only a sample result, recall that, among Banach spaces, Hilbert space is characterized by the Jordan-von Neumann (parallelogram) inequality

$$\forall x, y \in E \qquad \frac{\|x + y\|^2 + \|x - y\|^2}{2} \leq \|x\|^2 + \|y\|^2.$$

In other words, if we denote by

$$T: \ell_2^2(E) \longrightarrow \ell_2^2(E)$$

the operator defined by

$$T: (x, y) \longrightarrow \left(\frac{x+y}{\sqrt{2}}, \frac{x-y}{\sqrt{2}} \right),$$

then E is isometrically Hilbertian iff $\|T\| \leq 1$. In our theory, if E is an arbitrary operator space, the space $\ell_2^2(E)$ (denoted by $E \oplus_2 E$ in Remark 2.7.3) is also equipped with a natural operator space structure (this is a particular case of (7.22), with (Ω, μ) a two-point space). For that structure we have

Theorem 7.14. $\|T\|_{cb} \leq 1$ *iff E is completely isometric to $OH(I)$ for some set I.*

We end this chapter with a surprising estimate (improving a bound in Theorem 3.8) that uses the space OH. This is due to Éric Ricard.

Theorem 7.15. *([Ri1]) Let E be any n-dimensional operator space. Then*

$$\|I_E: \ \min(E) \to \max(E)\|_{cb} \leq n/2^{1/4}.$$

Equivalently, $\alpha(n) \leq n/2^{1/4}$.

Proof. By a well-known result on 2-summing operators (cf., e.g., [P4, p. 15]), I_E can be factorized as $E \xrightarrow{u} \ell_2^n \xrightarrow{u^{-1}} E$ with $\pi_2(u) = 1$ and $\|u^{-1}\| = \sqrt{n}$. By a well-known estimate, we have

$$\|u\|_{CB(\min(E), R_n)} \leq \pi_2(u) = 1,$$

and similarly

$$\|u\|_{CB(\min(E), C_n)} \leq 1;$$

hence

$$\|u\|_{CB(\min(E), R_n \cap C_n)} \leq 1.$$

On the other hand, the identity map on ℓ_2^n (denoted below by I) defines a completely contractive inclusion $I: \ R_n \cap C_n \to OH_n$. This follows from Corollary 7.11, but it can also be verified directly as follows. Let (T_i) denote an orthonormal basis in OH_n. Then, for any n-tuple (x_i) in $B(\ell_2)$, we have by (7.5) and (2.11.3)

$$\left\| \sum x_i \otimes T_i \right\| = \left\| \sum x_i \otimes \overline{x_i} \right\|^{1/2} \leq \left\| \sum x_i x_i^* \right\|^{1/4} \left\| \sum x_i^* x_i \right\|^{1/4};$$

hence a fortiori

$$\left\|\sum_1^n x_i \otimes T_i\right\| \leq \max\left\{\left\|\sum x_i x_i^*\right\|^{1/2}, \left\|\sum x_i^* x_i\right\|^{1/2}\right\},$$

which proves our claim that $\|I\colon R_n \cap C_n \to OH_n\|_{cb} \leq 1$. Furthermore, we also have (by (7.9) and Exercise 3.2) $\|I\colon OH_n \to \max(\ell_2^n)\|_{cb} = \|I\colon \min(\ell_2^n) \to \max(\ell_2^n)\|_{cb}^{1/2}$. Hence, by Theorem 3.8

$$\|I\colon OH_n \to \max(\ell_2^n)\|_{cb} \leq (n/2^{1/2})^{1/2}.$$

Finally, we have $\|u^{-1}\colon \max(\ell_2^n) \to \max(E)\|_{cb} = \|u^{-1}\| = n^{1/2}$. Thus we conclude

$$\|I_E\colon \min(E) \to \max(E)\|_{cb} \leq (n/2^{1/2})^{1/2} n^{1/2} = n/2^{1/4}. \qquad\blacksquare$$

Corollary 7.16. *([Ri1]) Any linear mapping $u\colon E \to F$ between operator spaces E, F with rank $\leq n$ satisfies*

$$\|u\|_{cb} \leq 2^{-1/4} n \|u\|.$$

The best constant (instead of $2^{-1/4}$) in the preceding bound is not yet known.

Remark. An important and classical fact in Banach space theory is that ℓ_2 embeds isometrically into L_1. This embedding can be obtained by mapping an orthonormal basis to a sequence of independent complex Gaussian random variables (see §9.8). It is natural to wonder about the o.s. analog of this fact for the space OH. This question was recently solved by Marius Junge [J2]. Curiously, we cannot use Gaussian variables, either "classical" or "free" (see §§9.8 and 9.9) to embed OH completely isomorphically into an L_1-space, either commutative or not. Nevertheless, Junge [J2] proved that OH embeds completely isomorphically into the predual M_* of a von Neumann algebra M. The completely isometric case remains open (at the time of this writing). Actually, Junge proved more generally that any quotient of $R \oplus C$ embeds into M_* for some M. Since, by Exercise 7.9, OH embeds completely isometrically in a quotient of $R \oplus C$ (this fact has been known for some time independently to Junge and the author), we recover the embedding of OH as a corollary. Junge's construction uses for M a von Neumann algebra "of type III," that is, one that does not admit any semi-finite and faithful trace. After learning of Junge's results, we were able to show (using [PiS]) that "type III" cannot be avoided: If $OH \subset M_*$ completely isomorphically, then M cannot be semifinite (see [P22]).

Exercises

Exercise 7.1. Let $E = OH(I)$. Prove that $E \simeq E^{op}$ and $E \simeq \overline{E}$ completely isometrically.

Exercise 7.2. In a suitable realization of $OH(I)$ there is an orthonormal basis $(\theta_i)_{i \in I}$ of $OH(I)$ formed of self-adjoint operators.

Exercise 7.3. For any orthonormal basis $(T_i)_{i \in I}$ of $OH(I)$ (with $|I| \neq 0$) we have

$$\sup_{\substack{J \subset I \\ |J| < \infty}} \left\| \sum_{i \in J} T_i \otimes \overline{T}_i \right\| = 1.$$

Exercise 7.4. Let $E = OH(I)$. Show that $\|u\| = \|u\|_{cb}$ for any $u \colon E \to E$, so that $CB(E, E) \simeq B(\ell_2(I))$ isometrically. Show that this identity is *not* completely isometric.

Exercise 7.5. Let $E = OH(I)$. Show that for all $x = (x_{ij})$ in $M_n(E)$ we have

$$\|x\|_{M_n(E)} = \|(\langle x_{ij}, x_{k\ell} \rangle)\|_{M_{n^2}}^{1/2}$$

or, more precisely,

$$= \left\| \sum_{i,j,k,\ell} \langle x_{ij}, x_{k\ell} \rangle e_{ij} \otimes e_{k\ell} \right\|_{\min}^{1/2}.$$

Exercise 7.6. Let E be any operator space. Then, for any $h_1, \ldots, h_n$ in OH and any $x_1, \ldots, x_n$ in E, we have

$$\left\| \sum h_i \otimes x_i \right\|_{OH \otimes_{\min} E} = \left\| \sum_{i,j} \langle h_i, h_j \rangle x_i \otimes \overline{x}_j \right\|_{\min}^{1/2}.$$

Exercise 7.7. Let $(T_1, \ldots, T_n)$ be an orthonormal basis in OH_n. Compute

$$\left\| \sum_1^n e_{i1} \otimes T_i \right\|_{C_n \otimes_{\min} OH_n} \quad \text{and} \quad \left\| \sum_1^n e_{1i} \otimes T_i \right\|_{R_n \otimes_{\min} OH_n}.$$

Let $u \colon C_n \to OH_n$ (resp. $v \colon R_n \to OH_n$) be any isometric map. Show that $\|u\|_{cb} = \|v\|_{cb} = n^{1/4}$.

Exercise 7.8. Let $E \subset B(H)$ be any n-dimensional operator space. For any $x = (x_i)_{i \leq n}$ in E^n we denote

$$\|x\|_{E(0)} = \left\| \sum x_i^* x_i \right\|^{1/2}, \quad \|x\|_{E(1)} = \left\| \sum x_i x_i^* \right\|^{1/2}.$$

Then we define for $0 < \theta < 1$

$$\|x\|_{E(\theta)} = \|x\|_{(E(0),E(1))_\theta}.$$

Show that there is a basis (x_i) in E with biorthogonal basis (ξ_i) in E^* such that

$$\|(x_i)\|_{E(\theta)} \|(\xi_i)\|_{E^*(\theta)} = \sqrt{n}.$$

In particular, for $\theta = 1/2$, deduce from this a new proof that $d_{cb}(E, OH_n) \le \sqrt{n}$.

Hint: Use Theorem 7.8 applied to $(E, E^{op})_\theta$, like in the proof of Corollary 7.9.

Exercise 7.9. Show that OH is completely isometric to a quotient of a subspace of $R \oplus C$.

Chapter 8. Group C^*-Algebras. Universal Algebras and Unitization for an Operator Space

We first recall some classical notation from noncommutative abstract harmonic analysis on an arbitrary discrete group G (we restrict ourselves to the discrete case for simplicity).

Let $\pi\colon G \to B(\mathcal{H})$ be a unitary representation on G. We denote by $C_\pi^*(G)$ the C^*-algebra generated by the range of π. Equivalently, $C_\pi^*(G)$ is the closed linear span of $\pi(G)$. In particular, this applies to the so-called universal representation of G, a notion that we now recall. Let $(\pi_j)_{j \in I}$ be a family of unitary representations of G, say,

$$\pi_j\colon G \to B(H_j),$$

in which every equivalence class of a cyclic unitary representation of G has an equivalent copy. Now one can define the *universal* representation $U_G\colon G \to B(\mathcal{H})$ of G by setting

$$U_G = \oplus_{j \in I}\pi_j \quad \text{on} \quad \mathcal{H} = \oplus_{j \in I}H_j.$$

Then the associated C^*-algebra $C_{U_G}^*(G)$ is simply denoted by $C^*(G)$ and is called the C^*-algebra of the group G (so this is the closed linear span of $\{U_G(t) \mid t \in G\}$). We will sometimes call it the *full* C^*-algebra of G to distinguish it from the *reduced* one, which is described in the following.

Let π be any unitary representation of G. By a classical argument, π is unitarily equivalent to a direct sum of cyclic representations; hence, for any finitely supported function $x\colon G \to \mathbb{C}$, we have

$$\left\|\sum x(t)\pi(t)\right\| \leq \left\|\sum x(t)U_G(t)\right\|.$$

Equivalently,

$$\left\|\sum x(t)U_G(t)\right\| = \sup\left\{\left\|\sum x(t)\pi(t)\right\|\right\}, \tag{8.1}$$

where the supremum runs over all possible unitary representations $\pi\colon G \to B(H)$ on an arbitrary Hilbert space H. More generally, for any finitely supported function $x\colon G \to B(\mathcal{H})$ we have

$$\left\|\sum x(t) \otimes U_G(t)\right\|_{\min} = \sup\left\{\left\|\sum x(t) \otimes \pi(t)\right\|_{\min}\right\}, \tag{8.2}$$

where the sup is the same as before. These formulas explain why $C^*(G)$ is sometimes called the "maximal" C^*-algebra of G.

We denote by

$$\lambda_G\colon G \to B(\ell_2(G)) \quad [\text{resp.} \quad \rho_G\colon G \to B(\ell_2(G))]$$

the left (resp. right) regular representation of G, which means that $\lambda_G(t)$ (resp. $\rho_G(t)$ is the unitary operator of left (resp. right) translation by t

(resp. t^{-1}) on $\ell_2(G)$. Explicitly, if we denote by $(\delta_t)_{t \in G}$ the canonical basis of $\ell_2(G)$, we have $\lambda_G(t)\delta_s = \delta_{ts}$ (resp. $\rho_G(t)\delta_s = \delta_{st^{-1}}$) for all t, s in G. Actually, we will often drop the index G and denote simply by U, λ, ρ the representations U_G, λ_G, ρ_G.

We denote by $C_\lambda^*(G)$ (resp. $C_\rho^*(G)$) the C^*-algebra generated in $B(\ell_2(G))$ by λ_G (resp. ρ_G). Equivalently, $C_\lambda^*(G) = \overline{\text{span}}\{\lambda(t) \mid t \in G\}$ and $C_\rho^*(G) = \overline{\text{span}}\{\rho(t) \mid t \in G\}$. Note that $\lambda(t)$ and $\rho(s)$ commute for all t, s in G.

We will study the maximal tensor product of two C^*-algebras A_1, A_2 in more detail in Chapter 11. For the moment we will content ourselves with its definition. For any x in $A_1 \otimes A_2$, we define

$$\|x\|_{\max} = \sup \|\pi(x)\|_{B(H)},$$

where the supremum runs over all possible Hilbert spaces H and all possible $*$-homomorphisms $\pi\colon A_1 \otimes A_2 \to B(H)$. The completion of $A_1 \otimes A_2$ with respect to this norm is a C^*-algebra called the maximal tensor product of A_1 and A_2 and is denoted $A_1 \otimes_{\max} A_2$.

Let G_1, G_2 be two discrete groups. It is easy to see (exercise) that

$$C^*(G_1) \otimes_{\max} C^*(G_2) \simeq C^*(G_1 \times G_2), \tag{8.3}$$

$$C_\lambda^*(G_1) \otimes_{\min} C_\lambda^*(G_2) \simeq C_\lambda^*(G_1 \times G_2), \tag{8.4}$$

and similarly for the free product $G_1 * G_2$ (the free product of C^*-algebras is defined before Theorem 5.13):

$$C^*(G_1) * C^*(G_2) \simeq C^*(G_1 * G_2).$$

These identities can be extended to arbitrary families $(G_i)_{i \in I}$ in place of the pair (G_1, G_2). In particular, we have

$$\underset{i \in I}{*}\, C^*(G_i) \simeq C^*\left(\underset{i \in I}{*}\, G_i\right). \tag{8.5}$$

The following very useful result is known as Fell's "absorption principle."

Proposition 8.1. *For any unitary representation* $\pi\colon G \to B(H)$, *we have*

$$\lambda_G \otimes \pi \simeq \lambda_G \otimes I \quad \text{(unitary equivalence)}.$$

Here I stands for the trivial representation of G in $B(H)$ (i.e., $I(t) = I_H$ $\forall t \in G$). In particular, for any finitely supported function $a\colon G \to \mathbb{C}$ and $b\colon G \to B(\ell_2)$, we have

$$\left\|\sum a(t)\lambda_G(t) \otimes \pi(t)\right\| = \left\|\sum a(t)\lambda_G(t)\right\|,$$

$$\left\|\sum b(t) \otimes \lambda_G(t) \otimes \pi(t)\right\| = \left\|\sum b(t) \otimes \lambda_G(t)\right\|.$$

Proof. Note that $\lambda_G \otimes \pi$ acts on the Hilbert space $K = \ell_2(G) \otimes_2 H \simeq \ell_2(G; H)$. Let $V \colon K \to K$ be the unitary operator taking $x = (x(t))_{t \in G}$ to $(\pi(t^{-1})x(t))_{t \in G}$. A simple calculation shows that

$$V^{-1}(\lambda_G(t) \otimes I_H)V = \lambda_G(t) \otimes \pi(t). \qquad \blacksquare$$

The next result illustrates the usefulness of this principle.

Theorem 8.2. *We have an isometric (C*-algebraic) embedding*

$$J \colon C^*(G) \subset C^*_\lambda(G) \otimes_{\max} C^*_\lambda(G)$$

taking $U_G(t)$ to $\lambda_G(t) \otimes \lambda_G(t)$ ($t \in G$). Moreover, there is a (completely) contractive projection P from $C^*_\lambda(G) \otimes_{\max} C^*_\lambda(G)$ onto the range of J.

Proof. Let $x \colon G \times G \to \mathbb{C}$ be finitely supported. It suffices to show the following claim:

$$\left\| \sum_t x(t,t)U_G(t) \right\|_{C^*(G)} \leq \left\| \sum_{s,t} x(s,t)\lambda_G(s) \otimes \lambda_G(t) \right\|_{\max}. \qquad (8.6)$$

Indeed, in the converse direction we have obviously

$$\left\| \sum x(t,t)\lambda_G(t) \otimes \lambda_G(t) \right\|_{\max} \leq \left\| \sum x(t,t)U_G(t) \right\|.$$

Hence (8.6) implies at the same time that J defines an isometric *-homomorphism and that the natural ("diagonal") projection onto its range is contractive. We will now prove this claim. Let $\pi \colon G \to B(H)$ be a unitary representation of G. We introduce a pair of commuting representations (π_1, π_2) as follows:

$$\pi_1(\lambda_G(t)) = \lambda_G(t) \otimes \pi(t) \quad \text{and} \quad \pi_2(\lambda_G(t)) = \rho_G(t) \otimes I.$$

Note that both π_1 and π_2 extend to continuous representations on $C^*_\lambda(G)$. For π_1, this follows from the preceding absorption principle. For π_2, it follows from the fact that $\rho_G \simeq \lambda_G$ (indeed, if $W \colon \ell_2(G) \to \ell_2(G)$ is the unitary taking δ_t to $\delta_{t^{-1}}$, then $W^*\lambda_G(\cdot)W = \rho_G(\cdot)$).

Since π_1 and π_2 have commuting ranges, we have

$$\left\| \sum_{s,t} x(s,t)\lambda_G(s)\rho_G(t) \otimes \pi(s) \right\| \leq \left\| \sum_{s,t} x(s,t)\lambda_G(s) \otimes \lambda_G(t) \right\|_{\max};$$

hence, restricting the left side to $\delta_e \otimes H \subset \ell_2(G) \otimes_2 H$, we obtain (note that $\langle \lambda_G(s)\rho_G(t)\delta_e, \delta_e \rangle = 1$ if $s = t$ and zero otherwise)

$$\left\| \sum_t x(t,t)\pi(t) \right\|_{B(H)} \leq \left\| \sum_{s,t} x(s,t)\lambda_G(s) \otimes \lambda_G(t) \right\|_{\max}.$$

Finally, taking the supremum over π, we obtain the announced claim (8.6). The same argument (using operator valued coefficients $(x(s,t))$ instead of scalar ones) shows that $\|P\|_{cb} \leq 1$. ∎

Theorem 8.3. *([BF, Jol]) Let G be a discrete group and H a Hilbert space. The following properties of a function $\varphi\colon G \to B(H)$ are equivalent:*

(i) *The linear mapping defined on* $\mathrm{span}[\lambda(t) \mid t \in G]$ *by*

$$M_\varphi(\lambda(t)) = \lambda(t) \otimes \varphi(t)$$

extends to a c.b. map $M_\varphi\colon C_\lambda^*(G) \to B(\ell_2(G) \otimes_2 H)$ *with* $\|M_\varphi\|_{cb} \leq 1$.

(ii) *There is a Hilbert space $\widehat{H}$ and bounded functions $x\colon G \to B(H, \widehat{H})$ and $y\colon G \to B(H, \widehat{H})$ with $\sup_{t \in G} \|x(t)\| \leq 1$ and $\sup_{s \in G} \|y(s)\| \leq 1$ such that*

$$\forall\, s, t \in G \qquad \varphi(st^{-1}) = y(s)^*x(t).$$

Proof. Assume (i). Then, by Theorem 1.6, there are a Hilbert space $\widehat{H}$, a representation $\pi\colon C_\lambda^*(G) \to B(\widehat{H})$, and operators $V_i\colon \ell_2(G) \otimes_2 H \to \widehat{H}$ $(i = 1, 2)$ with $\|V_1\|\|V_2\| \leq 1$ such that

$$\forall\, \theta \in G \qquad \lambda(\theta) \otimes \varphi(\theta) = M_\varphi(\lambda(\theta)) = V_2^*\pi(\lambda(\theta))V_1. \qquad (8.7)$$

We define $x(t) \in B(H, \widehat{H})$ and $y(s) \in B(H, \widehat{H})$ by $x(t)h = \pi(\lambda(t^{-1}))\, V_1(\delta_t \otimes h)$ and $y(s)k = \pi(\lambda(s^{-1}))V_2(\delta_s \otimes k)$. Note that

$$\langle (\lambda(\theta) \otimes \varphi(\theta))(\delta_t \otimes h), \delta_s \otimes k \rangle = 1_{\{\theta = st^{-1}\}}\langle \varphi(st^{-1})h, k \rangle;$$

hence (8.7) implies

$$\langle \varphi(st^{-1})h, k \rangle = \langle y(s)^*x(t)h, k \rangle,$$

and we obtain (ii). Conversely, assume (ii). Define $\pi\colon C_\lambda^*(G) \to B(\ell_2(G)\otimes_2\widehat{H})$ by $\pi(x) = x \otimes I_{\widehat{H}}$. Let $V_i\colon \ell_2(G)\otimes_2 H \to \ell_2(G)\otimes_2\widehat{H}$ be defined by $V_1(\delta_t\otimes h) = \delta_t \otimes x(t)h$ and $V_2(\delta_s \otimes k) = \delta_s \otimes y(s)k$. Then for any θ, t, s, h, k we have

$$\langle V_2^*\pi(\lambda(\theta))V_1(\delta_t \otimes h), \delta_s \otimes k \rangle = \langle \lambda(\theta)\delta_t, \delta_s \rangle\langle y(s)^*x(t)h, k \rangle$$
$$= \langle \lambda(\theta) \otimes \varphi(\theta)(\delta_t \otimes h), \delta_s \otimes k \rangle;$$

hence $M_\varphi(\lambda(\theta)) = V_2^* \pi(\lambda(\theta)) V_1$, so the converse part of Theorem 1.6 yields (ii) $\Rightarrow$ (i). ∎

Remark 8.4. Consider a complex-valued function $\varphi \colon G \to \mathbb{C}$. Then the preceding result applied with $\mathbb{C} = B(H)$ yields: $\|M_\varphi \colon C_\lambda^*(G) \to C_\lambda^*(G)\|_{cb} \le 1$ iff there are Hilbert space valued functions x, y with $\sup_t \|x(t)\| \le 1$ and $\sup_s \|y(s)\| \le 1$ such that

$$\forall\, s, t \in G \qquad \varphi(st^{-1}) = \langle y(s), x(t) \rangle.$$

Proposition 8.5. *Let G be a discrete group and let $\Gamma \subset G$ be a subgroup. Then the correspondence $\lambda_\Gamma(t) \to \lambda_G(t)$, $(t \in \Gamma)$ extends to an isometric (C^*-algebraic) embedding $J \colon C_\lambda^*(\Gamma) \to C_\lambda^*(G)$. Moreover, there is a completely contractive and completely positive projection P from $C_\lambda^*(G)$ onto the range of this embedding.*

Proof. Let $Q = G/\Gamma$ and let $G = \bigcup_{q \in Q} \Gamma g_q$ be the decomposition of G into disjoint right cosets. For convenience, let us denote by 1 the equivalence class of the unit element of G. Since $G \simeq \Gamma \times Q$, we have an identification

$$\ell_2(G) \simeq \ell_2(\Gamma) \otimes_2 \ell_2(Q)$$

such that

$$\forall\, t \in \Gamma \qquad \lambda_G(t) = \lambda_\Gamma(t) \otimes I.$$

This shows of course that J is an isometric embedding. Moreover, we have a natural (linear) isometric embedding $V \colon \ell_2(\Gamma) \to \ell_2(G)$ (note that the range of V coincides with $\ell_2(\Gamma) \otimes \delta_1$ in the above identification) such that

$$\forall\, t \in \Gamma \qquad \lambda_\Gamma(t) = V^* \lambda_G(t) V.$$

Let $u(x) = V^* x V$. Clearly we have $\forall\, t \in G$

$$u(\lambda_G(t)) = \lambda_\Gamma(t) \quad \text{if} \quad t \in \Gamma \quad \text{and} \quad u(\lambda_G(t)) = 0 \quad \text{if} \quad t \notin \Gamma.$$

Therefore $P = Ju$ is a completely positive and completely contractive projection from $C_\lambda^*(G)$ onto $JC_\lambda^*(\Gamma)$. ∎

In the case of the full C^*-algebras, the analogs of the preceding results are as follows.

Proposition 8.6. Let G be a discrete group. The following properties of a function $\varphi\colon G \to B(H)$ are equivalent:

(i) The linear mapping taking $U_G(t)$ to $\varphi(t)$ extends to a c.b. map $T_\varphi\colon$ $C^*(G) \to B(H)$ with $\|T_\varphi\|_{cb} \leq 1$.

(ii) The "multiplier" taking $U_G(t)$ to $U_G(t) \otimes \varphi(t)$ extends to a c.b. map $M_\varphi\colon C^*(G) \to C^*(G) \otimes_{\min} B(H)$ with $\|M_\varphi\|_{cb} \leq 1$.

(iii) There is a unitary representation $\pi\colon G \to B(H_\pi)$ and contractions $V, W\colon H \to H_\pi$ such that

$$\varphi(t) = V^*\pi(t)W. \qquad\qquad \forall\, t \in G$$

Proof. We will show (i) $\Rightarrow$ (iii) $\Rightarrow$ (ii) $\Rightarrow$ (i). Assume (i). Then (iii) follows by an immediate application of the factorization theorem for c.b. maps, namely, Theorem 1.6. Assume (iii). By maximality of $C^*(G)$, the mapping taking $U_G(t)$ to $U_G(t) \otimes \pi(t)$ extends to a representation on $C^*(G)$; hence, since $M_\varphi = (1 \otimes V^*)[U_G(t) \otimes \pi(t)](1 \otimes W)$, we have $\|M_\varphi\|_{cb} \leq \|1 \otimes V^*\| \cdot \|1 \otimes W\| \leq 1$, whence (ii). Assume (ii). Since the mapping T_1 taking $U_G(t)$ to 1 (the unit in $\mathbb{C}$) is clearly a representation on $C^*(G)$ (associated to the trivial representation on G), and since we have obviously

$$T_\varphi = (T_1 \otimes I)M_\varphi,$$

we obtain $\|T_\varphi\|_{cb} \leq 1$, whence (i). $\blacksquare$

Corollary 8.7. A function $\varphi\colon G \to \mathbb{C}$ defines a bounded multiplier $M_\varphi\colon$ $C^*(G) \to C^*(G)$ iff there are a unitary representation $\pi\colon G \to B(H_\pi)$ and ξ, η in H_π such that

$$\varphi(t) = \langle \pi(t)\xi, \eta \rangle. \qquad\qquad \forall\, t \in G$$

Moreover, we have

$$\|M_\varphi\| = \|M_\varphi\|_{cb} = \inf\{\|\xi\|\|\eta\|\},$$

where the infimum runs over all possible π, ξ, η for which this holds. Finally, if $\xi = \eta$, M_φ is completely positive.

Proof. The first assertion for c.b. maps and the equality $\|M_\varphi\|_{cb} = \inf\{\|\xi\|\|\eta\|\}$ follow from the equivalence (ii) $\Leftrightarrow$ (iii) in the preceding statement, using $V, W\colon \mathbb{C} \to H_\pi$ defined by $V1 = \eta$, $W1 = \xi$. Now, if M_φ is bounded, then (with the above notation) $T_\varphi\colon C^*(G) \to \mathbb{C}$ is bounded, and, since $T_\varphi = T_1 M_\varphi$, we have $\|T_\varphi\| \leq \|M_\varphi\|$; but since T_φ is scalar-valued, we have $\|T_\varphi\|_{cb} = \|T_\varphi\|$ and the proof can now be completed using (i) $\Leftrightarrow$ (ii) in the preceding statement. Note that M_φ is completely positive if $\xi = \eta$ since we then have $V = W$. $\blacksquare$

Proposition 8.8. *Let G be a discrete group and let $\Gamma \subset G$ be a subgroup. Then the correspondence $U_\Gamma(t) \to U_G(t)$, $(t \in \Gamma)$ extends to an isometric $(C^*$-algebraic) embedding J of $C^*(\Gamma)$ into $C^*(G)$. Moreover, there is a completely contractive and completely positive projection P from $C^*(G)$ onto the range of this embedding.*

Proof. We will need the classical fact that, for any unitary representation $\pi\colon \Gamma \to B(H)$, there is a Hilbert space $\widehat{H}$ with $H \subset \widehat{H}$ and a unitary representation $\widehat{\pi}\colon G \to B(\widehat{H})$ such that

$$\forall\, t \in \Gamma \qquad \widehat{\pi}(t)H \subset H \quad \text{and} \quad \pi(t) = \widehat{\pi}(t)_{|H}.$$

Here is a quick justification for this fact (see [Rie] for a more general framework). Let $\{s_j \mid j \in I\}$ be a set of elements of G chosen so that $G = \bigcup_{j \in I} s_j \Gamma$ is the decomposition of G into left cosets. For convenience, we assume that I contains a distinguished element, which we denote by 1, such that $s_1 = e$ (unit element of G). Let $\pi_0\colon G \to B(H)$ be the *trivial* extension of π defined for all x in G by

$$\pi_0(x) = \begin{cases} \pi(x) & \text{if } x \in \Gamma \\ 0 & \text{otherwise.} \end{cases}$$

We let $\widehat{H} = \bigoplus_{j \in I} H_j$, where each H_j is a copy of H. Then, for any x in G, we define $\widehat{\pi}(x)\colon \widehat{H} \to \widehat{H}$ as the operator associated to the matrix $(\pi_0(s_i^{-1}xs_j))_{ij}$ with entries in $B(H)$. We then leave as an exercise for the reader to check that $\widehat{\pi}$ has the announced properties. Following Mackey's classical terminology, $\widehat{\pi}$ is the representation "induced" by π. Using (8.1) and the preceding extension property, it is immediate that J is an isometric embedding.

Now, let $\pi_1\colon \Gamma \to B(\mathbb{C}) \simeq \mathbb{C}$ be the trivial representation ($\pi_1 \equiv 1$) and let $\widehat{\pi}_1\colon G \to B(\widehat{H})$ be the induced representation as above, with $\mathbb{C} \subset \widehat{H}$. Let $\xi \in \widehat{H}$ be the unit in $\mathbb{C}$ viewed as sitting in $\widehat{H}$. Then it is easy to check that the function φ defined by

$$\forall\, t \in G \qquad \varphi(t) = \langle \widehat{\pi}_1(t)\xi, \xi \rangle$$

is nothing but the indicator function of Γ. Therefore, by Corollary 8.7, the multiplier M_φ defines a completely contractive and completely positive projection from $C^*(G)$ onto $JC^*(\Gamma)$. ∎

Let F be a free group with generators $(g_i)_{i \in I}$. The first part of the next result is based on the classical observation that a unitary representation $\pi\colon F \to B(H)$ is entirely determined by its values $u_i = \pi(g_i)$ on the generators; and if we let π run over all possible unitary representations, then we obtain all possible families (u_i) of unitary operators. The second part is also well known (see, e.g., [Pa5]).

Lemma 8.9. Let F be a free group. Let $(U_i)_{i \in I}$ be the family composed of the unitary generators of $C^*(F)$ (i.e., these are the unitaries corresponding to the free generators of F in the universal unitary representation of F). Let $(x_i)_{i \in I}$ be a finitely supported family in $B(H)$. Consider the linear map $T \colon \ell_\infty(I) \to B(H)$ defined by $T((\alpha_i)_{i \in I}) = \sum_{i \in I} \alpha_i x_i$. Then we have

$$\left\| \sum_{i \in I} U_i \otimes x_i \right\|_{C^*(F) \otimes_{\min} B(H)} = \|T\|_{cb} = \sup\{\|\sum u_i \otimes x_i\|_{\min}\}, \qquad (8.8)$$

where the sup runs over all possible families of unitaries (u_i). Actually, the latter supremum remains the same if we let it run only over all possible families of finite-dimensional unitaries (u_i). Moreover, if $\dim(H) = \infty$, then

$$\left\| \sum_{i \in I} U_i \otimes x_i \right\|_{C^*(F) \otimes_{\min} B(H)} = \inf\{\left\| \sum y_i y_i^* \right\|^{1/2} \left\| \sum z_i^* z_i \right\|^{1/2}\}, \qquad (8.9)$$

where the infimum that runs over all possible factorizations $x_i = y_i z_i$ with y_i, z_i in $B(H)$ is actually attained. Moreover, all this remains true if we enlarge the family $(U_i)_{i \in I}$ by adding the unit element of $C^*(F)$.

Proof. It is easy to check by going back to the definitions that, on one hand

$$\|\sum U_i \otimes x_i\|_{\min} = \sup\{\|\sum u_i \otimes x_i\|_{\min}\},$$

where the sup runs over all possible families of unitaries (u_i), and on the other hand that

$$\|T\|_{cb} = \sup\{\|\sum t_i \otimes x_i\|_{\min}\},$$

where the sup runs over all possible families of contractions (t_i). By the Russo-Dye Theorem, any contraction is a norm limit of convex combinations of unitaries, so (8.8) follows by convexity. Actually, the preceding sup obviously remains unchanged (see Exercise 2.1.1) if we let it run only over all possible families of contractions (t_i) on a *finite-dimensional* Hilbert space. Thus it remains unchanged when restricted to families of *finite-dimensional* unitaries (u_i).

Now assume $\|T\|_{cb} = 1$. By the factorization of c.b. maps we can write $T(\alpha) = V^* \pi(\alpha) W$, where $\pi \colon \ell_\infty(I) \to B(\widehat{H})$ is a representation and where V, W are in $B(H, \widehat{H})$ with $\|V\| \|W\| = \|T\|_{cb}$. We can assume I finite and, since $\dim(H) = \infty$, $\widehat{H} = H$. Let $(e_i)_{i \in I}$ be the canonical basis of $\ell_\infty(I)$. We set

$$y_i = V^* \pi(e_i) \quad \text{and} \quad z_i = \pi(e_i) W.$$

It is then easy to check $\left\|\sum y_i y_i^*\right\|^{1/2} \left\|\sum z_i^* z_i\right\|^{1/2} \leq \|V\| \|W\| = \|T\|_{cb}$. Thus we obtain one direction of (8.9). The converse follows from the inequality

$$\left\|\sum b_i a_i\right\| \leq \left\|\sum b_i b_i^*\right\|^{1/2} \left\|\sum a_i^* a_i\right\|^{1/2}$$

(already proved in Remark 1.13) applied to $b_i = U_i \otimes y_i$ and $a_i = 1 \otimes z_i$. Finally, the last assertion follows from the next remark. ∎

Remark. Let $\{0\}$ be a singleton disjoint from the set I and let $J = I \cup \{0\}$. Then, for any finitely supported family $\{x_j \mid j \in J\}$ in $B(H)$ (H arbitrary), we have

$$\left\| I \otimes x_0 + \sum_{i \in I} U_i \otimes x_i \right\|_{\min} = \sup \left\{ \left\| \sum_{j \in J} u_j \otimes x_j \right\|_{\min} \right\}, \qquad (8.10)$$

where the supremum runs over all possible families $(u_j)_{j \in J}$ of unitaries.

Indeed, since

$$\left\| \sum_{j \in J} u_j \otimes x_j \right\| = \left\| I \otimes x_0 + \sum_{i \in I} u_0^{-1} u_i \otimes x_i \right\|,$$

the right side of (8.10) is the same as the supremum of $\left\| I \otimes x_0 + \sum_{i \in I} u_i \otimes x_i \right\|$ over all possible families of unitaries $(u_i)_{i \in I}$, but (recalling $U(g_i) = U_i$) the latter supremum coincides with

$$\left\| I \otimes x_0 + \sum_{i \in I} U_i \otimes x_i \right\|_{\min}. \qquad ∎$$

Remark 8.10. Let F be a free group with free generators (g_i). Then, for any finitely supported sequence of scalars (a_i), for any H, and for any family (u_i) of unitary operators in $B(H)$, we have

$$\left\| \sum a_i \lambda(g_i) \otimes u_i \right\| = \left\| \sum a_i \lambda(g_i) \right\|.$$

Indeed, this follows from the absorption principle (Proposition 8.1) applied to the function a defined by $a(g_i) = a_i$ and $= 0$ elsewhere, and to the unique unitary representation π on F such that $\pi(g_i) = u_i$.

We now return to the unital operator algebra of an operator space E, which we introduced in Chapter 6.

Recall that $OA_u(E)$ is defined as the completion of the unital tensor algebra $T_u(E)$ for the operator space structure determined by (6.3). Moreover,

$OA(E)$ is the subalgebra of $OA_u(E)$ formed by all elements with vanishing constant term.

By construction, the operator algebra $OA(E)$ (resp. $OA_u(E)$) is characterized by the following universal property (recall that "completely contractive" is abbreviated by c.c.): For any c.c. map $\sigma\colon E \to B(\mathcal{H})$, there is a unique c.c. morphism $\widehat{\sigma}\colon OA(E) \to B(\mathcal{H})$ (resp. unital morphism $\widehat{\sigma}\colon OA_u(E) \to B(\mathcal{H})$) extending σ.

This characterizes $OA(E)$ (resp. $OA_u(E)$) up to a completely isometric isomorphism of algebras (resp. unital algebras) among all (resp. unital) operator algebras containing E completely isometrically.

The next result is a reformulation of von Neumann's classical inequality.

Theorem 8.11. *If $E = \mathbb{C}$, then $OA_u(E)$ can be identified completely isometrically with the disc algebra*

$$A(D) = \overline{\operatorname{span}}\{z^n \mid n \geq 0\} \subset C(\partial D).$$

Here D denotes the open unit disc in $\mathbb{C}$.

Proof. Assume $E = \mathbb{C}$. Let us denote by $z \in E \subset OA_u(E)$ the element corresponding to the unit of $\mathbb{C}$. Then any element in $T_u(E)$ can be written as $P = \sum_0^d a_j z^{\otimes j}$ $(a_j \in \mathbb{C})$, which we identify with the polynomial $P(z) = \sum_0^d a_j z^j$. By (6.3) we have

$$\|P\|_{OA_u(E)} = \sup\left\{\left\|\sum_0^d a_j T^j\right\|\right\},$$

where the supremum runs over all contractive T in $B(H)$ (H arbitrary). Let $P(T) = a_0 I_H + \sum_1^d a_j T^j$. We now claim

$$\|P(T)\| \leq \sup\{|P(\omega)| \mid \omega \in \partial D\}. \tag{8.11}$$

This is nothing but von Neumann's classical inequality. Here is a quick proof: As already noted, for any contractive T in $B(H)$ there is a unitary V in $M_2(B(H)) \simeq B(H \oplus H)$ such that $T = V_{11}$. Indeed, we simply set

$$V = \begin{pmatrix} T & (1 - TT^*)^{1/2} \\ -(1 - T^*T)^{1/2} & T^* \end{pmatrix}.$$

Let j and p denote, respectively, the inclusion and the projection corresponding to the first coordinate in $H \oplus H$. Then $T = V_{11}$ can be rewritten as

$$T = p \begin{pmatrix} 1 & 0 \\ 0 & 0 \end{pmatrix} V \begin{pmatrix} 1 & 0 \\ 0 & 0 \end{pmatrix} j,$$

and more generally (since $pj = I$), for any polynomial P, we have trivially

$$P(T) = pP(U(0))j,$$

where we have set

$$U(\omega) = \begin{pmatrix} 1 & 0 \\ 0 & \omega \end{pmatrix} V \begin{pmatrix} 1 & 0 \\ 0 & \omega \end{pmatrix}.$$

Hence

$$\|P(T)\| = \|pP(U(0))j\| \leq \sup_{\omega \in \overline{D}} \|P(U(\omega))\|,$$

but since $\omega \to P(U(\omega))$ is obviously analytic in ω, the maximum principle yields

$$\|P(T)\| \leq \sup_{|\omega|=1} \|P(U(\omega))\|;$$

hence

$$\leq \sup\{\|P(U)\| \mid U \text{ unitary}\},$$

and, by the spectral theorem for unitary (or normal) operators, we obtain our claim:

$$\|P(T)\| \leq \sup_{z \in \partial D} |P(z)|.$$

Thus we conclude that

$$\|P\|_{OA_u(\mathbb{C})} = \|P\|_{A(D)},$$

or, equivalently, $OA_u(\mathbb{C}) \simeq A(D)$ isometrically. Using operator coefficients instead of scalars, it is easy to check that this identification actually is completely isometric. ∎

More generally, we have (cf. [Bo1]):

Theorem 8.12. *Let I be any set. Let F_I be the free group with generators $(g_i)_{i \in I}$ and let $U_i \in C^*(F_I)$ be the unitary element associated to g_i. We denote by e_i the canonical basis of $\ell_1(I)$. Then the correspondence $e_i \to U_i$ extends to a unital homomorphism that is a completely isometric embedding of $OA_u(\max(\ell_1(I)))$ into $C^*(F_I)$.*

Proof. Let $E = \max(\ell_1(I))$. We may as well assume by density that I is finite. Let $P \in \mathcal{K} \otimes T_u(E)$. We have $P = \sum_d P_d$ with $P_d \in \mathcal{K} \otimes E^{\otimes d}$ of the form

$$P_d = \sum \lambda_{i_1 \ldots i_d} \otimes (e_{i_1} \otimes \cdots \otimes e_{i_d})$$

with $\lambda_{i_1 \ldots i_d} \in \mathcal{K}$. Then, by (6.3), we have

$$\|P\|_{\mathcal{K} \otimes_{\min} OA_u(E)} = \sup \left\{ \left\| \sum_d \sum_{i_1 \ldots i_d} \lambda_{i_1 \ldots i_d} \otimes T_{i_1} \ldots T_{i_d} \right\|_{\min} \right\},$$

where the supremum runs over all H and over all families of contractions $(T_i)_{i \in I}$ in $B(H)$. By the same trick as in the preceding proof, the supremum remains the same if we restrict it to families of *unitaries* (cf. [Bo1]). Therefore we obtain, by the observation preceding Lemma 8.9,

$$\|P\|_{\mathcal{K} \otimes_{\min} OA_u(E)} = \left\| \sum_d \sum_{i_1 \ldots i_d} \lambda_{i_1 \ldots i_d} \otimes U_{i_1} U_{i_2} \ldots U_{i_d} \right\|_{\mathcal{K} \otimes_{\min} C^*(F_I)},$$

which shows that the correspondence $e_{i_1} e_{i_2} \ldots e_{i_d} \to U_{i_1} U_{i_2} \ldots U_{i_d}$ is a completely isometric homomorphism. ∎

Corollary 8.13. *Fix $d \geq 1$. With the same notation as before, we have a completely isometric isomorphism*

$$\max(\ell_1(I)) \otimes_h \cdots \otimes_h \max(\ell_1(I)) \simeq \overline{\mathrm{span}}[U_{i_1} \ldots U_{i_d}] \subset C^*(F_I)$$

taking $e_{i_1} \otimes \cdots \otimes e_{i_d}$ to $U_{i_1} \ldots U_{i_d}$.

Proof. We simply combine the preceding statement with Proposition 6.6. ∎

We will now introduce (following [Pes]) the universal C^*-algebra of an operator space E, which will be denoted by $C^*\langle E \rangle$. The definition stems from the next statement.

Theorem 8.14. *Let E be an operator space. There is a C^*-algebra A and a completely isometric embedding $j \colon E \to A$ with the following properties:*

(i) *For any C^*-algebra B and any completely contractive map $u \colon E \to B$ there is a representation $\pi \colon A \to B$ extending u, that is, such that $\pi j = u$.*

(ii) *A is the smallest C^*-algebra containing $j(E)$.*

Moreover, (ii) ensures that the representation π in (i) is unique.

Proof. The proof is immediate. Let I be the "collection" of all $u \colon E \to B_u$ with $\|u\|_{cb} \leq 1$. Let

$$\mathcal{B} = \bigoplus_{u \in I} B_u.$$

We define $j \colon E \to \mathcal{B}$ by

$$j(x) = \bigoplus_{u \in I} u(x).$$

It is easy to check that j is completely isometric. Then, if we define A to be the C^*-algebra generated by $j(E)$ in $\mathcal{B}$, it is very easy to check the announced universal property of A. ∎

Notation. We will denote by $C^*\langle E \rangle$ the C^*-algebra A appearing in the preceding statement, and we denote by $C_u^*\langle E \rangle$ its unitization.

Note that $C^*\langle E \rangle$ is essentially unique. Indeed, if $j_1\colon E \to A_1$ is another completely isometric embedding into a C^*-algebra A_1 with the property in Theorem 8.14, then the universal property of A (resp. A_1) implies the existence of a representation $\pi\colon A \to A_1$ (resp. $\pi_1\colon A_1 \to A$) such that $\pi j = j_1$ (resp. $\pi_1 j_1 = j$). Since C^*-representations are automatically contractive, we have $\|\pi\| \leq 1$, $\|\pi_1\| \leq 1$, and $\pi_1 = \pi^{-1}$ on $j_1(E)$ and hence on the $*$-algebra generated by $j_1(E)$, which is dense in A_1, by assumption. This implies that π is an isometric isomorphism from A onto A_1. ∎

Similarly, $C_u^*\langle E \rangle$ is characterized as the unique unital C^*-algebra C containing E completely isometrically in such a way that, for any unital C^*-algebra B (actually we may restrict to $B = B(H)$ with H arbitrary), any c.c. map $u\colon E \to B$ *uniquely* extends to a unital representation (i.e. $*$-homomorphism) from C to B.

Remark. If two operator spaces E, F are completely isometrically isomorphic, then E and F can be realized as "concrete" operator subspaces $E \subset A$ and $F \subset B$ of two C^*-algebras A and B for which there is an isometric $*$-homomorphism $\pi\colon A \to B$ such that $\pi(E) = F$.

Indeed, let $A = C^*\langle E \rangle$ and $B = C^*\langle F \rangle$, let $u\colon E \to F$ be a completely isometric isomorphism, let $\pi\colon C^*\langle E \rangle \to C^*\langle F \rangle$ be the (unique) extension of u (as in Theorem 8.10), and let $\sigma\colon C^*\langle F \rangle \to C^*\langle E \rangle$ be the (unique) extension of u^{-1}. Then clearly we must have (by unicity again) $\sigma\pi = I$ and $\pi\sigma = I$; hence $\sigma = \pi^{-1}$. ∎

To illustrate the use of this notion, we can state a nice relationship between direct sums of operator spaces and free products. Recall that we have defined the direct sum $E_1 \oplus_1 E_2$ of two operator spaces (in the ℓ_1-sense) in §2.6 and the free products in Chapter 5 (before Theorem 5.13). Using formula (5.7) together with Theorem 5.1 and comparing with (5.17), the following result is easy to check (we leave the details to the reader).

Theorem 8.15. *Let E_1, E_2 be arbitrary operator spaces. Then we have canonical (C^*-algebraic) isomorphisms*

$$C^*\langle E_1 \oplus_1 E_2 \rangle \simeq C^*\langle E_1 \rangle \dot{*} C^*\langle E_2 \rangle$$

and

$$C_u^*\langle E_1 \oplus_1 E_2 \rangle \simeq C_u^*\langle E_1 \rangle * C_u^*\langle E_2 \rangle.$$

Proof. It suffices to check that the free product C^*-algebras $C^*\langle E_1 \rangle \dot{*} C^*\langle E_2 \rangle$ and $C_u^*\langle E_1 \rangle * C_u^*\langle E_2 \rangle$ have the universal property characteristic, respectively, of $C^*\langle E\rangle$ and $C_u^*\langle E\rangle$ when $E = E_1 \oplus_1 E_2$ is embedded into the free product in the natural way. ∎

A similar result holds for N-tuples $E_1, ..., E_N$ (or actually for arbitrary families) of operator spaces.

Remark 8.16. It is easy to see that $OA(E)$ (resp. $OA_u(E)$) can be identified with the closed (resp. unital) subalgebra generated by E in $C^*\langle E\rangle$ (resp. $C_u^*\langle E\rangle$). Of course, $C^*\langle E\rangle$ can also be identified with a C^*-subalgebra of $C_u^*\langle E\rangle$.

For any c.c. map $\sigma\colon E \to B(H_\sigma)$, we denote by $\hat\sigma : OA_u(E) \to B(H_\sigma)$ the c.c. morphism extending σ and by

$$\pi_\sigma : C_u^*\langle E\rangle \to B(H_\sigma)$$

the unital representation extending σ. Note that $\hat\sigma$ can be identified with the restriction of π_σ to $OA_u(E)$ viewed as a subalgebra of $C_u^*\langle E\rangle$.

The following simple fact is sometimes useful. It shows the "residual finiteness" (see [Pes]) of the universal algebras of E.

Proposition 8.17. *Let E be any operator space. Let*

$$C = \{\sigma\colon E \to B(H) \mid \dim(H) < \infty, \|\sigma\|_{cb} \le 1\}.$$

Then the embedding
$$J\colon C_u^*\langle E\rangle \to \oplus_{\sigma \in C} B(H_\sigma)$$

defined by $J(x) = \oplus_{\sigma \in C} \pi_\sigma(x)$ is a completely isometric unital representation. A fortiori, when we restrict to either $C^*\langle E\rangle$ (resp. $OA_u(E)$ or $OA(E)$), we obtain a completely isometric representation (resp. morphism).

Proof. The fact that J is isometric is an easy consequence of the following elementary fact. Let $T_1, \ldots, T_n \in B(H)$. Let $P(\{T_i, T_i^*\})$ be a polynomial in $T_1, \ldots, T_n, T_1^*, \ldots, T_n^*$. Then

$$\|P(\{T_i, T_i^*\})\| \le \sup \|P(\{P_K T_{i|K}, P_K T_{i\ |K}^*\})\|,$$

where the supremum runs over all finite-dimensional subspaces $K \subset H$. This implies that, for any x in the $*$-algebra generated by E in $C_u^*\langle E\rangle$, we have

$$\|x\| \le \|J(x)\|,$$

and hence that J is isometric. The completely isometric case is proved similarly using polynomials with operator coefficients instead of scalar ones. ∎

Let (E_0, E_1) be a compatible couple of operator spaces and let $E_\theta = (E_0, E_1)_\theta$, with $0 < \theta < 1$. We have c.c. injective maps $E_0 \to E_0 + E_1$ and $E_1 \to E_0 + E_1$ that extend to c.c. morphisms $OA_u(E_0) \to OA_u(E_0 + E_1)$ and $OA_u(E_1) \to OA_u(E_0 + E_1)$. We claim that these morphisms are injective. Indeed, using Proposition 6.6, this can be reduced to the fact that, if a c.b. map $j\colon E \to F$ is injective, then, for any $N \geq 1$, $j \otimes \cdots \otimes j$ extends to an injective map from $E \otimes_h \cdots \otimes_h E$ to $F \otimes_h \cdots \otimes_h F$ (N times). The latter fact follows from Proposition 5.16.

Using this claim, we can view $(OA_u(E_0), OA_u(E_1))$ as a compatible couple. The next result shows that the interpolation functor commutes with the functor $E \to OA_u(E)$.

Theorem 8.18. *With the above notation we have completely isometric identities*

$$OA_u(E_\theta) \simeq (OA_u(E_0), OA_u(E_1))_\theta$$

and

$$OA(E_\theta) \simeq (OA(E_0), OA(E_1))_\theta.$$

Proof. We will prove only the second identity. The proof of the first one is exactly the same. By Propositions 6.6 and 2.7.6 we have a completely isometric embedding

$$v\colon E_\theta \subset (OA(E_0), OA(E_1))_\theta$$

obtained by interpolating between

$$E_0 \to OA(E_0) \quad \text{and} \quad E_1 \to OA(E_1).$$

Let $OA_\theta = (OA(E_0), OA(E_1))_\theta$. By the universal property of $OA(E_\theta), v$ extends to a c.c. morphism $\widehat{v}\colon OA(E_\theta) \to OA_\theta$. To show that $\widehat{v}$ is completely isometric, it suffices to show by Proposition 8.17 (applied to $OA(E_\theta)$) the following claim: For any n and any c.c. map $\sigma\colon E_\theta \to M_n$ there is a c.c. morphism $\widehat{\sigma}\colon OA_\theta \to M_n$ extending σ. Indeed, this claim allows us to extend the embedding $E_\theta \to OA(E_\theta)$ to a c.c. morphism from OA_θ to $OA(E_\theta)$, which must be the inverse of $\widehat{v}$.

To prove the preceding claim, we first assume that $E_0 \cap E_1$ is dense both in E_0 and E_1 and we make crucial use of the following identity:

$$\|\sigma\|_{CB(E_\theta, M_n)} = \|\sigma\|_{(CB(E_0, M_n), CB(E_1, M_n))_\theta},$$

which follows from Theorem 5.22 and the discussion of the duality in §2.7 (recall that $CB(E, M_n)$ can be identified with $C_n \otimes_h E^* \otimes_h R_n$ or, equivalently, with $(R_n \otimes_h E \otimes_h C_n)^*$). Finally, the restriction that $E_0 \cap E_1$ be dense both in E_0 and E_1 can be removed a posteriori using Lemma 2.7.2.

An alternate proof, perhaps more direct (not using duality), can be given using only Propositions 5.22 and 26.14.

Let E be an operator space. We define its unitization $\widetilde{E}$ as the linear span of the unit and E in $OA_u(E)$. It is easy to check that $\widetilde{E}$ can be characterized as the unique unital operator space containing E completely isometrically and such that any complete contraction $\sigma\colon E \to B(H)$ admits a *unique* unital completely contractive extension $\widetilde{\sigma}\colon \widetilde{E} \to B(H)$. But actually, the next result shows that this notion of unitization is essentially trivial.

Proposition 8.19. *The unitization $\widetilde{E}$ of an operator space E can be identified completely isometrically with the direct sum $\mathbb{C} \oplus_1 E$, with the unit corresponding to $(1,0)$.*

Proof. Let $j\colon \mathbb{C} \oplus_1 E \to \widetilde{E}$ be the mapping taking (λ, x) to $\lambda u + x$, where we have denoted by u the unit in $\mathbb{C} \subset OA_u(E)$. We claim that j is completely isometric. Let $e_0 = (1,0) \in \mathbb{C} \oplus E$, let $x_1, \ldots, x_n$ be elements of E, let $e_i = (0, x_i)$, $(1 \le i \le n)$ and let $a_0, \ldots, a_n$ be in $\mathcal{K}$. We have by (6.3)

$$\left\| \sum_{i=0}^n a_i \otimes j(e_i) \right\|_{\min} = \sup \left\{ \left\| a_0 \otimes I + \sum a_i \otimes \sigma(x_i) \right\|_{\min} \right\},$$

where the sup is over all complete contractions $\sigma\colon E \to B(H)$. On the other hand, we have, by definition (see §2.6),

$$\left\| \sum_{i=0}^n a_i \otimes e_i \right\|_{\min} = \sup \left\{ \left\| a_0 \otimes T + \sum_{i=1}^n a_i \otimes \sigma(x_i) \right\| \right\},$$

where the supremum runs over all c.c. maps $\sigma\colon E \to B(H)$ and all contractions T in $B(H)$. By the Russo-Dye Theorem, the sup is the same if we restrict T to be unitary, and then after multiplication by $I \otimes T^{-1}$ (on the right say) we obtain exactly the same supremum as above. Hence we conclude that

$$\left\| \sum_{i=0}^n a_i \otimes j(e_i) \right\|_{\min} = \left\| \sum_{i=0}^n a_i \otimes e_i \right\|_{\min},$$

which completes the proof by (2.1.8). ∎

Exercises

Exercise 8.1. Prove that any unital C^*-algebra (resp. separable C^*-algebra) C is isomorphic to a quotient of $C^*(\mathbb{F})$ for some free group $\mathbb{F}$ (resp. for $\mathbb{F} = \mathbb{F}_\infty$), so that $C \simeq C^*(\mathbb{F})/I$ for some (closed two-sided) ideal $I \subset C^*(\mathbb{F})$.

Exercise 8.2. Consider an operator space E_1 and a closed subspace $E_2 \subset E_1$. Let $j\colon E_2 \to E_1$ be the inclusion map. Then j extends to completely isometric embeddings $C^*\langle E_2 \rangle \to C^*\langle E_1 \rangle$ and $OA(E_2) \to OA(E_1)$ (and similarly in the unital case).

Exercise 8.3. Let E_1, E_2 be as above. Let $q\colon E_1 \to E_1/E_2$ be the quotient map. Then the unique representation $\pi\colon C^*\langle E_1 \rangle \to C^*\langle E_1/E_2 \rangle$ and the unique morphism $u\colon OA(E_1) \to OA(E_1/E_2)$ associated to q are complete metric surjections (and similarly in the unital case). (Thus the functors $E \to OA(E)$ and $E \to C^*\langle E \rangle$ are "projective"; they are also "injective" by the preceding exercise.)

Exercise 8.4. A discrete group G is called amenable if it admits an invariant mean, that is, a functional φ in $\ell_\infty(G)^*_+$ with $\varphi(1) = 1$ such that $\varphi(\delta_t * f) = \varphi(f)$ for any f in $\ell_\infty(G)$ and any t in G.

It is known (and the reader should use this as an alternate definition) that G is amenable iff there is a net (f_α) in the unit sphere of $\ell_2(G)$ such that $\|\lambda(t)f_\alpha - f_\alpha\|_2 \to 0$ for any t in G. Show that the following are equivalent:

(i) G is amenable.

(ii) $C^*(G) = C^*_\lambda(G)$.

(iii) There is a generating subset $S \subset G$ with $e \in S$ such that, for any finite subset $E \subset S$, we have

$$|E| = \left\| \sum_{t \in E} \lambda(t) \right\|.$$

Chapter 9 Examples and Comments

In Banach space theory, ever since Banach and Mazur's early work, several examples have played a privileged role, such as ℓ_p, c_0, L_p, and $C(K)$ (with K compact). These are usually called the *classical* Banach spaces. In the light of more recent developments, it is tempting to extend the list to the Orlicz, Sobolev, and Hardy spaces as well as the disc algebra and the Schatten p-classes, although in some sense all these examples are derived from those of the first generation. Analogously, one could make a list of all the classical C^*-algebras or von Neumann algebras (see e.g. [Da2]). In the present chapter, our aim is to describe the spaces that, in our opinion, are the best candidates to appear on a list of the *classical* operator spaces (we have already met some examples, such as R, C, $\min(\ell_2)$, $\max(\ell_2)$, and OH).

9.1. A concrete quotient: Hankel matrices

We start with a natural example of a *quotient operator space*, namely, the space

$$L_\infty/H_\infty^0$$

(here L_∞ is equipped with its natural o.s.s., i.e., the minimal one), which can be identified with the subspace of $B(\ell_2)$ formed of all the Hankel matrices, that is, all the matrices (a_{ij}) in $B(\ell_2)$ such that a_{ij} depends only on $i + j$.

To explain this identification, we need some specific notation and background from classical harmonic analysis. We denote by L_p the space L_p on the torus $\mathbb{T}$ equipped with the normalized Haar measure m. We denote by H_p the subspace of L_p formed of all functions φ with Fourier transform $\widehat{\varphi}$ vanishing on the negative integers and by $H_p^0 \subset H_p$ the subspace of those φ such that $\widehat{\varphi}(0) = 0$. More generally, given a Banach space X, we denote by $L_p(X)$, $H_p(X)$ and $H_p^0(X)$ the analogous spaces of X-valued functions on $\mathbb{T}$ ($1 \leq p < \infty$). To any φ in L_∞, one classically associates the Hankel matrix (a_{ij}) with entries

$$a_{ij} = \widehat{\varphi}(-i - j) \quad \forall i, j \geq 0.$$

It is easy to see (identifying ℓ_2 to H_2) that this matrix defines a bounded linear operator $u_\varphi \colon \ell_2 \to \ell_2$ such that $\|u_\varphi\| \leq \|\varphi\|_{L_\infty}$. Moreover, we have $u_\varphi = 0$ if $\varphi \in H_\infty^0$, so that u_φ depends only on the equivalence class of φ modulo H_∞^0. Let $\psi \in L_\infty/H_\infty^0$. We let $\mathcal{H}[\psi] = u_\varphi$, where φ is chosen arbitrarily in the equivalence class represented by ψ. By a famous theorem due to Nehari (1958), the correspondence $\psi \to \mathcal{H}[\psi]$ is an isometric linear embedding from L_∞/H_∞^0 into $B(\ell_2)$. Thus, Nehari's Theorem provides us with a realization of L_∞/H_∞^0 as an operator space. But actually much more is true: By a "vectorial" variant of Nehari's Theorem due to Sarason (see [Ni]), for each $n \geq 1$, the mapping

$$(\psi_{ij}) \to (\mathcal{H}[\psi_{ij}])$$

is an isometry from $L_\infty(M_n)/H_\infty^0(M_n)$ into $M_n(B(\ell_2)) = B(\ell_2^n(\ell_2))$. Hence, in the language of operator spaces, we can reformulate the "vectorial" Nehari Theorem as follows:

Theorem 9.1.1. *The correspondence $\psi \to \mathcal{H}[\psi]$ is a completely isometric embedding of L_∞/H_∞^0 into $B(\ell_2)$.*

This is a special case of Theorem 9.1.2, which is proved below.

Thus, Hankel matrices provide us with a very natural and concrete "realization" of the quotient L_∞/H_∞^0 as an operator space (the existence of which is guaranteed a priori by Ruan's Theorem). It would be interesting to have an analogous description in the general case for the quotient E_1/E_2 of two operator spaces with $E_2 \subset E_1$. Even in the particular case when $E_1 = L_\infty$ and E_2 is a translation invariant subspace, this does not seem to be known (see, however, Exercise 8.2).

We will prove in the following a generalization to Hankel matrices with entries in a von Neumann algebra. For that purpose, we need more notation. Let $M \subset B(H)$ be a von Neumann algebra on a Hilbert space H. For simplicity, we assume H separable.

We say that a function $\varphi\colon \mathbb{T} \to M$ is σ-measurable if for any x, y in H the function $\langle \varphi(\cdot)x, y \rangle$ is measurable. We will denote by $L_\infty(M)$ the tensor product (von Neumann sense) $L_\infty(\mathbb{T}) \overline{\otimes} M$. Equivalently, $L_\infty(M)$ is formed of all equivalence classes of bounded σ-measurable functions $\varphi\colon \mathbb{T} \to M$ equipped with the norm

$$\|\varphi\|_\infty = \operatorname*{ess\,sup}_{t \in \mathbb{T}} \|\varphi(t)\|_M.$$

Note that, for such a function φ, the Fourier transform $\widehat{\varphi}\colon \mathbb{Z} \to M$ can be defined by the formula:

$$\forall\, x, y \in H \qquad \langle \widehat{\varphi}(n)x, y \rangle = \int \langle \varphi(t)x, y \rangle e^{-int} dm(t).$$

We denote by $H_\infty^0(M)$ the subspace formed of all φ in $L_\infty(M)$ such that $\widehat{\varphi}(n) = 0\ \forall\, n \leq 0$.

For any φ in $L_\infty(M)$ let

$$M_\varphi\colon L_2(H) \to L_2(H)$$

be the operator of multiplication by the (operator-valued) function φ (taking $x \in L_2(H)$ to the function $t \to \varphi(t)x(t)$). Clearly

$$\|M_\varphi\| = \|\varphi\|_{L_\infty(M)}. \tag{9.1.1}$$

We will work with the subspaces $H_2(H) \subset L_2(H)$ and $H_2^0(H)^\perp = L_2(H) \ominus H_2^0(H)$.

The Hankel operator $\mathcal{H}[\varphi]$: $H_2(H) \to H_2^0(H)^\perp$ is defined by

$$\mathcal{H}[\varphi] = P_{H_2^0(H)^\perp} M_{\varphi|H_2(H)}.$$

The associated matrix is given by

$$\forall\, i, j \geq 0 \qquad (\mathcal{H}[\varphi])_{ij} = \widehat{\varphi}(-(i+j)).$$

Note that $\|\mathcal{H}[\varphi]\| \leq \|\varphi\|_\infty$. Here again $\mathcal{H}[\varphi] = 0$ if $\varphi \in H_0^\infty(M)$, so that $\varphi \to \mathcal{H}[\varphi]$ defines a contractive map from $L_\infty(M)/H_\infty^0(M)$ to $B(H_2(H), H_2^0(H)^\perp)$. Note that $L_\infty(M)/H_\infty^0(M)$ is the dual of $H_1(M_*)$.

Theorem 9.1.2. *The correspondence* $\varphi \to \mathcal{H}[\varphi]$ *defines a completely isometric embedding of* $L_\infty(M)/H_\infty^0(M)$ *into* $B(H_2(H), H_2^0(H)^\perp)$.

To prove this, we first analyze Toeplitz matrices indexed by $\mathbb{Z} \times \mathbb{Z}$, or indexed by $\mathbb{Z} \times \mathbb{Z}_+$. We denote here $\mathbb{Z}_+ = \{n \in \mathbb{Z} \mid n \geq 0\}$.

Proposition 9.1.3. *Consider* φ *in* $L_\infty(M)$.

(i) *Let* T_φ: $\ell_2(\mathbb{Z}; H) \to \ell_2(\mathbb{Z}; H)$ *be the operator defined by the matrix*

$$(T_\varphi)_{ij} = \widehat{\varphi}(i - j), \qquad (i, j \in \mathbb{Z}).$$

Then $\|T_\varphi\| = \|\varphi\|_\infty$.

(ii) *Let* τ_φ: $\ell_2(\mathbb{Z}_+; H) \to \ell_2(\mathbb{Z}; H)$ *be the operator defined by the matrix*

$$(\tau_\varphi)_{ij} = \widehat{\varphi}(i - j), \qquad (i \in \mathbb{Z},\ j \in \mathbb{Z}_+).$$

Then $\|\tau_\varphi\| = \|\varphi\|_\infty$. *(Note that* τ_φ *may be viewed as the restriction of* T_φ *to* $\ell_2(\mathbb{Z}_+; H)$.)

(iii) *Consider a function* b: $\mathbb{Z} \to M$. *Let* $\tau(b)$: $\ell_2(\mathbb{Z}_+; H) \to \ell_2(\mathbb{Z}; H)$ *be the operator defined by the matrix* $\tau(b)_{ij} = b(i - j), (i \in \mathbb{Z},\ j \in \mathbb{Z}_+)$. *Then* $\tau(b)$ *is bounded iff there is a (necessarily unique)* φ *in* $L_\infty(M)$ *such that* $\widehat{\varphi} = b$. *Moreover,* $\|\tau(b)\| = \|\varphi\|_\infty$.

Proof. (i) The Fourier transform is a unitary isomorphism from $L_2(H)$ to $\ell_2(\mathbb{Z}; H)$. Modulo this isomorphism, T_φ is the same as M_φ, so (i) follows from (9.1.1).

(ii) It suffices to show that $\|\varphi\|_\infty \leq \|\tau_\varphi\|$. We can write by (9.1.1) and by a density argument

$$\|\varphi\|_\infty = \|M_\varphi\| = \sup \left\{ \left| \int \langle \varphi(t)x(t), y(t) \rangle dm(t) \right| \right\},$$

where the supremum runs over all x, y in the unit ball of $L_2(H)$ such that $\widehat{x}$ and $\widehat{y}$ are finitely supported. In particular, we may assume $x = \sum_{j=-N}^{N} \widehat{x}(j) e^{ijt}$ with $\widehat{x}(j) \in H$. We then have

$$\langle M_\varphi x, y \rangle = \sum_{i,j \in \mathbb{Z}} \langle \widehat{\varphi}(i-j)\widehat{x}(j), \widehat{y}(i) \rangle = \sum_{i \in \mathbb{Z}, j \in \mathbb{Z}_+} \langle \widehat{\varphi}(i-j)\widehat{x}(j-N), \widehat{y}(i-N) \rangle;$$

hence

$$|\langle M_\varphi x, y \rangle| \le \|\tau_\varphi\| \|x\|_2 \|y\|_2,$$

which establishes (ii).

(iii) Assume first that there is φ in $L_\infty(M)$ with $\widehat{\varphi} = b$. Then $\|\tau(b)\| = \|\tau_\varphi\|$, and hence in this case (iii) follows from (ii).

In general, for any $0 < r < 1$, let $b_r \colon \mathbb{Z} \to M$ be defined by $b_r(n) = r^{|n|} b(n)$. Assume that $\tau(b)$ is bounded. Then clearly $\sup_n \|b(n)\| < \infty$, which implies that the series $\varphi_r = \sum_{n \in \mathbb{Z}} b_r(n) e^{int}$ is absolutely convergent. Hence by the first part we have $\|\varphi_r\|_\infty = \|\tau(b_r)\| \le \|\tau(b)\|$. Let φ be a weak-$*$ cluster point in $L_\infty(M)$ of $\{\varphi_r\}$ where $r \to 1$. It is easy to see that $\widehat{\varphi}(n) = b(n)$. ∎

Our main result will follow from the following beautiful lemma proved by several authors, including Parrott [Par1].

Lemma 9.1.4. *Consider a decomposition $K = K_1 \oplus K_2$ of a Hilbert space K. We view an operator on K as a matrix $\begin{pmatrix} a & b \\ c & d \end{pmatrix}$ with operator entries. Then*

$$\inf_{a \in B(K_1)} \left\| \begin{pmatrix} a & b \\ c & d \end{pmatrix} \right\| = \max \left\{ \left\| \begin{pmatrix} 0 & b \\ 0 & d \end{pmatrix} \right\|, \left\| \begin{pmatrix} 0 & 0 \\ c & d \end{pmatrix} \right\| \right\},$$

and this infimum is attained. Moreover, if $K_1 = K_2 = H$ and if b, c, d are all in a von Neumann algebra $M \subset B(H)$, then this infimum is attained on an element a of M.

Proof. It suffices to show that, if the right side is ≤ 1, then there is a so that $\left\| \begin{pmatrix} a & b \\ c & d \end{pmatrix} \right\| \le 1$. So assume the right side is ≤ 1. Equivalently, we have both $\|b^* b + d^* d\| \le 1$ and $\|cc^* + dd^*\| \le 1$ (see Remark 1.13), which implies $b^* b \le 1 - d^* d$ and $cc^* \le 1 - dd^*$. The latter ensure that we can find operators V, W with $\|V\| \le 1$, $\|W\| \le 1$ such that $b = V(1 - d^* d)^{1/2}$ and $c = (1 - dd^*)^{1/2} W$. We then set $a = -V d^* W$. Let

$$U = \begin{pmatrix} d & (1 - dd^*)^{1/2} \\ -(1 - d^* d)^{1/2} & d^* \end{pmatrix}.$$

It is easy to check that U is unitary and, moreover,

$$\begin{pmatrix} a & b \\ c & d \end{pmatrix} = \begin{pmatrix} 0 & -V \\ I & 0 \end{pmatrix} U \begin{pmatrix} 0 & I \\ W & 0 \end{pmatrix};$$

hence we conclude $\left\| \begin{pmatrix} a & b \\ c & d \end{pmatrix} \right\| \leq 1$. If b, c, d all belong to M, it is easy to see that, in the above argument, we can find V, W in M so that $a = -Vd^*W$ also lies in M.

∎

Theorem 9.1.5. *Let* $(a(n))_{n \geq 0}$ *be a sequence in* M. *Let* $\Gamma = (a(i+j))_{i,j \geq 0}$ *be the associated Hankel matrix. Then* $\|\Gamma\|_{B(\ell_2(H))} \leq 1$ *iff there is* φ *in* $L_\infty(M)$ *with* $\|\varphi\|_\infty \leq 1$ *such that*

$$\forall\, n \geq 0 \qquad a(n) = \widehat{\varphi}(-n).$$

Proof. If $a(n) = \widehat{\varphi}(-n)$ for all $n \geq 0$, then Γ is the matrix associated to $\mathcal{H}[\varphi]$; hence $\|\Gamma\| \leq \|\varphi\|_\infty$ and the "if" part is obvious.

Conversely, assume $\|\Gamma\| \leq 1$. We claim that we can find operators $a(-1)$, $a(-2), a(-3), \ldots$ in M such that the function $b \colon \mathbb{Z} \to M$ defined by $b(n) = a(-n)$ satisfies $\|\tau(b)\| \leq 1$.

We can first find $a(-1)$ in M so that the matrix

$$\begin{pmatrix} a(-1) & a(0) & a(1) & \cdots \\ a(0) & a(1) & \cdots & \cdots \\ a(1) & \cdots & \cdots & \cdots \end{pmatrix}$$

has norm ≤ 1. This is possible: Indeed, this matrix can be written as

$$\begin{pmatrix} a(-1) & b \\ c & d \end{pmatrix}$$

with $\left\| \begin{pmatrix} 0 & b \\ 0 & d \end{pmatrix} \right\| = \left\| \begin{pmatrix} 0 & 0 \\ c & d \end{pmatrix} \right\| = \|\Gamma\|$; hence the existence of $a(-1)$ follows from the preceding lemma, since we assume $\|\Gamma\| \leq 1$. Repeating the same argument, we can find $a(-2)$ in M so that

$$\begin{pmatrix} a(-2) & a(-1) & a(0) & \cdots \\ a(-1) & a(0) & \cdots & \cdots \\ a(0) & \cdots & \cdots & \cdots \end{pmatrix}$$

has norm ≤ 1. Continuing exactly in the same way, we find $a(-3), a(-4)$, and so on. This gives us an extension of a to the whole of $\mathbb{Z}$ such that, if we set $b(n) = a(-n)$, we have $\|\tau(b)\| \leq 1$. Then, by Proposition 9.1.3(iii), there is φ in $L_\infty(M)$ with $\|\varphi\|_\infty \leq 1$ such that $b = \widehat{\varphi}$. Hence $a(n) = \widehat{\varphi}(-n)$ for all $n \geq 0$.

∎

Proof of Theorem 9.1.2. By Theorem 9.1.5, for any ψ in $L_\infty(M)$ we have

$$\|\mathcal{H}[\psi]\| = \inf\{\|\varphi\|_\infty \mid \mathcal{H}[\varphi] = \mathcal{H}[\psi]\} = \inf\{\|\psi + h\|_\infty \mid h \in H^0_\infty(M)\}.$$

This shows that $\psi \to \mathcal{H}[\psi]$ defines an isometric embedding of $L_\infty(M)/H_\infty^0(M)$ into $B(H_2(H), H_2^0(H)^\perp)$. Replacing M by $M_n(M)$ $(n \geq 1)$, we obtain that this embedding is actually completely isometric. ∎

Corollary 9.1.6. Let $M_1 \subset B(H_1)$ and $M_2 \subset B(H_2)$ be two von Neumann algebras. Let $A = (A_{ij})$ and $B = (B_{ij})$ be Hankel matrices with entries, respectively, in M_1 and M_2. Assume that there is a bounded linear map $v \colon M_1 \to M_2$ with $\|v\| \leq 1$ such that $v(A_{ij}) = B_{ij}$ for all i, j. Then

$$\|B\|_{B(\ell_2(H_2))} \leq \|A\|_{B(\ell_2(H_1))}.$$

Proof. By the preceding result, for any bounded Hankel matrix (A_{ij}) with entries in M_1 there is φ in $L_\infty(M_1)$ with $\|\varphi\|_\infty = \|(A_{ij})\|$ such that $A_{ij} = \widehat{\varphi}(-i-j)$. We claim that

$$\|(v(\widehat{\varphi}(-i-j)))\| \leq \|(\widehat{\varphi}(-i-j))\|.$$

The following, slightly incorrect, argument is easier to understand on a first reading: Since $\|v \colon M_1 \to M_2\| \leq 1$, we have $\|I \otimes v \colon L_\infty(M_1) \to L_\infty(M_2)\| \leq 1$; hence, passing to the quotient v defines a contraction from $L_\infty(M_1)/H_\infty^0(M_1)$ to $L_\infty(M_2)/H_\infty^0(M_2)$, and the above claim now follows from the preceding result.

The incorrect point in the preceding argument lies in the fact that, since $t \to \varphi(t)$ is only σ-measurable, the σ-measurability of $t \to v(\varphi(t))$ is questionable when v is not weak-$*$ continuous, so that $I \otimes v$ is not a well-defined map from $L_\infty(M_1)$ to $L_\infty(M_2)$. However, this is easy to repair: Fix $0 < r < 1$. Then $\varphi_r(t) = \sum_{n \in \mathbb{Z}} r^{|n|} \widehat{\varphi}(n) e^{int}$ is strongly measurable, so the preceding objections do not apply and we have

$$\|(r^{i+j} v(A_{ij}))\| = \|(v(\widehat{\varphi}_r(-i-j)))\| \leq \|(\widehat{\varphi}_r(-i-j))\| \leq \|(A_{ij})\|.$$

Letting $r \uparrow 1$, we obtain the announced result $\|(v(A_{ij}))\| \leq \|(A_{ij})\|$. ∎

Remark. Actually Lemma 9.1.4 itself reflects a "concrete" realization of a quotient operator space, namely, M_2/S, where $S \subset M_2$ is the one-dimensional subspace spanned by e_{11}. Indeed, Lemma 9.1.4 implies that the mapping $M_2 \to M_2 \oplus M_2$ taking $\begin{pmatrix} a & b \\ c & d \end{pmatrix}$ to $\begin{pmatrix} 0 & b \\ 0 & d \end{pmatrix} \oplus \begin{pmatrix} 0 & 0 \\ c & d \end{pmatrix}$ defines (after passing to the quotient modulo its kernel) a completely isometric embedding of M_2/S into $M_2 \oplus M_2$. By a rearrangement of the basis, this also shows that the mapping

$$\begin{pmatrix} a & b \\ c & d \end{pmatrix} \to \begin{pmatrix} a & 0 \\ c & 0 \end{pmatrix} \oplus \begin{pmatrix} 0 & 0 \\ c & d \end{pmatrix}$$

defines a completely isometric embedding of $M_2/\mathbb{C}e_{12}$ into $M_2 \oplus M_2$. In that case, $\inf_b \left\| \begin{pmatrix} a & b \\ c & d \end{pmatrix} \right\|$ appears as the distance of the matrix $x = \begin{pmatrix} a & b \\ c & d \end{pmatrix}$ to the set of *strictly* upper triangular matrices (i.e., such that $a = c = d = 0$).

More generally, for any $n \geq 1$, let $T_n \subset M_n$ be the subspace of upper triangular matrices (a_{ij}), that is, such that $a_{ij} = 0$ for all $j < i$. Let P_k be the orthogonal projection onto $\mathrm{span}(e_1, \ldots, e_k)$. Then, for any matrix $x = (a_{ij})$, we have

$$\inf_{y \in T_n} \|x - y\| = \sup_{1 \leq k < n} \|(1 - P_k)xP_k\|.$$

Moreover, the mapping

$$x \to \bigoplus_{1 \leq k < n} (1 - P_k)xP_k$$

defines a completely isometric embedding of M_n/T_n into the $(n-1)$-fold direct sum $M_n \oplus \cdots \oplus M_n$. This can be proved by repeated applications of Lemma 9.1.3.

There is an analogous result for more general "nest algebras" (for which T_n is a prototypical case), called "Arveson's distance formula." See either [Da1] or [Pow] for more on all this.

Remark 9.1.7. Let us denote simply by C the subspace formed in L_∞ by the continuous functions, and let $A = H_\infty \cap C$ be the subspace of C formed by the boundary values of analytic functions. By the vectorial version of Hartman's Theorem (cf. [Ni]), the restriction to the continuous functions of the correspondence appearing in Theorem 9.1.1 is a complete isometry between the quotient operator space C/A and the space of all compact Hankel operators on ℓ_2. In particular, this implies that the operator space C/A is exact in the sense of §17 below. On the other hand (this remark is due to Junge), the results of [HP1, Theorem 2.2] suggest that the space L_∞/H_∞ might be exact, but this does not seem clear at the moment.

Until recently, we knew of no example of a quotient of a commutative C^*-algebra (by a closed subspace) that is not exact as an operator space, but an example of this kind has just been exhibited by N. Ozawa [Oz1]. An operator space that is (completely isometrically) a quotient of a minimal operator space is called a Q-space; equivalently, these are subspaces of quotients of commutative C^*-algebras. The class of Q-spaces seems to be quite interesting to study. See [Ri2] for an investigation of an analog of the Haagerup tensor product for Q-spaces.

9.2. Homogeneous operator spaces

We say that an operator space is homogeneous (resp. λ-homogeneous) if, for all $u\colon E \to E$, we have

$$\|u\|_{cb} = \|u\| \quad (\text{resp. } \|u\|_{cb} \le \lambda\|u\|).$$

In this case, every surjective isometry on E is a complete isometry. For example, it is easy to check that, for any Banach space E, the spaces $\min(E)$ and $\max(E)$ (defined in Chapter 3) are homogeneous.

Actually, the most interesting case seems to be the Hilbertian one: We will say that E is Hilbertian if it is isometric (as a normed space) to a Hilbert space. When E is only λ-isomorphic to a Hilbert space for some $\lambda > 1$, we will say that E is λ-Hilbertian.

The spaces R and C (introduced in (0.1) and (0.2)) as well as $\min(\ell_2)$, $\max(\ell_2)$, and OH are examples of homogeneous Hilbertian operator spaces. The homogeneous Hilbertian operator spaces reproduce the same operator space structure inside themselves; when a space E is a mixture of two distinct structures, for instance, $E = R \oplus C$, then it is not λ-homogeneous for any λ. More precisely, we have:

Proposition 9.2.1. Let $E \subset B(H)$ be a Hilbertian operator space. Then E is homogeneous iff, for any unitary operator $U\colon E \to E$ (here "unitary" refers to the Hilbert space structure of E), we have $\|U\|_{cb} \le 1$. In that case, whenever two subspaces $E_1 \subset E$ and $E_2 \subset E$ and have the same Hilbertian dimension, they are completely isometric.

Proof. The first assertion follows from the Russo-Dye Theorem: The unit ball of $B(E)$ is the closed convex hull of its unitaries ([Ped, p. 4]). Then the second assertion follows: Indeed, let $u\colon E_1 \to E_2$ be an isometric isomorphism. After comparing with the orthogonal projection onto E_1, we can extend u to $\widetilde{u}\colon E \to E$ such that $\|\widetilde{u}\| = 1$ and, if E is homogeneous, we have $\|\widetilde{u}\|_{cb} = \|\widetilde{u}\|$, whence $\|\widetilde{u}\|_{cb} \le 1$. Similarly (reversing the roles) $\|u^{-1}\|_{cb} \le 1$. Thus we conclude that u is completely isometric. ∎

Proposition 9.2.2. Every λ-homogeneous operator space is completely isomorphic to a homogeneous (i.e., 1-homogeneous) one $\widetilde{E}$ with $d_{cb}(E, \widetilde{E}) \le \lambda$.

Proof. Let β be the family of all the operators $u\colon E \to E$ with $\|u\| \le 1$. Let $B_u = B(H)$ for all u in β. Consider the direct sum $\oplus_{u\in\beta} B_u$ (equipped with the operator space structure defined in §2.6) and let $J\colon E \to \oplus_{u\in\beta} B_u$ be the isometric map defined by $J(x) = (u(x))_{u\in\beta}$. Let $\widetilde{E}$ be the range of J. Then J is an isometry from E onto $\widetilde{E}$. Moreover, $\|J\|_{cb} \le \lambda$ and $\|J^{-1}_{|\widetilde{E}}\|_{cb} \le 1$. Finally, by its very definition, the space $\widetilde{E}$ is homogeneous (i.e., 1-homogeneous). ∎

Remark. The consideration of a few simple examples like R, C, or OH gives the impression that a homogeneous Hilbertian operator space might be determined by its two-dimensional subspaces. But it is not so: Using [MP], C. Zhang [Z2] exhibited for each integer N a pair of homogeneous Hilbertian (infinite-dimensional) operator spaces that have the same (i.e. completely isometric) N-dimensional subspaces but which are not even completely isomorphic.

The examples that we will review in this chapter demonstrate the rich diversity of the homogeneous Hilbertian operator spaces appearing in the literature. Nevertheless, it seems reasonable to expect that a classification or a parametrization of this class of operator spaces will be available soon.

9.3. Fermions. Antisymmetric Fock space. Spin systems

Let I be any set. Consider a family of operators in $B(H)$ satisfying the following relations, called the canonical anticommutation relations (in short, CAR):

$$
\text{(CAR)} \quad \left\{ \begin{array}{ll} \forall i, j \in I & V_i V_j^* + V_j^* V_i = \delta_{ij} I \\ \forall i, j \in I & V_i V_j + V_j V_i = 0. \end{array} \right.
$$

Here, of course, we have set $\delta_{ij} = 1$ if $i = j$ and $\delta_{ij} = 0$ if $i \neq j$. The closed span in $B(H)$ of the family $[V_i \mid i \in I]$ gives us a very interesting example of an operator space. We will denote it by $\Phi(I)$. We set $\Phi = \Phi(\mathbb{N})$ and $\Phi_n = \Phi(\{1, 2, .., n\})$. Let us first justify the existence of such families. Let $\mathcal{H} = \ell_2(I)$ equipped with an orthonormal basis $(e_i)_{i \in I}$, let $H_0 = \mathbb{C}$, and let $H_n = \mathcal{H}^{\wedge n}$ (antisymmetric Hilbertian tensor product) for $n \geq 1$. We define the antisymmetric Fock space associated to $\mathcal{H}$ as follows:

$$
H = \oplus_{n \geq 0} H_n.
$$

For each h in $\mathcal{H}$, it is classical to denote by $c(h)$ (resp. $a(h)$) the so-called operator of "creation (resp. annihilation) of particle" defined by

$$
\forall x \in H \quad c(h)x = h \wedge x \quad (\text{resp. } a(h) = c(h)^*).
$$

Then we have the relations

$$
\forall h, k \in \mathcal{H} \quad a(h)a(k)^* + a(k)^* a(h) = \, <h, k> I \quad \text{et} \quad a(h)a(k) + a(k)a(h) = 0.
$$

If we let $V_i = a(e_i)$ (or if we set $V_i = c(e_i)$ for all i), we obtain a family with the announced properties. We refer the reader to [Gu2, Gu3, EvL, BR] for more information.

The next result is classical.

Theorem 9.3.1. *The operator space $\Phi(I)$ is (isometrically) Hilbertian and homogeneous. Up to a complete isometry, it only depends on the cardinality of I (and not on the particular family $(V_i)_{i \in I}$ chosen to define it).*

Proof. Let $(\alpha_i)_{i \in I}$ be a finitely supported family of scalars and let $T = \sum_{i \in I} \alpha_i V_i$. By the CAR we have

$$TT^* + T^*T = \sum |\alpha_i|^2 I.$$

Then

$$(T^*T)^2 = T^*(TT^* + T^*T)T = (\sum |\alpha_i|^2)T^*T.$$

Hence

$$\|T\|^4 = \|(T^*T)^2\| = \sum |\alpha_i|^2 \|T^*T\| = \sum |\alpha_i|^2 \|T\|^2,$$

whence

$$\|T\| = (\sum |\alpha_i|^2)^{1/2}. \tag{9.3.1}$$

Therefore, the space $\Phi(I) = \overline{\text{span}}[V_i \mid i \in I]$ is isometric to $\ell_2(I)$. Let H' be another Hilbert space and let $\{W_i \mid i \in I\}$ be a system in $B(H')$ satisfying the CAR. Let $\mathcal{A}$ (resp. $\mathcal{B}$) be the C^*-algebra with unit generated by $\{V_i \mid i \in I\}$ (resp. $\{W_i \mid i \in I\}$). By a classical argument (see [BR, p. 15]), one shows that there is a C^*-algebraic isomorphism $\pi \colon \mathcal{A} \to \mathcal{B}$ such that $\pi(V_i) = W_i$ for all i. A fortiori (since, by Proposition 1.5, π is completely isometric), the spaces $\overline{\text{span}}[V_i]$ and $\overline{\text{span}}[W_i]$ are completely isometric. This justifies the last assertion in Theorem 9.3.1.

Finally, let us show that $\Phi(I) = \overline{\text{span}}[V_i] \simeq \ell_2(I)$ is homogeneous: According to Proposition 9.2.1, it suffices to show that every unitary $U \colon \Phi(I) \to \Phi(I)$ is completely contractive. But, as shown by a simple computation, if U is unitary, then the operators $W_i = U(V_i)$ still satisfy the CAR; hence the mapping U extends to an isometric representation from $\mathcal{A}$ onto $\mathcal{B}$. A fortiori, U is a complete isometry and $\|U\|_{cb} = 1$.

Remark. We should mention that some variants of the space $\Phi(I)$ appear in the theory of Clifford algebras. On the other hand, a space very similar to $\Phi(I)$ also appears in the study of certain domains of holomorphy, called Cartan factors of type IV (cf. [Ha]). In the latter theory, the term "spin system" is used to designate a family of Hermitian and unitary operators $\{U_i \mid i \in I\}$ in $B(H)$ such that

$$\forall i \neq j \quad U_i U_j + U_j U_i = 0.$$

For instance, the classical Pauli matrices

$$\sigma_1 = \begin{pmatrix} 1 & 0 \\ 0 & -1 \end{pmatrix} \quad \sigma_2 = \begin{pmatrix} 0 & i \\ -i & 0 \end{pmatrix} \quad \sigma_3 = \begin{pmatrix} 0 & 1 \\ 1 & 0 \end{pmatrix}$$

form a spin system.

There is (essentially) a one-to-one correspondence between spin systems and systems satisfying the CAR. Indeed, if $(U'_j)_{j\in I}$ and $(U''_j)_{j\in I}$ are two systems such that their disjoint union forms a spin system, then the system $(V_j)_{j\in I}$, defined by $V_j = (U'_j + iU''_j)/2$, satisfies the CAR. In the converse direction, if $\{V_i \mid i \in I\}$ is a family satisfying the CAR, then the operators $U_i = V_i + V_i^*$ form a spin system. The subspace of $B(H)$ generated by a spin system $(U_i)_{i\in I}$ clearly is isomorphic to $\ell_2(I)$. To be more explicit, we must distinguish the real case from the complex one. We have

$$\forall(\alpha_i) \in \mathbb{R}^{(I)} \quad (\sum |\alpha_i|^2)^{1/2} = \|\sum \alpha_i U_i\|,$$

$$\forall(\alpha_i) \in \mathbb{C}^{(I)} \quad (\sum |\alpha_i|^2)^{1/2} \leq \|\sum \alpha_i U_i\| \leq 2^{1/2}(\sum |\alpha_i|^2)^{1/2}.$$

More precisely (see [Ha]), we have

$$\forall(\alpha_i) \in \mathbb{C}^{(I)} \quad \|\sum \alpha_i U_i\|^2 = \sum |\alpha_i|^2 + \left((\sum |\alpha_i|^2)^2 - |\sum \alpha_i^2|^2\right)^{1/2}.$$

Remarks. It is known that, if $I = \{1, ..., n\}$, the unital C^*-algebra generated by $\Phi(I) = \Phi_n$ is isomorphic to the algebra M_{2^n} of $2^n \times 2^n$ matrices (one can obtain an entirely explicit realization of Φ_n by using tensor products of the Pauli matrices). This remark shows that $\Phi(I)$ is an "exact" operator space with constant $= 1$ in the sense of §17. It is possible to show that, if I is infinite, for any completely isometric embedding $\Phi(I) \subset B(H)$, there is no c.b. projection from $B(H)$ onto $\Phi(I)$ (but there is a *bounded* one). Except for these few facts, there is very little available information on the operator space structure of $\Phi(I)$. It would be interesting to compute its dual $\Phi(I)^*$ as well as the iterated tensor products $\Phi(I) \otimes_{\min} ... \otimes_{\min} \Phi(I)$ (n times) in the style of [HP2] (note, however, that $\Phi(I)$ can be easily distinguished from all the examples reviewed in this chapter; see Chapter 10).

The field of fermionic analysis is a very active one at the moment. See, for example, [M] for the probabilistic viewpoint and also [BoS1, BoS2, CL, BCL, CK, PX]. It is only natural to expect that there will be further points of contacts with operator space theory.

9.4. The Cuntz algebra O_n

Let $s_1, ..., s_n \in B(H)$ be isometries such that

$$\sum_1^n s_i s_i^* = I.$$

The C^*-algebra generated by $s_1..., s_n$ is called the *Cuntz algebra* (cf. [Cu]) and is denoted by O_n. One can show that it does not depend on the particular choice of the sequence $\{s_i\}$ but only on n. It is tempting to study the operator space $E_n = \text{span}[s_1, .., s_n]$, but this is disappointing: A very simple calculation shows that for all $a_1, ..., a_n$ in $B(H)$, we have $\|\sum a_i \otimes s_i\| = \|\sum a_i^* a_i\|^{1/2}$. It follows that E_n is completely isometric to C_n. Similarly, $\text{span}[s_1^*, .., s_n^*]$ is completely isometric to R_n. Of course, similar remarks hold for the C^*-algebra O_∞ generated by a sequence of isometries (s_i) such that $\sum_1^\infty s_i s_i^* \leq I$ (this time the equality is not required for the unicity of O_∞).

Another very important family of isometries is provided by the creation operators on the full Fock space. To define this we need to introduce some notation.

Let H be a Hilbert space. We denote by $\mathcal{F}(H)$ (or simply by $\mathcal{F}$) the full Fock space associated to H, that is to say, we set $\mathcal{H}_0 = \mathbb{C}$, $\mathcal{H}_n = H^{\otimes n}$ (Hilbertian tensor product) and finally

$$\mathcal{F} = \oplus_{n \geq 0} \mathcal{H}_n.$$

We consider from now on $\mathcal{H}_n$ as a subspace of $\mathcal{F}$.

For every $h \in \mathcal{F}$, we denote by $\ell(h)$: $\mathcal{F} \to \mathcal{F}$ the operator defined by:

$$\ell(h)x = h \otimes x.$$

More precisely, if $x = \lambda 1 \in \mathcal{H}_0 = \mathbb{C}1$, we have $\ell(h)x = \lambda h$ and if $x = x_1 \otimes x_2 ... \otimes x_n \in \mathcal{H}_n$, we have $\ell(h)x = h \otimes x_1 \otimes x_2 ... \otimes x_n$. We will denote by Ω the unit element in $\mathcal{H}_0 = \mathbb{C}1$.

Now assume $H = \ell_2^n$ with its canonical basis (e_i). We denote $\mathcal{F}_n = \mathcal{F}(\ell_2^n)$ and we set

$$\ell_i = \ell(e_i) \qquad (i = 1, \ldots, n).$$

Then $\{\ell_i\}$ is a family of isometries on $\mathcal{F}_n$ with orthogonal ranges, and we have

$$\sum_1^n \ell_i \ell_i^* = I - P_\Omega,$$

where P_Ω is the orthogonal projection onto $\mathbb{C}\Omega$.

Now let $\{x_i\}$ be any n-tuple in $B(H)$ (H arbitrary). We will denote by $\mathcal{E}\{x_i\}$ the closed linear span of I and all the products of the form $x_{i_1} x_{i_2} \ldots x_{i_p} x_{j_1}^* \ldots x_{j_q}^*$ ($p \geq 0, q \geq 0$). In particular, taking $q = 0$, we see that

$$\text{span}[I, x_i] \subset \mathcal{E}\{x_i\}.$$

The space $\mathcal{E}\{x_i\}$ can also be viewed as the span of all products $P\{x_i\}Q\{x_i^*\}$, where $P\{X_i\}$ and $Q\{X_i\}$ are polynomials in noncommutative variables $\{X_i\}$. Note that our polynomials are allowed to have a "constant term" equal to a multiple of the unit in the free unital semi-group generated by $\{X_1, \ldots, X_n\}$. In a series of papers ([Pu1–3]) G. Popescu studied various extensions of von Neumann's inequality for n-tuples of operators; in particular, he proved the following.

Theorem 9.4.1. *([Pu3]) Assume* $\|\sum_1^n x_i x_i^*\| \leq 1$. *For any finite family* $P_k(x_1, \ldots, x_n)$, $Q_k(x_1, \ldots, x_n)$ *of polynomials in noncommuting variables* $\{X_1, \ldots, X_n\}$ *we have*

$$\left\| \sum_k P_k(x_1, \ldots, x_n) Q_k(x_1^*, \ldots, x_n^*) \right\| \leq \left\| \sum_k P_k(\ell_1, \ldots, \ell_n) Q_k(\ell_1^*, \ldots, \ell_n^*) \right\|.$$

More generally, the linear mapping

$$F \colon \mathcal{E}\{\ell_i\} \longrightarrow \mathcal{E}\{x_i\}$$

defined by

$$F(\ell_{i(1)} \ldots \ell_{i(p)} \ell_{j(1)}^* \ldots \ell_{j(q)}^*) = x_{i(1)} \ldots x_{i(p)} x_{j(1)}^* \ldots x_{j(q)}^*,$$

is completely contractive.

This shows that $\{\ell_1, \ldots, \ell_n\}$ is in some sense "maximal" among all the n-tuples $\{x_i\}$ with $\|\sum x_i x_i^*\| \leq 1$.

Remark. In [Pu3], Popescu also observed that for "analytic" polynomials, the Cuntz isometries $\{s_i\}$ can be substituted to $\{\ell_i\}$ in the preceding statement. Namely, for any polynomial P we have

$$\|P(\{\ell_i\})\| = \|P(\{s_i\})\|.$$

Moreover, the correspondence $\ell_i \to s_i$ extends to a completely isometric unital homomorphism on the (nonself-adjoint) algebra generated by $\{\ell_i\}$.

Corollary 9.4.2. *We have completely isometrically*

$$\mathrm{span}[I, \ell_1, \ldots, \ell_n] \simeq \mathbb{C} \oplus_1 C_n \quad \text{and} \quad \mathrm{span}[I, \ell_1^*, \ldots, \ell_n^*] \simeq \mathbb{C} \oplus_1 R_n,$$

and similarly with $\{s_i\}$ *substituted to* $\{\ell_i\}$.

Proof. Let us denote by e_0 the unit in $\mathbb{C}$. We will show that the correspondence $u \colon \mathbb{C} \oplus_1 C_n \longrightarrow \mathrm{span}[I, \ell_1, \ldots, \ell_n]$ defined by $u(e_0) = I$ and $u(e_{i1}) = \ell_i$ is completely isometric. Note that, by Remark 1.13, the restriction of u to C_n is completely contractive. Hence, by the definition of $\oplus_1$, we have $\|u\|_{cb} \leq 1$.

To show that $\|u^{-1}\|_{cb} \leq 1$, it suffices to establish the following claim: For any family of matrices $\{a_0, a_1, \ldots, a_n\}$ in M_N we have

$$\left\| a_0 \otimes e_0 + \sum a_i \otimes e_{i1} \right\| \leq \left\| a_0 \otimes I + \sum a_i \otimes \ell_i \right\|, \tag{9.4.1}$$

where the norm is the norm in $M_N(\mathbb{C} \oplus_1 C_n)$ and $M_N(B(\mathcal{F}_n))$. By Theorem 8.19, the left side of (9.4.1) is equal to the supremum of

$$\left\| a_0 \otimes I + \sum a_i \otimes v(e_{i1}) \right\|,$$

where $v: C_n \to B(H)$ runs over all possible maps with $\|v\|_{cb} \leq 1$. But, if we let $x_i = v(e_{i1})$, we clearly have (see Remark 1.13) $\|\sum x_i x_i^*\|^{1/2} \leq \|v\|_{cb} \leq 1$. Hence we find

$$\left\| a_0 \otimes e_0 + \sum a_i \otimes e_{i1} \right\| = \sup \left\| a_0 \otimes I + \sum a_i \otimes x_i \right\|,$$

and since $\|F\|_{cb} = 1$, we obtain

$$\leq \left\| a_0 \otimes I + \sum a_i \otimes \ell_i \right\|.$$

Thus we conclude $\|u^{-1}\|_{cb} \leq 1$ and the last assertion concerning $\{s_i\}$ follows from the preceding remark. ∎

Remark. It would be interesting to study more generally the relation between an operator space and the C^*-algebra that it generates. For some information in this direction, see [KiV, Z1, KiW, P11].

9.5. The operator space structure of the classical L_p-spaces

In this section, we describe the operator space structure of the usual (i.e. commutative) L_p-spaces. We will see that these spaces can be equipped with a specific structure, which we call their *natural* structure. We refer to [P2] for more details.

Let $(\Omega, \mathcal{A}, \mu)$ be a measure space. We will denote by $L_p(\mu)$ the associated L_p-space of complex-valued functions. Given a complex Banach space X, we will denote by $L_p(\mu; X)$ the L_p-space of X-valued measurable functions in Bochner's sense. It is well known that

$$L_p(\mu) = (L_\infty(\mu), L_1(\mu))_\theta \quad \text{and} \quad L_p(\mu; X) = (L_\infty(\mu; X), L_1(\mu; X))_\theta$$

with $\theta = 1/p$. A priori, this formula is an *isometric* identity that is only valid in the category of Banach spaces, but the complex interpolation of operator spaces will allow us to extend this formula to the operator space setting.

The case $p = \infty$. First, if A is a C^*-algebra, it possesses a privileged operator space structure associated to any realization of A as a C^*-subalgebra of $B(H)$. Indeed, by Proposition 1.5, if two C^*-algebras are isomorphic (as C^*-algebras), then they are completely isometric, and hence they are identical as operator spaces. We will say that this particular structure is the *natural* operator space structure on a C^*-algebra.

In particular, if $p = \infty$, we have a natural operator space structure on the space $L_\infty(\mu)$ or on the space $C(T)$ of all continuous functions on a compact set T. It is easy to check that this natural structure is determined by the isometric identity

$$C(T) \otimes_{\min} \mathcal{K} = C(T; \mathcal{K}),$$

where the space on the right-hand side is the space of $\mathcal{K}$-valued continuous functions equipped with its usual norm. This space coincides with the injective tensor product (in Grothendieck's sense) of $C(T)$ and $\mathcal{K}$. In other words, the natural operator space structure on $E = L_\infty(\mu)$ or $E = C(T)$ makes it identical to the space $\min(E)$, as defined in Chapter 3.

The case $p = 1$. If $p = 1$, again the choice is clear: The natural structure is defined as the one induced on $L_1(\mu)$ by the dual space $L_\infty(\mu)^*$ equipped with its dual operator space structure. Explicitly, this means that the norm of $M_n(L_1(\mu))$ is by definition the norm induced by $CB(L_\infty(\mu), M_n)$. (Note that $M_n(L_1(\mu))$ can be identified with the $\sigma(L_\infty(\mu), L_1(\mu))$-continuous linear maps from $L_\infty(\mu)$ to M_n.) For the resulting operator space structure on $L_1(\mu)$, we have $L_1(\mu)^* = L_\infty(\mu)$ completely isometrically.

Proposition 9.5.1. *The operator space obtained by equipping $L_1(\mu)$ with its natural o.s.s. (as defined above) coincides with $\max(L_1(\mu))$.*

Proof. Indeed, let $B = L_1(\mu)$ and let $E = \max(B)$. By (2.5.1) (or Corollary 5.11), the inclusion $E \to E^{**}$ is a complete isometry, and, by Exercise 3.2, we have $E^{**} = (\max(B))^{**} = \max(B^{**}) = (\min(B^*))^*$. Therefore the o.s.s. induced by $\max(B^{**})$ on B coincides with that of $\max(B)$. In the present case, we have $\max(B^{**}) = (\min(B^*))^* = L_\infty(\mu)^*$, so we conclude that the operator space structure of $\max(L_1(\mu))$ is the same as that induced by $L_\infty(\mu)^*$, where $L_\infty(\mu)$ is equipped with its natural (i.e. minimal) operator space structure. ∎

In particular, in the case $\Omega = \mathbb{N}$, we have a natural o.s.s. on ℓ_1. It is not hard to verify that this natural o.s.s. on ℓ_1 also coincides with the one obtained by considering ℓ_1 as the dual of c_0. We can describe the associated norm $\|\ \|_{\min}$ on $\mathcal{K} \otimes \ell_1$ in the following manner: Let (e_n) be the canonical basis of ℓ_1. For any finite sequence (a_n) in $\mathcal{K}$ (or in $B(\ell_2)$) we have

$$\left\| \sum a_n \otimes e_n \right\|_{\mathcal{K} \otimes_{\min} \ell_1} = \inf\{ \| \sum b_n b_n^* \|^{1/2} \| \sum c_n^* c_n \|^{1/2} \}, \qquad (9.5.1)$$

where the infimum runs over all possible decompositions $a_n = b_n c_n$ in $\mathcal{K}$ (resp. $B(\ell_2)$). We will give two more descriptions of the same structure, in (9.5.3) for $p = 1$ and in (9.6.1).

Analogously, we can describe the natural o.s.s. of $L_1(\mu)$ as follows. Let $f \in \mathcal{K} \otimes_{\min} L_1(\mu)$. We may consider f as a $\mathcal{K}$-valued function on Ω. We then have

$$\|f\|_{\mathcal{K} \otimes_{\min} L_1(\mu)} = \inf \left\{ \left\| \int g(t)g(t)^* d\mu(t) \right\|^{1/2} \left\| \int h(t)^* h(t) d\mu(t) \right\|^{1/2} \right\},$$
(9.5.2)

where the infimum runs over all possible decompositions of f as a product of (measurable) $\mathcal{K}$-valued functions.

The formula (9.5.1) (resp. (9.5.2)) is the *quantum* version of the fact that the unit ball of ℓ_1 (resp. $L_1(\mu)$) coincides with the set of all products of two elements in the unit ball of ℓ_2 (resp. $L_2(\mu)$). These two formulas (9.5.1) and (9.5.2) can be deduced from the fundamental factorization Theorem 1.6.

The case $1 < p < \infty$. For the general case, we use interpolation. Consider the operator space structure on $L_p(\mu)$ corresponding to $(L_\infty(\mu), L_1(\mu))_\theta$ as defined in §2.7. By definition, the norm on $M_n(L_p(\mu))$ is the one induced by the space $(M_n(L_\infty(\mu)), M_n(L_1(\mu)))_\theta$ with $\theta = 1/p$. Once again, we will say that this is the *natural* operator space structure on $L_p(\mu)$. We can describe more explicitly this structure using the Schatten classes S_p already introduced (see (7.19)). Indeed, let $f \in \mathcal{K} \otimes L_p(\mu)$ (viewed as a $\mathcal{K}$-valued measurable function). Then we have

$$\|f\|_{\mathcal{K} \otimes_{\min} L_p(\mu)} = \sup\{\|a \, f \, b\|_{L_p(\mu; S_p)}\},$$
(9.5.3)

where the supremum runs over all a, b in the unit ball of S_{2p}.

Since $L_p(\mu)$ has been equipped with an o.s.s., we may now unambiguously discuss completely bounded maps $u \colon L_p(\mu) \to L_p(\mu)$. These turn out to be easy to describe.

Proposition 9.5.2. *A linear map $u \colon L_p(\mu) \to L_p(\mu)$ is completely bounded (on $L_p(\mu)$ equipped with its natural o.s.s.) iff the mapping $u \otimes I_{S_p}$ is bounded on $L_p(\mu; S_p)$. Moreover, we have*

$$\|u\|_{CB(L_p(\mu), L_p(\mu))} = \|u \otimes I_{S_p}\|_{L_p(\mu; S_p) \to L_p(\mu; S_p)}.$$

It has been known for a long time that the Hilbert transform is bounded on the (vector-valued) L_p-space of S_p-valued functions for any $1 < p < \infty$ (see also in [Bou1] the analogous result for martingale transforms). This implies that, if $1 < p < \infty$, the Hilbert transform is completely bounded on L_p (on the torus or on $\mathbb{R}$). The same is valid for the Riesz transforms and their Gaussian analogs on the n-dimensional torus or on $\mathbb{R}^n$; see, for example, the discussion in [P15].

It is then tempting to compare boundedness and complete boundedness for maps on L_p. By Proposition 9.5.2, it is easy to check that, in the cases $p = 1, 2$ or ∞, every bounded map on L_p is automatically completely bounded.

However, if $1 < p \neq 2 < \infty$, there are examples of Fourier multipliers on L_p of the torus that are not c.b. Such examples (implicit in [P4, pp. 110–113]) can be obtained easily by observing that the norm in S_p is invariant under all transformations $(a_{ij}) \to (\varepsilon_i' a_{ij} \varepsilon_j'')$ for any choice of signs $\varepsilon_i' = \pm 1$, $\varepsilon_j'' = \pm 1$, but that there exist, if $1 < p \neq 2 < \infty$, mappings that are unbounded on S_p and of the form $(a_{ij}) \to (\varepsilon_{ij} a_{ij})$ with $\varepsilon_{ij} = \pm 1$ for $i \leq j$ and (say) $\varepsilon_{ij} = 0$ otherwise.

Fix (ε_{ij}) such that the latter mapping is unbounded on S_p. Then let $u \colon L_p \to L_p$ be the Fourier multiplier defined as follows: If n is of the form $n = 3^k + 3^l$ with $k \leq l$, then $u(e^{int}) = \varepsilon_{kl} e^{int}$ and $u(e^{int}) = 0$ otherwise. By a classical variant of the Khintchine inequalities (cf., e.g., [LR, p. 65]), it is known that u is bounded on L_p, but by the choice of (ε_{ij}) (and by the invariance property of the S_p-norm recalled above) u is unbounded on $L_p(S_p)$, and hence it is not c.b. Similarly, it follows from [HP1] that there are Fourier multipliers bounded on H^1 (of the torus) but not c.b.

The preceding definitions can be extended in the case of noncommutative L_p-spaces. Consider, for instance, the Schatten class S_p. It is well known that, if $\theta = 1/p$, we have an isometric identity

$$S_p = (\mathcal{K}, S_1)_\theta. \tag{9.5.4}$$

Here, it would be more appropriate to denote by S_∞ the space $\mathcal{K}$!

The space $\mathcal{K}$ can be equipped with its natural o.s.s. as a C^*-algebra and S_1 with the dual o.s.s. associated to the identity $S_1 = \mathcal{K}^*$, relative to the duality defined for $x \in S_1, y \in \mathcal{K}$ by $\langle x, y \rangle = \text{tr}(^t xy)$. Then, the formulas (2.7.3) and (9.5.4) allow us to define an o.s. structure (again called *natural*) on S_p. By Theorem 7.10, the space S_2 is then completely isometric to $OH(\mathbb{N} \times \mathbb{N})$ or, equivalently, to OH (see Remark 7.13 and the discussion preceding it).

As we already mentioned, given an operator space $E \subset B(H)$, exactly the same idea leads to the definition of a natural o.s.s. on the E-valued L_p-space that remains valid in the noncommutative case.

When $p = 1$, we define $S_1[E]$ as the "projective operator space tensor product" of S_1 with E, introduced in §2.8 and denoted by $S_1 \otimes^\wedge E$. By [ER8], we have

$$S_1[E^*] \simeq (\mathcal{K} \otimes_{\min} E)^* \tag{9.5.5}$$

completely isometrically. When $1 < p < \infty$ we define

$$S_p[E] = (\mathcal{K} \otimes_{\min} E, S_1[E])_\theta, \tag{9.5.6}$$

where $\theta = 1/p$. Then the following result holds (cf. [P2, Lemma 1.7])

Proposition 9.5.3. *Fix p with $1 \leq p < \infty$. Let $u \colon E \to F$ be a linear map between two operator spaces. Then u is c.b. (resp. completely isometric) iff* the map $I_{S_p} \otimes u$ defines a bounded (resp. an isometric) linear map from $S_p[E]$ to $S_p[F]$. Moreover, we have

$$\|u\|_{cb} = \|I_{S_p} \otimes u\|_{S_p[E] \to S_p[F]}.$$

Furthermore (see [P2, Corollary 1.4], if $1 \leq p_0, p_1 \leq \infty$ and $\frac{1}{p_\theta} = \frac{1-\theta}{p_0} + \frac{\theta}{p_1}$, then for any compatible couple of operator spaces (E_0, E_1) we have completely isometrically

$$S_{p_\theta}[(E_0, E_1)_\theta] = (S_{p_0}[E_0], S_{p_1}[E_1])_\theta. \tag{9.5.7}$$

We refer the reader to [P2] for more details.

9.6. The C^*-algebra of the free group with n generators

Let F_n (resp. F_∞) be the free group with n (resp. countably many) generators, and let $\{g_1, g_2, ...\}$) be the generators. Let $\pi \colon F_\infty \to B(H)$ be a unitary representation of the free group. We will see that, in several instances, the operator space $E(\pi)$ spanned in $B(H)$ by $\{\pi(g_i) \mid i = 1, 2, ...\}$ has interesting properties.

We will first illustrate this with the universal representation $U \colon F_\infty \to B(\mathcal{H})$, which generates the full C^*-algebra $C^*(F_\infty)$, as introduced in Chapter 8. The other classical choice for π is the left regular representation λ, which generates the "reduced" C^*-algebra $C^*_\lambda(F_\infty)$. This case is discussed in the next section, §9.7.

We let

$$E_U^n = \text{span}[U(g_i) \mid i = 1, 2, .., n]$$

and

$$E_U = \overline{\text{span}}[U(g_i) \mid i \geq 1].$$

By Lemma 8.9, for any finite sequence (a_i) in $B(\ell_2)$, we have

$$\| \sum a_i \otimes U(g_i) \|_{\min} = \sup\{\| \sum a_i \otimes u_i \|_{\min}\}, \tag{9.6.1}$$

where the supremum runs over all sequences (u_i) of unitary operators in $B(H)$ and over all possible Hilbert spaces H. Actually, the supremum remains unchanged if we restrict ourselves to $H = \ell_2$ or to H finite-dimensional with arbitrary dimension. But then formula (2.11.2) shows that, if we denote by (e_i^*) the dual basis to the canonical basis of c_0 (equipped with its natural o.s.s.), we also have

$$\| \sum a_i \otimes U(g_i) \|_{\min} = \| \sum a_i \otimes e_i^* \|_{B(\ell_2) \otimes_{\min} c_0^*}.$$

Therefore, the mapping $u \colon c_0^* \to E_U$ that takes e_i^* to $U(g_i)$ is a complete isometry. Hence, we have proved:

Theorem 9.6.1. *The operator space E_U spanned by the generators in $C^*(F_\infty)$ is completely isometric to ℓ_1 equipped with its natural operator space structure (or, equivalently, its o.s.s. as the dual of c_0). Similarly, E_U^n is completely isometric to ℓ_1^n.*

The formula (9.6.1) can be viewed as the "quantum" analog of the classical formula

$$\forall (\lambda_i) \in \mathbb{C}^{(I)} \quad \|(\lambda_i)\|_{\ell_1} = \sum |\lambda_i| = \sup\{|\sum \lambda_i z_i| \mid z_i \in \mathbb{C}, \ |z_i| = 1\}.$$
(9.6.2)

The space E_U gives us a concrete realization of the space ℓ_1 as an operator space. More generally, for any measure space (Ω, μ), one can describe the natural operator space structure of $L_1(\Omega, \mu)$ (induced by $L_\infty(\Omega, \mu)^*$) as follows. For all f in $L_1(\Omega, \mu) \otimes B(\ell_2)$, we have

$$\|f\|_{L_1(\Omega,\mu) \otimes_{\min} B(\ell_2)} = \sup \left\{ \left\| \int f(\omega) \otimes g(\omega) d\mu(\omega) \right\|_{B(\ell_2) \otimes_{\min} B(\ell_2)} \right\}, \quad (9.6.3)$$

where the supremum runs over all functions g in the unit ball of the space of L_∞-functions with values in $B(\ell_2)$.

9.7. Reduced C^*-algebra of the free group with n generators

Let G be a discrete group. We have defined in Chapter 8 the left regular representation $\lambda \colon G \to B(\ell_2(G))$. Recall that $\lambda(t)$ is the unitary operator of left translation by t on $\ell_2(G)$. We denote by $C_\lambda^*(G)$ the C^*-algebra generated in $B(\ell_2(G))$ by $\{\lambda(t) \mid t \in G\}$ or, equivalently, $C_\lambda^*(G) = \overline{\text{span}}\{\lambda(t) \mid t \in G\}$. Clearly, we have a C^*-algebra morphism

$$Q \colon C^*(G) \to C_\lambda^*(G)$$

that takes $U(t)$ to $\lambda(t)$. By elementary properties of C^*-algebras, it is onto and we have

$$C_\lambda^*(G) \simeq C^*(G)/\ker(Q).$$

In general, $\ker(Q) \neq \{0\}$, but one can show that $C_\lambda^*(G) = C^*(G)$ (i.e., $\ker(Q) = \{0\}$) iff G is amenable. The free groups are typical examples of nonamenable groups. The fact that the algebras $C_\lambda^*(G)$ and $C^*(G)$ are distinct in this case is manifestly visible on the generators. Indeed, if we let

$$E_\lambda^n = \text{span}[\lambda(g_i) \mid i = 1, .., n]$$

$$E_\lambda = \overline{\text{span}}[\lambda(g_i) \mid i \geq 1],$$

we can see that E_λ is a very different space from its analog in the full C^*-algebra, namely, the space E_U studied in §9.6. Indeed, as Banach spaces, we have $E_U \simeq \ell_1$ and $E_\lambda \simeq \ell_2$. The first isomorphism is elementary (see Theorem 9.6.1), while the second one is due to Leinert [Le]. Using Haagerup's ideas from [H2], one can describe the operator space structure of E_λ as follows (see [HP2] for more details).

Consider the space $B(\ell_2) \oplus B(\ell_2)$, equipped with the norm $\|(x \oplus y)\| = \max\{\|x\|, \|y\|\}$. In the subspace $R \oplus C \subset B(\ell_2) \oplus B(\ell_2)$, we consider the vectors δ_i ($i = 1, 2, ...$) defined by setting

$$\delta_i = e_{1i} \oplus e_{i1}.$$

We will denote by $R \cap C$ the closed subspace spanned in $R \oplus C$ by the sequence $\{\delta_i\}$.

This notation is compatible with the notion of "intersection" defined in §2.7, provided we view the pair (R, C) as a compatible pair using the transposition mapping $x \to {}^t x$ as a way to embed R into C. This means that we let $\mathcal{X} = C$, we use $x \to {}^t x$ to inject R into $\mathcal{X}$, and we use the identity map of C to inject C into $\mathcal{X}$.

Similarly, we will denote by $R_n \cap C_n$ the subspace of $R \cap C$ spanned by $\{\delta_i \mid i = 1, 2, ..., n\}$. It is easy to verify that, for any Hilbert space H and for any finite sequence (a_i) in $B(H)$, we have

$$\|\sum a_i \otimes \delta_i\|_{\min} = \max\{\|\sum a_i^* a_i\|^{1/2}, \|\sum a_i a_i^*\|^{1/2}\}.$$

Then, we can state (see [HP2]):

Theorem 9.7.1. *The space E_λ is completely isomorphic to $R \cap C$. More precisely, for any finite sequence (a_i) in $B(H)$, we have*

$$\|\sum a_i \otimes \delta_i\|_{\min} \le \|\sum a_i \otimes \lambda(g_i)\|_{\min} \le 2\|\sum a_i \otimes \delta_i\|_{\min}, \qquad (9.7.1)$$

so that the mapping $u\colon R \cap C \to E_\lambda$ defined by $u(\delta_i) = \lambda(g_i)$ is a complete isomorphism satisfying $\|u\|_{cb} = 2$ and $\|u^{-1}\|_{cb} = 1$. Moreover, the map $P\colon C_\lambda^*(F_\infty) \to E_\lambda$, defined by $P\lambda(t) = \lambda(t)$ if t is a generator and $P\lambda(t) = 0$ otherwise, is a c.b. projection from $C_\lambda^*(F_\infty)$ onto E_λ with norm $\|P\|_{cb} \le 2$. (Similar results hold for E_λ^n and $R_n \cap C_n$.)

Proof. Let $G = F_\infty$. Let $C_i^+ \subset G$ (resp. $C_i^- \subset G$) be the subset formed by all the reduced words that start by g_i (resp. g_i^{-1}). Note: Except for the empty word e, every element of G can be written as a reduced word in the generators admitting a well-defined "first" and "last" letter (where we read from left to right). Let P_i^+ (resp. P_i^-) be the orthogonal projection on $\ell_2(G)$

with range $\overline{\text{span}}[\delta_t \mid t \in C_i^+]$ (resp. $\overline{\text{span}}[(\delta_t \mid t \in C_i^-])$. Then it is easy to check that

$$\begin{aligned}
\lambda(g_i) &= \lambda(g_i)P_i^- + \lambda(g_i)(1 - P_i^-) \\
&= \lambda(g_i)P_i^- + P_i^+\lambda(g_i)(1 - P_i^-), \\
&= \lambda(g_i)P_i^- + P_i^+\lambda(g_i),
\end{aligned}$$

so that $\lambda(g_i) = x_i + y_i$ with (note $x_i^* x_i = P_i^-$ and $y_i y_i^* = P_i^+$)

$$\left\|\sum x_i^* x_i\right\| \le 1 \quad \text{and} \quad \left\|\sum y_i y_i^*\right\| \le 1.$$

Therefore, for any finite sequence (a_i) in $B(H)$ we have by (1.11) (note $a_i \otimes x_i = (a_i \otimes 1)(1 \otimes x_i)$ and similarly for $a_i \otimes y_i$)

$$\begin{aligned}
\left\|\sum a_i \otimes \lambda(g_i)\right\| &\le \left\|\sum a_i \otimes x_i\right\| + \left\|\sum a_i \otimes y_i\right\| \\
&\le \left\|\sum a_i a_i^*\right\|^{1/2} + \left\|\sum a_i^* a_i\right\|^{1/2} \\
&\le 2\left\|\sum a_i \otimes \delta_i\right\|.
\end{aligned}$$

The converse follows from a more general inequality valid for any discrete group G: For any finitely supported function $a: G \to B(H)$ we have

$$\max\left\{\left\|\sum a(t)^* a(t)\right\|^{1/2}, \left\|\sum a(t)a(t)^*\right\|^{1/2}\right\} \le \left\|\sum a(t) \otimes \lambda(t)\right\|_{\min}. \tag{9.7.2}$$

To check this, let $T = \sum a(t) \otimes \lambda(t)$. For any h in B_H we have $T(h \otimes \delta_e) = \sum a(t)h \otimes \delta_t$. Hence $\|T(h \otimes \delta_e)\| = \left(\sum_t \|a(t)h\|^2\right)^{1/2}$, and hence

$$\left\|\sum a(t)^* a(t)\right\|^{1/2} = \sup_{h \in B_H} \left(\sum \|a(t)h\|^2\right)^{1/2} \le \|T\|.$$

Similarly, since $T^* = \sum a(t^{-1})^* \otimes \lambda(t)$, we find

$$\left\|\sum a(t)a(t)^*\right\|^{1/2} \le \|T^*\| = \|T\|,$$

and we obtain (9.7.2).

If we now return to the case $G = \mathbb{F}_\infty$, we find that (9.7.2) implies the left side of (9.7.1). Moreover, if P is as in Theorem 9.7.1, we have

$$(I \otimes P)\left(\sum a(t) \otimes \lambda(t)\right) = \sum a(g_n) \otimes \lambda(g_n).$$

Hence, if $T = \sum a(t) \otimes \lambda(t)$, we have

$$\begin{aligned}
\|(I \otimes P)(T)\| &\le 2\max\left\{\left\|\sum a(g_n)^* a(g_n)\right\|^{1/2}, \left\|\sum a(g_n)a(g_n)^*\right\|^{1/2}\right\} \\
&\le 2\max\left\{\left\|\sum a(t)^* a(t)\right\|^{1/2}, \left\|\sum a(t)a(t)^*\right\|^{1/2}\right\},
\end{aligned}$$

and by (9.7.2) this is $\leq 2\|T\|$. Thus we conclude, as announced, that $\|P\|_{cb} \leq 2$. ∎

We will now describe the operator space generated by the free unitary generators $\{\lambda(g_i) \mid i = 1, 2, \ldots,\}$ in the noncommutative L_p-space $(1 \leq p < \infty)$ of the free group F_∞. Let us denote by τ the standard trace on $C_\lambda^*(F_\infty)$ defined by $\tau(\lambda(f)) = f(e)$. Let $L_p(\tau)$ denote for $1 \leq p < \infty$ the associated noncommutative L_p-space. The space $L_1(\tau)$ is the predual of the von Neumann algebra generated by $\lambda(F_\infty)$, which we will denote by $L_\infty(\tau)$. When $1 < p < \infty$ and $\theta = 1/p$, the space $L_p(\tau)$ can be identified with the complex interpolation space $(C_\lambda^*(F_\infty), L_1(\tau))_\theta$ or, equivalently, with $(L_\infty(\tau), L_1(\tau))_\theta$. Using §2.7, we may view $L_p(\tau)$ as an operator space. Let us denote by E_p the closed subspace of $L_p(\tau)$ generated by the free generators $\{\lambda(g_i) \mid i = 1, 2, \ldots\}$ (note that $E_\infty = E_\lambda$). We may view E_p as an operator space with the o.s.s. induced by $L_p(\tau)$. Clearly, the orthogonal projection P from $L_2(\tau)$ to E_2 is completely contractive (since, by Theorem 7.10, $L_2(\tau)$ can be identified with $OH(I)$ for a suitable set I). On the other hand, by Theorem 9.7.1, that same projection P is completely bounded from $C_\lambda^*(F_\infty)$ onto E_λ. Actually, the proof of Theorem 9.7.1 shows that P extends to a weak$-*$ continuous projection from $L_\infty(\tau)$ onto E_λ. By transposition, P also defines a c.b. projection from $L_1(\tau)$ onto E_1. Therefore, by interpolation, P defines a completely bounded projection from $L_p(\tau)$ onto E_p for any $1 < p < \infty$.

Moreover, by Proposition 2.7.6, the existence of this simultaneous c.b. projection ensures that the space E_p can be identified completely isomorphically with $(E_\lambda, E_1)_\theta$ with $\theta = 1/p$. In addition, $E_1 \simeq (E_\lambda)^*$. By Theorem 9.7.1, we have $E_\lambda \simeq R \cap C$ and by duality $E_1 \simeq R + C$; hence we have (completely isomorphically) $E_p \simeq (R \cap C, R + C)_\theta$. We will compute the latter space more explicitly in Theorem 9.8.7, but let us state what we just proved.

Corollary 9.7.2. *Let $L_p(\tau)$ denote the noncommutative L_p-space of the free group and let E_p be the closed subspace generated by the free generators $\{\lambda(g_i) \mid i = 1, 2, \ldots\}$. Then we have, completely isomorphically (with $\theta = 1/p$),*

$$E_p \simeq (R \cap C, R + C)_\theta, \tag{9.7.3}$$

where, as before, we use the transposition mapping as the continuous injection from R to C, which allows us to view (R, C) as a compatible couple. Moreover, the orthogonal projection from $L_2(\tau)$ onto E_2 defines a c.b. projection from $L_p(\tau)$ onto E_p for all $1 \leq p \leq \infty$. Moreover, the equivalence constants in (9.7.3) as well as $\|P: L_p(\tau) \to E_p\|_{cb}$ remain bounded when p runs aver the whole range $1 \leq p \leq \infty$.

We will see later (Theorem 10.4) that the space $R \cap C$ is very different from the Fermionic space Φ described in §9.3. Nevertheless, the following simple statement holds.

Proposition 9.7.3. *Let* (V_i) *be a sequence satisfying the CAR. For any finite sequence* (a_i) *in* $B(H)$, *we have*

$$\|\sum a_i \otimes \delta_i\|_{\min} \leq \sqrt{2}\|\sum a_i \otimes V_i\|_{\min},$$

so that the mapping $v \colon \Phi \to R \cap C$ *defined by* $v(V_i) = \delta_i$ *has norm* $\|v\|_{cb} \leq \sqrt{2}$.

Proof. It suffices to show the same result for Φ_n. We may assume that $V_1, ..., V_n$ are in a von Neumann algebra equipped with a normalized trace τ (for example, M_{2^n} equipped with its usual normalized trace). Then let $T = \sum a_i \otimes V_i$. The CAR and the trace property imply $\tau(V_j^* V_i) = \tau(V_i V_j^*) = 0$. Hence, if we set $T = \sum a_i \otimes V_i$, we obtain $\|\sum a_i^* a_i\| = 2\|(I \otimes \tau)(T^* T)\| \leq 2\|T\|_{\min}^2$, and similarly $\|\sum a_i a_i^*\| \leq 2\|T\|_{\min}^2$. We conclude that

$$\max\left\{\|\sum a_i^* a_i\|^{1/2}, \|\sum a_i a_i^*\|^{1/2}\right\} \leq \sqrt{2}\|\sum a_i \otimes V_i\|_{\min}. \qquad \blacksquare \qquad (9.7.4)$$

We will now generalize Theorem 9.7.1: Instead of the generators $\{g_i\}$ we consider all the reduced words of length d, d being a fixed integer. Recall that (after all possible cancellations have been made) any element t of a free group can be viewed as a reduced word in the generators and their inverses. We denote by $|t|$ the length (i.e., the number of letters) of this reduced word. For instance, for the empty word, we have $|e| = 0$, and also $|g_i| = |g_i^{-1}| = 1$, $|g_i g_j^{-1}| = 2 \; \forall i \neq j$, and so on. Let G be a free group, freely generated by $\{g_i \mid i \subset I\}$. Let

$$W_d = \{t \in G \mid |t| = d\}.$$

We will study the operator space spanned inside $C_\lambda^*(G)$ by $\{\lambda(t) \mid t \in W_d\}$.

When $d > 1$, the analog of $R \cap C$ is a bit more difficult to describe, because there are now $(d+1)$ ways (and not only two) to write an element t in W_d as a reduced product $L = \beta\gamma$ with $0 \leq |\beta| \leq d$ and $|\beta| + |\gamma| = d$. Indeed, we can take for β the word formed by the first j letters of t and for γ the one formed of its last $d - j$ letters. Actually, we will look at the $d + 1$ ways to decompose t as $t = \beta\gamma^{-1}$ instead of $t = \beta\gamma$, but this does not really matter here. This idea leads to a decomposition that we now describe precisely.

Let H be any Hilbert space. Let $a = \{a(t) \mid t \in G\}$ be a finitely supported family in $B(H)$. For any fixed $0 \leq j \leq d$, we consider the matrix $\{a_{\beta,\gamma}^{[d,j]} \mid \beta \in W_j, \gamma \in W_{d-j}\}$ defined by

$$a_{\beta,\gamma}^{[d,j]} = a(\beta\gamma^{-1})1_{\{|\beta\gamma^{-1}|=d\}}.$$

Note that here $|\beta\gamma^{-1}| = d$ expresses that there is no cancellation in the product $\beta\gamma^{-1}$. We denote by $\|a\|_{[d,j]}$ the norm of this matrix as an operator acting from $\ell_2(W_{d-j}; H)$ to $\ell_2(W_j; H)$.

The following extension of Theorem 9.7.1 is due to Buchholz [Buc1] and Haagerup (unpublished).

Theorem 9.7.4. *Let G be a free group and let $d \geq 1$. With the above notation we have*

$$\max_{0 \leq j \leq d}\{\|a\|_{[d,j]}\} \leq \left\| \sum_{t \in W_d} a(t) \otimes \lambda(t) \right\| \leq (d+1) \max_{0 \leq j \leq d}\{\|a\|_{[d,j]}\}. \qquad (9.7.5)$$

Moreover, the projection $P_d \colon C_\lambda^*(G) \to \overline{\text{span}}[\lambda(t) \mid t \in W_d]$ *defined by* $P_d\lambda(t) = \lambda(t)1_{\{t \in W_d\}}$ *is c.b. and satisfies* $\|P_d\| \leq d+1$ *and* $\|P_d\|_{cb} \leq 2d$.

Proof. Let $T = \sum_{t \in W_d} a(t) \otimes \lambda(t)$. We claim that there is a decomposition

$$T = T_0 + T_1 + \cdots + T_d$$

with each T_j of the form

$$T_j = \sum_{|\beta|=j,|\gamma|=d-j} a_{\beta,\gamma}^{[d,j]} \otimes u_\beta v_\gamma,$$

where $u_\beta, v_\gamma \in B(\ell_2(G))$ are such that

$$\left\| \sum u_\beta u_\beta^* \right\| \leq 1 \quad \text{and} \quad \left\| \sum v_\gamma^* v_\gamma \right\| \leq 1.$$

As we will see, this implies

$$\|T_j\| \leq \|(a_{\beta,\gamma}^{[d,j]})\| = \|a\|_{[d,j]},$$

and then the triangle inequality yields the right side of (9.7.5).

We now proceed to check this claim. Note that if we denote by $e_{s,t}(s,t \in G)$ the matrix units in $B(\ell_2(G))$, we have

$$T = \sum_{b,c \in G} a(bc^{-1})1_{\{|bc^{-1}|=d\}} \otimes e_{b,c}.$$

Note that $|bc^{-1}| = d$ iff we can write (uniquely) $b = \beta x$ (reduced) and $c = \gamma x$ (reduced) with $|\beta\gamma^{-1}| = |\beta| + |\gamma| = d$. Note that if $|\beta| = j$, after reduction the word bc^{-1} consists of the first j letters of b followed by the last $(d - j)$ letters of c^{-1}. Moreover, the equalities $|\beta x| = |\beta| + |x|$ and $|\gamma x| = |\gamma| + |x|$ are a convenient way to express that the words βx and γx are reduced. A simple calculation then shows that, if $|\beta| = j$, we have $|b| = j+|x|$ and $|c| = d-j+|x|$; hence $j = (|b| - |c| + d)/2$. Thus the set of all pairs (b,c) in $G \times G$ such that $|bc^{-1}| = d$ can be decomposed as a disjoint union $C_0 \cup C_1 \cup \cdots \cup C_d$, where

$$C_j = \{(b,c) \mid |bc^{-1}| = d, |b| - |c| + d = 2j\}.$$

Let $T_j = \sum_{(b,c) \in C_j} a(bc^{-1}) \otimes e_{b,c}$. By the preceding discussion we have $T = \sum_0^d T_j$ and moreover

$$T_j = \sum_{|\beta|=j, |\gamma|=d-j} a(\beta\gamma^{-1}) 1_{\{|\beta\gamma^{-1}|=d\}} \otimes \sum_{x \in G} e_{\beta x, \gamma x} 1_{\{|\beta x|=|\beta|+|x|\}} 1_{\{|\gamma x|=|\gamma|+|x|\}}.$$

Let $u_\beta = \sum_{x \in G} e_{\beta x, x} 1_{\{|\beta x|=|\beta|+|x|\}}$, and let $v_\gamma = u_\gamma^*$. Then

$$T_j = \sum_{|\beta|=j, |\gamma|=d-j} a(\beta\gamma^{-1}) 1_{\{|\beta\gamma^{-1}|=d\}} \otimes u_\beta v_\gamma.$$

Moreover, a simple calculation shows

$$\sum_{|\beta|=j} u_\beta u_\beta^* = \sum_{|\beta|=j} \sum_x e_{\beta x, \beta x} 1_{\{|\beta x|=|\beta|+|x|\}} = \sum_{b: |b| \geq j} e_{b,b}.$$

Hence $\left\| \sum u_\beta u_\beta^* \right\| \leq 1$ and similarly $\left\| \sum v_\gamma^* v_\gamma \right\| \leq 1$. This completes the proof of the above claim.

Now, an elementary verification shows that, if g is any fixed element in G (for instance $g = e$),

$$\sum a_{\beta,\gamma}^{[d,j]} \otimes u_\beta v_\gamma \otimes e_{g,g} = \left(I \otimes \sum_\beta u_\beta \otimes e_{g,\beta} \right) \left(\sum_{\beta,\gamma} a_{\beta,\gamma}^{[d,j]} \otimes I \otimes e_{\beta,\gamma} \right)$$
$$\left(I \otimes \sum_\gamma v_\gamma \otimes e_{\gamma,g} \right),$$

and hence

$$\|T_j\| = \|T_j \otimes e_{g,g}\| \leq \left\| \sum_{\beta,\gamma} a_{\beta,\gamma}^{[d,j]} \otimes I \otimes e_{\beta,\gamma} \right\| = \|a\|_{[d,j]}.$$

Then, by the triangle inequality, we have

$$\|T\| \leq \sum_0^d \|a\|_{[d,j]} \leq (d+1) \max\{\|a\|_{[d,j]} \mid 0 \leq j \leq d\},$$

which establishes the right side of (9.7.5).

To establish the left side, we will first study the projection P_d. Let $a: G \to B(H)$ be a finitely supported function. Let

$$Y = \sum_{t \in G} a(t) \otimes \lambda(t) \quad \text{and} \quad T = (I \otimes P_d)(Y) = \sum_{t \in W_d} a(t) \otimes \lambda(t).$$

Then, with the same notation as before (recall $\|T_j\| \leq \|a\|_{[d,j]}$), we will show that for any $0 < j < d$ we have $\|a\|_{[d,j]} \leq 2\|Y\|$ and that, if either $j = 0$ or $j = d$, the factor 2 can be removed.

Let $H_t = H \otimes \delta_t$. We have an orthogonal decomposition

$$H \otimes_2 \ell_2(G) = \bigoplus_{t \in G} H_t,$$

relative to which Y is represented by a matrix $(Y(s,t))$ with coefficients in $B(H)$ defined by

$$Y(s,t) = a(st^{-1}).$$

Now fix j with $0 \leq j \leq d$. If we restrict the above matrix to act from $\bigoplus_{|t|=d-j} H_t$ to $\bigoplus_{|t|=j} H_t$, we find

$$\left\| \sum_{|\beta|=j, |\gamma|=d-j} a(\beta\gamma^{-1}) \otimes e_{\beta,\gamma} \right\| \leq \|Y\|. \tag{9.7.6}$$

Now assume first that a is supported on W_d so that $Y = T$ and the sum in (9.7.6) can be restricted to $|\beta\gamma^{-1}| = d$. Then, the last estimate yields $\|a\|_{[d,j]} \leq \|Y\| = \|T\|$, whence the left side of (9.7.5).

We will now estimate the norm of P_d. We no longer assume that the coefficients $(a(t))$ are supported on W_d, but we assume that they are scalar. Then it is easy to check that

$$\|a\|_{[d,j]} \leq (\sum_{\beta,\gamma} |a^{[d,j]}(\beta\gamma^{-1})|^2)^{1/2} \leq (\sum_{t \in G} |a(t)|^2)^{1/2} \leq \left\| \sum_{t \in G} a(t)\lambda(t) \right\| = \|Y\|.$$

Thus, we obtain $\|T\| \leq \sum_0^d \|T_j\| \leq \sum_0^d \|a\|_{[d,j]} \leq (d+1)\|Y\|$ and hence $\|P_d\| \leq d+1$.

We will now estimate the c.b. norm of P_d. We return to the general case of a finitely supported function $a: G \to B(H)$. Note that if either $j = 0$ or $j = d$, (9.7.6) implies $\|a\|_{[d,j]} \leq \|Y\|$. Now assume $0 < j < d$. For any t in G of positive length we denote by $g(t)$ the last letter of t (equal to a generator or the inverse of one), and we define $x(t) = \delta_{g(t)}$, so that (assuming again $|s| > 0$) $\langle x(s), x(t) \rangle = 0$ iff the product st^{-1} is reduced (in other words, iff $|st^{-1}| = |s| + |t|$) and otherwise $\langle x(s), x(t) \rangle = 1$.

By Exercise 1.5 and the triangle inequality, we deduce from (9.7.6)

$$\left\| \sum_{|\beta|=j, |\gamma|=d-j} a(\beta\gamma^{-1})(\beta\gamma^{-1}) \otimes e_{\beta,\gamma}(1 - \langle x(\beta), x(\gamma) \rangle) \right\| \leq 2\|Y\|,$$

or, equivalently,

$$\|a\|_{[d,j]} \le 2\|Y\|. \tag{9.7.7}$$

This gives us

$$\|T\| \le \sum_0^d \|T_j\| \le \|a\|_{[0]} + \|a\|_{[d]} + \sum_1^{d-1} \|a\|_{[d,j]} \le (2 + 2(d-1))\|Y\|,$$

and hence $\|P_d\|_{cb} \le 2d$. ∎

Remark 9.7.5. Fix an integer $D \ge 1$. Let $a\colon G \to B(H)$ be a finitely supported function with support in $W_0 \cup \cdots \cup W_D$. Let $a_d\colon G \to B(H)$ be the function equal to a on W_d and equal to zero elsewhere, so that

$$\sum_{|t| \le D} a(t) \otimes \lambda(t) = \sum_{0 \le d \le D} a_d(t) \otimes \lambda(t).$$

We then have

$$2^{-1} \max_{0 \le d \le D} \max_{0 \le j \le d} \{\|a_d\|_{[d,j]}\} \le \left\| \sum_{|t| \le D} a(t) \otimes \lambda(t) \right\|$$
$$\le 2D(D+1) \max_{0 \le d \le D} \max_{0 \le j \le d} \{\|a_d\|_{[d,j]}\}. \tag{9.7.8}$$

Indeed, the left side is an immediate consequence of (9.7.7) while the right side follows from the triangle inequality and (9.7.5).

9.8. Operator space generated in the usual L_p-space by Gaussian random variables or by the Rademacher functions

Let $(\Omega, \mathcal{A}, P)$ be a probability space. We will say that a real-valued Gaussian random variable (in short r.v.) is standard if $E\gamma = 0$ and $E\gamma^2 = 1$. We will say that a complex-valued Gaussian r.v. $\widetilde{\gamma}$ is Gaussian standard if we can write $\widetilde{\gamma} = 2^{-1/2}(\gamma' + i\gamma'')$ with γ', γ'' real-valued, independent, standard Gaussian r.v.'s.

Let $\{\gamma_n \mid n = 1, 2, ...\}$ (resp. $\widetilde{\gamma}_n \mid n = 1, 2, ...\}$) be a sequence of real- (resp. complex-) valued independent standard Gaussian r.v.'s on $(\Omega, \mathcal{A}, P)$. As is well known, for any finitely supported sequence of real (resp. complex) scalars (α_n), the r.v. $S = \sum \alpha_i \gamma_i$ (resp. $\sum \alpha_i \widetilde{\gamma}_i$) has the same distribution as the variable $\widetilde{S} = (\sum |\alpha_i|^2)^{1/2} \gamma_1$ (resp. $(\sum |\alpha_i|^2)^{1/2} \widetilde{\gamma}_1$). In particular, we have for any finitely supported sequence of complex scalars

$$\left\| \sum \alpha_i \widetilde{\gamma}_i \right\|_p = \|\widetilde{\gamma}_1\|_p (\sum |\alpha_i|^2)^{1/2}. \tag{9.8.1}$$

Let $\mathcal{G}_p$ be the subspace of $L_p(\Omega, \mathcal{A}, P)$ generated by $\{\widetilde{\gamma}_n \mid n = 1, 2, ...\}$. Then, as a Banach space, $\mathcal{G}_p$ is isometric to ℓ_2 for all $1 \leq p < \infty$. (To simplify, we will discuss mostly the complex case in the sequel, although the real case is entirely similar provided we restrict ourselves to $\mathbb{R}$-linear transformations.) Moreover, for any isometry $U : \mathcal{G}_p \to \mathcal{G}_p$ the sequence $\{U(\widetilde{\gamma}_i) \mid i = 1, 2, ..\}$ has the same distribution as the sequence $\{\widetilde{\gamma}_i \mid i = 1, 2, ..\}$. If we equip $\mathcal{G}_p$ with the o.s.s. induced by $L_p(\Omega, \mathcal{A}, P)$, it follows (see Proposition 9.5.2) that U is a complete isometry from $\mathcal{G}_p$ to $\mathcal{G}_p$. Therefore, by Proposition 9.2.1, we have

Proposition 9.8.1. *For any $1 \leq p < \infty$, the space $\mathcal{G}_p$ is a homogeneous Hilbertian operator space.*

Let $\{\varepsilon_n \mid n = 1, 2, ...\}$ be a sequence of independent, identically distributed (in short i.i.d.) r.v.'s on $(\Omega, \mathcal{A}, P)$ with ± 1 values and such that $P\{\varepsilon_n = +1\} = P\{\varepsilon_n = -1\} = 1/2$. The reader who so wishes can replace $\{\varepsilon_n \mid n = 1, 2, ...\}$ by the classical Rademacher functions (r_n) on the Lebesgue interval; this does not make any difference in the sequel. Let $\mathcal{R}_p$ be the subspace generated in $L_p(\Omega, \mathcal{A}, P)$ by the sequence (ε_n).

The analog of (9.8.1) for the variables (ε_n) (or, equivalently, for the Rademacher functions) is given by the classical Khintchine inequalities (cf., e.g., [LT1, p. 66] or [DJT p. 10]), which say that, for $1 \leq p < \infty$, there are positive constants A_p and B_p such that, for any scalar sequence of coefficients (α_n) in ℓ_2, we have

$$A_p \left(\sum |\alpha_n|^2 \right)^{1/2} \leq \left\| \sum \alpha_n \varepsilon_n \right\|_p \leq B_p \left(\sum |\alpha_n|^2 \right)^{1/2}. \qquad (9.8.2)$$

(Note that we have trivially $B_p = 1$ if $p \leq 2$ and $A_p = 1$ if $p \geq 2$.) This implies that, as a Banach space, $\mathcal{R}_p$ is isomorphic to ℓ_2 for any $1 \leq p < \infty$. A fortiori, $\mathcal{R}_p$ and $\mathcal{G}_p$ are isomorphic Banach spaces if $1 \leq p < \infty$.

We now wish to describe the operator space structure induced by L_p on $\mathcal{G}_p$ (resp. $\mathcal{R}_p$). By (9.5.3), this can be reduced to the knowledge of the norm

$$\left\| \sum \gamma_n x_n \right\|_{L_p(\Omega, P; S_p)}$$

(resp. $\left\| \sum \varepsilon_n x_n \right\|_{L_p(\Omega, P; S_p)}$) when (x_n) is an arbitrary finite sequence of elements of S_p. In other words, to describe the o.s.s. of $\mathcal{G}_p$ (resp. $\mathcal{R}_p$) up to complete isomorphism, it suffices to produce two-sided inequalities analogous to (9.8.1) and (9.8.2) but with coefficients in S_p instead of scalar ones. The noncommutative versions of Khintchine's inequalities proved in [LuP] and [LPP] are exactly what is needed here. The case $1 < p < \infty$ is a remarkable result due to F. Lust-Piquard ([LuP]). The case $p = 1$ comes from the later paper [LPP], which also contains an alternate proof of the other cases.

Theorem 9.8.2. *(i) Assume $2 \leq p < \infty$. Then there is a constant B_p' such that, for any finite sequence (x_n) in S_p, we have*

$$(9.8.3) \quad \max\left\{ \left\| \left(\sum x_n^* x_n\right)^{1/2} \right\|_{S_p}, \left\| \left(\sum x_n x_n^*\right)^{1/2} \right\|_{S_p} \right\}$$

$$\leq \left\| \sum \varepsilon_n x_n \right\|_{L_p(\Omega, P; S_p)}$$

$$\leq B_p' \max\left\{ \left\| \left(\sum x_n^* x_n\right)^{1/2} \right\|_{S_p}, \left\| \left(\sum x_n x_n^*\right)^{1/2} \right\|_{S_p} \right\}.$$

(ii) Assume $1 \leq p \leq 2$. Then there is a positive constant A_p' such that, for any finite sequence (x_n) in S_p, we have

$$A_p' \||(x_n)\||_p \leq \left\| \sum \varepsilon_n x_n \right\|_{L_p(\Omega, P; S_p)} \leq \||(x_n)\||_p, \qquad (9.8.4)$$

where we have set

$$\||(x_n)\||_p = \inf\left\{ \left\| \left(\sum y_n^* y_n\right)^{1/2} \right\|_{S_p} + \left\| \left(\sum z_n z_n^*\right)^{1/2} \right\|_{S_p} \,\middle|\, x_n = y_n + z_n \right\}.$$

Moreover, similar inequalities are valid with a real or complex Gaussian i.i.d. sequence (γ_n) or $(\widetilde{\gamma}_n)$ in the place of (ε_n). Finally, the same inequalities are valid when S_p is replaced by any noncommutative L_p space associated to a semi-finite faithful normal trace on a von Neumann algebra.

Remark. (Observed independently by Marius Junge). We claim that there is a numerical constant C such that, for all $2 \leq p < \infty$,

$$B_p' \leq C\sqrt{p}.$$

Since this is only implicitly contained in [LPP] and it might be of independent interest, we will give the details explicitly. Let us denote by $P_1 \colon L_2 \to \mathcal{R}_2$ the orthogonal projection. Recall that the K-convexity constant of a Banach space X is defined as follows:

$$K(X) = \|P_1 \otimes I_X\|_{L_2(X) \to L_2(X)}.$$

By a standard averaging technique, one easily verifies that

$$\|P_1 \otimes I_X\|_{L_p(X) \to L_p(X)} \leq K(L_p(X)).$$

When $X = S_p$, since $\ell_p(S_p)$ embeds isometrically into S_p, we have $K(L_p(S_p)) = K(S_p)$. In [LPP] it is proved that the constants A_p' are uniformly bounded when $1 \leq p \leq 2$. By duality, it follows from the preceding estimate of $\|P_1 \otimes I_X\|_{L_p(X) \to L_p(X)}$ for $X = S_p$ that there is a numerical constant C' such that, for all $2 \leq p < \infty$, $B'_p \leq C' K(L_p(S_p)) = C' K(S_p)$. By [MaP, Remark 2.10], the latter constant is dominated by the type 2 constant, and by [TJ2] the type 2 constant of S_p is equal to the best constant in the classical (scalar) Khintchine inequalities B_p (at least when p is an even integer) that is of order $p^{1/2}$ when $p \to \infty$. Thus we obtain our claim that there is a numerical constant C such that, for all $2 \leq p < \infty$,

$$B'_p \leq C\sqrt{p}.$$

Remark. By Proposition 9.5.2, we can deduce from the known general results on Gaussian and Rademacher series in Banach spaces (cf. [MaP, Corollaire 1.3]) that the spaces $\mathcal{G}_p$ and $\mathcal{R}_p$ are completely isomorphic for any $1 \leq p < \infty$. Of course, in the case $p = 2$, $\mathcal{G}_2$ and $\mathcal{R}_2$ are completely isometric to OH since the space $L_2(\Omega, \mathcal{A}, P)$ itself is completely isometric to $OH(I)$, where the cardinal I is its Hilbertian dimension.

Let us now identify $\mathcal{G}_p$ and $\mathcal{R}_p$ as operator spaces. We start with the case $p = 1$, which is particularly interesting. Consider the direct sum $R \oplus_1 C$ (as defined in §2.6) and its subspace $\Delta \subset R \oplus_1 C$ defined by

$$\Delta = \{(x, -^t x) \mid x \in R\}.$$

In accordance with the definitions in §2.7, we will denote by $R+C$ the quotient operator space $(R \oplus_1 C)/\Delta$. Since $R \oplus_1 C$ is equipped with a natural o.s.s. (see §2.6), the space $R + C$ itself is thus equipped with a natural o.s.s. as a quotient space (see §2.4). It is easy to see (cf. §2.7) that

$$(R \cap C)^* = R + C \quad \text{completely isometrically.} \tag{9.8.5}$$

In particular, $R + C$ is isomorphic to ℓ_2 as a Banach space. We will denote by (σ_i) the natural basis of $R + C$ that is biorthogonal to the basis (δ_i) of $\Delta^\perp \simeq R \cap C$. Equivalently, if we denote by $q : R \oplus_1 C \to R + C$ the canonical surjection, then we have $\sigma_n = q(e_{1n} \oplus e_{n1})$. Similarly, we will denote by $R_n + C_n$ the quotient operator space $(R_n \oplus_1 C_n)/\Delta_n$, with $\Delta_n = \{(x, -^t x) \mid x \in R_n\}$. We also can identify $R_n + C_n$ with the subspace spanned in $R + C$ by $(\sigma_1, ..., \sigma_n)$. Moreover, we have $(R_n \cap C_n)^* = R_n + C_n$ completely isometrically.

Now we can reformulate the main result of [LPP] in the language of operator spaces as follows:

Theorem 9.8.3. *The space $\mathcal{R}_1$ (resp. $\mathcal{G}_1$) is completely isomorphic to $R+C$ via the isomorphism that takes ε_i (resp. $\tilde{\gamma}_i$) to σ_i.*

Proof. It is easy to verify that the norm $\||.\||_1$ appearing in Theorem 9.8.2 is the dual norm to the natural norm of the space $\mathcal{K} \otimes_{\min} (R \cap C)$. Equivalently, in the notation of §9.5 (using (9.5.5)) it coincides with the norm in the space $S_1[(R \cap C)^*] = S_1[R + C]$. Therefore, the linear mapping that takes ε_i to σ_i defines an isomorphism from $S_1[\mathcal{R}_1]$ onto $S_1[R + C]$. Thus, the announced result for $\mathcal{R}_1$ follows from Proposition 9.5.3 applied with $p = 1$. The case of $\mathcal{G}_1$ is analogous. ∎

By combining (9.8.5) with Theorem 9.7.1 we obtain a surprising connection between the standard Gaussian (or ± 1-valued) independent random variables and the generators of $C^*_\lambda(F_\infty)$:

Corollary 9.8.4. *We have* $(E_\lambda)^* \simeq \mathcal{G}_1 \simeq \mathcal{R}_1$ *completely isomorphically. More precisely, let us denote by* $(\lambda_*(g_i))$ *the system in* $(E_\lambda)^*$ *that is biorthogonal to* $(\lambda(g_i))$. *Then the mapping* $u : \mathcal{R}_1 \to (E_\lambda)^*$ *(resp.* $u : \mathcal{G}_1 \to (E_\lambda)^*$) *defined by* $u(\varepsilon_i) = \lambda_*(g_i)$ *(resp.* $u(\widetilde{\gamma}_i) = \lambda_*(g_i)$) *is a complete isomorphism.*

More generally, for all $1 \leq p \leq \infty$, let us denote by $R[p]$ (resp. $C[p]$) the operator space generated by the sequence $\{e_{1j} \mid j = 1, 2, ...\}$ (resp. $\{e_{i1} \mid i = 1, 2, ...\}$) in the operator space S_p equipped with its natural o.s.s. defined by interpolation (see (9.5.4)). Here again we set $S_\infty = \mathcal{K}$. Note that $R[\infty]$ (resp. $C[\infty]$) obviously coincides with the row (resp. column) space R (resp. C). A moment of thought (recall Exercise 2.3.5) shows that $R[1] \simeq R^*$ and $C[1] \simeq C^*$ completely isometrically; therefore we may identify $R[1]$ with C on one hand, and $C[1]$ with R on the other. Moreover, we obviously have a natural projection simultaneously completely contractive from S_1 to $R[1]$ and S_∞ to $R[\infty]$ (and similarly for columns). This implies, by Proposition 2.7.6, that if we make the couple (R, C) into a compatible one (as we did in Chapter 7) by using transposition to inject R into C, then we have completely isometric identifications (with $\theta = 1/p$)

$$R[p] = (R, C)_\theta \qquad C[p] = (C, R)_\theta.$$

Now, if we use (9.5.7), we find

$$S_p[R[p]] = (S_\infty[R], S_1[C])_\theta \qquad (9.8.6)$$

and $S_p[C[p]] = (S_\infty[C], S_1[R])_\theta$. If we view $S_\infty[R]$ (resp. $S_1[C]$) as a space of sequences of elements of S_∞ (resp. S_1), then the norm in $S_\infty[R]$ (resp. $S_1[C]$) is easily seen to be $(x_j) \to \left\| \left(\sum x_j x_j^* \right)^{1/2} \right\|_{S_\infty}$ (resp. $(x_j) \to \left\| \left(\sum x_j x_j^* \right)^{1/2} \right\|_{S_1}$). Therefore, since we have simultaneously contractive projections (see the discussion before Proposition 2.7.6) onto the corresponding subspaces of $S_\infty[S_\infty]$ and $S_1[S_1]$, we find that the norm in $(S_\infty[R], S_1[C])_\theta$ coincides with $(x_j) \to \left\| \left(\sum x_j x_j^* \right)^{1/2} \right\|_{S_p}$. In other words, for any sequence (x_j) in S_p we have, by (9.8.6),

$$\left\| \sum x_j \otimes e_{1j} \right\|_{S_p[R[p]]} = \left\| \left(\sum x_j x_j^* \right)^{1/2} \right\|_{S_p}. \qquad (9.8.7)$$

Similarly, we have

$$\left\|\sum x_i \otimes e_{i1}\right\|_{S_p[C[p]]} = \left\|\left(\sum x_i^* x_i\right)^{1/2}\right\|_{S_p}. \tag{9.8.8}$$

We denote by $R[p] \cap C[p]$ the subspace of $R[p] \oplus C[p]$ formed of all couples of the form $(x, {}^t x)$. On the other hand, we denote by $R[p] + C[p]$ the operator space that is the quotient of $R[p] \oplus_1 C[p]$ modulo the subspace formed of all couples of the form $(x, -{}^t x)$. Then, by (9.8.7) and (9.8.8), the norm appearing on the left in (9.8.3) is equivalent to the natural norm of the space $S_p[C[p]] \cap S_p[R[p]]$ or, equivalently, $S_p[R[p] \cap C[p]]$. Similarly, the norm $||| \ |||_p$ in Theorem 9.8.2 (case $p \leq 2$) is equivalent to the natural norm of the space $S_p[C[p]] + S_p[R[p]]$ or, equivalently, $S_p[R[p] + C[p]]$. This allows us to state:

Theorem 9.8.5. *Let* $1 < p < \infty$. *The space* $\mathcal{G}_p$ *(or the space* $\mathcal{R}_p$) *is completely isomorphic to* $R[p] + C[p]$ *if* $p \leq 2$ *and to* $R[p] \cap C[p]$ *if* $p \geq 2$.

Proof. First observe that the natural norm in the space $S_p[\mathcal{R}_p]$ is equal to the norm induced by $L_p(\Omega, \mathcal{A}, P; S_p)$, by the isometric case in proposition 9.5.2. Then, by (9.8.3) and the preceding discussion, the latter norm is equivalent to the natural norm of either the space $S_p[R[p] \cap C[p]]$ if $p \geq 2$ or the space $S_p[R[p] + C[p]]$ if $p \leq 2$, whence the announced complete isomorphisms by Proposition 9.5.3. ∎

Remark. Note that Theorem 9.8.3 is nothing but the natural extension of Theorem 9.8.5 to the case $p = 1$.

Remark 9.8.6. In the Banach space setting, it is well known that the orthogonal projection $P_2\colon L_2 \to \mathcal{G}_2$ (resp. $Q_2\colon L_2 \to \mathcal{R}_2$) extends to a bounded linear projection $P_p\colon L_p \to \mathcal{G}_p$ (resp. $Q_p\colon L_p \to \mathcal{R}_p$), provided $1 < p < \infty$, and this fails if $p = 1$ or $p = \infty$. (Warning: It is customary in harmonic analysis to consider that P_p and P_2 are the "same" operator, since they coincide on simple functions.)

In the operator space setting, the situation is analogous: For any $1 < p < \infty$, P_p (resp. Q_p) is a c.b. projection from L_p onto $\mathcal{G}_p$ (resp. onto $\mathcal{R}_p$). This can be seen easily using Proposition 9.5.2 and the fact that, when $1 < p < \infty$, S_p is a K-convex Banach space in the sense of [P16]. (See also [TJ1, p. 86] or [DJT, p. 258].) This can also be viewed as a corollary of Theorem 9.8.4, since the latter result implies that $(\mathcal{G}_p)^* \simeq \mathcal{G}_{p'}$ and $(\mathcal{R}_p)^* \simeq \mathcal{R}_{p'}$ (completely isomorphically) when $1 < p, p' < \infty$ with $\frac{1}{p} + \frac{1}{p'} = 1$. Indeed, the complete boundedness of the natural mapping

$$(\mathcal{G}_p)^* = L_{p'}/G_p^\perp \to \mathcal{G}_{p'}$$

is clearly equivalent to the complete boundedness of P_p, and similarly for Q_p.

Using the last assertion in Theorem 9.8.2, we obtain another striking isomorphism.

Theorem 9.8.7. *Let* $1 \leq p < \infty$. *Let* E_p *be as in Corollary 9.7.2. Then the correspondence* $\varepsilon_i \to \lambda(g_i)$ *(resp.* $\tilde{\gamma}_i \to \lambda(g_i)$*) is a complete isomorphism between the spaces* $\mathcal{R}_p$ *(resp.* $\mathcal{G}_p$*) and* E_p.

Proof. It suffices to prove this for $p \geq 2$. Let (x_i) be a finitely supported sequence of elements of S_p. Let

$$\|(x_i)\|_\varepsilon = \left\|\sum x_i \otimes \varepsilon_i\right\|_{S_p[\mathcal{R}_p]}$$

and

$$\|(x_i)\|_\lambda = \left\|\sum x_i \otimes \lambda(g_i)\right\|_{S_p[E_p]}.$$

By Proposition 9.5.3, it suffices to show that these two norms are equivalent for each $2 \leq p < \infty$. By Proposition 9.5.3 again (isometric case), the norm $\|\cdot\|_\varepsilon$ coincides with the norm induced by $S_p[L_p(\Omega, \mathcal{A}, P)]$, which, by "Fubini's Theorem" (cf. [P2, (5.6)]), coincides with that of $L_p(\Omega, \mathcal{A}, P; S_p)$, or, equivalently, with the norm in the middle of (9.8.3).

Similarly, the norm $\|\ \|_\lambda$ coincides with the norm induced by the space $S_p[L_p(\tau)]$. The latter space is identical (by Fubini again; cf. [P2, (5.6)]) to the noncommutative L_p-space associated to $B(\ell_2) \overline{\otimes} M$ equipped with the trace $\text{tr} \otimes \tau$, which we will denote simply by $L_p(\text{tr} \otimes \tau)$. Let (ξ_i) be any choice of signs $\xi_i = \pm 1$. By a well-known result, the linear map that takes $\lambda(g_i)$ to $\xi_i \lambda(g_i)$ $(i = 1, 2, \ldots)$ extends to an isometric C^*-representation from M to M and (by transposition) also defines a complete isometry from M_* to M_*. By interpolation, the same map defines a complete isometry on $L_p(\tau)$ for all $1 < p < \infty$. In particular, this implies that

$$\|(\xi_i x_i)\|_\lambda = \|(x_i)\|_\lambda \quad \text{for any} \quad \xi_i = \pm 1,$$

and, therefore, after averaging over $\xi_i = \pm 1$, we have

$$\|(x_i)\|_\lambda = \int_\Omega \|(\varepsilon_i x_i)\|_\lambda dP.$$

(Note: A slightly weaker result, namely, the equivalence of the last two expressions, follows from Corollary 9.7.2, and this suffices for the present argument.) If we set $y_i = x_i \otimes \lambda(g_i) \in S_p \otimes L_p(\tau)$, this yields

$$\|(x_i)\|_\lambda = \int \left\|\sum \varepsilon_i y_i\right\|_{L_p(\text{tr} \otimes \tau)} dP;$$

hence, by (9.8.3),

$$\|(x_i)\|_\lambda \leq B'_p \max\left\{\left\|\left(\sum y_i^* y_i\right)^{1/2}\right\|_{L_p(\text{tr} \otimes \tau)}, \left\|\left(\sum y_i y_i^*\right)^{1/2}\right\|_{L_p(\text{tr} \otimes \tau)}\right\}$$

$$\leq B'_p \|(x_i)\|_\lambda.$$

But since $y_i y_i^* = x_i x_i^* \otimes 1$ and $y_i^* y_i = x_i^* x_i \otimes 1$, we obtain finally

$$\|(x_i)\|_\lambda \leq B_p' \max\left\{\left\|\left(\sum x_i^* x_i\right)^{1/2}\right\|_{S_p}, \left\|\left(\sum x_i x_i^*\right)^{1/2}\right\|_{S_p}\right\}$$
$$\leq B_p'\|(x_i)\|_\lambda.$$

In other words, we conclude, that the norms $\|\cdot\|_\lambda$ and $\|\cdot\|_\varepsilon$ are equivalent. ∎

Curiously, the preceding result combined with the earlier Corollary 9.7.2 implies a result of independent interest on the couple (R, C), as follows.

Corollary 9.8.8. *For any $1 < p < \infty$ and $\theta = 1/p$ we have completely isomorphic identities*

$$(R \cap C, R + C)_\theta \simeq R[p] \cap C[p] \quad \text{if } p \geq 2$$
$$(R \cap C, R + C)_\theta \simeq R[p] + C[p] \quad \text{if } p \leq 2.$$

Proof. Indeed, by Corollary 9.7.2, the left side can be identified with E_p, and, by Theorem 9.8.5, the right side can be identified with $\mathcal{R}_p$. Thus the result follows from Theorem 9.8.7.   ∎

Remark. By a different argument (see [P2, p. 109-110]), one can show that the equivalence constants appearing in the preceding corollary remain bounded uniformly when p runs over the whole interval $)1, \infty($. Thus, together with (9.7.3), this implies completely isomorphic identities

$$E_p \simeq R[p] \cap C[p] \quad \text{if } p \geq 2$$
$$E_p \simeq R[p] + C[p] \quad \text{if } p \leq 2,$$

with equivalence constants independent of $1 < p < \infty$. When p is an even integer, the best possible constants are computed in the remarkable paper [Buc2].

Remark 9.8.9. Let $k \geq 1$ be a fixed integer. Let $(\varepsilon_n^1)_{n\geq 1}$, $(\varepsilon_n^2)_{n\geq 1}, \ldots,$ $(\varepsilon_n^k)_{n\geq 1}$ be independent copies of the original sequence (ε_n) as above, on a suitable probability space (Ω, A, P). Let us denote by $\mathcal{R}_p^k$ the subspace of $L_p(\Omega, A, P)$ spanned by the functions of the form $\varepsilon_{n_1}^1 \varepsilon_{n_2}^2 \ldots \varepsilon_{n_k}^k$ $(n_1 \geq 1, n_2 \geq 1, \ldots)$. Then, modulo a simple reformulation, the results of the paper [HP2] describe the operator space structure of the space $\mathcal{R}_1^k$ for any $k = 1, 2, \ldots$ and its dual. (The Gaussian case is similar by general arguments.)

The paper [HP2] also describes the space $E^\lambda \otimes_{\min} \cdots \otimes_{\min} E^\lambda$ (k times) and proves that $\mathcal{R}_1^k$ is completely isomorphic to $(E^\lambda \otimes_{\min} \cdots \otimes_{\min} E^\lambda)^*$. Here, of course, the isomorphism constants depend on k.

Concerning $\mathcal{R}_p^k$ for $1 < p < \infty$, it is easy to iterate the inequalities appearing in Theorem 9.8.2 to obtain (after successive integrations) two-sided inequalities describing the operator space structure of $\mathcal{R}_p^k$. To describe these iterated inequalities, assume for simplicity that $k = 2$. Let (x_{ij}) be a matrix with entries in S_p, with only finitely many of them nonzero. Then both $x = (x_{ij})$ and the transposed matrix ${}^tx = (x_{ji})$ can be viewed as elements of S_p on the Hilbert space $\ell_2 \oplus \ell_2 \oplus \cdots$, and we denote the corresponding norms simply by $\|x\|_{S_p}$ and $\|{}^tx\|_{S_p}$.

Then, after iteration, (9.8.3) becomes, when $2 \le p < \infty$,

$$\max\left\{ \|x\|_{S_p}, \|{}^tx\|_{S_p}, \left\|\left(\sum_{i,j} x_{ij}^* x_{ij}\right)^{1/2}\right\|_{S_p}, \left\|\left(\sum_{i,j} x_{ij} x_{ij}^*\right)^{1/2}\right\|_{S_p} \right\} \quad (9.8.9)$$

$$\le \left\|\sum_{i,j} \varepsilon_i^1 \varepsilon_j^2 x_{ij}\right\|_{L_p(\Omega,P;S_p)}$$

$$\le (B_p')^2 \max\left\{ \|x\|_{S_p}, \|{}^tx\|_{S_p}, \left\|\left(\sum_{i,j} x_{ij}^* x_{ij}\right)^{1/2}\right\|_{S_p}, \left\|\left(\sum_{i,j} x_{ij} x_{ij}^*\right)^{1/2}\right\|_{S_p} \right\}.$$

Moreover, the orthogonal projection induces a c.b. projection from $L_p(\Omega, P)$ onto $\mathcal{R}_p^k$, for any $k = 1, 2, \ldots$, so that (9.8.9) can be dualized to treat the case $1 < p < 2$. We do not spell out the corresponding inequality. Again, when $2 \le p < \infty$, these inequalities can be interpreted as describing $\mathcal{R}_p^2$ as completely isomorphic to the intersection of four operator spaces, as follows. First recall that, for any Hilbert space H, we denote by H_c (resp. H_r) the operator space obtained by equipping H with the o.s.s. of the column (resp. row) Hilbert space. Let us define $H_r[p] = (H_r, H_c)_{1/p}$ and $H_c[p] = (H_c, H_r)_{1/p} \simeq H_r[p']$ for $1 < p < \infty$. (Also set, by convention, $H_r[\infty] = H_r$, $H_c[\infty] = H_c$, $H_r[1] = H_c$, $H_c[1] = H_r$.)

Then (9.8.9) can be interpreted as saying that $\mathcal{R}_p^2$ is completely isomorphic to the intersection $S_p \cap S_p^{op} \cap (S_2)_c[p] \cap (S_2)_r[p]$. Thus we can extend essentially all the preceding discussion of $\mathcal{R}_p$ to the spaces $\mathcal{R}_p^k$. In particular, here is what becomes of Corollary 9.8.8 in the case $k = 2$:

Let $\theta = 1/p$, $1 < p < \infty$, and let us denote $\mathcal{K}$ by S_∞. Then the interpolation space

$$\left((S_2)_r \cap (S_2)_c \cap S_\infty \cap S_\infty^{op}, (S_2)_r + (S_2)_c + S_1 + S_1^{op}\right)_\theta$$

is completely isomorphic to the intersection

$$(S_2)_r[p] \cap (S_2)_c[p] \cap S_p \cap S_p^{op} \quad \text{if} \quad p \ge 2$$

and to the sum

$$(S_2)_r[p] + (S_2)_c[p] + S_p + S_p^{op} \quad \text{if} \quad p \leq 2.$$

In the case $p = 1$, the results of [HP2] show that $\mathcal{R}_1^2$ is completely isomorphic to the sum $(S_2)_r + (S_2)_c + S_1 + S_1^{op}$. The case of a general $k > 2$ can be handled similarly, and we obtain for $p \geq 2$ (resp. $p \leq 2$) the intersection (resp. the sum) of a family of 2^k operator spaces. We leave the details to the reader (see [HP2] for the cases $p = 1$ and $p = \infty$).

Remark. By a well-known symmetrization procedure, one can deduce from the Khintchine inequalities that, for any sequence $(Z_n)_{n \geq 1}$ of independent mean zero random variables in L_p $(1 \leq p < \infty)$, we have (for any n)

$$\frac{1}{2} A_p \left\| \left(\sum_1^n |Z_i|^2 \right)^{1/2} \right\|_p \leq \left\| \sum_{i=1}^n Z_i \right\|_p \leq 2 B_p \left\| \left(\sum_1^n |Z_i|^2 \right)^{1/2} \right\|_p.$$

Note that the partial sums $S_n = \sum_1^n Z_i$ form a very special class of martingales. The preceding inequalities were extended to the case of general martingales by Burkholder, Davis, and Gundy (see [Bur]).

For a noncommutative version of the Burkholder-Gundy inequalities, with an application to Clifford martingales and stochastic integrals, see [PX].

9.9. Semi-circular systems in Voiculescu's sense

In his recent and very beautiful theory of "free probability," Voiculescu discovered a "free" analog of Gaussian random variables; see [VDN] (see also [HiP]). This discovery gives a new insight into a remarkable limit theorem for random matrices, due to Wigner (1955). In Wigner's result, a particular probability distribution plays a crucial role, namely, the probability measure on $\mathbb{R}$ (actually supported by $[-2, 2]$) defined as follows:

$$\mu_W(dt) = 1_{[-2,2]} \sqrt{4 - t^2} \, dt/2\pi.$$

We will call it the standard Wigner distribution. We have

$$\int t \mu_W(dt) = 0 \qquad \int t^2 \mu_W(dt) = 1.$$

In classical probability theory, Gaussian random variables play a prominent role. They usually can be discussed in the framework attached to a family $(\gamma_i)_{i \in I}$ (resp. $(\widetilde{\gamma}_i)_{i \in I}$) of independent identically distributed (i.i.d. in short) real- (resp. complex-) valued Gaussian variables with mean zero and

L_2-norm equal to 1. When (say) $I = \{1, 2, \ldots, n\}$ the distribution of $(\gamma_i)_{i \in I}$ (resp. $(\widetilde{\gamma}_i)_{i \in I}$) is invariant under the orthogonal (resp. unitary) group $O(n)$ (resp. $U(n)$).

In Voiculescu's theory, stochastic independence of random variables is replaced by freeness of C^*-random variables. We will review the basic definitions below. After that, we will introduce a free family $(W_i)_{i \in I}$ of C^*-random variables, each distributed according to the standard Wigner distribution. These are called *free semi-circular* variables. The family $(W_i)_{i \in I}$ is the free analog of $(\gamma_i)_{i \in I}$ in classical probability; it satisfies a similar distributional invariance under the orthogonal group. But actually, since we work mostly with *complex* coefficients, we will also introduce a free family $(\widetilde{W}_i)_{i \in I}$ that is the free analog of $(\widetilde{\gamma}_i)_{i \in I}$; their "joint distribution" satisfies an analogous unitary invariance. Such variables are called *free circular* variables.

We now start reviewing the precise definitions of the basic concepts of "free probability," following [VDN].

Definitions. *A C^*-probability space is a unital C^*-algebra A equipped with a state φ (a state is a positive linear form of norm 1). We will say that an element x of A is a C^*-random variable (in short, C^*-r.v.). If x is self-adjoint, we will say that it is a real C^*-r.v. By definition, the distribution of a real C^*-r.v. x is the probability measure μ_x on $\mathbb{R}$ such that*

$$\forall k \geq 0 \quad \varphi(x^k) = \int t^k \mu_x(dt).$$

It follows that, for any continuous function $f : \mathbb{R} \to \mathbb{R}$, we have

$$\varphi(f(x)) = \int f(t) \mu_x(dt), \tag{9.9.1}$$

Indeed, we can approximate f by a sequence of polynomials uniformly on every compact subset. Hence, in particular, for all $0 < p < \infty$,

$$\varphi(|x|^p) = \int |t|^p \mu_x(dt). \tag{9.9.2}$$

Moreover, if φ is "faithful" on the C^*-algebra A_x generated by x (meaning that $\varphi(y) = 0$ for $y \geq 0$ implies $y = 0$), then the support of μ_x is exactly the spectrum of x, denoted by $\sigma(x)$. Therefore, we can record here the following fact:

Let (A, φ) and (B, ψ) be two C^*-probability spaces with φ and ψ faithful. Let $x \in A$ and $y \in B$ be two real C^*-r.v. with the same distribution, that is, such that $\mu_x = \mu_y$. Then we necessarily have $\|x\| = \|y\|$ (where $\|x\|$ is the norm in A and $\|y\|$ the norm in B).

This property is immediate, since

$$\|x\| = \sup\{|\lambda| \mid \lambda \in \sigma(x)\}. \tag{9.9.3}$$

It can also be obtained by letting p tend to infinity in (9.9.2). Note that it suffices that φ (resp. ψ) be faithful on the C^*-algebra generated by x (resp. y).

A probabilist will legitimately object that this theory is restricted to *bounded* variables and that the usual probability distributions (Gaussian, Poisson, etc.) have unbounded support. But, by a truncation, one can easily extend this viewpoint to the unbounded real case. Besides, it turns out that the free analog of Gaussian variables happens to be bounded (see below), although it is not so for the free stable distributions (see [BV]).

Example. Let $\omega \to a(\omega) \in M_n$ be a random $n \times n$ matrix defined on a standard probability space $(\Omega, \mathcal{A}, P)$. Then the space $A = L_\infty(\Omega, \mathcal{A}, P; M_n)$ can be viewed as a C^*-probability space once we equip it with the state φ defined by

$$\varphi(a) = \int \frac{1}{n} tr(a(\omega)) dP(\omega).$$

Assume moreover that $a(\omega) = a(\omega)^*$ almost surely. Let $(\lambda_1(\omega), ..., \lambda_n(\omega))$ be the eigenvalues of the matrix $a(\omega)$. Then the distribution μ_a of the real C^*-r.v. a is nothing but

$$\mu_a = \int \frac{1}{n} \sum_1^n \delta_{\lambda_i(\omega)} dP(\omega).$$

Definitions. Let (A_n, φ_n) *be a sequence of C^*-probability spaces, let $x_n \in A_n$ be a sequence of real C^*-r.v.'s, and let x be another real C^*-r.v. We will say that x_n tends to x in distribution if the distributions μ_{x_n} tend weakly to μ_x. By a classical criterion (since μ_x is compactly supported, the Stone-Weierstrass Theorem is applicable), this is equivalent to*

$$\forall k \geq 0 \quad \varphi_n(x_n^k) \to \varphi(x^k) \quad \text{when } n \to \infty.$$

More generally, we can define the joint distribution of a family $x = (x_i)_{i \in I}$ of real C^*-r.v.'s, but it is no longer a measure: We consider the set of all polynomials in a family of noncommuting variables $(X_i)_{i \in I}$. First we define

$$F(X_{i_1} X_{i_2} ... X_{i_k}) = \varphi(x_{i_1} x_{i_2} ... x_{i_k}),$$

then we extend F linearly to a linear form on $\mathcal{P}(I)$. We will say that F is the "joint distribution" of the family $x = (x_i)_{i \in I}$. If we give ourselves for each n such a family $(x_i^n)_{i \in I}$ with distribution F^n, we say that $(x_i^n)_{i \in I}$ converges in distribution to $(x_i)_{i \in I}$ if F^n converges pointwise to F.

Let (A, φ) be a C^*-probability space and let $(A_i)_{i \in I}$ be a family of sub-algebras of A. We say that $(A_i)_{i \in I}$ is free if $\varphi(a_1 a_2 ... a_n) = 0$ every time we have $a_j \in A_{i_j}, i_1 \neq i_2 \neq ... \neq i_n$ and $\varphi(a_j) = 0 \; \forall j$.

Let $(x_i)_{i \in I}$ be a family of C^*-r.v.'s in A. Let A_i be the unital algebra (resp. C^*-algebra) generated by x_i inside A. We say that the family $(x_i)_{i \in I}$ is free (resp. $*$-free) if $(A_i)_{i \in I}$ is free.

The preceding definitions are restricted to C^*-r.v.'s, which correspond to bounded r.v.'s in the commutative case, but, by convention, we will use the same definition whenever we are dealing with a sequence in a noncommutative L_1-space. So, from now on, convergence in distribution or weak convergence means the convergence of all moments just like in the preceding definitions (of course, for this to make sense, we implicitly assume that all moments exist, like in the Gaussian case, for instance). When the limit distribution is determined by its moments (in particular, if it is compactly supported, or in the Gaussian case) this coincides with the usual notion of convergence in distribution.

We can now reformulate Wigner's Theorem in Voiculescu's language. Fix $n \geq 1$. We introduce the random (real symmetric) $n \times n$ matrix

$$G^n = (g_{ij})_{1 \leq ij \leq n}$$

with entries defined as follows: $\{g_{ij} \mid i \leq j\}$ is a collection of independent Gaussian real-valued r.v.'s with distribution $N(0, 1/n)$ (i.e., $E(g_{ij}) = 0$ and $E|g_{ij}|^2 = 1/n$) and $g_{ij} = g_{ji} \; \forall i > j$. We assume these (classical sense) random variables defined on a sufficiently rich probability space (for instance, the Lebesgue interval). Let $A_n = L_\infty(\Omega, \mathcal{A}, P; M_n)$ and let φ_n be the state defined on A_n by setting

$$\forall x \in A_n \quad \varphi_n(x) = \int \frac{1}{n} tr(x(\omega)) dP(\omega).$$

Then, Voiculescu's reformulation of Wigner's Theorem is

Theorem 9.9.1. *If we consider G^n as a real C^*-r.v. relative to (A_n, φ_n), then we have the weak convergence of probability measures:*

$$\mu_{G_n} \to \mu_W \quad \text{when } n \to \infty.$$

More generally, Voiculescu showed:

Theorem 9.9.2. *Let $(G^n_i)_{i \in I}$ be a family of independent copies (in the usual sense) of the random variable G^n. Then, when $n \to \infty$, the family $(G^n_i)_{i \in I}$ converges in distribution to a free family $(W_i)_{i \in I}$ of real C^*-r.v.'s each with the same distribution equal to μ_W.*

We will say that a real C^*-r.v. x is semi-circular if there exists $\lambda > 0$ such that the distribution of λx is equal to μ_W. If $\lambda = 1$, and if x admits exactly

μ_W for its distribution, then we will say that x is semi-circular standard. We then have $\varphi(x) = 0$, $\varphi(x^2) = 1$. (We should warn the reader that our standard normalization differs from that of [VDN].)

Actually we even have an almost sure result as follows:

Theorem 9.9.3. *Let $(G_i^n)_{i \in I}$ be as above. Then, for almost all ω in Ω, we have*

$$\sup_n \|G_i^n(\omega)\|_{M_n} < \infty$$

for each i and, moreover, the distribution of $(G_i^n(\omega))_{i \in I}$ (on M_n equipped with its normalized trace) tends to that of $(W_i)_{i \in I}$ when $n \to \infty$.

Proof. This is based on a (so-called) "concentration of measure" argument. Actually, the most basic form of Sobolev's inequality in the Gaussian setting will suffice for our purposes. Let $\{g_m\}$ be any collection of independent standard (i.e., $N(0,1)$) real-valued Gaussian random variables. Then it is well known (cf., e.g., [Che]) that, for any polynomial $F = F(g_1, g_2, \ldots)$, we have

$$\|F - \mathbb{E}F\|_2 \leq \|\nabla F\|_2, \tag{9.9.4}$$

where

$$\|\nabla F\|_2^2 = \sum_m \left\| \frac{\partial F}{\partial x_m}(g_1, g_2, \ldots) \right\|_2^2.$$

Now let $P = P(\{X_i\})$ be a polynomial in noncommutative variables $\{X_i\}$, and let

$$F_n = \tau_n(P(\{G_i^n\})).$$

As we already mentioned, we have almost surely

$$\limsup_{n \to \infty} \|G_i^n\|_{M_n} < \infty$$

and also for any $p < \infty$

$$\limsup_{n \to \infty} \mathbb{E}\|G_i^n\|_{M_n}^p < \infty.$$

Using this, it is rather easy to deduce from (9.9.4) that there is a constant C_P depending only on P such that

$$\|F_n - \mathbb{E}F_n\|_2 \leq C_P n^{-1}. \tag{9.9.5}$$

Indeed, it suffices to majorize $\|\nabla F\|_2$ accordingly when P is a *monomial*, that is a finite product $X_{i_1} X_{i_2} \ldots X_{i_k}$, and in that case the preceding observations and the specific normalizations of τ_n and G_i^n yield (9.9.5). Then (9.9.5) implies for any $\varepsilon > 0$

$$\sum P(|F_n - \mathbb{E}F_n| > \varepsilon) < \infty,$$

and hence for almost all ω

$$\limsup_{n \to \infty} |F_n - \mathbb{E}F_n| = 0.$$

On the other hand, by Voiculescu's central limit theorem (i.e., Theorem 9.9.2) we know that

$$\mathbb{E}F_n \to \tau(P(\{W_i\})).$$

Hence, we conclude that for almost all ω we have

$$\tau_n(P(\{G_i^n(\omega)\})) \to \tau(P(\{W_i\})). \tag{9.9.6}$$

Finally, taking a suitable countable intersection, we find a subset $\Omega' \subset \Omega$ with $P(\Omega') = 1$ such that, for any ω in Ω', (9.9.6) is satisfied by *all* polynomials P with (say) rational coefficients and moreover $\sup_n \|G_i^n(\omega)\| < \infty$. This yields the announced result. ∎

Remark. In the preceding proof the passage from convergence in distribution to the almost sure one is based on a very general principle called the *concentration of measure* phenomenon, which has many important applications in geometry and analysis (see [MS]). As pointed out by Voiculescu in [Vo2, pp. 216–217], the random matrix situation (either in the Gaussian or unitary case, for instance) is usually so "extremely concentrated" that results such as the preceding one are then easy to derive using this principle. Note in passing that we chose to use only the elementary Sobolev bound (9.9.4), but more refined estimates are actually available (see, e.g., [MS] or [P8, pp. 44–48]). For other methods to obtain similar almost sure results see [HiP, T].

In Voiculescu's theory, the analog of an *independent* family of standard real Gaussian variables is a *free* family of standard semi-circular C^*-r.v.'s. Such a family can be realized on the full Fock space, as follows. Let $H = \ell_2(I)$. Recall we denote by $\mathcal{F}(H)$(or simply by $\mathcal{F}$) the full Fock space associated to H; that is to say, we set $\mathcal{H}_0 = \mathbb{C}$, $\mathcal{H}_n = H^{\otimes n}$ (Hilbertian tensor product) and finally

$$\mathcal{F} = \oplus_{n \geq 0}\mathcal{H}_n.$$

We consider from now on $\mathcal{H}_n$ as a subspace of $\mathcal{F}$. For every $h \in \mathcal{F}$, we denote by $\ell(h)\colon \mathcal{F} \to \mathcal{F}$ the operator defined by:

$$\ell(h)x = h \otimes x.$$

More precisely, if $x = \lambda 1 \in \mathcal{H}_0 = \mathbb{C}1$, we have $\ell(h)x = \lambda h$, and if $x = x_1 \otimes x_2 \ldots \otimes x_n \in \mathcal{H}_n$, we have $\ell(h)x = h \otimes x_1 \otimes x_2 \ldots \otimes x_n$. We will denote by Ω the unit element in $\mathcal{H}_0 = \mathbb{C}1$. The C^*-algebra $B(\mathcal{F})$ is equipped with the state φ defined by

$$\varphi(T) = <T\Omega, \Omega>.$$

Let $(e_i)_{i \in I}$ be an orthonormal basis of H.

The pair $(B(\mathcal{F}), \varphi)$ is an example of a C^*-probability space. Moreover, φ is tracial on the C^*-algebra generated by the operators $\ell(e_i) + \ell(e_i)^*$ $(i \in I)$, that is, we have $\varphi(xy) = \varphi(yx)$ for all x, y in this subalgebra. (Note however that $\varphi(\ell(h)^*\ell(h)) = \langle h, h \rangle$ and $\varphi(\ell(h)\ell(h)^*) = 0$, so that φ is not tracial on the whole of $B(\mathcal{F})$!)

In this subalgebra, let

$$W_i = \ell(e_i) + \ell(e_i)^*.$$

Then the family $(W_i)_{i \in I}$ is an example of a free family of standard semi-circular C^*-r.v.'s, or, in short, a standard semi-circular free family. This family enjoys properties very much analogous to those of a standard independent Gaussian family $(g_i)_{i \in I}$. Indeed, for every family $(\alpha_i)_{i \in I} \in \mathbb{R}^{(I)}$ with $\sum \alpha_i^2 = 1$ the real C^*-r.v. $S = \sum_{i \in I} \alpha_i W_i$ admits μ_W as its distribution. This is analogous to the rotational invariance of the usual Gaussian distributions. More explicitly, this means that, for every continuous function $f \colon \mathbb{R} \to \mathbb{R}$, we have

$$\varphi(f(S)) = \int f(t)\mu_W(dt).$$

In particular, by the fact preceding (9.9.3), for all finitely supported families of real scalars we have

$$\left\| \sum_{i \in I} \alpha_i W_i \right\| = 2\left(\sum \alpha_i^2\right)^{1/2}.$$

Thus, the operator space $\mathbb{R}$-linearly generated by $(W_i)_{i \in I}$ is isometric to a real Hilbert space.

We now pass to the complex case. Let $(Z_i)_{i \in I}$ be a family of (not necessarily self-adjoint) C^*-r.v.'s. We can then consider the distribution F of the family of real C^*-r.v.'s obtained by forming the disjoint union of the family of real parts and that of imaginary parts of $(Z_i)_{i \in I}$. We will say that F is the joint $*$-distribution of the family $(Z_i)_{i \in I}$. Of course, if the family is reduced to one variable Z, we will say that F is the $*$-distribution of Z. Note that the data of the $*$-distribution of $(Z_i)_{i \in I}$ are equivalent to those of all possible moments of the form

$$\varphi(X_{i_1} X_{i_2} \ldots X_{i_n}),$$

where $X_i = $ either Z_i or Z_i^* and where $i_1, i_2, \ldots, i_n$ are arbitrary in I.

We now come to the analog of complex Gaussian random variables. Let (W', W'') be a standard semi-circular free family (with two elements). We set $\widetilde{W} = \frac{1}{\sqrt{2}}(W' + iW'')$.

Every C^*-r.v. having the same $*$-distribution as $\widetilde{W}$ (resp. as $\lambda\widetilde{W}$ for some $\lambda > 0$) will be called "standard circular" (resp. "circular").

Suppose we are given a (partitioned) orthonormal basis $\{e_i \mid i \in I\} \cup \{f_i \mid i \in I\}$ of H. Then, one can show that $\widetilde{W}_i = \ell(e_i) + \ell(f_i)^*$ is a $*$-free family of standard circular C^*-r.v.'s (in short, a standard circular $*$-free family).

Now, let $(\widetilde{W}_i)_{i \in I}$ be any $*$-free family formed of standard circular variables. Then, for any finitely supported family $(\alpha_i)_{i \in I}$ of complex scalars with $\sum |\alpha_i|^2 = 1$, the variable $\widetilde{S} = \sum_{i \in I} \alpha_i \widetilde{W}_i$ has the same $*$-distribution as $\widetilde{W}$. As above, we have

$$\Big\| \sum \alpha_i \widetilde{W}_i \Big\| = 2\Big(\sum |\alpha_i|^2\Big)^{1/2} \tag{9.9.7}$$

(one can verify that $\|\widetilde{W}\| = 2$).

Let $\mathcal{V}_I$ be the operator space spanned by this family $\{\widetilde{W}_i \mid i \in I\}$. By (9.9.7), $\mathcal{V}_I$ is isometrically Hilbertian and $(\widetilde{W}_i)_{i \in I}$ is an orthonormal basis. Moreover (see [VDN, p. 56]) for any isometric transformation $U \colon \mathcal{V}_I \to \mathcal{V}_I$ the family $(U(\widetilde{W}_i))_{i \in I}$ has the same $*$-distribution as $(\widetilde{W}_i)_{i \in I}$. In particular, by the next lemma, this implies that the operator space $\mathcal{V}_I$ is homogeneous.

Lemma 9.9.4. *Let (A, φ) and (B, ψ) be two C^*-probability spaces with φ and ψ faithful. Let $(Z_i)_{i \in I}$ and $(Y_i)_{i \in I}$ be two families of C^*-r.v.'s in A and in B, respectively, admitting the same joint $*$-distribution. Let A_Z (resp. B_Y) be the C^*-algebra generated by $(Z_i)_{i \in I}$ (resp. $(Y_i)_{i \in I}$), and let $E_Z \subset A_Z$ (resp. $E_Y \subset B_Y$) be the operator space spanned by the families. Then the linear mapping U defined by $U(Z_i) = Y_i$ extends to a complete isometry from E_Z onto E_Y and actually to an isometric representation from A_Z onto B_Y.*

Proof. Without restricting the generality, we may replace the family $(Z_i)_{i \in I}$ by the disjoint union of the families $(Z_i)_{i \in I}$ and $(Z_i^*)_{i \in I}$, and similarly for the family $(Y_i)_{i \in I}$. Then let $P = \sum \alpha_{i_1 i_2 \ldots i_k} Z_{i_1} \ldots Z_{i_k}$ be a polynomial with complex coefficients in the noncommutative variables $(Z_i)_{i \in I}$. We set

$$\pi(P) = \sum \alpha_{i_1 i_2 \ldots i_k} Y_{i_1} \ldots Y_{i_k}.$$

Then, since (Z_i) and (Y_i) have the same joint $*$-distribution, P^*P and $\pi(P)^*\pi(P)$ have the same distribution; hence, by the fact stated before (9.9.3), since φ and ψ are faithful, $\|P\| = \|\pi(P)\|$. In particular, π extends to an isometric representation from A_Z onto B_Y. A fortiori (cf. Proposition 1.5), the restriction U of π to E_Z is completely isometric. ∎

Actually, it is very easy to identify the operator space $\mathcal{V}_I$ (up to complete isomorphism), as the next result shows (see [HP2] for some refinements).

Theorem 9.9.5. _The operator space $\mathcal{V}_I$ generated by a standard circular $*$-free family $(\widetilde{W}_i)_{i \in I}$ is Hilbertian and homogeneous. Similarly, the closed span of a free semi-circular family $\{(W_i)_{i \in I}\}$ is 2-Hilbertian and 2-homogeneous. Moreover, if (say) $I = \mathbb{N}$, each of these spaces is completely isomorphic to $R \cap C$ or, equivalently, to E_λ._

Proof. We already saw that $\mathcal{V}_I$ is Hilbertian and homogeneous.

Let $(a_i)_{i \in I}$ be a finitely supported family in $B(H)$. The identity $W_i = \ell(e_i) + \ell(e_i)^*$ together with $\sum \ell(e_i)\ell(e_i)^* \leq I$ yields

$$\left\| \sum a_i \otimes W_i \right\|_{\min} \leq \left\| \sum a_i \otimes \ell(e_i) \right\|_{\min} + \left\| \sum a_i \otimes \ell(e_i)^* \right\|_{\min}$$
$$\leq \left\| \sum a_i^* a_i \right\|^{1/2} + \left\| \sum a_i a_i^* \right\|^{1/2}$$

whence

$$\left\| \sum a_i \otimes W_i \right\|_{\min} \leq 2 \left\| \sum a_i \otimes \delta_i \right\|_{\min}. \qquad (9.9.8)$$

Conversely, it is easy to check that $\varphi(W_i^* W_j) = \varphi(W_j W_i^*) = 0$ if $i \neq j$ and $= 1$ otherwise. Hence, letting $T = \sum a_i \otimes W_i$, we have $\| \sum a_i^* a_i \| = \|(I \otimes \varphi)(T^*T)\| \leq \|T\|_{\min}^2$, and similarly we have $\| \sum a_i a_i^* \| \leq \|T\|_{\min}^2$. It follows that

$$\max \left\{ \left\| \sum a_i^* a_i \right\|^{1/2}, \left\| \sum a_i a_i^* \right\|^{1/2} \right\} \leq \left\| \sum a_i \otimes W_i \right\|_{\min}. \qquad (9.9.9)$$

The inequalities (9.9.8) and (9.9.9) imply that $\overline{\text{span}}[W_i \mid i \in I]$ is 2-Hilbertian and 2-homogeneous.

For simplicity, we assume $I = \mathbb{N}$ in the rest of the proof. By (9.7.3) and Theorem 9.7.1, the last two inequalities imply that the closed span of $(W_i)_{i \in I}$ is completely isomorphic to E_λ or, equivalently, to $R \cap C$. Finally, as the variables $\widetilde{W}_j = (W_j' + iW_j'')2^{-1/2}$ appear as a sequence of "blocks" (normalized in ℓ_2) on a standard semi-circular system, the same inequalities (9.9.8) and (9.9.9) remain valid if we replace $(W_i)_{i \in I}$ by $(\widetilde{W}_i)_{i \in I}$. Therefore, we conclude that $\mathcal{V}_I$ itself is completely isomorphic to E_λ or to $R \cap C$. This last point can also be deduced from the concrete realization $\widetilde{W}_i = \ell(e_i) + \ell(f_i)^*$ already mentioned for a standard circular $*$-free system. ∎

Remark 9.9.6. Let M be the von Neumann algebra generated by a free semi-circular family $(W_i)_{i \in I}$. We assume $I = \mathbb{N}$ for simplicity. Recall a classical notation: For any x in M, we define $x\varphi \in M_*$ by $x\varphi(y) = \varphi(yx)$ for all y in M. Thus we obtain a continuous injection $M \to M_*$ that allows us to consider the interpolation spaces $(M, M_*)_\theta$ for $0 < \theta < 1$. Let us denote for simplicity $L_\infty(\varphi) = M$, $L_1(\varphi) = M_*$, and $L_p(\varphi) = (M, M_*)_\theta$ with $\theta = 1/p$.

Let us denote by $\mathcal{W}_p$ the closed linear span of $(W_i)_{i \in I}$ in $L_p(\varphi)$. In analogy with Corollary 9.7.2, we claim that the orthogonal projection $\mathcal{P}$ from $L_2(\varphi)$ onto $\mathcal{W}_2$ defines a completely bounded projection from $L_p(\varphi)$ onto $\mathcal{W}_p$ for any $1 \leq p \leq \infty$. Here again, the case $p = 2$ is clear since, by Theorem 7.10, $L_2(\varphi)$ is $OH(I)$ for some set I. Therefore, by interpolation and transposition, it suffices to prove this claim for $p = \infty$. The latter case can be justified as follows: Given a Hilbert space H, let us denote by H_r (resp. H_c) the space H equipped with the row (resp. column) operator space structure associated to $B(H^*, \mathbb{C})$ (resp. $B(\mathbb{C}, H)$). It is easy to check that the natural inclusion map $M \to L_2(\varphi)$ is completely contractive from M to $L_2(\varphi)_r$ (resp. $L_2(\varphi)_c$). Hence (recall that W_i is normalized in $L_2(\varphi)$) $\mathcal{P}$ induces a completely contractive mapping $T\colon M \to R \cap C$ defined by

$$\forall x \in M \qquad T(x) = \sum_i \delta_i \, \varphi(x W_i^*).$$

Let $V\colon R \cap C \to M$ be the mapping defined by $V(\delta_i) = W_i$. By (9.9.8), the composition $VT\colon M \to M$ satisfies $\|VT\|_{cb} \leq \|V\|_{cb} \leq 2$. Moreover, VT is the adjoint of an operator on M_* and VT "coincides" with $\mathcal{P}$ on the $*$-algebra generated by $(W_i)_{i \in I}$. Therefore, VT is a completely bounded projection from M onto $\mathcal{W}_\infty$, which naturally extends $\mathcal{P}$. By transposition, we obtain a c.b. projection from $L_1(\varphi)$ onto $\mathcal{W}_1$ and by interpolation from $L_p(\varphi)$ onto $\mathcal{W}_p$ for all $1 \leq p \leq \infty$. This establishes the above claim.

Therefore, exactly as in Corollary 9.7.2, we conclude that $\mathcal{W}_1 \simeq \mathcal{W}_\infty^*$ and $\mathcal{W}_p \simeq (\mathcal{W}_\infty, \mathcal{W}_1)_\theta$ (completely isomorphically) with $\theta = 1/p$. But, by Theorem 9.9.5, we already know that $\mathcal{W}_\infty \simeq R \cap C$; hence, by duality $\mathcal{W}_1 \simeq R + C$ and, consequently, by Corollary 9.8.8, we obtain again $\mathcal{W}_p \simeq R[p] + C[p]$ if $p \leq 2$ and $\mathcal{W}_p \simeq R[p] \cap C[p]$ if $p \geq 2$. The case of circular variables can be treated by the same argument; thus, to recapitulate, we can state

Theorem 9.9.7. *For simplicity, let $I = \mathbb{N}$. Let $(W_i)_{i \in I}$ (resp. $(\widetilde{W}_i)_{i \in I}$) be a standard free semi-circular (reps. $*$-free circular) family. For $1 \leq p \leq \infty$, let $\mathcal{W}_p$ (resp. $\widetilde{\mathcal{W}}_p$) be the closed span of $(W_i)_{i \in I}$ (resp. $(\widetilde{W}_i)_{i \in I}$) in $L_p(\varphi)$. Then, for any $p < \infty$, $\mathcal{W}_p$ and $\widetilde{\mathcal{W}}_p$ are completely isomorphic to the Gaussian subspace $\mathcal{G}_p$ (or to the space $\mathcal{R}_p$) considered in §9.8. The correspondences $W_i \to \gamma_i$, $\widetilde{W}_i \to \widetilde{\gamma}_i$ (and also $W_i \to \widetilde{W}_i$), or $W_i \to \varepsilon_i$ all define complete isomorphisms between the corresponding L_p-subspaces. Moreover, the orthogonal projection defines a c.b. projection from $L_p(\varphi)$ onto $\mathcal{W}_p$ (or onto $\widetilde{\mathcal{W}}_p$) for any $1 \leq p \leq \infty$.*

We refer the reader to [VDN] for a description of the various forms of Voiculescu's central limit theorem which is a generalization of Theorem 9.9.1.

Note that there is a Fermionic variant in which a free semi-circular family appears as the limit of suitably normalized matrices with entries satisfying the CAR. On the other hand, the reader will find in [Sk] a description of the applications of Voiculescu's theory to von Neumann algebras.

9.10. Embeddings of von Neumann algebras into ultraproducts

In this section, we discuss ultraproducts in the von Neumann sense. This is a slightly different notion from the usual one considered in §2.8. We will prove in detail that the von Neumann algebra of a free group embeds into a (von Neumann sense) ultraproduct of *finite-dimensional* matrix algebras. We will also describe several important related open questions for which operator space theory might be useful.

Let $\{M(n) \mid n \geq 1\}$ be a sequence of von Neumann algebras equipped with normal faithful traces $\{\tau_n \mid n \geq 1\}$ with $\tau_n(1) = 1$. Let $B = \ell_\infty(\{M(n) \mid n \geq 1\})$, and let $\mathcal{U}$ be a free ultrafilter on $\mathbb{N}$. We define a functional $f \in B^*$ by setting for all $t = (t_n)_{n\geq 1}$ in B

$$f_{\mathcal{U}}(t) = \lim_{\mathcal{U}} \tau_n(t_n).$$

Clearly $f_{\mathcal{U}}$ is a tracial state on B. Let H_n be the Hilbert space associated to the GNS construction for $(M(n), \tau_n)$. Note that H_n can be viewed as the noncommutative version of the L_2-space over a probability space. (Indeed, if $M(n)$ is commutative, τ_n can be identified with a probability on its spectrum and $H_n \simeq L_2(\tau_n)$). Let ξ_n be the (cyclic) unit vector of H_n corresponding to the identity of $M(n)$. We then have

$$\forall\ x_n \in M(n) \qquad \tau_n(x_n) = \langle x_n \xi_n, \xi_n \rangle.$$

Let $\widehat{H}_{\mathcal{U}} = \Pi H_n / \mathcal{U}$. We will denote by $(\widehat{h_n})$ the equivalence class in $\widehat{H}_{\mathcal{U}}$ of a bounded sequence (h_n) with $h_n \in H_n$ for all n. Clearly, $M(n)$ acts by left and right multiplication on H_n. Therefore, passing to the ultraproduct we obtain representations

$$\widehat{L} \colon B \to B(\widehat{H}_{\mathcal{U}}) \quad \text{and} \quad \widehat{R} \colon B^{op} \to B(\widehat{H}_{\mathcal{U}})$$

defined by

$$\widehat{L}((b_n))(\widehat{h_n}) = (\widehat{b_n h_n}) \quad \text{and} \quad \widehat{R}((b_n))(\widehat{h_n}) = (\widehat{h_n b_n}).$$

But actually, this Hilbert space $\widehat{H}_{\mathcal{U}}$ is "too large." We need to reduce it and to consider the restriction of $\widehat{L}$ and $\widehat{R}$ to a smaller space $H_{\mathcal{U}}$, which we now describe. The space $H_{\mathcal{U}}$ can be defined as the Hilbert space associated to the tracial state $f_{\mathcal{U}}$ in the GNS construction applied to B. We denote by

$$L \colon B \to B(H_{\mathcal{U}}) \quad \text{and} \quad R \colon B^{op} \to B(H_{\mathcal{U}})$$

the representations of B corresponding to left and right multiplication by an element of B. More precisely, let $p(x) = \lim_{\mathcal{U}} \tau_n(x_n^* x_n)^{1/2}$. Clearly p is a Hilbertian semi-norm on B. Let

$$I_{\mathcal{U}} = \ker(p).$$

Then $I_{\mathcal{U}}$ is a closed two-sided ideal, and $H_{\mathcal{U}}$ is defined as the completion of $B/I_{\mathcal{U}}$ equipped with the Hilbertian norm associated to p.

For any $t = (t_n)_n$ in B we denote by $\dot{t}$ the equivalence class of t in $B/I_{\mathcal{U}}$. Then L and R are defined by $L(x)\dot{t} = \widehat{xt}$ and $R(t)\dot{t} = \widehat{tx}$. Clearly, since $f_{\mathcal{U}}$ is tracial, these are contractive representations of B on $H_{\mathcal{U}}$.

Note that we have a natural isometric embedding

$$H_{\mathcal{U}} \hookrightarrow \widehat{H}_{\mathcal{U}}, \tag{9.10.1}$$

which takes $\dot{t}$ to $\hat{t}$. Clearly L and R are nothing but the restrictions of $\widehat{L}$ and $\widehat{R}$ to the invariant subspace $H_{\mathcal{U}}$.

The next result seems to go back to Mac Duff's early work (see [Sa]). Note that the kernel $I_{\mathcal{U}}$ is *not* weak-* closed in B, so the fact that the quotient $B/I_{\mathcal{U}}$ is a von Neumann algebra is a priori somewhat surprising. We call $B/I_{\mathcal{U}}$ the (von Neumann) ultraproduct of the family $(M(n), \tau_n)$ with respect to $\mathcal{U}$. Note that a priori, this is a quotient of the (Banach space) ultraproduct described in §2.8.

Theorem 9.10.1. *The kernels of L and R coincide with the set*

$$I_{\mathcal{U}} = \{t \in B \mid \lim_{\mathcal{U}} \tau_n(t_n^* t_n) = 0\}.$$

After passing to the quotient, L and R define isometric representations

$$L_{\mathcal{U}} \colon B/I_{\mathcal{U}} \to B(H_{\mathcal{U}}) \quad \text{and} \quad R_{\mathcal{U}} \colon B^{op}/I_{\mathcal{U}} \to B(H_{\mathcal{U}})$$

with commuting ranges. Moreover, $f_{\mathcal{U}}$ defines a faithful trace $\tau_{\mathcal{U}}$ on $B/I_{\mathcal{U}}$ such that, if $q \colon B \to B/I_{\mathcal{U}}$ denotes the quotient map, we have

$$\tau_{\mathcal{U}}(q(t)) = f_{\mathcal{U}}(t) \qquad \forall\, t \in B.$$

Finally, the commutants satisfy

$$[L_{\mathcal{U}}(B/I_{\mathcal{U}})]' = R_{\mathcal{U}}(B^{op}/I_{\mathcal{U}}) \quad \text{and} \quad [R_{\mathcal{U}}(B^{op}/I_{\mathcal{U}})]' = L_{\mathcal{U}}(B/I_{\mathcal{U}}).$$

In particular, the ranges $L_{\mathcal{U}}(B/I_{\mathcal{U}})$ and $R_{\mathcal{U}}(B^{op}/I_{\mathcal{U}})$ are von Neumann sub-algebras of $B(H_{\mathcal{U}})$, and they are factors (i.e., their center is reduced to the scalars) if all the $M(n)$ are themselves factors.

Proof. Let 1_n be the unit of $M(n)$ and let $\xi = (1_n)_n \in B$. We have

$$f_{\mathcal{U}}(t) = \langle L(t)\dot{\xi}, \dot{\xi} \rangle = \langle R(t)\dot{\xi}, \dot{\xi} \rangle.$$

If $L(t) = 0$, then $L(t^*t) = 0$, which, by the preceding line, implies $f_{\mathcal{U}}(t^*t) = 0$; hence $t \in I_{\mathcal{U}}$. Conversely, if $t \in I_{\mathcal{U}}$, then $x^*t^*tx \in I_{\mathcal{U}}$ for any x in B and hence (Cauchy-Schwarz) $f_{\mathcal{U}}(x^*t^*tx) = 0$, which means $\overset{\frown}{tx} = 0$ for all x in B or, equivalently, $L(t) = 0$.

A similar argument applies for R, so we obtain that $\ker(L) = \ker(R) = I_{\mathcal{U}}$. Then, after passing to the quotient by $I_{\mathcal{U}}$, L and R define the isometric representations $L_{\mathcal{U}}$ and $R_{\mathcal{U}}$ with the same respective ranges. Therefore, $L_{\mathcal{U}}$ and $R_{\mathcal{U}}$ still have commuting ranges.

Finally, let $T \in B(H_{\mathcal{U}})$ be an operator commuting with $L_{\mathcal{U}}(B/I_{\mathcal{U}})$, that is, $T \in L_{\mathcal{U}}(B/I_{\mathcal{U}})'$. We will show that T must be in the range of $R_{\mathcal{U}}$. Let

$$\beta = T(\dot{\xi}) \in H_{\mathcal{U}}.$$

We claim that there is $b = (b_n)$ in B such that $\beta = \dot{b}$ and that

$$T = R(b) = R_{\mathcal{U}}(\dot{b}).$$

Indeed, we have for any $t = (t_n)$ in B

$$TL(t)\dot{\xi} = L(t)T\dot{\xi} = L(t)\beta; \qquad (9.10.2)$$

hence

$$\|L(t)\beta\|_{H_{\mathcal{U}}} \le \|T\| \, \|L(t)\dot{\xi}\|_{H_{\mathcal{U}}} = \|T\| \, \|\dot{t}\|_{H_{\mathcal{U}}}. \qquad (9.10.3)$$

We now use the embedding (9.10.1). Let (β_n) with $\beta_n \in H_n$ and $\sup \|\beta_n\|_{H_n} < \infty$ be a representative of β in $\widehat{H}_{\mathcal{U}}$. Then (9.10.3) implies for any t in B

$$\lim_{\mathcal{U}} \tau_n(\beta_n\beta_n^*t_n^*t_n) \le \|T\|^2 \lim_{\mathcal{U}} \tau_n(t_n^*t_n). \qquad (9.10.4)$$

Let $\beta_n = h_n v_n$ be the polar decomposition of β_n in H_n (see the next remark) with $h_n \in H_n$, $h_n \ge 0$, v_n partial isometry in $M(n)$, and $h_n = (\beta_n\beta_n^*)^{1/2}$. Fix $\varepsilon > 0$. Let p_n be the spectral projection of h_n associated to $)\|T\| + \varepsilon, \infty($. Note that $\beta_n\beta_n^*p_n = h_n^2 p_n \ge (\|T\| + \varepsilon)^2 p_n$. A priori (p_n) defines an element of $\widehat{H}_{\mathcal{U}}$, but actually, since $\beta \in H_{\mathcal{U}}$, it is rather easy to show that (p_n) also corresponds to an element in $H_{\mathcal{U}}$. We leave this point to the reader (see the next remark for some clarification). Hence (9.10.4) implies (with $t_n = p_n$)

$$(\|T\| + \varepsilon)^2 \lim_{\mathcal{U}} \tau_n(p_n) \le \lim_{\mathcal{U}} \tau_n(\beta_n\beta_n^*p_n) \le \|T\|^2 \lim_{\mathcal{U}} \tau_n(p_n).$$

This forces $\lim_{\mathcal{U}} \tau_n(p_n) = 0$, and hence $\lim_{\mathcal{U}} \tau_n(\beta_n\beta_n^*p_n) = 0$.

Therefore, if we set finally $b_n = (1 - p_n)h_n v_n$, we find $\|b_n\| \le \|(1 - p_n)h_n\| \le \|T\| + \varepsilon$ and $\|\beta_n - b_n\|_{H_n}^2 \le \|p_n h_n v_n\|_{H_n}^2 \le \tau_n(\beta_n\beta_n^*p_n)$; hence $\lim_{\mathcal{U}} \|\beta_n - b_n\|_{H_n} = 0$. Let $b = (b_n)$. Note that $b \in B$ with $\|b\|_B \le \|T\| + \varepsilon$. Then, going back to (9.10.2), we obtain finally

$$TL(t)\dot{\xi} = L(t)\beta = L(t)\dot{b} = (\overset{\frown}{t_n b_n}) = R_{\mathcal{U}}(b)L(t)\dot{\xi}.$$

This shows that $T = R_{\mathcal{U}}(b)$, which completes the proof that $L_{\mathcal{U}}(B/I_{\mathcal{U}})' = R_{\mathcal{U}}(B^{op}/I_{\mathcal{U}})$. The same argument clearly yields $R_{\mathcal{U}}(B^{op}/I_{\mathcal{U}})' = L_{\mathcal{U}}(B/I_{\mathcal{U}})$, and hence $L_{\mathcal{U}}(B/I_{\mathcal{U}})'' = L_{\mathcal{U}}(B/I_{\mathcal{U}})$, which proves (von Neumann's bicommutant theorem) that $L_{\mathcal{U}}(B/I_{\mathcal{U}})$ is a von Neumann algebra. ∎

Remark. In the above, we invoked the polar decomposition in H_n, viewing H_n as the noncommutative L_2-space relative to τ_n and using its structure as a bimodule over $M(n)$. But actually, we will apply the preceding result only in the case when $M(n)$ is a matrix algebra. In that case, all the points left to the reader in the preceding proof are especially easy to verify and the polar decomposition is then done in $M(n)$ itself.

Remark 9.10.2. By construction, $H_{\mathcal{U}}$ appears as the closure in $\widehat{H}_{\mathcal{U}}$ of the subspace of all elements of the form $(\widehat{b_n})$ with $\sup_n \|b_n\|_{M(n)} < \infty$. Alternatively, $H_{\mathcal{U}} \subset \widehat{H}_{\mathcal{U}}$ can also be described as the subspace corresponding to the "uniformly square integrable" sequences. More precisely, let $\widehat{\beta} = (\widehat{\beta_n})_n$ be an element of $\widehat{H}_{\mathcal{U}}$, with $\sup_n \|\beta_n\|_{H_n} < \infty$. Then $\widehat{\beta}$ belongs to $H_{\mathcal{U}}$ iff

$$\lim_{c \to \infty} \lim_{\mathcal{U}} \tau_n(\beta_n \beta_n^* 1_{\{\beta_n \beta_n^* > c\}}) = 0$$

or iff

$$\lim_{c \to \infty} \lim_{\mathcal{U}} \tau_n(\beta_n^* \beta_n 1_{\{\beta_n^* \beta_n > c\}}) = 0,$$

where we have denoted (abusively) by $1_{\{h > c\}}$ the spectral projection of the Hermitian operator h for the interval $)c, \infty($.

A group G is called *residually finite* if there exists a collection of finite groups (G_i) and homomorphisms $\varphi_i \colon G \to G_i$ separating the points of G; that is, for any finite subset $S \subset G$ there is an i for which the restriction of φ_i to S is injective. Without loss of generality, we may assume that $G_i = G/\Gamma_i$, where each $\Gamma_i \subset G$ is a normal subgroup with finite index and φ_i is the canonical quotient map. Thus, G is residually finite iff it admits a family of normal subgroups with finite index (Γ_i), directed by (downward) inclusion and such that $\bigcap_{i \in I} G_i = \{e\}$.

Corollary 9.10.2. *([Wa1]) Let G be any countable residually finite discrete group. Then $VN(G)$ embeds into an ultraproduct of the form $B/I_{\mathcal{U}}$ as above with all the algebras $M(n)$ finite-dimensional.*

Proof. Let Γ_n be a decreasing sequence of normal subgroups of finite index with intersection reduced to the unit e, and let $G_n = G/\Gamma_n$. Let $\varphi_n \colon G \to G_n$ denote the quotient morphism and let $M(n) = VN(G_n)$ equipped with its normalized trace τ_n. We consider $B = \ell_\infty(\{M(n) \mid n \geq 1\})$ and its quotient $B/I_{\mathcal{U}}$ as above. For any t in G, we denote by $y(t)$ the equivalence class modulo $I_{\mathcal{U}}$ of $(\lambda_{G_n}(t))_{n \geq 1} \in B$.

Let τ denote the normalized trace on $VN(G)$. We then have

$$\forall\, t \in G \qquad \lim_{\mathcal{U}} \tau_n(\lambda_{G_n}(t)) = \tau(\lambda_G(t)).$$

This implies that the family $\{y(t) \mid t \in G\}$ has the same $*$-distribution with respect to $\tau_{\mathcal{U}}$ as the family $\{\lambda_G(t) \mid t \in G\}$ with respect to τ. Therefore, by Lemma 9.9.4, the von Neumann algebras that they generate are isomorphic, via the isomorphism taking $\lambda_G(t)$ to $y(t)$.   ∎

The following fact is classical.

Lemma 9.10.3. *Free groups are residually finite.*

Proof. Let $G = F_I$. Let $\{g_i \mid i \in I\}$ be the (free) generators. Let $C \subset G$ be a finite subset. It suffices to produce a (group) homomorphism $h\colon G \to \Gamma$ into a finite group Γ such that, for any c in C, we have $h(c) \neq e_\Gamma$ if $c \neq e$, where e_Γ denotes the unit in Γ and e the unit in G.

We may assume that $C \subset G'$, where G' is the subgroup generated by a finite subset $\{g_i \mid i \in J\}$ of the generators. Let $k = \max\{|c| \mid c \in C\}$ (recall $|c|$ is the length of c). We then set

$$S = \{t \in G' \mid |t| \leq k\}.$$

We will take for Γ the (finite) group of all permutations of the (finite) set S. For any i in J, we introduce

$$S_i = \{t \in S \mid g_i t \in S\}.$$

Then clearly $S_i \subset S$ and $g_i S_i \subset S$. Hence (since $|S_i| = |g_i S_i|$ and S is finite) there is a permutation $\sigma_i\colon S \to S$ such that $\sigma_i(s) = g_i s$ for any s in S_i. Thus if $s, t \in S$ and if $g_i t = s$ (or, equivalently, $t = g_i^{-1} s$), we have $\sigma_i(t) = s$ (or, equivalently, $t = \sigma_i^{-1}(s)$). Thus it is easy to check that if a reduced word $t = g_{i_1}^{\varepsilon_1} g_{i_2}^{\varepsilon_2} \ldots g_{i_m}^{\varepsilon_m}$ ($m \leq k$ $\varepsilon_i = \pm 1$) lies in S (note that, by definition of S, e and all the subwords of t also lie in S), we have

$$\sigma_{i_1}^{\varepsilon_1} \sigma_{i_2}^{\varepsilon_2} \ldots \sigma_{i_m}^{\varepsilon_m}(e) = t.$$

Therefore, if we define $h\colon G \to \Gamma$ as the unique homomorphism such that $h(g_i) = \sigma_i \ \forall i \in J$ and $\sigma(g_i) = e_\Gamma \ \forall i \notin J$, we find, for t as before, $h(t) = \sigma_{i_1}^{\varepsilon_1} \sigma_{i_2}^{\varepsilon_2} \ldots \sigma_{i_m}^{\varepsilon_m}$ and $h(t)(e) = t$ in particular, we have $h(t) \neq e_\Gamma$ whenever $t \in S$ and $t \neq e$. Since $C \subset S$, we obtain the announced result.   ∎

Consequently:

Theorem 9.10.4. *([Wa1]) The von Neumann algebra of the free groups $\mathbb{F}_n$ or $\mathbb{F}_\infty$ embeds into a von Neumann ultraproduct of matrix algebras.*

We wish to give a second proof of the preceding result using Voiculescu's "matrix model" for free semi-circular systems. This approach is well known to specialists. See also [T] for more along this line.

Theorem 9.10.5. *Let M be the von Neumann algebra generated by a free semi-circular system $(W_i)_{i\in I}$ (as in Theorem 9.9.2) with I countable. Then M embeds into a von Neumann ultraproduct of matrix algebras.*

Proof. Let τ_n (resp. τ) be the normalized trace on M_n (resp. on M). Let $B = \ell_\infty(\{M_n \mid n \geq 1\})$. Let $\mathcal{U}$ be any free ultrafilter on $\mathbb{N}$ and let $I_\mathcal{U}$ be as before. By Theorem 9.9.3 (using the same notation), we know that, for almost all ω, $(G_i^n(\omega))_{n\geq 1}$ is in B. Moreover, if we denote by $\widehat{G}_i(\omega)$ the equivalence class modulo $I_\mathcal{U}$ associated to $(G_i^n(\omega))_{n\geq 1}$, then the family $\{\widehat{G}_i(\omega) \mid i \in I\}$ has the same distribution with respect to $(B/I_\mathcal{U}, \tau_\mathcal{U})$ as the system $\{W_i \mid i \in I\}$ with respect to τ. Thus, by Lemma 9.9.4, the von Neumann algebras they generate must be isomorphic. ∎

In [Co1], Alain Connes observes in passing the above Theorem 9.10.4 and casually asks whether the same embedding is valid for any II_1 factor. This has become one of the main open problems in von Neumann algebra theory. Actually, since, by desintegration, any finite von Neumann algebra (on a separable Hilbert space) can be seen as the "direct integral" of a field of factors, one may reformulate Connes's question as follows:

Problem. *Let (M, τ) be a von Neumann algebra (on a separable Hilbert space) equipped with a faithful, normal, and normalized trace τ. Is there always a (trace preserving) embedding of M (as a von Neumann subalgebra) in an ultraproduct of matrix algebras?*

9.11. Dvoretzky's Theorem

It is natural to wonder whether there is an analog for operator spaces of the famous Dvoretzky Theorem on the spherical sections of convex bodies. This fundamental result can be formulated as follows: Every infinite-dimensional Banach space X contains, for each $\varepsilon > 0$, a sequence of subspaces $E_n \subset X$ ($n = 1, 2, ...$) such that E_n is $(1 + \varepsilon)$-isomorphic to ℓ_2^n for every n.

Vitali Milman (see [MS] and [FLM]) gave a remarkable proof based on Paul Lévy's isoperimetric inequality on the unit sphere of $\mathbb{R}^n$. It is then tempting to try and find in every (infinite-dimensional) operator space X a sequence of subspaces (E_n) uniformly completely isomorphic to OH_n. Unfortunately, this is absurd: For a counterexample it suffices to consider a homogeneous Hilbertian X different from OH (for instance, R or C). Then, by Proposition 9.2.1, every n-dimensional subspace $E_n \subset X$ is completely isometric to the n-dimensional version of X (for instance, R_n if $X = R$). Therefore, one cannot find inside X anything else but "copies" of X. The best that one can hope to find in an arbitrary (infinite-dimensional) operator space seems to be a sequence of almost Hilbertian and almost homogeneous subspaces (E_n) (here "almost" means up to ε) with $\dim E_n = n$ for all n. In this form, it is rather easy to adapt Milman's ideas (the concentration of measure phenomenon; see [MS]) to the *quantum* situation. There is a slight difficulty due to the noncompactness of the unit ball of the *quantum scalar coefficients*, that is, the unit ball of $\mathcal{K}$ (see the discussion in §2.11). To circumvent this difficulty, we choose the language of ultraproducts, which is convenient in the present situation. See §2.8 for precise definitions.

With the ultraproduct terminology, we can reformulate Dvoretzky's Theorem as follows:

Theorem 9.11.1. *Let $(X_i)_{i \in I}$ be a family of infinite-dimensional Banach spaces (or merely such that $\lim_{\mathcal{U}} \dim(X_i) = \infty$). Then their ultraproduct $\Pi_{i \in I} X_i / \mathcal{U}$ contains a subspace isometric to ℓ_2.*

The "quantum" version of this statement is then (cf. [P9]):

Theorem 9.11.2. *Let $(X_i)_{i \in I}$ be a family of infinite-dimensional operator spaces (or merely such that $\lim_{\mathcal{U}} \dim(X_i) = \infty$). Then their ultraproduct $\Pi_{i \in I} X_i / \mathcal{U}$ contains a subspace completely isometric to an infinite-dimensional Hilbertian homogeneous operator space.*

This statement explains perhaps why Hilbertian homogeneous operator spaces appear so often in "nature."

Chapter 10. Comparisons

In the previous chapters, we have met the following examples of homogeneous (or λ-homogeneous) Hilbertian operator spaces:

$$OH, \quad \Phi, \quad R, \quad C, \quad R \cap C, \quad R + C, \quad \min(\ell_2), \quad \max(\ell_2).$$

We will now show that all the spaces in that list are mutually completely nonisomorphic. More precisely, let E, F be any two spaces in the preceding list, and let E_n, F_n be their n-dimensional version. We will show that $d_{cb}(E_n, F_n) \to \infty$, and we will give an asymptotic estimate of $d_{cb}(E_n, F_n)$ when $n \to \infty$.

Our aim is to illustrate what can be done with the ideas of operator space theory, by elementary tools. In the very early days of Banach space theory, methods were discovered to distinguish, up to isomorphism, the so-called classical Banach spaces, such as the spaces ℓ^p or L^p (see [Ba, Chapter 12]) and also to estimate the Banach-Mazur distances $d(\ell_n^p, \ell_n^q)$ with $1 \le p \ne q \le \infty$. Our aim in this chapter is to initiate an analogous study for operator spaces.

Many estimates in this chapter come from C. Zhang's PhD thesis ([Z3]) or can be easily derived by his methods. (The only exception may be the fact that Φ and $\max(\ell_2)$ are not isomorphic, which was only established in [JP].)

The following result, from [Z3], will be very convenient throughout this chapter (see also [Z1, Z2]).

Proposition 10.1. *Let E, F be two Hilbertian homogeneous operator spaces. Assume for simplicity that E and F are isometric to ℓ_2. Let (e_n) (resp. (f_n)) be an orthonormal basis of E (resp. F). Let $E_n = \mathrm{span}[e_1, ..., e_n]$ (resp. $F_n = \mathrm{span}[f_1, ..., f_n]$).*

(i) *Let $u_n \colon E_n \to F_n$ (resp. $u \colon E \to F$) be the (isometric) linear map defined by $u_n(e_i) = f_i$ (resp. $u(e_i) = f_i$). Then we have*

$$d_{cb}(E_n, F_n) = \|u_n\|_{cb} \|u_n^{-1}\|_{cb} \tag{10.1}$$

and

$$d_{cb}(E, F) = \|u\|_{cb} \|u^{-1}\|_{cb} = \sup_n \|u_n\|_{cb} \|u_n^{-1}\|_{cb}. \tag{10.2}$$

(ii) *We also have*

$$d_{cb}(E_n, \overline{E_n^*}) = d_{cb}(E_n, OH_n)^2. \tag{10.3}$$

Proof. Using the polar decomposition of an $n \times n$ matrix, it is easy to see that $d_{cb}(E_n, F_n) = \inf\{\|v\|_{cb} \|v^{-1}\|_{cb}\}$, where the infimum runs over all maps $v \colon E_n \to F_n$ of the form $v(e_i) = \lambda_i f_i$ with $\lambda_i > 0$. Let σ be a permutation of $[1, 2, .., n]$. Let U_σ (resp. V_σ) be the unitary transformation defined on E_n by

$U_\sigma(e_i) = e_{\sigma(i)}$ (resp. $V_\sigma(f_i) = f_{\sigma(i)}$). Since E_n, F_n are homogeneous, we can write:

$$\|\frac{1}{n!} \sum_\sigma V_{\sigma^{-1}} \, v \, U_\sigma\|_{cb} \le \|v\|_{cb}.$$

Since $\frac{1}{n!} \sum_\sigma V_{\sigma^{-1}} \, v \, U_\sigma = (1/n)(\sum \lambda_i) u_n$, we deduce $\|u_n\|_{cb}(1/n)(\sum \lambda_i) \le \|v\|_{cb}$. Similarly, exchanging the roles of E_n and F_n, we find $\|u_n^{-1}\|_{cb}(1/n)(\sum \lambda_i^{-1}) \le \|v^{-1}\|_{cb}$. By Cauchy-Schwarz, we have $1 \le n^{-1}(\sum \lambda_i)n^{-1}(\sum \lambda_i^{-1})$, whence

$$\|u_n\|_{cb}\|u_n^{-1}\|_{cb} \le \|v\|_{cb}\|v^{-1}\|_{cb}.$$

Taking the infimum over all possible v, we find

$$\|u_n\|_{cb}\|u_n^{-1}\|_{cb} \le d_{cb}(E_n, F_n).$$

Since the inverse inequality is obvious, we obtain (10.1).

Let $w \colon E \to F$ be an arbitrary isomorphism. By Proposition 9.2.1, $w(E_n)$ is completely isometric to F_n. Therefore we have $d_{cb}(E_n, F_n) \le d_{cb}(E_n, w(E_n)) \le \|w\|_{cb}\|w^{-1}\|_{cb}$, whence $d_{cb}(E_n, F_n) \le d_{cb}(E, F)$. This implies

$$\sup_n \|u_n\|_{cb}\|u_n^{-1}\|_{cb} \le d_{cb}(E, F). \tag{10.2$'$}$$

Note then that $\|u\|_{cb}$ (resp. $\|u^{-1}\|_{cb}$) is the nondecreasing limit of $\|u_n\|_{cb}$ (resp. $\|u_n^{-1}\|_{cb}$) when $n \to \infty$.

From the obvious inequality $d_{cb}(E, F) \le \|u\|_{cb}\|u^{-1}\|_{cb}$, we deduce the converse of $(10.2)'$ and obtain (10.2). To show (ii), observe that if $u \colon \overline{E_n^*} \to E_n$ denotes the map defined by the canonical isometry from $\overline{E_n^*}$ onto E_n, and if $v \colon OH_n \to E_n$ is an arbitrary isometry, we have $u = v^t \overline{v}$ hence $\|u\|_{cb} = \|v\|_{cb}^2$ by (7.9). ∎

Remark. In [Z2], Zhang made the interesting observation that, for any normed space E,

$$d_{cb}(\min(E), \max(E)) = \|\min(E) \to \max(E)\|_{cb},$$

or, equivalently, $d_{cb}(\min(E), \max(E))$ is equal to Paulsen's constant $\alpha(E)$ described in Chapter 3. The argument is quite simple: Let $i_E \colon \min(E) \to \max(E)$ be the identity. Then, for any complete isomorphism $u \colon \min(E) \to \max(E)$ we have $i_E = (i_E u^{-1})u$, but since $\max(E)$ is homogeneous, we have $\|i_E u^{-1}\|_{cb} = \|i_E u^{-1}\| = \|u^{-1}\|$; hence

$$\|i_E\|_{cb} \le \|i_E u^{-1}\|_{cb}\|u\|_{cb} \le \|u^{-1}\| \, \|u\|_{cb} \le \|u^{-1}\|_{cb}\|u\|_{cb},$$

which implies

$$\|i_E\|_{cb} \le d_{cb}(\min(E), \max(E)).$$

The converse is obvious. ∎

Convention. Throughout the sequel, we will write $a_n \simeq b_n$ if there exist positive constants c_1 and c_2 such that $c_1 a_n \leq b_n \leq c_2 a_n$ for all n.

Notation. Let E_n and F_n be two (isometrically) Hilbertian operator spaces of dimension n. Assume F_n homogeneous. If $u: E_n \to F_n$ and $v: E_n \to F_n$ are two isometries; then $\|u\|_{cb} = \|v\|_{cb}$. Indeed, $uv^{-1}: F_n \to F_n$ and $vu^{-1}: F_n \to F_n$ are isometric; hence by the homogeneity of F_n we have $\|uv^{-1}\|_{cb} = \|vu^{-1}\|_{cb} = 1$, which obviously implies $\|u\|_{cb} \leq \|uv^{-1}\|_{cb}\|v\|_{cb} = \|v\|_{cb}$ and conversely $\|v\|_{cb} \leq \|u\|_{cb}$. We will simply denote by

$$\|E_n \to F_n\|_{cb}$$

the c.b. norm of *any* isometry from E_n into F_n.

By the triangle inequality, this c.b. norm is $\leq n$, whence

$$1 \leq \|E_n \to F_n\|_{cb} \leq n. \tag{10.4}$$

Therefore we can rewrite (10.1) as follows:

$$d_{cb}(E_n, F_n) = \|E_n \to F_n\|_{cb}\|F_n \to E_n\|_{cb}. \tag{10.1}'$$

We start by estimating the distance $d_{cb}(OH_n, E_n)$ when E is any one of the spaces in the above list. Recall that we know (Corollary 7.7) that, if E_n is an arbitrary n-dimensional operator space, we have $d_{cb}(OH_n, E_n) \leq \sqrt{n}$.

Proposition 10.2. *The following estimates hold:*

$$d_{cb}(OH_n, R_n) = d_{cb}(OH_n, C_n) = \sqrt{n}, \tag{10.5}$$

$$d_{cb}(OH_n, \min(\ell_2^n)) = d_{cb}(OH_n, \max(\ell_2^n)) \simeq \sqrt{n}, \tag{10.6}$$

$$d_{cb}(OH_n, \Phi_n) = d_{cb}(OH_n, \overline{\Phi_n^*}) \simeq \sqrt{n}, \tag{10.7}$$

$$d_{cb}(OH_n, R_n \cap C_n) = d_{cb}(OH_n, R_n + C_n) = n^{1/4}. \tag{10.8}$$

Proof. We have

$$\|OH_n \to R_n\|_{cb} = \left\|\sum_1^n e_{1i} \otimes \overline{e_{1i}}\right\|_{\min}^{1/2} = n^{1/4}, \tag{10.9}'$$

$$\|R_n \to OH_n\|_{cb} = \|OH_n^* \to R_n^*\|_{cb} = \left\|\sum_1^n e_{i1} \otimes \overline{e_{i1}}\right\|_{\min}^{1/2} = n^{1/4}, \tag{10.9}''$$

and similarly with C_n instead of R_n; whence (10.5). We could also deduce this from (10.3) and the identity $d_{cb}(C_n, R_n) = n$ already mentioned above.

By Theorem 3.8

$$\| \min(\ell_2^n) \to \max(\ell_2^n) \|_{cb} \simeq n, \tag{10.10}$$

whence (10.6) by (10.3) and (10.1)'.

Proof. To show (10.7), let $\{V_i\}$ be as in §9.3. Note that $\|OH_n \to \Phi_n\|_{cb} = \| \sum V_i \otimes \overline{V_i} \|_{\min}^{1/2}$. Then (with the same notation as in the proof of Proposition 9.7.3) $\| \sum V_i \otimes \overline{V_i} \|_{\min} \geq \sum_1^n \tau(V_i V_i^*) = n/2$, and hence $\|OH_n \to \Phi_n\|_{cb} \geq (n/2)^{1/2}$. Conversely, we have obviously (by the triangle inequality) $\| \sum V_i \otimes \overline{V_i} \|_{\min}^{1/2} \leq \sqrt{n}$, and hence

$$\|OH_n \to \Phi_n\|_{cb} \simeq \sqrt{n}. \tag{10.11}$$

By Proposition 9.7.3 and by (2.11.4), we have $\|\Phi_n \to OH_n\|_{cb} \leq \sqrt{2}$. This establishes (10.7), taking into account (10.1), and Theorem 9.3.1. Finally, to check (10.8), we observe

$$\|OH_n \to R_n \cap C_n\|_{cb} = \| \sum \delta_i \otimes \overline{\delta_i} \|_{\min}^{1/2}$$
$$= \max\{\| \sum \delta_i^* \delta_i \|^{1/4}, \| \sum \delta_i \delta_i^* \|^{1/4}\} = n^{1/4}.$$

In the converse direction, the inequality (2.11.4) implies $\|R_n \cap C_n \to OH_n\|_{cb} = 1$. Then by (10.1) we have (10.8). ∎

Note in passing that, since $\|R_n \cap C_n \to OH_n\|_{cb} = 1$ and (see Proposition 9.7.3)

$$\|\Phi_n \to R_n \cap C_n\|_{cb} \simeq 1, \tag{10.12}$$

we have (taking (10.4) into account)

$$\|\Phi_n \to OH_n\|_{cb} \simeq 1, \tag{10.13}$$

and hence $\|\Phi_n \to \overline{\Phi_n^*}\|_{cb} \simeq 1$. By (10.11) and (10.3) we therefore have

$$\|\overline{\Phi_n^*} \to \Phi_n\|_{cb} \simeq n. \tag{10.14}$$

Moreover, since we have trivially $\|OH_n \to \min(\ell_2^n)\|_{cb} = \| \max(\ell_2^n) \to OH_n\|_{cb} = 1$, (10.6) implies

$$\| \min(\ell_2^n) \to OH_n\|_{cb} = \|OH_n \to \max(\ell_2^n)\|_{cb} \simeq \sqrt{n}. \tag{10.15}$$

By Proposition 10.1(ii) we immediately deduce from the preceding statement:

Corollary 10.3. *The following estimates hold:*

$$d_{cb}(C_n, R_n) \simeq d_{cb}(\min(\ell_2^n), \max(\ell_2^n)) \simeq d_{cb}(\Phi_n, \overline{\Phi_n^*}) \simeq n,$$

$$d_{cb}(R_n \cap C_n, R_n + C_n) = \sqrt{n}.$$

In the next statement we compare Φ_n to the other spaces in the list.

Theorem 10.4. *The following estimates hold:*

$$d_{cb}(\Phi_n, R_n) \simeq d_{cb}(\Phi_n, C_n) \simeq \sqrt{n}, \tag{10.16}$$

$$d_{cb}(\Phi_n, R_n \cap C_n) \simeq \sqrt{n}, \tag{10.17}$$

$$d_{cb}(\Phi_n, R_n + C_n) \simeq \sqrt{n}, \tag{10.18}$$

$$d_{cb}(\Phi_n, \min(\ell_2^n)) \simeq n, \tag{10.19}$$

$$d_{cb}(\Phi_n, \max(\ell_2^n)) \simeq \sqrt{n}. \tag{10.20}$$

Proof. It is easy to check that $\| \sum V_i^* V_i \| \simeq \| \sum V_i V_i^* \| \simeq n$ (in fact $= n$), so that

$$\|R_n \to \Phi_n\|_{cb} \simeq \|C_n \to \Phi_n\|_{cb} \simeq \sqrt{n}. \tag{10.21}$$

Conversely, we have (10.12); hence a fortiori (recalling (10.4))

$$\|\Phi_n \to R_n\|_{cb} \simeq \|\Phi_n \to C_n\|_{cb} \simeq 1, \tag{10.22}$$

whence (10.16).

To check (10.17), let $c = \|R_n \cap C_n \to \Phi_n\|_{cb}$. We have clearly

$$c \le \|R_n \cap C_n \to R_n\|_{cb}\|R_n \to \Phi_n\|_{cb} = \|R_n \to \Phi_n\|_{cb} \simeq \sqrt{n}.$$

Conversely, we can show (cf. [HP2, Remark 1.2]) that there exists a decomposition $V_i = a_i + b_i$ with $\| \sum a_i^* a_i \|^{1/2} \le c$ and $\| \sum b_i b_i^* \|^{1/2} \le c$. This yields (with the notation in the proof of Proposition 22.3, using $\tau(b_i b_i^*) = \tau(b_i^* b_i)$)

$$(n/2)^{1/2} = (\sum \tau(V_i^* V_i))^{1/2} \le (\sum \tau(a_i^* a_i))^{1/2} + (\sum \tau(b_i^* b_i))^{1/2} \le 2c.$$

Hence finally $c \ge \sqrt{n}(2\sqrt{2})^{-1}$. Thus we have $c = \|R_n \cap C_n \to \Phi_n\|_{cb} \simeq \sqrt{n}$. By (10.12) and (10.1)′ we then obtain (10.17).

To check (10.18), note that by (10.21) and (10.22) we have (a fortiori)

$$\|R_n + C_n \to \Phi_n\|_{cb} \simeq \sqrt{n} \quad \text{and} \quad \|\Phi_n \to R_n + C_n\|_{cb} \simeq 1,$$

whence (10.18).

To check (10.19), note that (trivially) $\|\Phi_n \to \min(\ell_2^n)\|_{cb} = 1$. Then we can write by (10.14)

$$n \simeq \|\overline{\Phi_n^*} \to \Phi_n\|_{cb} \le \|\overline{\Phi_n^*} \to \min(\ell_2^n)\|_{cb} \, \| \min(\ell_2^n) \to \Phi_n\|_{cb} = \| \min(\ell_2^n)$$
$$\to \Phi_n\|_{cb},$$

whence (by (10.4)) $\| \min(\ell_2^n) \to \Phi_n\|_{cb} \simeq n$. Thus we obtain (10.19).

To check (10.20), note that (trivially) $\| \max(\ell_2^n) \to \Phi_n\|_{cb} = 1$. Then (by (10.13) and (10.15))

$$\|\Phi_n \to \max(\ell_2^n)\|_{cb} \le \|\Phi_n \to OH_n\|_{cb}\|OH_n \to \max(\ell_2^n)\|_{cb} \simeq \sqrt{n},$$

whence $d_{cb}(\Phi_n, \max(\ell_2^n)) \le C\sqrt{n}$ with C independant of n. In the converse direction, the results of [JP] show that $\max(\ell_2))$ is not exact, contrary to Φ. The corresponding estimate from [JP] (see Exercise 19.2 below) is $d_{SK}(\max(\ell_2^n)) \ge 4^{-1}\sqrt{n}$, whence $d_{cb}(\Phi_n, \max(\ell_2^n)) \ge 4^{-1}\sqrt{n}$. Thus we obtain (10.20). ∎

It is now easy to complete these results. First we can compare R and C to the other spaces in the list (recall $d_{cb}(R_n, C_n) = n$):

Theorem 10.5. *The following estimates hold:*

$$d_{cb}(R_n, R_n \cap C_n) = d_{cb}(C_n, R_n \cap C_n) = \sqrt{n},$$

$$d_{cb}(C_n, R_n + C_n) = d_{cb}(R_n, R_n + C_n) = \sqrt{n},$$

$$d_{cb}(R_n, \min(\ell_2^n)) = d_{cb}(C_n, \min(\ell_2^n)) = d_{cb}(R_n, \max(\ell_2^n))$$
$$= d_{cb}(C_n, \max(\ell_2^n)) = \sqrt{n}.$$

Proof. By (1.5) we have $\|R_n \to C_n\|_{cb} = \|C_n \to R_n\|_{cb} = \sqrt{n}$, from which it immediately follows that $\|R_n \to R_n \cap C_n\|_{cb} = \|C_n \to R_n \cap C_n\|_{cb} = \sqrt{n}$. On the other hand, we can verify easily (using (1.11) and (3.1))

$$\| \min(\ell_2^n) \to R_n\|_{cb} = \| \min(\ell_2^n) \to C_n\|_{cb} = \sqrt{n}. \tag{10.23}$$

It is then an easy task to prove the missing estimates using the duality. ∎

Let us now compare $R_n \cap C_n$ and $R_n + C_n$ to the remaining spaces.

Theorem 10.6. *The following estimates hold:*

$$d_{cb}(R_n \cap C_n, \min(\ell_2^n)) \simeq d_{cb}(R_n + C_n, \min(\ell_2^n)) \simeq \sqrt{n},$$

$$d_{cb}(R_n \cap C_n, \max(\ell_2^n)) \simeq d_{cb}(R_n + C_n, \max(\ell_2^n)) \simeq \sqrt{n}.$$

Proof. From (10.23) we deduce that

$$\| \min(\ell_2^n) \to R_n \cap C_n \|_{cb} = \sqrt{n}.$$

A fortiori, we have $\| \min(\ell_2^n) \to R_n + C_n \|_{cb} \leq \sqrt{n}$, whence, by duality, $\|R_n + C_n \to \max(\ell_2^n)\|_{cb} = \sqrt{n}$. Therefore, we can write by (10.10)

$$n \simeq \| \min(\ell_2^n) \to \max(\ell_2^n) \|_{cb} \leq \| \min(\ell_2^n) \to R_n + C_n \|_{cb} \| R_n + C_n$$
$$\to \max(\ell_2^n) \|_{cb}$$

$$\leq \sqrt{n} \| \min(\ell_2^n) \to R_n + C_n \|_{cb},$$

and we obtain

$$\| \min(\ell_2^n) \to R_n + C_n \|_{cb} \simeq \sqrt{n}.$$

Since we have trivially $\|R_n \cap C_n \to \min(\ell_2^n)\|_{cb} = \|R_n + C_n \to \min(\ell_2^n)\|_{cb} = 1$, we obtain the estimates on the first line and those on the second one follow by duality. ■

We can now state, as a recapitulation:

Theorem 10.7. *The operator spaces*

$$OH, \quad \Phi, \quad R, \quad C, \quad R \cap C, \quad R + C, \quad \min(\ell_2), \quad \max(\ell_2)$$

are mutually completely nonisomorphic.

PART II

OPERATOR SPACES AND C^*-TENSOR PRODUCTS

Chapter 11. C^*-Norms on Tensor Products. Decomposable Maps. Nuclearity

Let A_1, A_2 be two C^*-algebras. Their algebraic tensor product $A_1 \otimes A_2$ is an involutive algebra for the natural operations defined by

$$(a_1 \otimes a_2) \cdot (b_1 \otimes b_2) = a_1 b_1 \otimes a_2 b_2$$

and

$$(a_1 \otimes a_2)^* = a_1^* \otimes a_2^*.$$

Then a norm $\| \ \|$ on $A_1 \otimes A_2$ is called a C^*-norm if it satisfies

$$\|x\| = \|x^*\|, \qquad \|xy\| \le \|x\| \, \|y\| \quad \text{and} \quad \|x^* x\| = \|x\|^2$$

for any x, y in $A_1 \otimes A_2$.

It can be shown that we then automatically have

$$\|a_1 \otimes a_2\| = \|a_1\| \, \|a_2\| \quad \forall a_1 \in A_1, \forall a_2 \in A_2. \tag{11.0}$$

This subject was initiated in the 1950s by Turumaru in Japan. Later work by Takesaki [Ta1] and Guichardet [Gu1] lead to the following striking result.

Theorem 11.1. *There is a minimal C^*-norm $\| \ \|_{\min}$ and a maximal one $\| \ \|_{\max}$, so that any C^*-norm $\| \cdot \|$ on $A_1 \otimes A_2$ must satisfy*

$$\forall x \in A_1 \otimes A_2 \qquad \|x\|_{\min} \le \|x\| \le \|x\|_{\max}.$$

We denote by $A_1 \otimes_{\min} A_2$ (resp. $A_1 \otimes_{\max} A_2$) the completion of $A_1 \otimes A_2$ for the norm $\| \ \|_{\min}$ (resp. $\| \ \|_{\max}$).

The maximal C^*-norm is easy to describe. We simply write

$$\|x\|_{\max} = \sup \|\pi(x)\|_{B(H)},$$

where the supremum runs over all possible Hilbert spaces H and all possible $*$-homomorphisms $\pi \colon A_1 \otimes A_2 \to B(H)$. It is easy to see that, for any such π, there is a pair of (necessarily contractive) $*$-homomorphisms $\pi_i \colon A_i \to B(H)$ $(i = 1, 2)$ *with commuting ranges* such that

$$\pi(a_1 \otimes a_2) = \pi_1(a_1) \pi_2(a_2) \qquad \forall a_1 \in A_1 \quad \forall a_2 \in A_2.$$

Conversely, any such pair $\pi_i \colon A_i \to B(H)$ $(i = 1, 2)$ of $*$-homomorphisms with commuting ranges determines uniquely a $*$-homomorphism $\pi \colon A_1 \otimes A_2 \to B(H)$ by setting $\pi(a_1 \otimes a_2) = \pi_1(a_1) \pi_2(a_2)$. Thus, we can write for any $x = \sum a_k^1 \otimes a_k^2$ in $A_1 \otimes A_2$

$$\|x\|_{\max} = \sup \left\{ \left\| \sum \pi_1(a_k^1) \pi_2(a_k^2) \right\| \right\},$$

where the supremum runs over all possible such pairs.

Then the inequality $\| \ \|\leq \| \ \|_{\max}$ follows by considering the Gelfand-Neumark embedding of the completion of $(A_1 \otimes A_2, \| \ \|)$ into $B(H)$ for some H. All this goes back to [Gu1]. The lower bound $\| \ \|_{\min} \leq \| \ \|$ is due to Takesaki [Ta1] and is much more delicate. For a proof, see either [Ta3] or [KaR].

The minimal norm can be described as follows: Embed A_1 and A_2 as C^*-subalgebras of $B(H_1)$ and $B(H_2)$, respectively. Then, for any $x = \sum a_i^1 \otimes a_i^2$ in $A_1 \otimes A_2$, $\|x\|_{\min}$ coincides with the norm induced by the space $B(H_1 \otimes_2 H_2)$, that is, we have an embedding (i.e. an isometric $*$-homomorphism) of the completion, denoted by $A_1 \otimes_{\min} A_2$, into $B(H_1 \otimes_2 H_2)$.

In other words, the minimal tensor product of operator spaces, considered in §2.1, when restricted to two C^*-algebras coincides with the minimal C^*-tensor product. Let (B_1, B_2) be another pair of C^*-algebras and consider c.b. maps $u_i \colon A_i \to B_i$ $(i = 1, 2)$. Then (as emphasized already in §2.1) $u_1 \otimes u_2$ defines a c.b. map from $A_1 \otimes_{\min} A_2$ to $B_1 \otimes_{\min} B_2$ with $\|u_1 \otimes u_2\|_{cb} = \|u_1\|_{cb}\|u_2\|_{cb}$.

In sharp contrast, the analogous property does *not* hold for the max-tensor products. However, it does hold if we moreover assume that u_1 and u_2 are *completely positive*, and then (see e.g. [Ta3, p. 218], [Pa1, p. 164], or [Wa2, p. 11]) the resulting map $u_1 \otimes u_2$ is also completely positive (on the max-tensor product), and we have

$$\forall x \in A_1 \otimes A_2 \quad \|u_1 \otimes u_2(x)\|_{B_1 \otimes_{\max} B_2} \leq \|u_1\|\|u_2\| \ \|x\|_{A_1 \otimes_{\max} A_2}. \quad (11.1)$$

This follows from the next two statements.

Theorem 11.2. Let $\varphi \colon A_1 \otimes A_2 \to \mathbb{C}$ be a linear form and let $u_\varphi \colon A_1 \to A_2^*$ be the corresponding linear map. The following are equivalent:

(i) φ extends to a positive linear form in the unit ball of $(A_1 \otimes_{\max} A_2)^*$.

(ii) $u_\varphi \colon A_1 \to A_2^*$ is a contractive c.p. map, that is, $\|u_\varphi\| \leq 1$, and

$$\sum_{i,j} \langle u_\varphi(x_{ij}), y_{ij} \rangle \geq 0 \quad \forall n \ \forall x \in M_n(A_1)_+ \ \forall y \in M_n(A_2)_+.$$

Proof. Assume (i). Recall (11.0). Then clearly $\|u_\varphi\| \leq 1$. Moreover (by the GNS construction) there are a representation $\pi \colon A_1 \otimes_{\max} A_2 \to B(H)$ and ξ in the unit ball of H such that $\varphi(\cdot) = \langle \pi(\cdot)\xi, \xi \rangle$. We may assume that $\pi = \pi_1 \cdot \pi_2$ as above. Let x, y be as in (ii). Let $z = y^{1/2}$, so that $y_{ij} = \sum_k z_{ki}^* z_{kj}$ (note $z_{ik} = z_{ki}^*$). We claim that the matrix $(\pi_1(x_{ij})\pi_2(y_{ij}))$ is positive. Indeed, for each fixed k the matrix

$$(\pi_2(z_{ki})^* \pi_1(x_{ij})\pi_2(z_{kj}))_{ij}$$

is positive and, since π_1, π_2 have commuting ranges, we have

$$\pi_1(x_{ij})\pi_2(y_{ij}) = \sum_k \pi_2(z_{ki})^* \pi_1(x_{ij})\pi_2(z_{kj}).$$

This proves our claim. Let $\widetilde{\xi} \in H \oplus \cdots \oplus H$ (n times) be defined by $\widetilde{\xi} = \xi \oplus \cdots \oplus \xi$. Then we have

$$\sum \langle u_\varphi(x_{ij}), y_{ij} \rangle = \langle [\pi_1(x_{ij})\pi_2(y_{ij})]\widetilde{\xi}, \widetilde{\xi} \rangle \geq 0,$$

which shows that (i) $\Rightarrow$ (ii). Conversely, assume (ii). Consider $t = \sum a_j \otimes b_j$ in $A_1 \otimes A_2$. Then $t^*t = \sum_{i,j} a_i^* a_j \otimes b_i^* b_j$, and hence by (ii)

$$\varphi(t^*t) = \sum_{i,j} \langle u_\varphi(a_i^* a_j), b_i^* b_j \rangle \geq 0,$$

which shows that φ extends to a positive linear form on $A_1 \otimes A_2$. By the GNS construction, there is a $*$-homomorphism $\pi \colon A_1 \otimes A_2 \to B(H)$ and ξ in H such that $\varphi(\cdot) = \langle \pi(\cdot)\xi, \xi \rangle$. Therefore φ extends to a positive linear form on $A_1 \otimes_{\max} A$ of norm $\leq \|\xi\|$. Moreover, if we assume A_1, A_2 and π unital, we have $\|\xi\|^2 = \langle \pi(1 \otimes 1)\xi, \xi \rangle = \langle u_\varphi(1), 1 \rangle \leq \|u_\varphi\|$; hence we obtain (i). In the nonunital case it is easy to modify this argument using an approximate unit (we leave the details to the reader). ∎

Corollary 11.3. *Let* $u_i \colon A_i \to B_i$ ($i = 1, 2$) *be c.p. maps between* C^*-*algebras. Then* $u_1 \otimes u_2 \colon A_1 \otimes A_2 \to B_1 \otimes B_2$ *extends to a c.p. map from* $A_1 \otimes_{\max} A_2$ *to* $B_1 \otimes_{\max} B_2$ *and (11.1) holds.*

Proof. We may assume $\|u_i\| \leq 1$. We will first show that $u_1 \otimes u_2$ extends to a continuous positive map on the max-tensor products. Consider φ in the unit ball of $(B_1 \otimes_{\max} B_2)_+^*$, and let $u_\varphi \colon B_1 \to B_2^*$ be the corresponding contractive linear map. Let $\psi(t) = \varphi((u_1 \otimes u_2)(t)) \; \forall t \in A_1 \otimes A_2$. Then $u_\psi \colon A_1 \to A_2^*$ is given by $u_\psi = (u_2)^* u_\varphi u_1$, and hence is c.p., which shows by Theorem 11.2 that $\psi \in (A_1 \otimes_{\max} A_2)_+^*$ and $\|\psi\|_{(A_1 \otimes_{\max} A_2)^*} = \|u_\psi\| \leq 1$. Since any element in $(A_1 \otimes_{\max} A_2)^*$ decomposes as a linear combination of positive elements, it is clear that $u_1 \otimes u_2$ must be bounded from $(A_1 \otimes A_2, \| \; \|_{\max})$ to $(B_1 \otimes B_2, \| \; \|_{\max})$.

Let $t \in A_1 \otimes_{\max} A_2$ and $\theta = (u_1 \otimes u_2)(t)$. Since $\theta \in (B_1 \otimes_{\max} B_2)_+$ iff $\varphi(\theta) \geq 0$ for all φ in $(B_1 \otimes_{\max} B_2)_+^*$, we clearly have $t \geq 0 \Rightarrow \varphi(\theta) = \psi(t) \geq 0$; hence $t \geq 0 \Rightarrow \theta \geq 0$, so that $u_1 \otimes u_2$ is a positive map on $A_1 \otimes_{\max} A_2$. This shows the positivity of $u_1 \otimes u_2$. Replacing A_1 by $M_n(A_1)$, we obtain its complete positivity. By Exercise 11.5(iii), if A_1, A_2 are unital, we have

$$\|u_1 \otimes u_2 \colon A_1 \otimes_{\max} A_2 \to B_1 \otimes_{\max} B_2\| \leq \|u_1(1)\| \cdot \|u_2(1)\| \leq 1.$$

In the nonunital case, we obtain the same conclusion using approximate units. ∎

Let A, B be C^*-algebras. We will denote by $CP(A, B)$ the set of all completely positive maps $u: A \to B$. Recall (see Exercise 11.5 (iii)) that, for any c.p. map $u: A \to B$, we have $\|u\|_{cb} = \|u\|$ and in the unital case $\|u\| = \|u(1)\|$. Thus, by Corollary 1.8, a unital map $u: A \to B$ is completely positive iff it is completely contractive.

More generally, we denote by $D(A, B)$ the set of all decomposable maps $u: A \to B$, that is, maps that can be written as $u = u_1 - u_2 + i(u_3 - u_4)$ with $u_1, \ldots, u_4 \in CP(A, B)$.

This space can be normed in a fairly straightforward way by setting

$$\|u\|_{[d]} = \inf\{\sum_1^4 \|u_i\|\},$$

where the infimum runs over all the above possible decompositions of u with $u_i \in CP(A, B)$. It is easy to see that $D(A, B)$ equipped with this norm is a Banach space. However, we will use an equivalent norm, the "decomposable norm" $\|u\|_{dec}$, which was introduced by Haagerup in [H1]. This norm allows us to suppress various unnecessary numerical factors (such as 2 or 4) that would appear were we to use the norm $\|u\|_{[d]}$ instead. The *decomposable norm* $\|u\|_{dec}$ is defined as follows. Consider all possible mappings S_1, S_2 in $CP(A, B)$ such that the map $v: A \to M_2(B)$ defined by

$$v(x) = \begin{pmatrix} S_1(x) & u(x^*)^* \\ u(x) & S_2(x) \end{pmatrix}$$

is completely positive. Then we set

$$\|u\|_{dec} = \inf\{\max\{\|S_1\|, \|S_2\|\}\},$$

where the infimum runs over all possible such mappings.

See Chapter 14 for more on $D(A, B)$.

We will use the following list of basic results (from [H3, Proposition 1.3 and Theorem 1.6]):

$$\forall\, u \in D(A, B) \qquad \|u\|_{cb} \le \|u\|_{dec}. \tag{11.2}$$

If u and v are as above, and if, say, $B \subset B(H)$, we can write $u(x) = p_2 v(x) p_1^*$, where $p_i: H \oplus H \to H$ is the projection to the i-th coordinate:

$$\forall\, u \in CP(A, B) \qquad \|u\| = \|u\|_{cb} = \|u\|_{dec}. \tag{11.3}$$

If C is any C^*-algebra, $u \in D(A, B)$, and $v \in D(B, C)$, then $vu \in D(A, C)$ and

$$\|vu\|_{dec} \le \|v\|_{dec}\|u\|_{dec}, \tag{11.4}$$

$$\forall\, u \in CB(A, B(\mathcal{H})) \qquad \|u\|_{cb} = \|u\|_{dec}. \tag{11.5}$$

Now let A_1, A_2, B_1, B_2 be arbitrary C^*-algebras, and let $u_1 \in D(A_1, B_1)$, $u_2 \in D(A_2, B_2)$. We claim that Haagerup's results imply that $u_1 \otimes u_2$ extends to a decomposable map, denoted by $u_1 \otimes_{\max} u_2$, from $A_1 \otimes_{\max} A_2$ into $B_1 \otimes_{\max} B_2$ satisfying

$$\|u_1 \otimes_{\max} u_2\|_{dec} \le \|u_1\|_{dec}\|u_2\|_{dec}. \tag{11.6}$$

By (11.2) we have a fortiori

$$\|u_1 \otimes u_2\|_{CB(A_1 \otimes_{\max} A_2, B_1 \otimes_{\max} B_2)} \le \|u_1\|_{dec}\|u_2\|_{dec}. \tag{11.7}$$

All this extends to tensor products of n-tuples $A_1, A_2, \ldots, A_n$ of C^*-algebras, but the resulting tensor products $A_1 \otimes_{\min} \cdots \otimes_{\min} A_n$ and $A_1 \otimes_{\max} \cdots \otimes_{\max} A_n$ are associative, so that we have, for instance, when $\alpha =$ either min or max,

$$(A_1 \otimes_\alpha A_2) \otimes_\alpha A_3 = A_1 \otimes_\alpha A_2 \otimes_\alpha A_3 = A_1 \otimes_\alpha (A_2 \otimes_\alpha A_3).$$

Therefore the theory of multiple products reduces, by iteration, to that of products of pairs.

We have obviously a bounded $*$-homomorphism $q\colon A_1 \otimes_{\max} A_2 \to A_1 \otimes_{\min} A_2$, which (as all C^*-representations) has a closed range; hence $A_1 \otimes_{\min} A_2$ is C^*-isomorphic to the quotient $(A_1 \otimes_{\max} A_2)/\ker(q)$. The observation that in general q is not injective is at the basis of the theory of nuclear C^*-algebras:

Definition 11.4. *A C^*-algebra A is called nuclear if, for any C^*-algebra B, we have $\|\ \|_{\min} = \|\ \|_{\max}$ on $A \otimes B$ or, in short, if $A \otimes_{\min} B = A \otimes_{\max} B$. In that case, by Theorem 11.1, there is only one C^*-norm on $A \otimes B$.*

This notion was introduced (under a different name) by Takesaki and was especially investigated by Lance [La1].

For example, if $\dim(A) < \infty$, A is nuclear, because $A \otimes B$ (there is no need to complete it!) is already a C^*-algebra; hence it admits a unique C^*-norm. Moreover, all *commutative* C^*-algebras are nuclear and $\mathcal{K} = K(\ell_2)$ is nuclear (see Exercise 11.7), but $B(\ell_2)$ is *not* nuclear (cf. [Wa1]; we will reprove this later at least twice in Chapter 17 and in Chapter 22). Also, the Cuntz algebra (see §9.4) is nuclear.

Other examples or counterexamples can be given among group C^*-algebras. For any discrete group G, one can define the full C^*-algebra $C^*(G)$ and the reduced one $C_\lambda^*(G)$ (see §§9.6 and 9.7 for precise definitions). Then $C^*(G)$ (or $C_\lambda^*(G)$) is nuclear iff G is amenable. So, for instance, if G is the free group on two generators, $C^*(G)$ and $C_\lambda^*(G)$ are not nuclear. (Note that for continuous groups the situation is quite different: Connes [Co1] proved that, for any separable connected locally compact group G, $C^*(G)$ and $C_\lambda^*(G)$ are nuclear.)

Moreover, as already mentioned in Chapter 8, if G_1, G_2 are two discrete groups, we have

$$C^*(G_1) \otimes_{\max} C^*(G_2) \simeq C^*(G_1 \times G_2) \text{ and } C^*_\lambda(G_1) \otimes_{\min} C^*_\lambda(G_2) \simeq C^*_\lambda(G_1 \times G_2).$$

The basic characterizations of nuclear C^*-algebras are incorporated in the next statements.

Theorem 11.5. *Let A be a C^*-algebra. The following are equivalent.*

(i) *A is nuclear.*

(ii) *There is a net (u_i) of finite rank completely positive maps on A, converging pointwise to the identity. (We then say that A has the "completely positive approximation property," in short CPAP.)*

(iii) *There is a net of maps (u_i) admitting a contractive completely positive factorization through matrix algebras of the form*

$$
\begin{array}{ccc}
 & M_{n_i} & \\
v_i \nearrow & & \searrow w_i \\
A & \xrightarrow{\;\;u_i\;\;} & A
\end{array}
$$

(that is, $u_i = w_i v_i$ and v_i, w_i are contractive completely positive) and converging pointwise to the identity.

(iv) *There is a net of maps (u_i) admitting a completely contractive factorization through matrix algebras of the form*

$$
\begin{array}{ccc}
 & M_{n_i} & \\
v_i \nearrow & & \searrow w_i \\
A & \xrightarrow{\;\;u_i\;\;} & A
\end{array}
$$

(i.e., $u_i = w_i v_i$ and v_i, w_i are completely contractive) and converging pointwise to the identity.

The equivalence of (i) and (ii) was obtained independently in [Ki3] and [CE1]. (i) $\Leftrightarrow$ (iii) was obtained in [CE2] and (i) $\Leftrightarrow$ (iv) in [Sm1].

Moreover, using Connes's deep work [Co1], the following striking result was shown in [CE2]. (See §2.5 for a description of the C^*-algebra structure of the bidual A^{**}.)

Theorem 11.6. *A C^*-algebra A is nuclear iff its bidual A^{**} is injective, that is, there is a completely contractive projection P from $B(H)$ onto A^{**}. By Corollary 1.8, P is automatically completely positive.*

The next notion was introduced in [EL].

Definition 11.7. *A von Neumann algebra M is called semi-discrete if the identity map on the predual M_* is the limit of a net of finite rank contractive completely positive maps in the pointwise topology (here we say that a map on M_* is c.p. if its adjoint is c.p. on M). Equivalently, this means that there is a net of finite rank contractive completely positive normal maps $u_i \colon M \to M$ such that, for all x in M, $u_i(x)$ tends to x in the $\sigma(M, M_*)$-topology.*

Remark. In [CE1, p. 75], it is proved that if M is semi-discrete, there exists a net of maps (u_i) admitting a contractive completely positive normal factorization through matrix algebras of the form

$$
\begin{array}{ccc}
 & M_{n_i} & \\
v_i \nearrow & & \searrow w_i \\
M & \xrightarrow{\quad u_i \quad} & M
\end{array}
$$

(i.e., $u_i = w_i v_i$ and v_i, w_i are contractive, completely positive, and normal) such that, for all x in M, $u_i(x)$ tends to x in the $\sigma(M, M_*)$-topology.

Effros and Lance [EL] proved that semi-discreteness implies injectivity. Relying heavily on Connes's work, Choi and Effros ([CE4]) proved the converse on a separable Hilbert space (see [Wa5] for a simpler proof covering the nonseparable case). The key to these developments is Connes's proof that all injective factors on a separable Hilbert space are hyperfinite. A von Neumann algebra is called hyperfinite if there is a directed net of finite-dimensional subalgebras $M_\alpha \subset M$ the union of which is weak-∗ dense in M. It is not too hard to show (but, beware, it is easy to give a false proof of this; see Remark 11.13 later) that hyperfinite implies injective (see [To2] for a proof based on the fact that M is injective iff its commutant is). As a consequence of Connes's work on factors, one finally gets the converse for von Neumann algebras (on a separable Hilbert space). Thus, on a separable Hilbert space, injectivity, semi-discreteness, and hyperfiniteness are all equivalent properties for von Neumann algebras. (See Elliott's papers [Ell1–2] for the nonseparable case, where hyperfiniteness has to be replaced by a different notion of "approximately finite-dimensional.") See [Tor] for an exposition. Simpler proofs of Connes's most difficult results were obtained more recently in [H5] and [Pol–2].

It is rather easy to show that nuclearity passes to ideals (see Corollary 11.9 later). It is known (and nontrivial) that nuclearity passes to quotients (cf. [CE2]), but not to subspaces ([Ch2, Bla1]). Also (cf. [CE2]) nuclearity is preserved under "extensions." This means that if $I \subset A$ is an ideal in a C^*-algebra, and if both I and A/I are nuclear, then A is nuclear. These assertions are easy consequences of Theorem 11.6 and the fact that, for any ideal $I \subset A$, we have a C^*-isomorphism

$$A^{**} \simeq (A/I)^{**} \oplus I^{**}.$$

Indeed, the latter isomorphism shows that A^{**} is injective iff both $(A/I)^{**}$ and I^{**} are injective.

A major difference between the two "extreme" tensor products is that, while the maximal one is projective (see Exercise 11.2), the minimal one is not. On the other hand, the minimal tensor product is injective, but the maximal one is not. More precisely, it can happen that, when B, A_1, A_2 are C^*-algebras with $B \subset A_1$ (as a C^*-subalgebra), the norm induced by $A_1 \otimes_{\max} A_2$ on $B \otimes A_2$ does not coincide with the maximal norm on $B \otimes A_2$. Of course, this difficulty disappears when either B or A_2 is nuclear, since we then have a unique C^*-norm on $B \otimes A_2$. Since this "defect" is a key to the subsequent theory, we now review, for emphasis, some elementary sufficient conditions for the injectivity of the maximal tensor product.

Proposition 11.8. *Let A_1, A_2 be two C^*-algebras, let $B \subset A_1$ be a C^*-subalgebra, and let $E \subset B$ be a subspace. Each of the following conditions is sufficient to ensure that*

$$\forall x \in E \otimes A_2 \qquad \|x\|_{B \otimes_{\max} A_2} = \|x\|_{A_1 \otimes_{\max} A_2}.$$

(i) *There is a contractive completely positive (in short c.p.) projection $P \colon A_1 \to B$ from A_1 onto B.*

(ii) *There is a net of contractive c.p. maps $T_i \colon A_1 \to B$ such that $T_i(x) \to x$ for all x in E.*

(iii) *There is a net of decomposable maps $T_i \colon A_1 \to B$ such that $\|T_i\|_{\mathrm{dec}} \to 1$ and $T_i(x) \to x$ for all x in E.*

Proof. This is an immediate consequence of (11.1) and (11.7). (Recall that, for any c.p. map u between C^*-algebras, we have $\|u\|_{cb} = \|u\| = \|u\|_{\mathrm{dec}}$.)

Corollary 11.9. *Let A_1, A_2 be C^*-algebras. For any (closed, two-sided, self-adjoint) ideal $I \subset A$ we have an isometric embedding*

$$I \otimes_{\max} A_2 \subset A_1 \otimes_{\max} A_2.$$

Proof. As is well known, I satisfies (ii) in the preceding statement (here we take $B = E = I$). Indeed, I has an approximate unit (cf. [Ta3] or Lemma 2.4.4), namely, there is a net (a_i) in I such that $\|a_i\| \leq 1$ with $a_i x \to x$ and $x a_i \to x$ for any x in I. Thus, if we let $T_i(x) = a_i x a_i$, we find (iii). ∎

Remark. Another useful situation of the same kind as in the preceding two statements is provided by the canonical inclusion $A_1 \subset A_1^{**}$. For any C^*-algebra A_2, we have an *isometric* inclusion

$$A_1 \otimes_{\max} A_2 \subset A_1^{**} \otimes_{\max} A_2. \tag{11.8}$$

For the proof, see the solution to Exercise 11.6.

It will be useful to record here the following fact.

Proposition 11.10. *Let $u\colon A_1 \to A_2$ be a decomposable linear map between two C*-algebras, and let B be another C*-algebra. If $I_B \otimes u\colon B \otimes_{\min} A_1 \to B \otimes_{\max} A_2$ is bounded, then*

$$\|I_B \otimes u\|_{B \otimes_{\min} A_1 \to B \otimes_{\max} A_2} \leq \|u\|_{dec}.$$

In particular, this holds for any finite rank map u. Moreover, for any finite rank c.p. map u, we have

$$\|I_B \otimes u\|_{B \otimes_{\min} A_1 \to B \otimes_{\max} A_2} \leq \|u\|.$$

Proof. Let us denote by $I_B \otimes_{\max} u$ the mapping $I_B \otimes u\colon B \otimes_{\max} A_1 \to B \otimes_{\max} A_2$. By (11.7), we have

$$\|I_B \otimes_{\max} u\| \leq \|u\|_{dec}.$$

Observe that the natural morphism $q\colon B \otimes_{\max} A_1 \to B \otimes_{\min} A_1$ is surjective (since it has dense range); hence it defines an isometric isomorphism from $B \otimes_{\max} A_1 / \ker(q)$ to $B \otimes_{\min} A_1$. Let $\tilde{u} = I_B \otimes u\colon B \otimes_{\min} A_1 \to B \otimes_{\max} A_2$. Since $\tilde{u}$ is assumed bounded, we must have $\tilde{u}\, q = I_B \otimes_{\max} u$; hence $(I_B \otimes_{\max} u)(\ker(q)) \subset \{0\}$, and hence $I_B \otimes_{\max} u$ defines a mapping of norm $\leq \|u\|_{dec}$ from $B \otimes_{\max} A_1/\ker(q)$ to $B \otimes_{\max} A_2$, which yields the announced result.

Now, if u has finite rank, it is rather easy to verify that $\tilde{u}$ is bounded. Indeed, it suffices to verify this for a map u of rank 1, but then u factors through $\mathbb{C}$ as a product of decomposable maps $u\colon A_1 \xrightarrow{v} \mathbb{C} \xrightarrow{w} A_2$, so that $\tilde{u}$ can be seen as the composition

$$B \otimes_{\min} A_1 \xrightarrow{I_B \otimes v} B \otimes_{\min} \mathbb{C} = B \otimes_{\max} \mathbb{C} \xrightarrow{I_B \otimes w} B \otimes_{\max} A_2,$$

and hence $\tilde{u}$ is bounded.

Since $\|u\|_{dec} = \|u\|$ when u is c.p., we obtain the last assertion. ∎

Remark 11.11. Let C be a nuclear C*-algebra, and let $A \subset C$ be a C*-subalgebra. Consider a C*-embedding $C \subset B(H)$. Then, for any C*-algebra B, we have an isometric embedding

$$A \otimes_{\min} B \longrightarrow B(H) \otimes_{\max} B. \tag{11.9}$$

Indeed, $C \otimes_{\min} B \to C \otimes_{\max} B$ is isometric since C is nuclear; moreover, $A \otimes_{\min} B \to C \otimes_{\min} B$ and $C \otimes_{\max} B \to B(H) \otimes_{\max} B$ are both clearly contractive. Thus (11.9) is contractive, and hence isometric. We will see in Chapter 17 that this property *characterizes* the (separable) C*-algebras that can be embedded in a nuclear one.

Remark 11.12. The C^*- or von Neumann algebra setting leads to all sorts of implications that are surprising to a Banach space theorist. For instance, the fact that A nuclear (which, by Theorem 11.5, is a kind of approximation property for A) implies A^{**} semi-discrete (which, by definition, is one for A^*) is in sharp contrast with the fact that there exist a Banach space X (separable and with separable dual) with Grothendieck's Banach space approximation property but with dual X^* failing it; see [LT1, p. 34].

Concerning injectivity, Connes's results are formally analogous to part of the theory of $\mathcal{L}_\infty$-spaces (see [LiR]). A Banach space X is called $\mathcal{L}_\infty$ if there is a net of finite-dimensional subspaces $X_\alpha \subset X$ uniformly isomorphic to ℓ^n_∞-spaces of the same dimension with union dense in X. In other words, X can be written as a norm dense directed union of subspaces uniformly isomorphic to injective and finite-dimensional spaces. By [LiR], this holds iff the bidual X^{**} is isomorphic to an injective Banach space, that is, iff X^{**} is isomorphic to a complemented subspace of an L_∞-space. Thus, this notion appears as the Banach space analog of nuclearity. Note that the difficulty in Connes's proof that injective implies hyperfinite, when compared to the analogous fact in [LiR], is that, starting from an injective infinite-dimensional object, one must construct a (suitably dense) directed net of finite-dimensional and injective $*$-*subalgebras*; not surprisingly, it is much easier to construct (finite-dimensional injective) *subspaces* than $*$-*subalgebras*.

Remark 11.13. In yet another direction, we wish to point out that the proof that hyperfiniteness implies injectivity cannot be too simple in view of the following example (kindly pointed out to me by W. B. Johnson): There exists a constant C and a dual Banach space X^*, with a weak-$*$ closed subspace $M \subset X^*$ and a directed net of weak-$*$ closed subspaces $M_\alpha \subset M$ onto which there is a projection $P_\alpha \colon X^* \to M_\alpha$ with $\|P_\alpha\| \leq C$ and with $\cup M_\alpha$ weak-$*$ dense in M but such that there is no bounded projection from X^* onto M.

As we will see in the next chapter, the properties of nuclear C^*-algebras are valid more generally for a certain class of mappings between C^*-algebras. It would have been natural to call these mappings "nuclear," but the risk of confusion with Grothendieck's notion of nuclearity would be too great. Still, in various discussions, the corresponding term is lacking, so (rather daringly!) we propose the following terminology.

Definition 11.14. Let $u \colon X \to A$ be a mapping from an operator space X to a C^*-algebra A. Let B be another C^*-algebra. We will say that u is B-maximizing if $I_B \otimes u$ defines a bounded map from $B \otimes_{\min} X$ to $B \otimes_{\max} A$. We will say that u is maximizing if this happens for all C^*-algebras B.

With this terminology, by definition A is nuclear iff its identity map is maximizing. More generally, we will see later (see Theorem 17.11) that A is "exact" iff the inclusion map $A \subset B(H)$ is maximizing. In the next chapter (see Corollary 12.6) we show that the maximizing maps $u\colon X \to A$ are characterized by a certain approximation property by (finite rank) maps factoring through matrix algebras.

Note that the analogous notion of "maximizing" map makes perfectly good sense in the Banach space category: A map $u\colon X \to Y$ between Banach spaces could be called maximizing if, for any Banach space Z, the map $I_Z \otimes u\colon Z \overset{\vee}{\otimes} X \to Z \overset{\wedge}{\otimes} Y$ is bounded. It is then an easy (and instructive) exercise to show that this holds iff u is the pointwise limit of a net of finite rank maps $u_\alpha\colon X \to Y$ with uniformly bounded nuclear norms. Equivalently, u is an integral operator in Grothendieck's sense; that is, if we view u as acting into Y^{**}, then u can be factorized as

$$X \longrightarrow L_\infty(\mu) \overset{J}{\longrightarrow} L_1(\mu) \longrightarrow Y^{**},$$

where $J\colon L_\infty(\mu) \to L_1(\mu)$ is the inclusion map relative to a probability space (Ω, μ).

Exercises

Exercise 11.1. Let A, B be C^*-algebras, and let $\mathcal{I} \subset B$ be a (closed two-sided) ideal. Prove that the sequence

$$\{0\} \to \mathcal{I} \otimes_{\max} A \to B \otimes_{\max} A \to (B/\mathcal{I}) \otimes_{\max} A \to \{0\}$$

is exact.

More precisely, prove that any x in $(B/\mathcal{I}) \otimes A$ with $\|x\|_{\max} < 1$ admits a lifting $\widetilde{x}$ in $B \otimes A$ such that $\|\widetilde{x}\|_{\max} < 1$.

Exercise 11.2. Let A_1, A_2 be C^*-algebras, and let $I \subset A_1$ be an ideal. Prove the following (isometric) identity

$$(A_1/I) \otimes_{\max} A_2 \simeq (A_1 \otimes_{\max} A_2)/(I \otimes_{\max} A_2).$$

Exercise 11.3. Let A_1, A_2 be two C^*-algebras, and let $\|\ \|$ be a C^*-norm on $A_1 \otimes A_2$. The goal is to show that $\|a_1 \otimes a_2\| \le \|a_1\|\,\|a_2\|$ for all $a_i \in A_i$ ($i = 1, 2$). Prove that $0 \le a_1 \le b_1$ and $a_2 \ge 0$ implies $\|a_1 \otimes a_2\| \le \|b_1 \otimes a_2\|$. Then, using $a_1^* x^* x a_1 \le \|x\|^2 a_1^* a_1$, prove that $\|x a_1 \otimes a_2\| \le \|x\|\,\|a_1 \otimes a_2\|$, and deduce $\|x a_1 \otimes y a_2\| \le \|x\|\,\|y\|\,\|a_1 \otimes a_2\|$ for $x \in A_1, y \in A_2$. Finally, choosing $x = a_1^*$ and $y = a_2^*$, prove that $\|a_1 \otimes a_2\| \le \|a_1\|\,\|a_2\|$.

Exercise 11.4. Let A_1, A_2 be two C^*-algebras, and let $\pi\colon A_1 \otimes A_2 \to B(H)$ be a $*$-homomorphism. The goal is to show that there are representations $\pi_i\colon A_i \to B(H)$ with commuting ranges such that

$$\pi(a_1 \otimes a_2) = \pi_1(a_1)\pi_2(a_2) \qquad (a_i \in A_i).$$

One can proceed as follows:

(i) Let $K = \overline{\pi(A_1 \otimes A_2)H}$. Show that π decomposes as $\pi(\cdot)_{|K} \oplus 0$ with respect to $H = K \oplus K^{\perp}$. Thus the problem reduces to the case $K = H$.

(ii) Let $\{x_\alpha\}$ (resp. $\{y_\beta\}$) be an approximate unit in the unit ball of A_1 (resp. A_2), so that $x_\alpha x \to x$ (resp. $y_\beta y \to y$) for any x in A_1 (resp. y in A_2). Assuming $K = H$ and using the preceding exercise, prove that, for any a_i in A_i, the nets $\pi(a_1 \otimes y_\beta)$ and $\pi(x_\alpha \otimes a_2)$ converge in the strong operator topology of $B(H)$. We denote by $\pi_1(a_1)$ and $\pi_2(a_2)$ the respective limits. Show that these limits do not depend on the choice of the approximate units.

(iii) Show that π_1, π_2 satisfy the announced properties.

Exercise 11.5.

(i) Let p, q, a be operators on H. Assume $p, q \geq 0$. Show that

$$\begin{pmatrix} p & a \\ a^* & q \end{pmatrix} \geq 0$$

iff $|\langle ax, y \rangle|^2 \leq \langle py, y \rangle \langle qx, x \rangle \ \forall x, y \in H$.

(ii) In particular,

$$\begin{pmatrix} 1 & a \\ a^* & 1 \end{pmatrix} \geq 0 \Leftrightarrow \|a\| \leq 1.$$

(iii) Let $u \colon A \to B$ be a c.p. map between C^*-algebras. If A is unital, show that $\|u\| = \|u(1)\|$. If $a_\alpha \geq 0$ is a net in A such that $a_\alpha^{1/2} a a_\alpha^{1/2} \to a$ for any a in A, show that $\|u\| \leq \sup_\alpha \|u(a_\alpha)\|$. Conclude that $\|u\|_{cb} = \|u\|$. In any case note that

$$\|u\| = \sup\{\|u(a)\| \mid a \geq 0, \|a\| \leq 1\}.$$

Exercise 11.6.

(i) For any C^*-algebras A_1, A_2 we have an isometric embedding

$$A_1 \otimes_{\max} A_2 \to A_1^{**} \otimes_{\max} A_2.$$

(ii) More generally, we have an isometric embedding

$$A_1 \otimes_{\max} A_2 \to A_1^{**} \otimes_{\max} A_2^{**}.$$

(iii) The canonical inclusion $j \colon A_1 \otimes_{\max} A_2 \to (A_1 \otimes_{\max} A_2)^{**}$ extends to a (contractive) representation

$$\pi \colon A_1^{**} \otimes_{\max} A_2^{**} \to (A_1 \otimes_{\max} A_2)^{**}.$$

Exercise 11.7.

(i) Let A be a C^*-algebra. Consider a directed net (A_α) of C^*-subalgebras with dense union. Show that if all the A_α are nuclear, then A is nuclear.

(ii) Deduce from this that $L_\infty(\Omega, \mathcal{A}, \mu)$ is nuclear (for any measure space $(\Omega, \mathcal{A}, \mu)$) or, equivalently, that any commutative von Neumann algebra is nuclear.

(iii) Show that any commutative C^*-algebra is nuclear.

Exercise 11.8. Show that, if $\| \ \|_{\min} = \| \ \|_{\max}$ either on $C^*_\lambda(G) \otimes C^*_\lambda(G)$ or on $C^*_\lambda(G) \otimes C^*(G)$, then G is amenable.

Hint: Use Exercise 8.4.

Chapter 12. Nuclearity and Approximation Properties

The aim of this chapter is to give a reasonably direct proof of the fact that a C^*-algebra A is nuclear iff it has the approximation properties mentioned in Theorem 11.5. Actually, we will show the following more general statement: Let A_1, A_2 be two C^*-algebras, and let $u\colon A_1 \to A_2$ be any linear map. Fix a constant $C > 0$. Then the following assertions are equivalent:

(i) For any C^*-algebra B,

$$\|I_B \otimes u\|_{B \otimes_{\min} A_1 \to B \otimes_{\max} A_2} \leq C.$$

(ii) There is a net $u_i\colon A_1 \to A_2$ of finite rank maps tending pointwise to u and such that $\sup_i \|u_i\|_{dec} \leq C$.

This will be proved in Corollary 12.6. The proof will require a dual description of the maximal C^*-norm (see Theorem 12.1) for which we need to introduce a specific notation.

Notation. Let E, F be operator spaces, and let $\sigma\colon E \to B(H)$ and $\pi\colon F \to B(H)$ be linear mappings. We denote by

$$\sigma \cdot \pi\colon E \otimes F \to B(H)$$

the linear mapping defined by

$$\sigma \cdot \pi \left(\sum x_i \otimes y_i \right) = \sum \sigma(x_i)\pi(y_i).$$

The main ingredient for this chapter is the following.

Theorem 12.1. *Let $A \subset B(\mathcal{H})$ be a closed unital subalgebra, and let E be an operator space. Consider an element y in $E \otimes A$. Let*

$$\Delta(y) = \sup\{\|\sigma \cdot \pi(y)\|_{B(H)}\},$$

where the supremum runs over all Hilbert spaces H and all pairs (σ, π), where $\pi\colon A \to B(H)$ is a completely contractive unital homomorphism and $\sigma\colon E \to \pi(A)'$ is a complete contraction. On the other hand, let

$$\delta(y) = \inf \left\{ \|x\|_{M_n(E)} \left\| \sum a_i a_i^* \right\|^{1/2} \left\| \sum b_j^* b_j \right\|^{1/2} \right\}, \tag{12.1}$$

where the infimum runs over all possible n and all possible representations of y of the form

$$y = \sum_{ij=1}^{n} x_{ij} \otimes a_i b_j. \tag{12.2}$$

Then $\Delta(y) = \delta(y)$.

Remark. Let q: $A \otimes E \otimes A \to E \otimes A$ be the linear mapping taking $a_1 \otimes x \otimes a_2$ to $x \otimes a_1 a_2$. Then, by definition of the norm on $A \otimes_h E \otimes_h A$, it is easy to check that

$$\delta(y) = \inf\{\|\widehat{y}\|_{A \otimes_h E \otimes_h A} \mid \widehat{y} \in A \otimes E \otimes A, \ q(\widehat{y}) = y\}.$$

Note in particular that this shows that δ is a norm. Equivalently, if we denote by $E \otimes_\delta A$ the completion of $E \otimes A$ for the norm δ, then the mapping q extends to a norm 1 mapping from $A \otimes_h E \otimes_h A$ onto $E \otimes_\delta A$, and the resulting mapping (which we still denote – abusively – by q) is a metric surjection from $A \otimes_h E \otimes_h A$ onto $E \otimes_\delta A$, that is, it induces an isometric isomorphism from $(A \otimes_h E \otimes_h A)/\ker(q)$ onto $E \otimes_\delta A$. The preceding remark (which I had overlooked) is due to C. Le Merdy. It allows us to shorten the proof of Theorem 12.1.

Proof of Theorem 12.1. We first show the easy inequality $\Delta(y) \leq \delta(y)$. We may assume by homogeneity that $\delta(y) = 1$. Let (σ, π) be as in the definition of $\Delta(y)$. Then we have, assuming (12.2),

$$(\sigma \cdot \pi)(y) = \sum \sigma(x_{ij})\pi(a_i)\pi(b_j) = \sum \pi(a_i)\sigma(x_{ij})\pi(b_j)$$

and hence

$$
\begin{aligned}
\|(\sigma \cdot \pi)(y)\| &= \sup_{\xi,\eta \in B_H} \left| \sum \langle \sigma(x_{ij})\pi(b_j)\xi, \pi(a_i)^*\eta \rangle \right| \\
&\leq \|(\sigma(x_{ij}))\|_{M_n(B(H))} \sup_{\xi \in B_H} \left(\sum \|\pi(b_j)\xi\|^2 \right)^{1/2} \\
&\quad \sup_{\eta \in B_H} \left(\sum \|\pi(a_i)^*\eta\|^2 \right)^{1/2} \\
&\leq \|x\|_{M_n(E)} \left\| \sum \pi(b_j)^*\pi(b_j) \right\|^{1/2} \left\| \sum \pi(a_i)\pi(a_i)^* \right\|^{1/2} \\
&\leq \|x\|_{M_n(E)} \left\| \sum b_j^* b_j \right\|^{1/2} \left\| \sum a_i a_i^* \right\|^{1/2},
\end{aligned}
$$

whence $\Delta(y) \leq \delta(y)$.

To show the converse, since δ and Δ are norms (see the preceding remark), it suffices to show that

$$\Delta^* \leq \delta^*.$$

So let ξ: $E \otimes A \to \mathbb{C}$ be a linear form such that $\delta^*(\xi) \leq 1$, or, equivalently, such that

$$\forall y \in E \otimes A \qquad |\xi(y)| \leq \delta(y).$$

By the preceding remark, we can associate to ξ an element $\widehat{\xi}$ in the unit ball of $(A \otimes_h E \otimes_h A)^*$ determined by the identity

$$\forall x \in E \ \ \forall a_1 \in A \ \ \forall a_2 \in A \quad \widehat{\xi}(a_1 \otimes x \otimes a_2) = \xi(x \otimes a_1 a_2).$$

Now, by Corollary 5.4 and Remark 5.5 applied to $\widehat{\xi}$: $A \otimes_h E \otimes_h A \to \mathbb{C}$, we can find two representations π_1: $B(H) \to B(H_1)$ and π_2: $B(H) \to B(H_2)$ together with a completely contractive map v: $E \to B(H_2, H_1)$ and unit vectors $\xi_1 \in H_1$ and $\xi_2 \in H_2$ such that, for any a, b in A and any x in E, we have

$$\xi(x \otimes ab) = \widehat{\xi}(a \otimes x \otimes b) = \langle \pi_1(a)v(x)\pi_2(b)\xi_2, \xi_1 \rangle. \tag{12.3}$$

If we now replace $\pi_2(\cdot)$ by $\pi_2(\cdot)|_{\overline{\pi_2(A)\xi_2}}$ and $\pi_1(\cdot)$ by the homomorphism $a \to (\pi_1(a^*)|_{\overline{\pi_1(A^*)\xi_1}})^*$, we may as well assume that $\pi_2(A)\xi_2$ and $\{\pi_2(a^*)\xi_1 \mid a \in A\}$ are dense in H_2 and H_1, respectively.

Then, writing $(ac)b = a(cb)$ into (12.3) and using the density just mentioned, we find for all c in A

$$\forall x \in E \quad v(x)\pi_2(c) = \pi_1(c)v(x). \tag{12.4}$$

Let π: $A \to B(H_1 \oplus H_2)$ and $\widetilde{v}$: $E \to B(H_1 \oplus H_2)$ be defined by

$$\pi(a) = \begin{pmatrix} \pi_1(a) & 0 \\ 0 & \pi_2(a) \end{pmatrix} \quad \text{and}$$

$$\widetilde{v}(x) = \begin{pmatrix} 0 & v(x) \\ 0 & 0 \end{pmatrix}.$$

Then (12.4) implies $\widetilde{v}(x)\pi(c) = \pi(c)\widetilde{v}(x)$ for $x \in E$, $c \in A$, and π is a completely contractive unital homomorphism on A. Finally, letting $\eta_2 = \begin{pmatrix} 0 \\ \xi_2 \end{pmatrix}$ and $\eta_1 = \begin{pmatrix} \xi_1 \\ 0 \end{pmatrix}$, we find

$$\xi(x \otimes ab) = \langle \pi(a)\widetilde{v}(x)\pi(b)\eta_2, \eta_1 \rangle.$$

Hence (taking for a the unit of A) for all y in $E \otimes A$

$$\xi(y) = \langle \widetilde{v} \cdot \pi(y)\eta_2, \eta_1 \rangle;$$

therefore

$$|\xi(y)| \le \|\widetilde{v} \cdot \pi(y)\| \le \Delta(y).$$

This completes the proof that $\Delta^* \le \delta^*$ and hence $\delta \le \Delta$. ∎

We will mostly restrict attention here to the case when A is a C^*-algebra. We will return to the non-self-adjoint case in Chapter 25.

Proposition 12.2. *Consider an operator space E. We view E as embedded into $C^*\langle E \rangle$ (as defined in Chapter 8). Then, if A is a unital C^*-algebra, we have*

$$\delta(y) = \|y\|_{C^*\langle E \rangle \otimes_{\max} A}. \tag{12.5}$$

Proof. Note that any (completely) contractive unital homomorphism $\pi \colon A \to B(H)$ is necessarily a $*$-homomorphism, or, equivalently, a (C^*-algebraic) representation. Then $\pi(A)'$ is also a C^*-algebra, and any completely contractive map $\sigma \colon E \to \pi(A)'$ extends to a representation $\widehat{\sigma} \colon C^*\langle E \rangle \to \pi(A)'$. Hence we clearly have $\forall y \in E \otimes A$

$$\Delta(y) \leq \sup \|\widehat{\sigma} \cdot \pi(y)\| \leq \|y\|_{C^*\langle E \rangle \otimes_{\max} A}.$$

The converse is clear. ∎

Corollary 12.3. *In the situation of Theorem 12.1, assume that A is a C^*-algebra and let F be another operator space and B another C^*-algebra. Consider $u_1 \in CB(E, F)$ and $u_2 \in D(A, B)$. Then, for all y in $E \otimes A$, we have*

$$\delta((u_1 \otimes u_2)(y)) \leq \|u_1\|_{cb} \|u_2\|_{dec} \delta(y). \tag{12.6}$$

Proof. Assume $\|u_1\|_{cb} = 1$. Note that $u_1 \colon E \to F$ extends to a C^*-representation from $C^*\langle E \rangle$ to $C^*\langle F \rangle$. Then (12.6) is an immediate consequence of (11.6) and (12.5). ∎

In the particular case when $E = M_N^*$, Theorem 12.1 becomes:

Corollary 12.4. *In Theorem 12.1, assume A is a C^*-algebra and let $E = M_N^*$ for some $N \geq 1$. Then, for all y in $E \otimes A$, with associated linear map $\widetilde{y} \colon M_N \to A$, we have*

$$\delta(y) = \|\widetilde{y}\|_{D(M_N, A)}. \tag{12.7}$$

Proof. Let $t \in M_N^* \otimes M_N$ be associated to the identity of M_N. Let (ξ_{ij}) be biorthogonal to the standard basis (e_{ij}) in M_N, so that $t = \sum \xi_{ij} \otimes e_{ij}$. Then we can write

$$t = \sum \xi_{ij} \otimes e_{i1} e_{1j},$$

and hence

$$\delta(t) \leq \|(\xi_{ij})\|_{M_n(E)} \left\| \sum e_{i1} e_{1i} \right\|^{1/2} \left\| \sum e_{j1} e_{1j} \right\|^{1/2} \leq \|(\xi_{ij})\|_{M_n(E)} = 1.$$

Hence, by Corollary 12.3, we have

$$\delta(y) = \delta((I \otimes \widetilde{y})(t)) \leq \|\widetilde{y}\|_{dec} \delta(t) \leq \|\widetilde{y}\|_{dec}.$$

On the other hand, if we have (12.2), then let $v\colon E^* \to M_n$ be the map defined by $v(\xi) = (\xi(x_{ij}))$ ($\xi \in E^*$) and let $w\colon M_n \to A$ be defined by $w(e_{ij}) = a_i b_j$. Then it is fairly easy to verify (see Exercise 12.2) that

$$\|w\|_{dec} \le \left\|\sum a_i a_i^*\right\|^{1/2} \left\|\sum b_j^* b_j\right\|^{1/2}, \tag{12.8}$$

whence, if we let $\widetilde{y} = wv$ by (11.4) and (11.5),

$$\|\widetilde{y}\|_{dec} \le \|w\|_{dec}\|v\|_{dec} = \|w\|_{dec}\|v\|_{cb} \le \left\|\sum a_i a_i^*\right\|^{1/2} \left\|\sum b_j^* b_j\right\|^{1/2} \|x\|_{M_n(E)}.$$

Taking the infimum over all representations of the form (12.2), we obtain

$$\|\widetilde{y}\|_{dec} \le \delta(y). \qquad \blacksquare$$

Corollary 12.5. *In the situation of Theorem 12.1, assume A is a C^*-algebra. Consider an element y in $E \otimes A$. Let $\widetilde{y}\colon E^* \to A$ be the associated linear map. Then*

$$\delta(y) = \inf\{\|v\|_{cb}\|w\|_{dec}\},$$

where the infimum runs over all possible factorizations of $\widetilde{y}$ of the form

$$E^* \xrightarrow{\ v\ } M_n \xrightarrow{\ w\ } A,$$

with v weak$-*$ continuous. Moreover, if $E = F^*$ for some operator space F, let $\widehat{y}\colon F \to A$ be the restriction of $\widetilde{y}$ to $F \subset F^{**}$. Then again

$$\delta(y) = \inf\{\|v\|_{cb}\|w\|_{dec}\},$$

where the infimum runs over all possible factorizations of $\widehat{y}$ of the form

$$F \xrightarrow{\ v\ } M_n \xrightarrow{\ w\ } A.$$

Proof. First observe that the weak-$*$ continuity of v implies that we have $v = (v_*)^*$ for some map $v_*\colon M_n^* \to E$. Then, if $\widetilde{y} = wv$ with v, w as above, we can write

$$y = (v_* \otimes I_A)(\theta),$$

where $\theta \in M_n^* \otimes A$ is associated to w. Hence we have by (12.6) and (12.7)

$$\delta(y) \le \|v_*\|_{cb}\delta(\theta) = \|v\|_{cb}\|w\|_{dec},$$

whence $\delta(y) \le \inf\{\|v\|_{cb}\|w\|_{dec}\}$. For the converse inequality, we argue as in the preceding proof.

In the case $E = F^*$, the argument is similar, and we leave the details to the reader. ∎

Corollary 12.6. *Let λ be a positive constant. Consider two C^*-algebras A_1 and A_2 and an operator subspace $X \subset A_1$. Let $u: X \to A_2$ be a linear mapping. The following assertions are equivalent.*

(i) *For any C^*-algebra B, $I_B \otimes u$ defines a bounded linear map from $B \otimes_{\min} X$ to $B \otimes_{\max} A_2$ with norm $\leq \lambda$.*

(ii) *Same as (i) with $B = C^*\langle E^* \rangle$ for all finite-dimensional operator subspaces $E \subset A_1$.*

(iii) *For any finite-dimensional subspace $E \subset X$, the restriction $u_{|E}$ admits, for any $\varepsilon > 0$, a factorization of the form $E \xrightarrow{V} M_n \xrightarrow{w} A_2$ with $\|V\|_{cb} \|w\|_{dec} \leq \lambda + \varepsilon$.*

(iv) *There is a net of finite rank maps $u_\alpha: X \to A_2$ admitting factorizations through matrix algebras of the form*

$$M_{n_\alpha}$$

$$v_\alpha \nearrow \qquad \searrow w_\alpha$$

$$X \qquad \xrightarrow{u_\alpha} \qquad A_2$$

with

$$\|v_\alpha\|_{cb} \|w_\alpha\|_{dec} \leq \lambda$$

such that $u_\alpha = w_\alpha v_\alpha$ converges pointwise to u.

(v) *There is a net $u_\alpha: A_1 \to A_2$ of finite rank maps with $\sup \|u_\alpha\|_{dec} \leq \lambda$ that tends pointwise to u when restricted to X.*

Proof. (i) $\Rightarrow$ (ii) is trivial. Assume (ii). Let $E \subset X$ be an arbitrary finite-dimensional subspace, and let $i_E \in E^* \otimes X$ be the tensor associated to the inclusion of E into X. Let $B = C^*\langle E^* \rangle$. By (ii) we have $\|(I_B \otimes u)(i_E)\|_{B \otimes_{\max} A_2} \leq \lambda \|i_E\|_{B \otimes_{\min} X}$. But by the injectivity of the min-norm, we have

$$\|i_E\|_{B \otimes_{\min} X} = \|i_E\|_{E^* \otimes_{\min} X} = \|i_E\|_{CB(E,X)} = 1.$$

Hence we have $\|(I_B \otimes u)(i_E)\|_{B \otimes_{\max} A_2} \leq \lambda$. By (12.5) and Corollary 12.5, this implies that, for any $\varepsilon > 0$, there is a factorization of $u_{|E}$ of the following form:

$$M_n$$

$$V \nearrow \qquad \searrow w$$

$$E \qquad \xrightarrow{u_{|E}} \qquad A_2$$

with $\|V\|_{cb} \le 1$ and $\|w\|_{dec} \le 1+\varepsilon$. This shows (ii) $\Rightarrow$ (iii). Assume (iii). By the extension property of M_n, we can extend V to a mapping $v\colon X \to M_n$ with $\|v\|_{cb} \le 1$. Thus, if we take for index set I the set of all finite-dimensional subspaces $E \subset X$ (directed by inclusion), we obtain nets $v_\alpha\colon X \to M_{n_\alpha}$ and $w_\alpha\colon M_{n_\alpha} \to A_2$ such that $\|w_\alpha v_\alpha(x) - u(x)\| \to 0$ for all x in X and such that (after a suitable renormalization) $\sup \|v_\alpha\|_{cb} \le 1$, $\sup \|w_\alpha\|_{dec} \le 1$. This completes the proof that (iii) $\Rightarrow$ (iv). We may clearly assume (by Corollary 1.7) that v_α is extended to A_1 with the same c.b. norm; thus, recalling (11.4) and (11.5), (iv) $\Rightarrow$ (v) is immediate. Finally, the proof that (v) $\Rightarrow$ (i) is an immediate consequence of Proposition 11.10 in the preceding chapter. ∎

Using an earlier version of the present work, Marius Junge and C. Le Merdy [JLM] obtained the following striking result.

Theorem 12.7. *([JLM]) Let $u\colon B \to A$ be a finite rank map between two C^*-algebras. Then, for any $\varepsilon > 0$, there is an integer n and a factorization $u = vw$ of the form*

$$B \xrightarrow{v} M_n \xrightarrow{w} A$$

with $(\|v\|_{cb}\|w\|_{cb} \le)\ \|v\|_{cb}\|w\|_{dec} \le \|u\|_{dec}(1 + \varepsilon)$. Therefore, if $y \in B^* \otimes A$ is the tensor associated to $u\colon B \to A$, we have

$$\|u\|_{dec} = \delta(y).$$

Remark. The following question seems interesting: Fix $\varepsilon > 0$ and let k be the rank of u. Can we obtain the above factorization with $n \le f(k, \varepsilon)$ for some function f? In other words, can we control n by a function f depending *only* on k and ε?

The proof will use an original application of Kaplansky's density theorem discovered by M. Junge [J1], as follows.

Lemma 12.8. *Let A, B be arbitrary C^*-algebras. Then any c.b. map $\sigma\colon B^* \to A$ can be approximated in the point-norm topology by a net of weak*-continuous finite rank maps $\sigma_\alpha\colon B^* \to A$ with $\|\sigma_\alpha\|_{cb} \le \|\sigma\|_{cb}$.*

Proof. Let $M = A^{**}\overline{\otimes}B^{**}$. By Theorem 2.5.2, $CB(B^*, A^{**})$ can be identified isometrically with M in a natural way. Let $t \in A^{**}\overline{\otimes}B^{**}$ be the tensor associated to σ composed with the inclusion $A \to A^{**}$. Then, by Kaplansky's density theorem (see any standard book, such as [Di1, Sa, Ta3, Ped, KaR]), there is a net (t_α) in $A \otimes B$ with $\|t_\alpha\|_{\min} \le \|t\|_M$ such that $t_\alpha\ \sigma(M, M_*)$-tends to t. Let $\sigma_\alpha\colon B^* \to A$ be the finite rank map associated to t_α. We have $\|\sigma_\alpha\|_{cb} = \|t_\alpha\|_{\min} \le \|t\|_M = \|\sigma\|_{cb}$. Moreover, for any ξ in B^*, $\sigma_\alpha(\xi)$ must $\sigma(A^{**}, A^*)$-tend to $\sigma(\xi)$. But, since $\sigma_\alpha(\xi)$ and $\sigma(\xi)$ both lie in A, this means that $\sigma_\alpha(\xi)$ tends to $\sigma(\xi)$ weakly in A. Passing to suitable convex hulls, we obtain a net (σ_α) such that, for any ξ, $\sigma_\alpha(\xi)$ tends to $\sigma(\xi)$ in norm. ∎

Remark. The same argument shows that, for any von Neumann algebra R, any c.b. map $\sigma \colon R_* \to A$ can be approximated by a net of finite rank maps $\sigma_\alpha \colon R_* \to A$ with $\|\sigma_\alpha\|_{cb} \le \|\sigma\|_{cb}$.

Proof of Theorem 12.7. If $u = wv$ as in Theorem 12.7, then, by (11.5), we have $\|u\|_{dec} \le \|v\|_{dec}\|w\|_{dec} = \|v\|_{cb}\|w\|_{dec}$, hence by Corollary 12.5,

$$\|u\|_{dec} \le \delta(y).$$

We now turn to the converse. We may assume $u(x) = \sum_1^k \xi_i(x)a_i$ with $\xi_i \in B^*$ and $a_i \in A$, or equivalently $y = \sum \xi_i \otimes a_i$. Assume $\|u\|_{dec} \le 1$. By Corollary 12.5 (and the equality $\delta = \Delta$) it suffices to show that, for any representation $\pi \colon A \to B(H)$ and any completely contractive map $\sigma \colon B^* \to \pi(A)'$, we have

$$\left\| \sum \sigma(\xi_i)\pi(a_i) \right\| \le 1.$$

By Lemma 12.8, there is a net of finite rank weak*-continuous maps $\sigma_\alpha \colon B^* \to \pi(A)'$ with $\|\sigma_\alpha\|_{cb} \le 1$ tending pointwise to σ. Hence

$$\left\| \sum_1^k \sigma_\alpha(\xi_i)\pi(a_i) \right\| \to \left\| \sum_1^k \sigma(\xi_i)\pi(a_i) \right\|.$$

Let $t_\alpha \in B \otimes \pi(A)'$ be the tensor representing σ_α. By Proposition 11.10, since u has finite rank, we have

$$\|(u \otimes I)(t_\alpha)\|_{A \otimes_{\max} \pi(A)'} \le \|u\|_{dec}\|t_\alpha\|_{\min} = \|u\|_{dec}\|\sigma_\alpha\|_{cb} \le \|u\|_{dec}.$$

Since $(u \otimes I)(t_\alpha) = \sum a_i \otimes \sigma_\alpha(\xi_i)$, this yields

$$\left\| \sum \sigma_\alpha(\xi_i)\pi(a_i) \right\| \le \left\| \sum a_i \otimes \sigma_a(\xi_i) \right\|_{\max} \le \|u\|_{dec}$$

hence in the limit

$$\left\| \sum \sigma(\xi_i)\pi(a_i) \right\| \le \|u\|_{dec}.$$

Thus, by Theorem 12.1, we obtain

$$\Delta \left(\sum \xi_i \otimes a_i \right) = \delta(y) \le \|u\|_{dec},$$

so we conclude $\delta(y) = \|u\|_{dec}$. Then, the first assertion follows from the second part of Corollary 12.5 (applied with $F = B$). ∎

We now return to complete positivity. Recall (see Exercise 11.5(iii)) that a c.p. map $u \colon A_1 \to A_2$ between two C^*-algebras is necessarily continuous and that $\|u\|_{cb} = \|u\| = \sup\{\|u(x)\| \mid x \ge 0, \|x\| \le 1\}$. Moreover, if A_1 has a unit, we have simply $\|u\| = \|u(1)\|$. To derive the known results on the "c.p. approximation property," the following result, which is a simple adaptation of well-known ideas, will be useful.

Lemma 12.9. Let $E \subset B(H)$ be a finite-dimensional operator space. Assume moreover that E is an operator system, that is, E is self-adjoint and unital. Let A be a C^*-algebra. Consider a unital self-adjoint mapping $u\colon E \to A$ associated to a tensor $t \in E^* \otimes A$. Fix $\varepsilon > 0$. Then, if $\delta(t) < 1 + \varepsilon$, we can decompose u as $u = \varphi - \psi$ with φ, ψ c.p. such that $\|\psi\| \leq \varepsilon$ and φ admits for some n a factorization of the form

$$M_n$$

$$V \nearrow \qquad\qquad \searrow W$$

$$E \qquad \xrightarrow{\ \ \varphi\ \ } \qquad A$$

where V, W are c.p. maps with

$$\|V\| \leq 1 + \varepsilon \text{ and } \|W\| \leq 1.$$

Proof. By the definition of the norm δ and by Theorem 1.6, we can assume that $u = wv$, where $v\colon E \to M_n$ and $w\colon M_n \to A$ are as follows:

$$\forall x \in E \quad v(x) = v_1^* \pi(x) v_2,$$

where $\pi\colon E \to B(\widehat{H})$ is the restriction of a representation, v_1, v_2 are operators in $B(\ell_2^n, \widehat{H})$ with $\|v_1\| = \|v_2\| < (1 + \varepsilon)^{1/2}$, and w is defined by

$$w(e_{ij}) = a_i b_j$$

with

$$\left\| \sum a_i a_i^* \right\| = \left\| \sum b_j^* b_j \right\| < 1.$$

Let

$$b = \begin{pmatrix} b_1 \\ \vdots \\ b_n \end{pmatrix}$$

and $a^* = (a_1, \ldots, a_n)$ so that, in matrix notation, we have

$$u(x) = a^* \cdot v_1^* \pi(x) v_2 \cdot b.$$

Since u is self-adjoint, we have

$$u(x) = u(x^*)^* = b^* v_2^* \pi(x) v_1 a;$$

hence we have (by "polarization")

$$u(x) = \varphi(x) - \psi(x)$$

with

$$\varphi(x) = \tfrac{1}{4}[(v_1 a + v_2 b)^* \pi(x)(v_1 a + v_2 b)]$$

and

$$\psi(x) = \tfrac{1}{4}[(v_1 a - v_2 b)^* \pi(x)(v_1 a - v_2 b)].$$

Clearly φ and ψ are c.p. and

$$1 = u(1) = \varphi(1) - \psi(1);$$

hence $\varphi(1) \geq 1$ and $\|\varphi(1)\| \leq \frac{\|v_1 a + v_2 b\|^2}{4} \leq (1+\varepsilon)$. This implies that $\|\varphi(1) - 1\| \leq \varepsilon$, and hence we obtain (see Exercise 11.5 (iii)) $\|\psi\| = \|\psi(1)\| \leq \varepsilon$. It remains to show that φ admits the announced factorization. Let (again in matrix notation)

$$V(x) = (1/2) \begin{pmatrix} v_1^* \pi(x) v_1 & v_1^* \pi(x) v_2 \\ v_2^* \pi(x) v_1 & v_2^* \pi(x) v_2 \end{pmatrix} = (1/2) \begin{pmatrix} v_1^* \\ v_2^* \end{pmatrix} \pi(x)(v_1, v_2).$$

Clearly V is c.p. from E to M_{2n} and

$$\|V\| \leq \tfrac{1}{2}(\|v_1\|^2 + \|v_2\|^2) \leq 1 + \varepsilon.$$

Moreover, if we define $W \colon M_{2n} \to A$ by

$$W(t) = \tfrac{1}{2}(a^*, b^*) t \begin{pmatrix} a \\ b \end{pmatrix},$$

then W is c.p. with $\|W\| \leq 1$ and we have

$$\varphi(x) = W(V(x)). \qquad \blacksquare$$

We now recover a result due to Choi and Effros and Kirchberg [CE1, Ki3] as follows. (This is the analog of Corollary 12.6 for c.p. maps.)

Corollary 12.10. Let $u \colon A_1 \to A_2$ be a completely positive and unital linear mapping between two unital C^*-algebras. The following assertions are equivalent.

(i) For any C^*-algebra B, $I_B \otimes u$ defines a completely positive linear map from $B \otimes_{\min} A_1$ to $B \otimes_{\max} A_2$ with norm $= 1$.

(ii) Same as (i) with $B = C^*\langle E^* \rangle$ for all finite-dimensional operator subspaces $E \subset A_1$.

(iii) There is a net of finite rank maps (u_α) admitting factorizations through matrix algebras of the form

$$M_{n_\alpha}$$

$$v_\alpha \nearrow \qquad \searrow w_\alpha$$

$$A_1 \xrightarrow{\ u_\alpha\ } A_2$$

with c.p. maps v_α and w_α satisfying $\|v_\alpha\|\|w_\alpha\| \leq 1$ such that $u_\alpha = w_\alpha v_\alpha$ converges pointwise to u.

(iv) There is a net $u_\alpha \colon A_1 \to A_2$ of finite rank c.p. maps that tends pointwise to u.

Proof. (i) $\Rightarrow$ (ii) is trivial. Assume (ii). Let E be a finite-dimensional operator system inside A_1. Let $B = C^*\langle E^*\rangle$. Let $y \in E^* \otimes A_1$ be the element associated to the inclusion i_E map of E into A_1, so that $\|y\|_{\min} = \|i_E\|_{cb} = 1$. By (ii) we have $\|(I_B \otimes u)(y)\|_{\max} \leq 1$. Note that the linear map from E to A_2 associated to the tensor $(I_B \otimes u)(y)$ is nothing but $u_{|E}$. Hence, by (12.5) and by Lemma 12.9, for any $\varepsilon > 0$ we can write $u_{|E} = \varphi - \psi$ with $\varphi = WV$ as in Lemma 12.9. By the extension property of c.p. maps we may as well assume that V is a c.p. mapping of norm $\leq 1 + \varepsilon$ from A_1 to M_n. Then, using the net (directed by inclusion) formed by the finite-dimensional operator systems $E \subset A_1$ and letting $\varepsilon \to 0$, we obtain (iii) (after a suitable renormalization of V). Then (iii) $\Rightarrow$ (iv) is trivial.

Finally assume (iv). Note that since $\|u_\alpha\| = \|u_\alpha(1)\|$ and $\|u_\alpha(1)\| \to \|u(1)\| = 1$, we have "automatically" $\|u_\alpha\| \to 1$. By Proposition 11.10, we have for any C^*-algebra B

$$\|I_B \otimes u_\alpha\|_{B\otimes_{\min}A_1 \to B\otimes_{\max}A_2} \leq \|u_\alpha\|;$$

hence, since $u_\alpha \to u$ pointwise,

$$\|I_B \otimes u\|_{B\otimes_{\min}A_1 \to B\otimes_{\max}A_2} \leq 1.$$

This shows that (iv) $\Rightarrow$ (i). ∎

Remark 12.11. In the preceding situation, every unital c.p. map of finite rank satisfies the equivalent four properties in Corollary 12.10.

Corollary 12.12. *A unital C^*-algebra A is nuclear iff there is a net of finite rank maps of the form $A\xrightarrow{v_\alpha}M_{n_\alpha}\xrightarrow{w_\alpha}A$, where v_α, w_α are c.p. maps with $\|v_\alpha\| \leq 1$, $\|w_\alpha\| \leq 1$, which tends pointwise to the identity.*

Proof. Apply Corollary 12.10 with $u = I_A$. ∎

Remark 12.13. A C^*-algebra A is nuclear iff its unitization is nuclear. This remark allows us to extend the preceding results to the nonunital case. (Indeed, A is an ideal in its unitization $\widetilde{A}$, so that $\widetilde{A}/A = \mathbb{C}$. By Corollary 11.9, an ideal in a nuclear C^*-algebra is nuclear. Thus $\widetilde{A}$ nuclear implies A nuclear. The converse is easy and left as an exercise for the reader.)

Recently, A. Sinclair and R. Smith [SS3] have obtained a characterization of injective von Neumann algebras by decomposability properties of multilinear maps with values in the algebra. In the case of nuclear C^*-algebras, the property they consider has the following analog.

Theorem 12.14. *Let A be a nuclear C^*-algebra. Let E_1, E_2, E_3 be operator spaces. Then the product map of A defines a complete metric surjection from $(E_1 \otimes_{\min} A) \otimes_h E_2 \otimes_h (E_3 \otimes_{\min} A)$ into $(E_1 \otimes_h E_2 \otimes_h E_3) \otimes_{\min} A$. In particular, if $E_2 = \mathbb{C}$, we have a complete metric surjection from $(E_1 \otimes_{\min} A) \otimes_h (E_3 \otimes_{\min} A)$ onto $(E_1 \otimes_h E_3) \otimes_{\min} A$.*

Proof. Let $E = E_1 \otimes_h E_2 \otimes_h E_3$. By Theorem 12.1 and the remark after it, if A is nuclear, the mapping $q\colon A \otimes_h E \otimes_h A \to E \otimes_{\min} A$ is a complete metric surjection. The result then follows from the fact that q factorizes completely contractively through $(E_1 \otimes_{\min} A) \otimes_h E_2 \otimes_h (E_3 \otimes_{\min} A)$. Indeed, we have by associativity $A \otimes_h E \otimes_h A = (A \otimes_h E_1) \otimes_h E_2 \otimes_h (E_3 \otimes_h A)$, and we have canonical complete contractions $A \otimes_h E_1 \to E_1 \otimes_{\min} A$ and $E_3 \otimes_h A \to E_3 \otimes_{\min} A$. Moreover, the product map defines (cf. Exercise 6.1) a complete contraction from $(E_1 \otimes_{\min} A) \otimes_h E_2 \otimes_h (E_3 \otimes_{\min} A)$ to $(E_1 \otimes_h E_2 \otimes_h E_3) \otimes_{\min} A$. Thus we obtain the announced result. ■

Exercises

Exercise 12.1. Let A be a C^*-algebra. Fix a, b in A. Let $u\colon A \to A$ be defined by $u(x) = axb^*$. Note that u is c.p. when $a = b$. Then show that for any a, b

$$\|u\|_{\mathrm{dec}} \le \|a\| \, \|b\|.$$

Exercise 12.2. Let $u\colon M_n \to A$ be a linear mapping into a C^*-algebra. Assume that we can write $u(e_{ij}) = a_i b_j^*$ with a_i, b_j in A. Note that u is c.p. if $a_i = b_i$ for all i. Then show that

$$\|u\|_{\mathrm{dec}} \le \left\| \sum a_i a_i^* \right\|^{1/2} \left\| \sum b_j b_j^* \right\|^{1/2}.$$

Chapter 13. $C^*(\mathbb{F}_\infty) \otimes B(H)$

Recently, E. Kirchberg [Ki2–5] revived the study of pairs of C^*-algebras A, B such that there is only one C^*-norm on the algebraic tensor product $A \otimes B$, or, equivalently, such that $A \otimes_{\min} B = A \otimes_{\max} B$. He constructed the first example of a nonnuclear C^*-algebra such that $A \otimes_{\min} A^{op} = A \otimes_{\max} A^{op}$. He also proved the following striking result [Ki5]. A simpler proof is given in [P11], which we follow rather closely in the following.

Theorem 13.14. *([Ki5]) Let F be any free group, $C^*(F)$ the (full) C^*-algebra of F, and H any Hilbert space. Then*

$$C^*(F) \otimes_{\min} B(H) = C^*(F) \otimes_{\max} B(H).$$

This is particularly striking if one recalls that $C^*(F)$ and $B(H)$ both are universal objects: Every C^*-algebra is a quotient of $C^*(F)$ and a subalgebra of $B(H)$ for suitable choices of F and H (see §2.12).

More generally, we will prove

Theorem 13.2. *([P11]) Let $(A_i)_{i \in I}$ be a family of C^*-algebras (resp. unital C^*-algebras). Assume that for each i in I*

$$A_i \otimes_{\min} B(H) = A_i \otimes_{\max} B(H). \tag{13.1}$$

We denote by $\dot{*}_{i \in I} A_i$ (resp. by $*_{i \in I} A_i$) their free product in the category of C^*-algebras (resp. in the category of unital C^*-algebras). Then we have

$$\left(\mathop{\dot{*}}_{i \in I} A_i \right) \otimes_{\min} B(H) = \left(\mathop{\dot{*}}_{i \in I} A_i \right) \otimes_{\max} B(H), \tag{13.2'}$$

and in the unital case

$$\left(\mathop{*}_{i \in I} A_i \right) \otimes_{\min} B(H) = \left(\mathop{*}_{i \in I} A_i \right) \otimes_{\max} B(H). \tag{13.2}$$

Corollary 13.3. *Let $(G_i)_{i \in I}$ be a family of discrete amenable groups, and let $G = *_{i \in I} G_i$ be their free product. Then*

$$C^*(G) \otimes_{\min} B(H) = C^*(G) \otimes_{\max} B(H).$$

Remark. Kirchberg's theorem for $F = F_I$ corresponds to $A_i = C^*(\mathbb{Z})$ for all i in I (in the unital case).

The main idea of our proof of Theorem 13.1 is that, if E is the linear span of 1 and the free unitary generators of $C^*(F)$, then it suffices to check that the min- and max-norms coincide on $E \otimes B(H)$. More generally, we will prove

Theorem 13.4. *Let A_1, A_2 be unital C^*-algebras. Let $(u_i)_{i \in I}$ (resp. $(v_j)_{j \in J}$) be a family of unitary operators that generate A_1 (resp. A_2). Let E_1 (resp. E_2) be the closed span of $(u_i)_{i \in I}$ (resp. $(v_j)_{j \in J}$). Assume $1 \in E_1$ and $1 \in E_2$. Then the following assertions are equivalent:*

(i) The inclusion map $E_1 \otimes_{\min} E_2 \to A_1 \otimes_{\max} A_2$ is completely isometric.

(ii) $A_1 \otimes_{\min} A_2 = A_1 \otimes_{\max} A_2$.

We will use several elementary facts, as follows. The first one is a well-known property of unitary dilations.

Lemma 13.5. *Let $u \in B(\mathcal{H}), \widehat{u} \in B(\widehat{H})$ be unitaries. Assume there is an isometry $S \colon \mathcal{H} \to \widehat{H}$ be an isometry with range $K \subset \widehat{H}$, such that*

$$u = S^* \widehat{u} S.$$

Then $K = S(\mathcal{H})$ is necessarily invariant under $\widehat{u}$ and $\widehat{u}^*$, so that $\widehat{u}$ commutes with P_K.

Proof. For simplicity we assume $\mathcal{H} \subset \widehat{H}$ (S being the inclusion map) and we decompose $\widehat{H}$ as $\mathcal{H} \oplus \mathcal{H}^\perp$. Then $\widehat{u}$ can be represented as a matrix

$$\widehat{u} = \begin{pmatrix} u & b \\ c & d \end{pmatrix},$$

where both u and $\widehat{u}$ are unitary. Computing the $(1,1)$-entry of $\widehat{u}^* \widehat{u}$ and $\widehat{u} \widehat{u}^*$, we find $c^* c = 0$ and $bb^* = 0$; hence $b = c = 0$, which means that $\widehat{u}$ commutes with $P_{\mathcal{H}}$. ∎

The following simple fact is essential in our argument.

Proposition 13.6. *Let A, B be two unital C^*-algebras. Let $(u_i)_{i \in I}$ be a family of unitary elements of A generating A as a unital C^*-algebra (i.e., the smallest unital C^*-subalgebra of A containing them is A itself). Let $E \subset A$ be the linear span of $(u_i)_{i \in I}$ and 1_A. Let $T \colon E \to B$ be a linear operator such that $T(1_A) = 1_B$ and taking each u_i to a unitary in B. Then, $\|T\|_{cb} \leq 1$ suffices to ensure that T extends to a (completely) contractive representation (i.e. $*$-homomorphism) from A to B.*

Proof. Consider B as embedded in $B(\mathcal{H})$. Then, it clearly suffices to prove this statement for $B = B(\mathcal{H})$, which we now assume. By the Arveson-Wittstock extension theorem (see Corollary 1.7), T extends to a complete contraction $\widehat{T} \colon A \to B(\mathcal{H})$. Since T is assumed unital, $\widehat{T}$ is unital; hence by Corollary 1.8 we can write

$$\widehat{T}(x) = S^* \widehat{\pi}(x) S,$$

where $\widehat{\pi}\colon A \to B(\widehat{H})$ is a unital representation (i.e. $*$-homomorphism) and $S\colon \mathcal{H} \to \widehat{H}$ is an isometry. Now, for any unitary U in the family $(u_i)_{i \in I}$, we have $T(U) = \widehat{T}(U) = S^*\widehat{\pi}(U)S$. Hence, by Lemma 13.5, since $T(U)$ is unitary by assumption, if $K = S(\mathcal{H})$, then P_K commutes with $\widehat{\pi}(U)$. Now since these operators U generate A, this implies that P_K commutes with $\widehat{\pi}(A)$, so that $\widehat{T}$ is actually a $*$-homomorphism. Thus, $\widehat{T}$ is an extension of T and a (contractive) $*$-homomorphism. This completes the proof. [Note that, a posteriori, the Arveson-Wittstock completely contractive extension $\widehat{T}$ is unique. In short, the proof reduces to this: The multiplicative domain (see Lemma 14.2 below) of $\widehat{T}$ is a unital C^*-algebra (cf. [Ch1]) and contains $(u_i)_{i \in I}$; hence it is equal to A.] ∎

Remark. We will apply Proposition 13.6 in the following particular situation. Let $\mathcal{A} \subset A$ be the (dense) unital $*$-algebra generated by E. Consider a unital $*$-homomorphism $u\colon \mathcal{A} \to B$. Then $\|u_{|E}\|_{cb} \leq 1$ suffices to ensure that u extends to a (completely) contractive representation (i.e. $*$-homomorphism) from the whole of A to B.

Remark 13.7. In the same situation as in Proposition 13.6, note that, if T is a complete isometry, then $\widehat{T}$ is a faithful representation onto the C^*-algebra B_1 generated by the range of T. Indeed, by Proposition 13.6 applied to T^{-1}, $\widehat{T}\colon A \to B_1$ is left invertible. This can be used to give a very simple proof of the fact due to Choi ([Ch2]) that the full C^*-algebra of any free group admits a faithful representation into a direct sum of matrix algebras. By Proposition 13.6, it suffices to check this on the free generators, and this is quite easy (see Exercise 13.1).

Proof of Theorem 13.4. The implication (ii) $\Rightarrow$ (i) is trivial, so we prove only the converse. Assume (i). Let $E = E_1 \otimes_{\min} E_2$. We view E as a subspace of $A = A_1 \otimes_{\min} A_2$. By (i), we have an inclusion map $T\colon E_1 \otimes_{\min} E_2 \to A_1 \otimes_{\max} A_2$ with $\|T\|_{cb} \leq 1$. By Proposition 13.6, T extends to a (contractive) representation $\widehat{T}$ from $A_1 \otimes_{\min} A_2$ to $A_1 \otimes_{\max} A_2$. Clearly $\widehat{T}$ must preserve the algebraic tensor products $A_1 \otimes 1$ and $1 \otimes A_2$, and hence also $A_1 \otimes A_2$. Thus we obtain (ii). ∎

Remark 13.8. Let us denote by $E_1 \otimes 1 + 1 \otimes E_2$ the linear subspace spanned by elements of $A_1 \otimes A_2$ of the form $\{a_1 \otimes 1 + 1 \otimes a_2\}$. Then, in the situation of Theorem 13.4, $E_1 \otimes 1 + 1 \otimes E_2$ generates $A_1 \otimes_{\min} A_2$, so that it suffices for the conclusion of Theorem 13.4 to assume that the operator space structures induced on $E_1 \otimes 1 + 1 \otimes E_2$ by the min- and max-norms coincide.

Proof of Kirchberg's Theorem 13.1. Let $A_1 = C^*(F)$, $A_2 = B(H)$. We may clearly assume $\dim(H) = \infty$. For simplicity, this time we denote by $(U_i)_{i \in J}$ the free unitary generators $(U_i \mid i \in I)$ of $C^*(F)$ augmented of the unit (so J is I plus one more point). We take $E_2 = B(H)$ and let E_1 be the linear span of the unit and $(U_i)_{i \in J}$.

with the remark after it), we have $\|u\|_{\text{dec}} = \Delta(y)$, and by definition

$$\Delta(y) = \sup \|\sum v_i \pi(x_i)\|,$$

where the supremum runs over all representations $\pi\colon A \to B(H)$ and all families (v_i) of contractions in $\pi(A)' \subset B(H)$. By the Russo-Dye Theorem ([Ped, p. 4]), the latter supremum remains unchanged if we let it run only over all the families (v_i) of *unitaries* in $\pi(A)'$. Equivalently, this means:

$$\Delta(y) = \left\|\sum_{i \in I} U_i \otimes x_i\right\|_{C^*(F) \otimes_{\max} A},$$

which establishes (13.9). ∎

In [Ki5], Kirchberg also proves a general result on the tensor products $C \otimes N$ when N is an *arbitrary* von Neumann algebra. In that case (but with C an arbitrary C^*-algebra), we can define a C^*-norm $\| \ \|_{\text{nor}}$ on $C \otimes N$ as follows:

$$\left\|\sum a_i \otimes b_i\right\|_{\text{nor}} = \sup \left\{\left\|\sum \sigma(a_i)\pi(b_i)\right\|\right\},$$

where the supremum runs over all pairs of representations $\sigma\colon C \to B(\mathcal{H})$ $\pi\colon N \to B(\mathcal{H})$ with commuting ranges and with π *normal*. We denote by $C \otimes_{\text{nor}} N$ the completion of $C \otimes N$ for this norm. (See [EL] for more information.) The preceding method also yields this result, but we only state it:

Theorem 13.14. ([Ki5]) Let F be any free group and let $C = C^*(F)$. Let N be any von Neumann algebra. Then

$$C \otimes_{\text{nor}} N = C \otimes_{\max} N.$$

The reader will find in Chapter 15 a characterization of the C^*-algebras A such that

$$C^*(F_\infty) \otimes_{\min} A = C^*(F_\infty) \otimes_{\max} A$$

and in Chapter 16 a characterization of those such that

$$A \otimes_{\min} B(H) = A \otimes_{\max} B(H).$$

Exercise

Exercise 13.1. Let G be any free group with free generators $\{g_i \mid i \in I\}$. Show that $C^*(G)$ can be embedded into a direct sum of matrix algebras ([Ch2]).

Hint: Let $\{\pi \mid \pi \in \widehat{G}_0\}$ denote the collection of all the finite-dimensional unitary representations of G (without repetitions). We define

$$\forall \, t \in G \qquad \sigma(t) = \oplus\{\pi(t) \mid \pi \in \widehat{G}_0\}.$$

Clearly, σ extends to a (contractive) representation $\widetilde{\sigma}\colon C^*(G) \to \bigoplus_{\pi \in \widehat{G}_0} B(H_\pi)$. We claim that $\widetilde{\sigma}$ is isometric. Let $\mathcal{A}$ be the (incomplete) $*$-algebra generated by $\sigma(G)$. To prove this claim, it clearly suffices to show that $(\widetilde{\sigma})^{-1}_{|\mathcal{A}}$ is (completely) contractive. By Proposition 13.6, for the latter it suffices to show that the restriction of $(\widetilde{\sigma})^{-1}$ to the linear span E of I and $\{\sigma(g_i) \mid i \in I\}$ is completely contractive, which indeed follows from Theorem 8.9.

Chapter 14. Kirchberg's Theorem on Decomposable Maps

We have seen (see (11.1) and (11.6)) that c.p. maps (and more generally decomposable maps) "tensorize" for the maximal tensor product. Actually, as Kirchberg proved, the converse also holds, and this is the main result of this section:

Theorem 14.1. *([Ki6]) Let A, B be C^*-algebras and let $u\colon A \to B$ be a linear map. Let us denote by $i_B\colon B \to B^{**}$ the canonical inclusion map of B into B^{**} viewed as a von Neumann algebra as usual. The following are equivalent:*

(i) *The map $i_B u\colon A \to B^{**}$ is decomposable with $\|i_B u\|_{\mathrm{dec}} \leq 1$.*

(ii) *For any C^*-algebra D, we have*

$$\forall\, x \in D \otimes A \qquad \|(I_D \otimes u)(x)\|_{D \otimes_{\max} B} \leq \|x\|_{D \otimes_{\max} A}.$$

Remark. It is easy to check that (i) holds iff $\|u^{**}\|_{\mathrm{dec}(A^{**}, B^{**})} \leq 1$.

The proof uses several results on c.p. maps relative to "multiplicative domains" ([Ch1]), which have not yet been discussed here, so we start by a brief review.

Lemma 14.2. *Let $u\colon A \to B$ be a c.p. map between C^*-algebras with $\|u\| \leq 1$.*

(i) *Then, if $a \in A$ satisfies $u(a^*a) = u(a)^*u(a)$, we have necessarily*

$$\forall x \in A \qquad u(xa) = u(x)u(a),$$

and the set of such a forms an algebra.

(ii) *Let $C_u = \{a \in A \mid u(a^*a) = u(a)^*u(a) \text{ and } u(aa^*) = u(a)u(a)^*\}$. Then C_u is a C^*-subalgebra of A satisfying*

$$\forall a, b \in C_u \quad \forall x \in A \qquad u(axb) = u(a)u(x)u(b).$$

Proof. First observe that u satisfies the following kind of Cauchy-Schwarz inequality (see [Kad, Ch1] for similar results for positive or merely 2-positive maps);

$$\forall x \in A \qquad u(x^*x) \geq u(x)^*u(x). \tag{14.1}$$

This is easy for c.p. maps. Indeed, by Arveson's Theorem, we can write u as $u(\cdot) = V^*\pi(\cdot)V$ for some representation $\pi\colon A \to B(\widehat{H})$ and $V\colon H \to \widehat{H}$ with $B \subset B(H)$. Then we have for all T with $0 \leq T \leq 1$

$$u(x^*x) = V^*\pi(x)^*\pi(x)V \geq V^*\pi(x)^*T\pi(x)V.$$

Hence, choosing $T = VV^*$, we obtain (14.1). This implies that the "defect"

$$\varphi(x, y) = u(x^*y) - u(x)^*u(y)$$

behaves like a B-valued scalar product. In particular, we clearly have (by Cauchy-Schwarz)

$$\forall \xi \in H \quad \forall x, y \in A \quad |\langle \varphi(x, y)\xi, \xi\rangle| \leq \langle \varphi(x, x)\xi, \xi\rangle^{1/2} \langle \varphi(y, y)\xi, \xi\rangle^{1/2}.$$

This shows (taking $y = a$) that, if $\varphi(a, a) = 0$, we have $\langle \varphi(x, a)\xi, \xi\rangle = 0$ for all ξ; hence $\varphi(x, a) = 0$ for all x in A. Changing x to x^* (and recalling that a c.p. map is self-adjoint) we obtain

$$u(xa) = u(x)u(a) \quad \text{for all} \quad x \text{ in } A.$$

Thus we have proved

$$\{a \in A \mid u(a^*a) = u(a)^*u(a)\} = \{a \in A \mid u(xa) = u(x)u(a) \quad \forall x \in A\}.$$
$$(14.2)$$

It is easy to see that the right side of this equality is an algebra. This proves (i). To check (ii), we note that by reversing the roles of a and a^* in (i) we have $u(aa^*) = u(a)u(a)^*$ iff

$$u(ay) = u(a)u(y) \quad \text{for all} \quad y \text{ in } A.$$

Note that $a \in C_u$ iff both a and a^* belong to the set (14.2). Therefore, C_u is a C^*-algebra and we have for any a, b in C_u and any x in A

$$u(axb) = u(a)u(xb) = u(a)[u(x)u(b)]. \qquad \blacksquare$$

As a consequence, we have

Lemma 14.3. *Let C be a C^*-algebra and let $\pi\colon C \to \pi(C) \subset B(H)$ be a representation. Assume $C \subset B(K)$. Then any contractive c.p. map (in particular any unital c.p. map) $u\colon B(K) \to B(H)$ extending π must satisfy*

$$\forall c_1, c_2 \in C \quad \forall x \in B(K) \quad u(c_1 x c_2) = \pi(c_1)u(x)\pi(c_2),$$

that is, u must be a C-bimodule map (for the action defined by π).

Proof. Indeed, with the notation in Lemma 14.2 we have clearly $C \subset C_u$; hence this follows from the second part of Lemma 14.2. $\qquad \blacksquare$

A basic (elementary) fact used in the sequel is that for any $x \in B(H)$ we have (see Exercise 11.5 (ii))

$$\|x\| \leq 1 \quad \text{iff} \quad \begin{pmatrix} 1 & x \\ x^* & 1 \end{pmatrix} \geq 0.$$

The next two lemmas are extensions of this to the case when x is replaced by a c.p. map (resp. a bimodule map) from an operator space (resp. an operator bimodule) into $B(H)$.

Lemma 14.4. Let $v\colon E \to B(H)$ be a map defined on an operator space $E \subset B(K)$. Let $S \subset M_2(B(K))$ be the operator system consisting of all matrices $\begin{pmatrix} \lambda 1 & a \\ b^* & \mu 1 \end{pmatrix}$ with $\lambda, \mu \in \mathbb{C}$, $a, b \in E$, and let $V\colon S \to M_2(B(H))$ be the mapping defined by

$$V\left(\begin{pmatrix} \lambda 1 & a \\ b^* & \mu 1 \end{pmatrix}\right) = \begin{pmatrix} \lambda 1 & v(a) \\ v(b)^* & \mu 1 \end{pmatrix}.$$

Then $\|v\|_{cb} \leq 1$ iff V is c.p.

Actually, we will use a generalization of this lemma in the setting of "operator modules," as follows (this result appears in [Su]; see also [PaSu]).

Lemma 14.5. Let $C \subset B(K)$ be a C^*-algebra given with a representation $\pi\colon C \to B(H)$. Let $E \subset B(K)$ be a C-bimodule, that is, an operator space stable by (left and right) multiplication by any element of C. Consider a bimodule map $w\colon E \to B(H)$, that is, a map satisfying $w(c_1 x c_2) = \pi(c_1) w(x) \pi(c_2)$, $(c_1, c_2 \in C, x \in E)$. Let $S \subset M_2(B(K))$ be the operator system consisting of all matrices of the form $\begin{pmatrix} \lambda & a \\ b^* & \mu \end{pmatrix}$ with $\lambda, \mu \in C$, $a, b \in E$. Let $W\colon S \to M_2(B(H))$ be defined by

$$W\left(\begin{pmatrix} \lambda & a \\ b^* & \mu \end{pmatrix}\right) = \begin{pmatrix} \pi(\lambda) & w(a) \\ w(b)^* & \pi(\mu) \end{pmatrix}.$$

Then $\|w\|_{cb} \leq 1$ iff W is c.p.

Proof. The easy direction is W c.p. $\Rightarrow \|w\|_{cb} \leq 1$. Indeed, if W is c.p., we have $\|W\|_{cb} = \|W(1)\| = 1$, and a fortiori $\|w\|_{cb} \leq 1$.

We now turn to the converse. Assume $\|w\|_{cb} \leq 1$. Consider an element $s \in M_n(S)$, say, $s = \begin{pmatrix} \lambda & a \\ b^* & \mu \end{pmatrix}$, $\lambda, \mu \in M_n(C)$, $a, b \in M_n(E)$. Assume $s \geq 0$. Then necessarily $\lambda, \mu \in M_n(C)_+$ and $a = b$. Fix $\varepsilon > 0$ and let $\lambda_\varepsilon = \lambda + \varepsilon 1$ and $\mu_\varepsilon = \mu + \varepsilon 1$ (invertible perturbations of λ and μ). Let $s_\varepsilon = s + \varepsilon 1$. Let us denote $x_\varepsilon = \lambda_\varepsilon^{-1/2} a \mu_\varepsilon^{-1/2}$ and let $W_n = I_{M_n} \otimes W$, $w_n = I_{M_n} \otimes w$. We then have

$$\begin{pmatrix} 1 & x_\varepsilon \\ x_\varepsilon^* & 1 \end{pmatrix} = \begin{pmatrix} \lambda_\varepsilon^{-1/2} & 0 \\ 0 & \mu_\varepsilon^{-1/2} \end{pmatrix} s_\varepsilon \begin{pmatrix} \lambda_\varepsilon^{-1/2} & 0 \\ 0 & \mu_\varepsilon^{-1/2} \end{pmatrix}; \qquad (14.3)$$

hence the left side of the preceding equation is ≥ 0, which implies by the basic fact recalled above that $\|x_\varepsilon\| \leq 1$. Therefore, if $\|w\|_{cb} \leq 1$, we have $\|w_n(x_\varepsilon)\| \leq 1$, which implies that

$$\begin{pmatrix} 1 & w_n(x_\varepsilon) \\ w_n(x_\varepsilon)^* & 1 \end{pmatrix} \geq 0.$$

But this last matrix is the same as $W_n \begin{pmatrix} 1 & x_\varepsilon \\ x_\varepsilon^* & 1 \end{pmatrix}$. Now, applying W_n to both sides of (14.3) and using the fact that w is a C-bimodule map, we find

$$W_n \begin{pmatrix} 1 & x_\varepsilon \\ x_\varepsilon^* & 1 \end{pmatrix} = \begin{pmatrix} \lambda_\varepsilon^{-1/2} & 0 \\ 0 & \mu_\varepsilon^{-1/2} \end{pmatrix} W_n(s_\varepsilon) \begin{pmatrix} \lambda_\varepsilon^{-1/2} & 0 \\ 0 & \mu_\varepsilon^{-1/2} \end{pmatrix}.$$

Since the left side is ≥ 0, we have $W_n(s_\varepsilon) \geq 0$, and letting $\varepsilon \to 0$ we conclude that $W_n(s) \geq 0$, whence that W is c.p. ∎

Kirchberg's Proof of Theorem 14.1. We assume A unital for simplicity. The implication (i) $\Rightarrow$ (ii) is easy using the isometric embedding $C \otimes_{\max} B \subset C \otimes_{\max} B^{**}$ (see Exercise 11.6). Hence the main point is the converse. Assume (ii). Let $M = B^{**}$ viewed as a von Neumann algebra embedded in $B(H)$ for some H. We have $B \subset B^{**}$. We will use $D = M'$. We will work with the mapping

$$w \colon M' \otimes_{\max} A \to B(H)$$

defined by $w\left(\sum c_i \otimes a_i\right) = \sum c_i u(a_i) = \sum u(a_i) c_i$.

Note that $\|w\|_{cb} \leq 1$. Indeed, replacing D by $M_n(D)$, we see that (if (ii) holds) $I_D \otimes u \colon D \otimes_{\max} A \to D \otimes_{\max} B$ is completely contractive and w is the composition of this map with the $*$-homomorphism $\sigma \colon M' \otimes_{\max} B \to B(H)$ defined by $\sigma(c \otimes b) = cb = bc$.

Let K be a suitable Hilbert space so that $M' \otimes_{\max} A \subset B(K)$. Let $C = M' \otimes 1 \subset B(K)$, and let $E = M' \otimes_{\max} A$. Note that E is a C-bimodule and that $w \colon E \to B(H)$ is a C-bimodule map. Hence, with the same notation as in Lemma 14.5, the mapping $W \colon S \to M_2(B(H))$ must be completely positive (and unital).

Recall (see Corollary 1.8) that for a unital mapping on a C^*-algebra, "completely contractive" is equivalent to "completely positive"; hence, by Corollary 1.7, any unital c.p. map $V \colon A_1 \to B(H)$ on a (unital) C^*-subalgebra $A_1 \subset A_2$ admits a (unital) c.p. extension $\widetilde{V} \colon A_2 \to B(H)$. Thus, let $\widetilde{W} \colon M_2(B(K)) \to M_2(B(H))$ be a completely positive extension of W, and let $T \colon M_2(A) \to M_2(B(H))$ be its restriction to $M_2(A)$. (Here we identify A and $1 \otimes A \subset B(K)$.) Let $u_*(x) = u(x^*)^*$.

We then claim that T is of the following form:

$$\forall x \in M_2(A) \qquad T(x) = \begin{pmatrix} T_{11}(x_{11}) & u(x_{12}) \\ u_*(x_{21}) & T_{22}(x_{22}) \end{pmatrix},$$

where T_{11} and T_{22} are c.p. maps from A into $M = B^{**}$, with $\|T_{11}\|, \|T_{22}\| \leq 1$. Taking this claim for granted, it is easy to conclude the proof since we can then define

$$\forall a \in A \qquad R(a) = T \begin{pmatrix} a & a \\ a & a \end{pmatrix} = \begin{pmatrix} T_{11}(a) & u(a) \\ u_*(a) & T_{22}(a) \end{pmatrix}.$$

Since T is c.p. and the mapping $a \to \begin{pmatrix} a & a \\ a & a \end{pmatrix}$ is clearly c.p., R must be c.p. Hence, by the definition of the dec-norm, we have

$$\|u\|_{\mathrm{dec}} \leq \max\{\|T_{11}\|, \|T_{22}\|\} \leq 1.$$

Thus it only remains to prove the claim. For this purpose, observe that, by its definition, W is a $*$-homomorphism on the algebra of matrices $\begin{pmatrix} c_1 & 0 \\ 0 & c_2 \end{pmatrix}$ with c_1, c_2 in $C = M' \otimes 1$. Indeed, let $\mathcal{C}$ denote the set of all such matrices, and let $\pi \colon \mathcal{C} \to M_2(M')$ be defined by

$$\pi \begin{pmatrix} c_1 \otimes 1 & 0 \\ 0 & c_2 \otimes 1 \end{pmatrix} = \begin{pmatrix} c_1 & 0 \\ 0 & c_2 \end{pmatrix}.$$

Then the definition of W shows that $W_{|\mathcal{C}} = \pi$. Therefore, since $\widetilde{W}$ is a c.p. extension of π, Lemma 14.3 implies that $\widetilde{W}$ must be a $\mathcal{C}$-bimodule map; that is, we have

$$\forall y_1, y_2 \in \mathcal{C} \quad \forall x \in M_2(B(H)) \qquad \widetilde{W}(y_1 x y_2) = \pi(y_1)\widetilde{W}(x)\pi(y_2). \quad (14.4)$$

Applying this with c_1, c_2 scalars, we find (by linear algebra) that $\widetilde{W}$ is necessarily such that $\widetilde{W}(x)_{ij}$ depends only on x_{ij}. A fortiori the same is true for $T = \widetilde{W}_{|M_2(A)}$. Thus we can write a priori

$$\forall x \in M_2(A) \qquad T(x) = \begin{pmatrix} T_{11}(x_{11}) & T_{12}(x_{12}) \\ T_{21}(x_{21}) & T_{22}(x_{22}) \end{pmatrix}.$$

Taking $x_{11} = x_{22} = 0$ and recalling that $\widetilde{W}$ extends W, we find $T_{12} = u$ and $T_{21} = u_*$.

Moreover, since T is unital and c.p., the same is true for T_{11} and T_{22}, hence, we have $\|T_{11}\| \leq 1$ and $\|T_{22}\| \leq 1$ by Exercise 11.5 (iii).

Finally, it remains to check that T_{11} and T_{22} take their values into $B^{**} = M$. For that it suffices to check that $T_{11}(x_{11})$ and $T_{22}(x_{22})$ commute with M'. But this is an easy consequence of (14.4). Indeed, let $x_{11} \in A$, $x_{22} \in A$ and let

$$x = \begin{pmatrix} 1 \otimes x_{11} & 0 \\ 0 & 1 \otimes x_{22} \end{pmatrix} \quad \text{and} \quad c = \begin{pmatrix} c_1 \otimes 1 & 0 \\ 0 & c_2 \otimes 1 \end{pmatrix}.$$

Then, with the same notation as in (14.4), we have $cx = xc$, and hence $\pi(c)T(x) = T(x)\pi(c)$, which implies that $T_{11}(x_{11})$ and $T_{22}(x_{22})$ commute with M'; hence they take their values in $M'' = M = B^{**}$. This ends the proof of the claim and of Theorem 14.1. ∎

The same proof yields an extension theorem as follows:

Theorem 14.6. *Consider two C^*-algebras A, B, a subspace $X \subset A$, and a linear map $u\colon X \to B$. The following are equivalent:*

(i) The map u admits a decomposable extension $\tilde{u}\colon A \to B^{**}$ with $\|\tilde{u}\|_{\mathrm{dec}} \leq 1$.

(ii) For any C^*-algebra D, we have

$$\forall\, x \in D \otimes X \qquad \|(I_D \otimes u)(x)\|_{D \otimes_{\max} B} \leq \|x\|_{D \otimes_{\max} A}.$$

Proof. We simply repeat the preceding proof with $E = M' \otimes X \subset M' \otimes_{\max} A$. ∎

Chapter 15. The Weak Expectation Property (WEP)

In this short chapter we discuss a property introduced by Lance [La1-2] as a generalization of nuclearity, namely, the weak expectation property (WEP).

Definition 15.1. A C^*-algebra A has the WEP (or "is WEP") if the inclusion map $i_A: A \to A^{**}$ factors completely positively and completely contractively through $B(H)$ for some H; that is, we have completely positive complete contractions $T_1: B(H) \to A^{**}$ and $T_2: A \to B(H)$ such that $T_1 T_2 = i_A$.

Remark 15.2. (i) If we assume A^{**} embedded as a von Neumann subalgebra in $B(H)$. Then the WEP implies that there is a completely positive and completely contractive mapping $T: B(H) \to A^{**}$ such that $T(a) = a$ for all a in a. Thus we can take $T_1 = T$ and T_2 equal to the inclusion mapping of A into $B(H)$. Note that in general T will *not* be a projection. But of course, if A^{**} is injective, then we can find a projection from $B(H)$ onto A^{**} so that a fortiori A^{**}, and hence A, has the WEP.

(ii) A von Neumann algebra M has the WEP iff it is injective. Indeed, if M has the WEP, it is injective since there is a projection $P: M^{**} \to M$, which is actually a normal $*$-homomorphism (P can be described as the canonical normal $*$-homomorphism extending the identity $M \to M$). Conversely, if $M \subset B(H)$ is injective, there is a c.p. projection from $B(H)$ onto it, so M is WEP.

Since we know that A^{**} is injective whenever A is nuclear (see Theorem 11.6), we find that nuclearity implies the WEP. A more direct proof can be given using the following result.

Proposition 15.3. Let A be a C^*-algebra. The following are equivalent:

(i) A has the WEP.

(ii) For any embedding $j: A \subset B$ into a C^*-algebra B and for any C^*-algebra C, the associated morphism $A \otimes_{\max} C \to B \otimes_{\max} C$ is isometric.

(iii) There is an embedding $A \subset B(H)$ such that, if $C = C^*(\mathbb{F}_\infty)$, the associated morphism $A \otimes_{\max} C \to B(H) \otimes_{\max} C$ is isometric.

(iv) There is an embedding $A \subset B(H)$ such that the associated morphism $A \otimes_{\max} C \to B(H) \otimes_{\max} C$ is isometric for any C^*-algebra C.

Proof. Assume (i). Let T_1 and T_2 be as above. By the extension property (see Corollary 1.7), T_2 admits a completely contractive extension $\widetilde{T}_2: B \to B(H)$ and by (11.5) we have $\|\widetilde{T}_2\|_{dec} = \|\widetilde{T}_2\|_{cb} \leq 1$. Then, by (11.3) and (11.4), the mapping $\widetilde{T} = T_1 \widetilde{T}_2: B \to A^{**}$ is decomposable with $\|\widetilde{T}\|_{dec} \leq 1$ and satisfies $\widetilde{T}_{|A} = i_A$. By (11.7) we have

$$\|\widetilde{T} \otimes I_C: B \otimes_{\max} C \to A^{**} \otimes_{\max} C\| \leq 1.$$

Hence, for any t in $A \otimes C$, we have

$$\|t\|_{A^{**}\otimes_{\max}C} \leq \|t\|_{B\otimes_{\max}C};$$

but since the inclusion $A\otimes_{\max}C \to A^{**}\otimes_{\max}C$ is isometric (cf. Exercise 11.6), this implies $\|t\|_{A\otimes_{\max}C} \leq \|t\|_{B\otimes_{\max}C}$, and the converse inequality is obvious. This shows that (i) $\Rightarrow$ (ii). The implication (ii) $\Rightarrow$ (iii) is trivial.

Assume (iii). We first claim that (iii) remains true when $\mathbb{F}_\infty$ is replaced by an arbitrary free group $\mathbb{F}$. Indeed, it is easy to see that, for any finite-dimensional subspace $E \subset C^*(\mathbb{F})$, there is a C^*-subalgebra D with $E \subset D \subset C^*(F)$ such that $D \simeq C^*(\mathbb{F}_\infty)$ and there is a completely positive contractive projection from $C^*(F)$ onto D. (*Hint:* Any element in $C^*(\mathbb{F})$ can be described using only countably many elements in $\mathbb{F}$, and hence only countably many generators. Then we can invoke Proposition 8.8.) By (11.1), our claim follows. Now, let C be any C^*-algebra. Let $\mathbb{F}$ be a large enough free group so that, if $B = C^*(\mathbb{F})$, we can write $C = B/\mathcal{I}$ for some ideal $\mathcal{I}$ (see Exercise 8.1). By Exercise 11.2, we have

$$(B/\mathcal{I}) \otimes_{\max} A \simeq \frac{B \otimes_{\max} A}{\mathcal{I} \otimes_{\max} A},$$

from which it is easy to see that, if (iii) holds for $C = B$, it also holds for $C = B/\mathcal{I}$. This shows that (iii) implies (iv).

Finally, to show that (iv) implies (i), one way is to apply Theorem 14.6 to our embedding $A \subset B(H)$. Assuming A unital, it follows that there is a completely contractive map $T \colon B(H) \to A^{**}$ extending the inclusion $A \to A^{**}$. Since T is unital, it must be c.p. by Corollary 1.8. For the convenience of the reader, we will give a more explicit argument based on the same idea but avoiding decomposable maps. Let $\mathcal{H}$ be large enough so that A^{**} can be realized as a von Neumann subalgebra of $B(\mathcal{H})$. Let $\pi \colon A^{**} \to B(\mathcal{H})$ be the inclusion map, and let $C = \pi(A^{**})'$. Let $u \colon A \otimes_{\max} C \to B(\mathcal{H})$ be the representation defined by

$$u(a \otimes c) = \pi(a)c.$$

Since $A \otimes_{\max} C \subset B(H) \otimes_{\max} C$, there is a (unital) complete contraction $\tilde{u} \colon B(H) \otimes_{\max} C \to B(\mathcal{H})$ extending u with $\|\tilde{u}\| = \|u\| = 1$. By Corollary 1.8, $\tilde{u}$ is c.p., and by Lemma 14.3, $\tilde{u}$ must be an $(A \otimes C)$-bimodule map. In particular, if we define $T_1 \colon B(H) \to B(\mathcal{H})$ by $T_1(b) = \tilde{u}(b \otimes 1)$ for any b in $B(H)$, then we must have for any c in C

$$T_1(b)c = T_1(b \otimes c) \quad \text{and} \quad cT_1(b) = T_1(b \otimes c).$$

Therefore $T_1(b)$ commutes with $C = \pi(A^{**})'$, and hence $T_1(b) \in \pi(A^{**})'' = \pi(A^{**})$. Thus we obtain a c.p. map $T_1 \colon B(H) \to A^{**}$ with $\|T_1\| = 1$ such that $T_1|A = i_A$, which proves that A is WEP. This shows that (iv) implies (i). ∎

Corollary 15.4. *Any nuclear C^*-algebra has the WEP.*

Proof. With the same notation as in Proposition 15.3(ii), if A is nuclear, there is a unique C^*-norm on $A \otimes C$, but the norm induced by $B \otimes_{\max} C$ on $A \otimes C$ is a C^*-norm; hence (ii) in Proposition 15.3 must hold. ∎

The preceding statement can also be derived from the next one, which gives Kirchberg's striking characterization of the WEP (cf. [Ki2]).

Theorem 15.5. *Let $C = C^*(\mathbb{F}_\infty)$. Then a C^*-algebra A has the WEP iff*

$$C \otimes_{\min} A = C \otimes_{\max} A.$$

Proof. Assume $C \otimes_{\min} A = C \otimes_{\max} A$. We claim that (iii) in Proposition 15.3 holds. Assume $A \subset B(H)$. Consider $t \in C \otimes A$. Then we have

$$\|t\|_{C \otimes_{\max} B(H)} \geq \|t\|_{C \otimes_{\min} B(H)} = \|t\|_{C \otimes_{\min} A} = \|t\|_{C \otimes_{\max} A}.$$

Hence (since $\|t\|_{C \otimes_{\max} A} \geq \|t\|_{C \otimes_{\max} B(H)}$ is trivial) we obtain our claim, so that A has the WEP by Proposition 15.3. Conversely, assume A WEP. Then, by Proposition 15.3, the inclusion

$$C \otimes_{\max} A \to C \otimes_{\max} B(H)$$

is isometric. On the other hand, of course, by the injectivity of the min-norm, the inclusion

$$C \otimes_{\min} A \to C \otimes_{\min} B(H)$$

is isometric as well. But now, by Theorem 13.1 we know that $C \otimes_{\max} B(H) = C \otimes_{\min} B(H)$, hence we must have equality of the norms induced on $C \otimes A$, which yields $C \otimes_{\max} A = C \otimes_{\min} A$. ∎

Remark. The same argument shows that if A has the WEP, then for any free group F we have

$$C^*(F) \otimes_{\min} A = C^*(F) \otimes_{\max} A.$$

The WEP does not seem to have been studied much in the operator space context, although its definition extends rather naturally, as follows.

Definition. *Let X be an operator space. Let $i_X \colon X \to X^{**}$ be the canonical inclusion. Let $\lambda \geq 1$. We say that X has the λ-WEP if there are maps $T_1 \colon B(H) \to X^{**}$ and $T_2 \colon X \to B(H)$ such that $i_X = T_1 T_2$ and satisfying*

$$\|T_1\|_{cb} \|T_2\|_{cb} \leq \lambda.$$

Note that $1 \leq \|T_1\|_{cb}\|T_2\|_{cb}$; hence, when $\lambda = 1$, this implies that we may choose T_1, T_2 so that $\|T_1\|_{cb} = \|T_2\|_{cb} = 1$.

It is not at all obvious that, if a C^*-algebra A has the λ-WEP for some $\lambda \geq 1$, then it has the WEP in the sense of Definition 15.1, but fortunately this is indeed true. It follows from an unpublished result due to Haagerup [H4] (see Remark 15.12). Thus we may say that an operator space has the WEP if it has the λ-WEP for some $\lambda \geq 1$ or, equivalently, if i_X factors completely boundedly through $B(H)$.

Haagerup's result [H4] is as follows.

Theorem 15.6. *The following properties of a C^*-algebra A are equivalent:*

(i) A has the WEP.

(ii) For any n and any $a_1, \ldots, a_n$ in A we have

$$\left\| \sum a_i \otimes \overline{a_i} \right\|_{A \otimes_{\min} \overline{A}} = \left\| \sum a_i \otimes \overline{a_i} \right\|_{A \otimes_{\max} \overline{A}}.$$

(iii) There is a constant C such that, for all n and all $a_1, \ldots, a_n$ in A, we have

$$\left\| \sum a_i \otimes \overline{a_i} \right\|_{A \otimes_{\max} \overline{A}} \leq C \left\| \sum a_i \otimes \overline{a_i} \right\|_{A \otimes_{\min} \overline{A}}.$$

We only give indications on the proof. In particular we will omit the proof of the most delicate implication (ii) $\Rightarrow$ (i).

Remark. It is worthwhile to recall:

$$\overline{A} \simeq A^{op} \quad \text{via the correspondence} \quad a \to a^*.$$

Hence we have in particular

$$\left\| \sum a_i \otimes \overline{a_i} \right\|_{A \otimes_{\max} \overline{A}} = \left\| \sum a_i \otimes a_i^* \right\|_{A \otimes_{\max} A^{op}}.$$

The following key result plays an important role in the proof.

Theorem 15.7. *Let A be any C^*-algebra. Fix an integer $n \geq 1$. Let E_0 (resp. E_1) be the space of n-tuples $(a_1, \ldots, a_n)$ in A equipped with the norm*

$$\|(a_i)\|_{E_0} = \left\| \sum a_i^* a_i \right\|^{1/2} \quad \left(\text{resp. } \|(a_i)\|_{E_1} = \left\| \sum a_i a_i^* \right\|^{1/2}\right).$$

We then have

$$\|(a_i)\|_{(E_0, E_1)_{1/2}} = \left\| \sum a_i \otimes \overline{a_i} \right\|_{A \otimes_{\max} \overline{A}}^{1/2}. \tag{15.1}$$

Note that here either E_0 or E_1, or the complex interpolation space $(E_0, E_1)_\theta$, are the *same* spaces (namely A^n) equipped with different norms. The preceding equality computes the "interpolated norm" for $\theta = 1/2$. Haagerup's unpublished results extend those of the previous paper [P7] restricted to the case when A is a semi-finite von Neumann algebra, as follows.

Theorem 15.8. *Let A be a von Neumann algebra, equipped with a normal, faithful semi-finite trace τ, so that we can define in a standard way the space $L_2(\tau)$. Then, if we denote by $a \to L(a)$ (resp. $a \to R(a)$) the operator of left (resp. right) multiplication by a acting on $L_2(\tau)$, then for any $a_1, \ldots, a_n$ in A we have*

$$\|(a_i)\|_{(E_0,E_1)_{1/2}} = \left\|\sum L(a_i)R(a_i^*)\right\|_{B(L_2(\tau))}^{1/2}. \tag{15.2}$$

Remark. Note that (15.2) immediately implies that $\|(a_i)\|_{(E_0,E_1)_{1/2}} \leq \|\sum a_i \otimes \overline{a_i}\|_{\max}^{1/2}$. Conversely, we have by (1.11)

$$\left\|\sum a_i \otimes \overline{b_i}\right\|_{\max} \leq \|(a_i)\|_{E_0}\|(b_i)\|_{E_1},$$

and also since $a_i \otimes 1$ and $1 \otimes b_i$ commute

$$\left\|\sum a_i \otimes \overline{b_i}\right\|_{\max} \leq \|(a_i)\|_{E_1}\|(b_i)\|_{E_0}.$$

Hence, by the complex interpolation theorem ([BeL, p. 96]) for bilinear maps (noticing that $(E_0, E_1)_{1/2} = (E_1, E_0)_{1/2}$), we must have

$$\left\|\sum a_i \otimes \overline{b_i}\right\|_{\max} \leq \|(a_i)\|_{(E_0,E_1)_{1/2}}\|(b_i)\|_{(E_0,E_1)_{1/2}}.$$

Finally, taking $b_i = \overline{a_i}$ we obtain

$$\left\|\sum a_i \otimes \overline{a_i}\right\|_{\max} \leq \|(a_i)\|_{(E_0,E_1)_{1/2}}^2,$$

which yields half of (15.1).

Now, in the case $A = B(H)$, by Corollary 7.12 we have $\|(a_i)\|_{(E_0,E_1)_{1/2}} = \|\sum a_i \otimes \overline{a_i}\|_{\min}^{1/2}$. Hence when $A = B(H)$ we obtain

$$\left\|\sum a_i \otimes \overline{a_i}\right\|_{\max} = \left\|\sum a_i \otimes \overline{a_i}\right\|_{\min}. \tag{15.3}$$

More generally, (15.3) holds if A is WEP. Indeed, if

$$T_1: B(H) \to A^{**} \quad \text{and} \quad T_2: A \to B(H)$$

are unital c.p. maps as in Definition 15.1, we have for any b_i in $B(H)$

$$\left\|\sum T_1(b_i) \otimes \overline{T_1(b_i)}\right\|_{\max} \leq \left\|\sum b_i \otimes \overline{b_i}\right\|_{\max} = \left\|\sum b_i \otimes \overline{b_i}\right\|_{\min};$$

and if $b_i = T_2(a_i)$ with $a_i \in A$, we have $T_1 T_2(a_i) = a_i$ and

$$\left\|\sum b_i \otimes \overline{b_i}\right\|_{\min} \leq \left\|\sum a_i \otimes \overline{a_i}\right\|_{\min}.$$

Hence we conclude that (15.3) holds in any WEP C^*-algebra A. This proves the implication (i) $\Rightarrow$ (ii) in Theorem 15.6.

Remark. The proof that (iii) $\Rightarrow$ (ii) in Theorem 15.6 is elementary: Let $t = \sum a_i \otimes \overline{a_i}$. Since, for any C^*-norm, we have $\|t\|^{2k} = \|(t^*t)^k\|$ and since $(t^*t)^k$ is again of the form $\sum \alpha_j \otimes \overline{\alpha_j}$, we have

$$\|t\|_{\max}^{2k} = \|(t^*t)^k\|_{\max} \le C\|(t^*t)^k\|_{\min} = C\|t\|_{\min}^{2k};$$

hence $\|t\|_{\max} \le C^{\frac{1}{2k}}\|t\|_{\min}$ and letting k go to infinity we obtain (ii).

Remark. The implication that $A \otimes_{\min} \overline{A} = A \otimes_{\max} \overline{A} \Rightarrow A$ has the WEP had been previously observed by Kirchberg in [Ki2].

The next two corollaries follow from Theorem 15.6.

Corollary 15.9. *Let* $u\colon A \to B$ *be a linear map between* C^**-algebras. Assume that* u *is c.b. (actually, up to the constant, it suffices to assume that* $I_G \otimes u\colon G \otimes_{\min} A \to G \otimes_{\min} B$ *is bounded when* $G = R \oplus C$*). Then, for any finite sequence* $a_1, \ldots, a_n$ *in* A*, we have*

$$\left\| \sum u(a_i) \otimes \overline{u(a_i)} \right\|_{\max} \le \|u\|_{cb}^2 \left\| \sum a_i \otimes \overline{a_i} \right\|_{\max}. \tag{15.4}$$

Proof. Let $C = \|u\|_{cb}$. We have

$$\left\| \sum u(a_i)^*u(a_i) \right\|^{1/2} \le C\|(a_i)\|_{E_0} \quad \text{and} \quad \left\| \sum u(a_i)u(a_i)^* \right\|^{1/2} \le C\|(a_i)\|_{E_1}.$$

Hence, by interpolation (applying Theorem 15.7 both in A and in B), we deduce (15.4). ∎

Corollary 15.10. *In the situation of Corollary 15.9, assume that* u *is a complete isomorphism. Then, if* A *has the WEP, so does* B*.*

Proof. The analog of (15.4) for the minimal C^*-norm is clear since, if (T_i) is any orthonormal basis in OH, we have

$$\left\| \sum a_i \otimes \overline{a_i} \right\|_{\min}^{1/2} = \left\| \sum a_i \otimes T_i \right\|_{A \otimes_{\min} OH}.$$

Hence, if A has the WEP, we find

$$\left\| \sum u(a_i) \otimes \overline{u(a_i)} \right\|_{\max} \le \|u\|_{cb}^2 \left\| \sum a_i \otimes \overline{a_i} \right\|_{\max} = \|u\|_{cb}^2 \left\| \sum a_i \otimes \overline{a_i} \right\|_{\min}$$

$$\le \|u\|_{cb}^2 \|u^{-1}\|_{cb}^2 \left\| \sum u(a_i) \otimes \overline{u(a_i)} \right\|_{\min},$$

and by (iii) $\Rightarrow$ (i) in Theorem 15.6 we conclude that B has the WEP. ∎

The next result was originally proved, independently and differently, in [P3, CS3].

Corollary 15.11. *Let $M \subset B(H)$ be a von Neumann algebra. Assume that there is a projection P from $B(H)$ onto M that is c.b. (or merely such that $I_G \otimes P$: $G \otimes_{\min} B(H) \to G \otimes_{\min} M$ is bounded when $G = R \oplus C$). Then M is injective.*

Proof. By Corollary 15.9 applied to P: $B(H) \to M$, for any $a_1, \ldots, a_n$ in M, we can write

$$\left\| \sum a_i \otimes \overline{a_i} \right\|_{M \otimes_{\max} \overline{M}} = \left\| \sum P(a_i) \otimes \overline{P(a_i)} \right\|_{M \otimes_{\max} \overline{M}}$$
$$\leq C \left\| \sum a_i \otimes \overline{a_i} \right\|_{B(H) \otimes_{\max} \overline{B(H)}}.$$

Hence, by (15.3),

$$\leq C \left\| \sum a_i \otimes \overline{a_i} \right\|_{\min}.$$

Thus, by (iii) $\Rightarrow$ (i) in Theorem 15.6, M is WEP, and, for a von Neumann algebra, WEP $\Rightarrow$ injective by Remark 15.2. ∎

Remark 15.12. Assume that the maps T_1, T_2 appearing in Definition 15.1 are merely c.b. (actually, for T_1 it suffices even to assume that $I_G \otimes T_1$: $G \otimes_{\min} B(H) \to G \otimes_{\min} A^{**}$ is bounded when $G = R \oplus C$). Then, A necessarily has the WEP. Indeed, by Corollary 15.9 (and the validity of (15.3) when $A = B(H)$), for any a_i in A we have

$$\left\| \sum a_i \otimes \overline{a_i} \right\|_{A^{**} \otimes_{\max} \overline{A^{**}}} \leq \|T_1\|_{cb}^2 \|T_2\|_{cb}^2 \left\| \sum a_i \otimes \overline{a_i} \right\|_{A \otimes_{\min} \overline{A}}.$$

But by Exercise 11.6 (applied twice), the left side is equal to $\left\| \sum a_i \otimes \overline{a_i} \right\|_{A \otimes_{\max} \overline{A}}$; hence we conclude by Theorem 15.6 that A has the WEP.

Remark. Ozawa proved recently that $B(\ell_2) \otimes_{\min} B(\ell_2)$ fails the WEP (see [Oz5]).

We discuss several important open problems about the WEP at the end of Chapter 16.

Exercises

Exercise 15.1. Let $C = C^*(F_\infty)$. Show that a C^*-algebra A is nuclear iff

$$A^{**} \otimes_{\min} C = A^{**} \otimes_{\max} C.$$

Exercise 15.2. Consider the following property of a C^*-algebra A: For any pair of C^*-algebras B_1, B_2 with $B_1 \subset B_2$, we have $A \otimes_{\max} B_1 \subset A \otimes_{\max} B_2$. Show that this property characterizes nuclear C^*-algebras.

Hint: By Theorem 11.6, it suffices to show that this property implies the injectivity of A^{**}. To do that, use the well-known fact ([To2]) that a von Neumann algebra $M \subset B(H)$ is injective iff its commutant M' is injective.

Exercise 15.3. Let $\mathcal{M}$ be a C^*-algebra that is a quotient of a WEP C^*-algebra by an ideal (QWEP in short). Let $M \subset \mathcal{M}$ be a C^*-subalgebra such that there is a completely positive contractive projection P from $\mathcal{M}$ onto M. Show that M is QWEP.

Exercise 15.4. Using the preceding exercise, show that if $\mathcal{M}$ is any (von Neumann sense) ultraproduct of matrix algebras as in Theorem 9.10.1, then any von Neumann subalgebra $M \subset \mathcal{M}$ is QWEP.

Hint: Use the existence of a conditional expectation from $\mathcal{M}$ onto M.

Exercise 15.5. Let G be a discrete group. If either $C_\lambda^*(G)$ or $VN(G) = C_\lambda^*(G)''$ is WEP, then G is amenable.

Hint: Use Exercise 8.4 and Theorem 15.6 (see also Exercise 11.8).

Chapter 16. The Local Lifting Property (LLP)

In Banach space theory, the *lifting property* of ℓ_1 is classical: For any bounded linear map u from ℓ_1 into a quotient Banach space X/Y and for any $\varepsilon > 0$, there is a lifting $\widetilde{u}$: $\ell_1 \to X$ with $\|\widetilde{u}\| \leq (1 + \varepsilon)\|u\|$. We will discuss in Chapter 24 the operator space analogs of this (see also [KyR]). But in this chapter we concentrate on the C^*-algebraic analog of the lifting property as introduced by Kirchberg. Following [Ki2], we will say that a unital C^*-algebra C has the lifting property (LP in short) if any unital c.p. map u from C to a quotient A/I (here A is a C^*-algebra and I a two-sided closed ideal in A) admits a unital c.p. lifting $\widetilde{u}$: $C \to A$.

Kirchberg (see [Ki2]) proved that, for any countable free group F, $C^*(F)$ (and also $M_n(C^*(F))$ for any $n \geq 1$) has the LP. But actually very little is known about the LP, and it is rather the local version defined in the following that has proved fruitful.

Definition 16.1. *We say that a unital C^*-algebra C has the "local lifting property" (LLP in short) if the following holds for any C^*-algebra A and any (closed two-sided) ideal $I \subset A$: For any unital c.p. map u: $C \to A/I$ and for any finite-dimensional subspace $E \subset C$, there is a complete contraction $\widetilde{u}$: $E \to A$ that lifts $u_{|E}$: $E \to A/I$. In other words, although a priori u is not liftable, it roughly "locally lifts." When A is not unital, we say that it has the LLP if its unitization has it.*

While the LP remains a rather elusive property up to now, Kirchberg [Ki2] discovered the following nice characterization for the LLP.

Theorem 16.2. *The following properties of a C^*-algebra C are equivalent:*

(i) *C has the LLP.*

(ii) *$C \otimes_{\min} B(H) = C \otimes_{\max} B(H)$ for any Hilbert space H.*

(iii) *$C \otimes_{\min} B(H) = C \otimes_{\max} B(H)$ for $H = \ell_2$.*

(iv) *$C \otimes_{\min} B = C \otimes_{\max} B$ for any C^*-algebra B with the WEP.*

Proof. We will first prove (iii) $\Rightarrow$ (i).

Assume (iii). Let $A, I, E \subset C$ and u: $E \to A/I$ be as in Definition 16.1. Let $t \in C \otimes E^*$ be the tensor corresponding to the inclusion map $E \to C$, and let $s \in (A/I) \otimes E^*$ be the one corresponding to $u_{|E}$: $E \to A/I$, so that

$$s = (u \otimes I)(t).$$

We assume $E^* \subset B(H)$. Then by (2.3.2) we have

$$\|t\|_{C \otimes_{\min} B(H)} = \|I_E\|_{cb} = 1.$$

Therefore, since we assume (iii),

$$\|t\|_{C\otimes_{\max}B(H)} = 1,$$

and therefore by (11.1)

$$\|s\|_{(A/I)\otimes_{\max}B(H)} = \|(u \otimes I)(t)\|_{(A/I)\otimes_{\max}B(H)} \leq 1$$

But now, by Exercise 11.2, we have

$$(A/I) \otimes_{\max} B(H) = (A \otimes_{\max} B(H))/(I \otimes_{\max} B(H)).$$

Hence, by Lemma 2.4.6, there is an element $\widehat{s}$ in the unit ball of $A \otimes_{\max} B(H)$ such that, if $q \colon A \to A/I$ denotes the quotient map, we have

$$s = (q \otimes I)(\widehat{s}),$$

and of course a fortiori $\|\widehat{s}\|_{A\otimes_{\min}B(H)} \leq 1$. By Lemmas 2.4.8 and 2.4.7, this implies that there is an element $\widetilde{s}$ in the unit ball of $A \otimes_{\min} E^*$ such that $s = (q \otimes I)(\widetilde{s})$. Let $\widetilde{u} \colon E \to A$ be the linear map defined by $\widetilde{s}$. Note that by (2.3.2) again

$$\|\widetilde{u}\|_{cb} = \|\widetilde{s}\|_{\min} \leq 1,$$

and $s = (q \otimes I)(\widetilde{s})$ equivalently means that $u_{|E} = q\widetilde{u}$. Thus we conclude that C has the LLP. This completes the proof that (iii) $\Rightarrow$ (i).

Note that, by Theorem 13.1, this proves in particular that $C^*(\mathbb{F})$ has the LLP for any free group $\mathbb{F}$.

The proof that (i) $\Rightarrow$ (ii) is based on Theorem 13.1. Assume (i). Consider $t \in C \otimes B(H)$. We can assume $t \in E \otimes B(H)$ with $E \subset C$ finite-dimensional. Let F be a free group such that $C \simeq C^*(F)/I$ for some ideal $I \subset C^*(F)$. Let us denote $A = C^*(F)$, let $q \colon A \to A/I \simeq C$ be the quotient map, and let $u \colon E \to A/I$ denote the natural inclusion. By the LLP of C, u admits a completely contractive lifting $\widetilde{u} \colon E \to A$, so that $q\widetilde{u} = u$. We then have $t = (q \otimes I)(\widetilde{u} \otimes I)t$, and hence

$$
\begin{aligned}
\|t\|_{C\otimes_{\max}B(H)} &\leq \|(\widetilde{u} \otimes I)(t)\|_{A\otimes_{\max}B(H)} \\
&= \|(\widetilde{u} \otimes I)(t)\|_{A\otimes_{\min}B(H)} \quad \text{(by Theorem 13.1)} \\
&\leq \|t\|_{E\otimes_{\min}B(H)} = \|t\|_{C\otimes_{\min}B(H)}.
\end{aligned}
$$

Hence we obtain (i) $\Rightarrow$ (ii).

(ii) $\Rightarrow$ (iii) is trivial.

Assume (iii). Let B be as in (iv). We may as well assume $B \subset B(H)$ for some H. Now, if B has the WEP, we have

$$C \otimes_{\max} B \subset C \otimes_{\max} B(H) \quad \text{(isometrically)}$$

and obviously also

$$C \otimes_{\min} B \subset C \otimes_{\min} B(H) \quad \text{(isometrically)}.$$

Hence, if (iii) holds, the induced norms on $C \otimes B$ must coincide so that $C \otimes_{\min} B = C \otimes_{\max} B$. This establishes (iii) $\Rightarrow$ (iv). Since (iv) $\Rightarrow$ (iii) is obvious, and we already showed (iii) $\Rightarrow$ (i), the proof is complete. ∎

Remark 16.3. (i) The preceding proof of (iii) $\Rightarrow$ (i) shows that, if C has the LLP, any decomposable map $u: C \to A/I$ with $\|u\|_{dec} \le 1$ locally lifts in the following sense: For any finite-dimensional $E \subset C$, $u_{|E}: E \to A/I$ admits a completely contractive lifting $\widetilde{u}: E \to A$.

(ii) The preceding proof of (i) $\Rightarrow$ (ii) shows that C has the LLP iff there is a free group F and a surjective representation $q: C^*(F) \to C$ such that the resulting isomorphism $u: C \to C^*(F)/\ker(q)$ locally lifts in the preceding sense. Thus, if we restrict ourselves to unital representations $u: C \to A/I$ in Definition 16.1, the resulting property is the same.

(iii) For a useful characterization of complete metric surjections that "locally lift" in the preceding sense, see Exercise 2.4.1.

(iv) Consider a separable C^*-algebra A with the metric CBAP (that is, there is a sequence of completely contractive finite rank maps tending pointwise to the identity). Then, if A has the LLP, by Exercise 2.4.2 it must have the LP.

(v) By Theorem 16.2, all nuclear C^*-algebras have the LLP. Actually, in the separable case, they have the LP by the preceding point and by Theorem 11.5.

Remark. Fix $\lambda \ge 1$. We could say that C has the λ-LLP if, given any unital c.p. map $u: C \to A/I$ and any finite-dimensional subspace $E \subset C$, the restriction $u_{|E}: E \to A/I$ admits a lifting $\widetilde{u}: E \to A$ such that $\|\widetilde{u}\|_{cb} \le \lambda$. But this is not needed: Indeed, the preceding argument shows that if C has this λ-LLP, then

$$\| \cdot \|_{C \otimes_{\max} B(H)} \le \lambda \| \cdot \|_{C \otimes_{\min} B(H)},$$

and since there is only one C^*-norm on a C^*-algebra, we must have $C \otimes_{\max} B(H) = C \otimes_{\min} B(H)$ isometrically. In other words, for C^*-algebras, this λ-LLP implies the LLP (or, equivalently, the 1-LLP).

We will now turn to the operator space version of the LLP introduced in [Oz3].

Definition 16.4. *Let $\lambda \ge 1$. We say that an operator space X has the λ-OLLP if, for any unital C^*-algebra A, for any ideal $I \subset A$, for any complete contraction $u: X \to A/I$, and for any finite-dimensional subspace $E \subset X$, the restriction $u_{|E}: E \to A/I$ admits a lifting $\widetilde{u}: E \to A$ such that $\|\widetilde{u}\|_{cb} \le \lambda$.*

We say that X has the λ-OLP if we can always take $E = X$ in the preceding definition.

Remark. By Exercise 2.4.2, if X has the $(\lambda + \varepsilon)$-OLLP (resp. X has the $(\lambda + \varepsilon)$-OLP and is separable) for any $\varepsilon > 0$, then it also has it for $\varepsilon = 0$.

Moreover, if X is separable and has the metric CBAP (that is, there is a sequence of completely contractive finite rank maps tending pointwise to the identity), then (again by Exercise 2.4.2) X has the λ-OLP if it has the λ-OLLP.

Remark. The spaces $\max(\ell_1)$ and S_1^n have the 1-OLP (and a fortiori the 1-OLLP). Indeed, it is easy to check that they are "projective," that is, they satisfy a more general lifting property valid for all quotients of operator spaces (see Chapter 24) and not only for C^*-algebra quotients.

Theorem 16.5. *([Oz3]) Let X be an operator space. Then X has the 1-OLLP iff $C_u^*\langle X \rangle$ has the LLP.*

Proof. By the universal property of the embedding $X \hookrightarrow C_u^*\langle X \rangle$, any complete contraction $u: X \to A/I$ extends to a representation $U: C_u^*\langle X \rangle \to A/I$. Thus, if $C_u^*\langle X \rangle$ has the LLP (resp. LP), it is clear that X must have the 1-OLLP (resp. 1-OLP).

Conversely assume that X has the 1-OLLP. Let $\pi: C_u^*\langle X \rangle \to A/I$ be a unital representation. Let $\{E_i \mid i \in I\}$ denote the family of all the finite-dimensional subspaces of X directed by inclusion. Recall (see Exercise 8.2) that the inclusion $E_i \subset X$ extends to an isometric unital representation $C_u^*\langle E_i \rangle \to C_u^*\langle X \rangle$. Since by assumption $C_u^*\langle X \rangle$ has the LLP, for each i, the restriction $\pi_{|E_i}: E_i \to A/I$ admits a lifting $\widetilde{\pi}: E_i \to A$ with $\|\widetilde{\pi}\|_{cb} \leq 1$, which extends to a unital representation $\pi_i: C_u^*\langle E_i \rangle \to A$. Let $q: A \to A/I$ denote the quotient map. Clearly $q\pi_i$ coincides with the restriction of π to $C_u^*\langle E_i \rangle \subset C_u^*\langle X \rangle$. Since the (directed) union of the spaces $C_u^*\langle E_i \rangle$ is clearly dense in $C_u^*\langle X \rangle$, we have obtained a dense subspace $M \subset C_u\langle X \rangle$ such that, for any finite-dimensional $E \subset M$, the restriction $\pi_{|E}: E \to A/I$ admits a completely contractive lifting from E to A. By a simple perturbation argument (see Lemma 2.13.1), this implies that π locally lifts; hence we conclude, by Remark 16.3(ii), that $C_u^*\langle X \rangle$ has the LLP. ∎

Remark. If X is separable, then X has the 1-OLP iff $C_u^*\langle X \rangle$ has the LP. Indeed, if X is separable, then so is $C_u^*\langle X \rangle$; hence (see Exercise 8.1) we can write

$$C_u^*\langle X \rangle \simeq A/I$$

for some ideal I in $A = C^*(\mathbb{F}_\infty)$. Let $\pi: C_u^*\langle X \rangle \to A/I$ be the corresponding isomorphism, and let u denote its restriction to X. Assume that X has the 1-OLP. Then, since $\|u\|_{cb} = 1$, u admits a completely contractive lifting $\widetilde{u}\colon X \to A$. Let $\widehat{\pi}\colon C_u^*\langle X\rangle \to A$ be the unital representation extending $\widetilde{u}$. We obtain a factorization of the identity of $C_u^*\langle X\rangle$ as follows:

$$C_u^*\langle X\rangle \xrightarrow{\ \widehat{\pi}\ } A \xrightarrow{\ q\ } A/I.$$

But since $A = C^*(\mathbb{F}_\infty)$ has the LP (cf. [Ki5]), we conclude from this factorization that $C_u^*\langle X\rangle$ must also have the LP. This proves the "only if" part. For the "if" part (which does not require separability), see the above proof of Theorem 16.5. $\blacksquare$

Theorem 16.6. *([Oz3]) Let $\lambda \geq 1$. The following properties of an operator space X are equivalent.*

(i) *X has the λ-OLLP.*

(ii) *$X \otimes_{\min} B(H) = X \otimes_\delta B(H)$ for any H and $\delta(t) \leq \lambda\|t\|_{\min}$ for any t in $X \otimes B(H)$.*

(ii)' *Same as (ii) for $H = \ell_2$.*

(iii) *For any finite-dimensional subspace $E \subset X$ and any $\varepsilon > 0$, there is, for some $N \geq 1$, a subspace $G \subset M_N$ and a factorization $E \xrightarrow{\ v\ } G^* \xrightarrow{\ w\ } X$ of the inclusion map $E \subset X$ such that $\|v\|_{cb}\|w\|_{cb} < \lambda + \varepsilon$.*

Proof. Assume (i). Let F be a free group so that $C_u^*\langle X\rangle \simeq A/I$ with $A = C^*(F)$ (see Exercise 8.1). Let $E \subset X$ be a finite-dimensional subspace and let $u\colon E \to A/I$ be the inclusion map. Since we assume (i), there is a lifting $\widetilde{u}\colon E \to A$ with $\|u\|_{cb} \leq \lambda$. Hence $\forall\, t \in E \otimes B(H)$ and we can write $t = (q \otimes I)(\widetilde{u} \otimes I)t$, and hence

$$\begin{aligned}
\delta(t) = \|t\|_{C_u^*\langle X\rangle \otimes_{\max} B(H)} &\leq \|(\widetilde{u} \otimes I)t\|_{A \otimes_{\max} B(H)}\\
\text{hence by Theorem 13.1} \quad &= \|(\widetilde{u} \otimes I)t\|_{A \otimes_{\min} B(H)}\\
\text{hence by (2.3.2)} \quad &\leq \|\widetilde{u}\|_{cb}\|t\|_{\min},
\end{aligned}$$

so that we obtain $\delta(t) \leq \lambda\|t\|_{\min}$ and (ii) holds.

(ii) $\Rightarrow$ (ii)' is obvious. We now prove (ii)' $\Rightarrow$ (iii). Assume (ii)'. Let E be as in (iii). Since it is separable, we may embed E^* (completely isometrically) into $B(\ell_2)$. Let $t \in E \otimes B(\ell_2)$ be the tensor associated to the inclusion map $E^* \to B(\ell_2)$. By (ii)' we have

$$\|t\|_{X \otimes_\delta B(\ell_2)} \leq \lambda\|t\|_{\min} = \lambda\|I_E\|_{cb} = \lambda.$$

Hence, by Corollary 12.5, we have for any $\varepsilon > 0$ a factorization of the form

$$\begin{array}{ccc}
X^* & \longrightarrow & E^* \subset B(\ell_2)\\
{\scriptstyle \alpha} \searrow & & \nearrow {\scriptstyle \beta}\\
& M_N &
\end{array}$$

with $\|\alpha\|_{cb}\|\beta\|_{cb} < \lambda + \varepsilon$ and $\alpha^*(M_N{}^*) \subset X$. Hence, letting $G = \alpha(X^*)$ and denoting by α_1 and β_1 the restrictions of α and β, we find a factorization

$$X^* \longrightarrow E^*$$
$$\alpha_1 \searrow \quad \nearrow \beta_1$$
$$G$$

with $\|\alpha_1\|_{cb}\|\beta_1\|_{cb} < \lambda + \varepsilon$. Taking the adjoint diagram, we obtain (iii) with $v = \beta_1^*$ and $w = \alpha_1^*$.

Assume (iii). Let $u\colon X \to A/I$ be a complete contraction and let $E \subset X$ be finite-dimensional. Let G, v, w be as in (iii). Let $U = uw\colon G^* \to A/I$. Note that $\|U\|_{cb} \le \|u\|_{cb}\|w\|_{cb} \le \|w\|_{cb}$. By Lemma 2.4.8, since $G \subset M_N$, we have

$$G \otimes_{\min} A/I = (G \otimes_{\min} A)/(G \otimes_{\min} I) \qquad \text{(isometrically)}.$$

Consequently, by (2.3.2) and by Lemma 2.4.7, U admits a lifting $\widetilde{U}\colon G^* \to A$ with $\|\widetilde{U}\|_{cb} \le \|w\|_{cb}$. Then the mapping $\widetilde{u} = \widetilde{U}v\colon E \to A$ lifts $u_{|E}$ and satisfies $\|\widetilde{u}\|_{cb} \le \|v\|_{cb}\|w\|_{cb}$. This shows that X has the λ-OLLP. ∎

The next statement is immediate (take $E = X$ in Theorem 16.6(iii)).

Corollary 16.7. *A finite-dimensional operator space E has the λ-OLP (or, equivalently, the λ-OLLP) iff for any $\varepsilon > 0$ there is, for some $N \ge 1$, a subspace $G \subset M_N$ such that*

$$d_{cb}(E^*, G) < \lambda + \varepsilon.$$

Remark 16.8. By a classical result (see, e.g., [HWW, p. 59]) if X is a separable Banach space with the metric approximation property, then any contraction $u\colon X \to A/I$ admits a contractive lifting $\widetilde{u}\colon X \to A$. Obviously this implies that the operator space $\max(X)$ has the 1-OLP. It is likely that this becomes false without assuming some kind of approximation property, but no couterexample is known (see Proposition 18.14 later). Here is another example:

Proposition 16.9. *([Oz3]) Let M be a von Neumann algebra with predual M_*. Then M_* has the 1-OLLP, and even the 1-OLP if it is separable.*

Proof. Let $q\colon A \to A/I$ be as before. We have a canonical surjective representation $M \otimes_{\min} A \to M \otimes_{\min} (A/I)$. By Lemma 2.4.6 and (2.3.2), for any finite rank map $u\colon M_* \to A/I$ there is a lifting $\widetilde{u}\colon M_* \to A$ with $\|\widetilde{u}\|_{cb} = \|u\|_{cb}$. (Indeed, let $t \in M \otimes (A/I)$ be the tensor associated to u. By Lemma 2.4.6, t has a lifting $\widetilde{t}$ in $M \otimes_{\min} A$ with $\|\widetilde{t}\|_{\min} = \|t\|_{\min}$, and thus we can take for $\widetilde{u}\colon M_* \to A$ the linear map associated to $\widetilde{t}$.)

Now, by the remark following Lemma 12.8, any complete contraction $u: M_* \to A/I$ can be approximated pointwise by completely contractive finite rank maps $u_\alpha: M_* \to A/I$. By the preceding observation, each u_α admits a completely contractive lifting $\widetilde{u}_\alpha: M_* \to A$, so that $q\widetilde{u}_\alpha \to u$ pointwise. Now, by Exercise 2.4.2, if M_* is separable, this implies that u admits a completely contractive lifting; hence M_* has the 1-OLP. In general, if we restrict to a separable subspace $E \subset M_*$, then Exercise 2.4.2 implies that $u_{|E}$ admits a completely contractive lifting, and hence a fortiori M_* has the 1-OLLP. $\blacksquare$

Let us denote $\mathcal{B} = B(\ell_2)$. We now turn to the lifting property relative to the "Calkin algebra," that is, the quotient $\mathcal{B}/\mathcal{K}$.

The following result comes from Ozawa's thesis.

Theorem 16.10. *([Oz6]) Let $\lambda \geq 1$. A separable operator space X has the λ-OLLP iff every complete contraction $u: X \to \mathcal{B}/\mathcal{K}$ admits a c.b. lifting $\widetilde{u}: X \to \mathcal{B}$ with $\|u\|_{cb} \leq \lambda$.*

Proof. Assume that X has the λ-OLLP and let $u \in CB(X, \mathcal{B}/\mathcal{K})$. By the OLLP and the injectivity of $\mathcal{B}$, there is a net $u_\alpha: X \to \mathcal{B}$ of maps with $\|u_\alpha\|_{cb} \leq \lambda$ such that qu_α tends pointwise to u, where $q: \mathcal{B} \to \mathcal{B}/\mathcal{K}$ denotes the quotient map. Then, by Exercise 2.4.2, we conclude that there is a "global" lifting $\widetilde{u}$ with $\|\widetilde{u}\|_{cb} \leq \lambda$. This proves the "only if" part.

To prove the "if" part, we will need the following lemma, for which we first introduce some notation. Consider a finite-dimensional operator space $E \subset B(H)$, a separable unital C^*-algebra B, and an ideal $\mathcal{J} \subset B$. Let $\theta: B \to \mathcal{B}$ be a unital c.p. map such that $\theta(\mathcal{J}) \subset \mathcal{K}$. Let

$$\dot{\theta}: B/\mathcal{J} \longrightarrow \mathcal{B}/\mathcal{K}$$

and

$$\overset{\vee}{\theta}: (E \otimes_{\min} B)/E \otimes_{\min} \mathcal{J} \to (E \otimes_{\min} \mathcal{B}/(E \otimes_{\min} \mathcal{K})$$

denote the associated complete contractions. We can now state:

Lemma 16.11. *([Oz6]) For any y in $E \otimes (B/\mathcal{J})$ we have*

$$\|y\|_{E\otimes_{\min}(B/\mathcal{J})} = \sup_{\theta \in \mathcal{C}} \|(I \otimes \dot{\theta})(y)\|_{E\otimes_{\min}(\mathcal{B}/\mathcal{K})} \qquad (16.1)$$

and

$$\|y\|_{(E\otimes_{\min}B)/(E\otimes_{\min}\mathcal{J})} = \sup_{\theta \in \mathcal{C}} \|\overset{\vee}{\theta}(y)\|_{(E\otimes_{\min}\mathcal{B})/(E\otimes_{\min}\mathcal{K})}, \qquad (16.2)$$

where $\mathcal{C}$ denotes the set of all unital c.p. maps $\theta: B \to \mathcal{B}$ such that $\theta(\mathcal{J}) \subset \mathcal{K}$.

Proof. This argument is inspired by an idea of Kirchberg. We first observe that in both cases the norm of y is $\geq$ the supremum because $\|\theta\|_{cb} \leq 1$ for each θ. We now turn to the converse inequality. We will only prove (16.2), which is the only one needed in the sequel (the proof of the first one is similar). We will show that, if the left side of (16.2) is > 1, then the right side is also > 1. Let $v \in E \otimes_{\min} B$ be such that

$$d(v, E \otimes_{\min} \mathcal{J}) > 1. \qquad (16.3)$$

We claim that there is a θ in $\mathcal{C}$ such that

$$d((I \otimes \theta)(v) + E \otimes_{\min} \mathcal{K}) > 1.$$

From this claim, (16.2) follows easily (modulo our initial observation). Thus it remains to prove this claim. Since $\mathcal{J}$ is separable, there is an element h in $\mathcal{J}$ such that $0 \le h \le 1$ and $\mathcal{J} = \overline{h \mathcal{J} h}$ (when $\mathcal{J} = \mathcal{K}$ any strictly positive element of $\mathcal{K}$ will do). Let g_n be the indicator function of the interval $[n^{-1}, 1]$ and let $p_n = g_n(h) \in \mathcal{J}^{**}$. Then p_n is a projection, and, since $\lim \|(1 - p_h)h\| = 0$, we have:

$$\lim \|(1 - p_n)x(1 - p_n)\| = 0 \qquad \forall\, x \in \mathcal{J}. \qquad (16.4)$$

Let $f_n \in C_0((0,1])$ be the function equal to g_n on $[n^{-1}, 1]$ and such that

$$f_n(t) = nt \quad \text{if} \quad 0 < t \le n^{-1}.$$

Then $h_n = f_n(h) \in \mathcal{J}$. Since $(1 - p_n) \ge (1 - h_n) \ge 0$, for each n we have

$$\|[1 \otimes (1 - p_n)]v[1 \otimes (1 - p_n)]\|_{E \otimes_{\min} B^{**}} \ge \|[1 \otimes (1 - h_n)]v[1 \otimes (1 - h_n)]\|_{E \otimes_{\min} B} > 1$$

(by our assumption (16.3) because

$$[1 \otimes h_n]v, \quad v[1 \otimes h_n] \quad \text{and} \quad [1 \otimes h_n]v[1 \otimes h_n]$$

are all in $E \otimes_{\min} \mathcal{J}$).

Note: Here the left and right multiplications are done in $B(H) \otimes B^{**}$, but in any case $v \in E \otimes_{\min} B$ implies $[1 \otimes (1 - p_n)]v[1 \otimes (1 - p_n)] \in E \otimes B^{**}$.

Hence (see Proposition 2.12.1), for each n, for some suitable $N(n) \ge 1$, there is a unital c.p. map $T_n \colon (1 - p_n)B(1 - p_n) \to M_{N(n)}$ such that

$$\|(I \otimes T_n)[1 \otimes (1 - p_n)]v[1 \otimes (1 - p_n)]\| > 1. \qquad (16.5)$$

(Note: Here $(1 - p_n)B(1 - p_n)$ is viewed as a C^*-algebra with unit $1 - p_n$; hence we have $T_n(1 - p_n) = 1$.) Then we define unital c.p. maps $\theta_n \colon B \to M_{N(n)}$ and $\theta \colon B \to \bigoplus_n M_{N(n)}$ by

$$\theta_n(b) = T_n((1 - p_n)b(1 - p_n))$$

and

$$\theta(b) = \bigoplus_n \theta_n(b).$$

We will view $\bigoplus_n M_{N(n)}$ as an embedded block diagonally into

$$\mathcal{B} \simeq B\left(\bigoplus_n \ell_2^{N(n)}\right),$$

so that $c_0(\{M_{N(n)}\}) \subset \mathcal{K}$. By (16.4), we have $\theta(J) \subset \mathcal{K}$, and using the fact that (16.5) holds for each n, we can easily verify that

$$\text{dist}((I \otimes \theta)(v), E \otimes_{\min} \mathcal{K}) > 1.$$

Thus, we have proved our claim. ∎

End of the Proof of Theorem 16.10. It remains to prove the "if" part. Assume that the lifting property into $\mathcal{B}/\mathcal{K}$ holds as in Theorem 16.9. Consider a complete contraction $u \colon X \to B/I$, as in Definition 16.4. Since X is separable, we may assume (by an elementary argument) that B is also separable. Let $F \subset X$ be finite-dimensional. Let $y \in F^* \otimes (B/I)$ be associated to $u_{|F} \colon F \to B/I$. Let $E = F^*$. Applying the lifting property for $\mathcal{B}/\mathcal{K}$ to the tensor $[I \otimes \dot{\theta}](y) \in F^* \otimes \mathcal{B}/\mathcal{K}$ (for each θ in $\mathcal{C}$), we find (recall Lemma 2.4.7)

$$\|\overset{\vee}{\theta}(y)\|_{(F^* \otimes_{\min} \mathcal{B})/(F^* \otimes_{\min} \mathcal{K})} \le \lambda.$$

Hence, by (16.2),

$$\|y\|_{(F^* \otimes_{\min} B)/(F^* \otimes_{\min} I)} \le \lambda,$$

which means (again recall Lemma 2.4.7) that $u_{|F}$ admits an extension $\widetilde{u} \colon F \to B$ with $\|\widetilde{u}\|_{cb} \le \lambda$. Thus we conclude that the X has the λ-OLLP. ∎

Kirchberg's conjecture. The following conjecture due to Kirchberg [Ki2] is equivalent to many fundamental open questions about von Neumann algebras. In particular (see [Ki2]) it is equivalent to the Connes problem mentioned at the end of §9.10. Its solution would be a major step forward.

Conjecture 16.12. ([Ki2])

(i) The C^*-algebra $C^*(\mathbb{F}_\infty)$ has the WEP.

The next statement gives several equivalent reformulations (see [Ki2] for more).

Proposition 16.13. Each of the following assertions is equivalent to the preceding conjecture.

(ii) If $C = C^*(F_\infty)$, we have $C \otimes_{\min} C = C \otimes_{\max} C$.

(iii) Every C^*-algebra is a quotient of a C^*-algebra with the WEP (such algebras are called QWEP).

(iv) LLP ⇒ WEP.

(v) If C is any C^*-algebra with the LLP, we have $C \otimes_{\min} C = C \otimes_{\max} C$.

(vi) For any finite-dimensional operator space $E \neq \{0\}$ with $d_{SK}(E^*) = 1$ (as defined in (0.2)), $C^*\langle E \rangle$ has the WEP.

(vii) The C^*-algebra $C_u^*\langle \mathbb{C} \rangle$ has the WEP.

Remark 16.14. For C^*-algebras with the LLP, QWEP ⇒ WEP. Indeed, if A has the LLP and if $A \simeq W/I$ with W having the WEP, the isomorphism $A \to W/I$ locally lifts up through W, from which it follows (say, by Theorem 15.5) that A has the WEP.

Proof of Proposition 16.13. By Theorem 15.5, (i) ⇔ (ii). Moreover, (i) holds iff, for any free group F, $C^*(F)$ has the WEP. Since any C^*-algebra A is a quotient of $C^*(F)$ for a suitable F (see Exercise 8.1), (i) implies that it is QWEP, and hence (i) ⇒ (iii). Moreover, (iii) ⇒ (iv) by Remark 16.14. Then (iv) ⇒ (v) follows from Theorem 16.2 and (v) ⇒ (ii) is clear since $C^*(\mathbb{F}_\infty)$ has the LLP. Thus we have proved that (i)–(v) are equivalent. Then (iv) ⇒ (vi) is clear since (by Theorems 16.5 and 16.6) $C_u^*\langle \mathbb{C} \rangle$ has the LLP; (vi) ⇒ (vii) is trivial. Thus it remains to show (vii) ⇒ (i). It is easy to see that $C_u^*\langle \mathbb{C} \rangle$ is the universal unital C^*-algebra of a contraction T. Any singly generated unital C^*-algebra is a quotient of $C_u^*\langle \mathbb{C} \rangle$. Let $C = C[-1, 1]$. We claim that the full unital free product $C * C$ is a quotient of $C_u^*\langle \mathbb{C} \rangle$. Indeed, since C is singly generated by a self-adjoint contraction x, $C * C$ is generated by a pair $\{x_1, x_2\}$ of self-adjoint contractions, hence $C * C$ is singly generated by $(x_1 + ix_2)/2$, and therefore is a quotient of $C_u^*\langle \mathbb{C} \rangle$. A fortiori $C_u^*\langle \mathbb{C} \rangle$ admits as a quotient $C * C$, where C is any finite-dimensional commutative C^*-algebra. Thus, for any pair G_1, G_2 of finite Abelian groups, $C^*(G_1 * G_2)$ is a quotient of $C_u^*\langle \mathbb{C} \rangle$, and hence (vii) implies that $C^*(G_1 * G_2)$ must have the WEP. But now it is well known that $\mathbb{F}_\infty$ is a subgroup of $\mathbb{Z}_3 * \mathbb{Z}_3$; hence, by Proposition 8.8, $C^*(\mathbb{F}_\infty)$ itself must have the WEP. ∎

Remark. By adapting the proof of Lemma 3.3 and Theorem 3.4 in [EH], one can show (as pointed out by Kirchberg [Ki2, Proposition 2.2(iv)]) that, whenever J is a WEP ideal in $C = C^*(\mathbb{F}_\infty)$, for the quotient C/J the LLP implies "automatically" the LP. See also [Har1, Corollary 3.3] for a related useful fact. Note that obviously J has the WEP if C does. Thus, if Conjecture 16.12 is true, we must have LLP ⇒ LP (hence also OLLP ⇒ OLP) in the separable case.

We discuss several important open problems concerning the OLLP in Chapter 18.

Chapter 17. Exactness

The notion of exactness was introduced by Kirchberg as early as 1977 [Ki4], but his major contributions on that topic did not circulate until the late 1980s; cf. [Ki1–2].

Definition. *An operator space X is called exact if, for any C^*-algebra B and any (closed two-sided) ideal $\mathcal{I} \subset B$, we have an exact sequence*

$$\{0\} \to \mathcal{I} \otimes_{\min} X \to B \otimes_{\min} X \to (B/\mathcal{I}) \otimes_{\min} X \to \{0\}. \tag{17.1}$$

Remark. The analogous property for the max-tensor product holds if X is replaced by an arbitrary C^*-algebra A. Indeed, first $\mathcal{I} \otimes_{\max} A$ embeds isometrically into $B \otimes_{\max} A$ and second, by maximality, the max-norm on $(B/\mathcal{I}) \otimes A$ must dominate the C^*-norm induced by $(B \otimes_{\max} A)/(\mathcal{I} \otimes_{\max} A)$ (see Exercise 11.1 for more details). Hence

$$(B/\mathcal{I}) \otimes_{\max} A \simeq \frac{B \otimes_{\max} A}{\mathcal{I} \otimes_{\max} A}$$

and consequently the sequence

$$\{0\} \to \mathcal{I} \otimes_{\max} A \to B \otimes_{\max} A \to (B/\mathcal{I}) \otimes_{\max} A \to \{0\} \tag{17.2}$$

is always exact. This goes back to [Gu1].

In particular, this remark shows that all nuclear C^*-algebras are exact.

By the properties of C^*-representations, if X is a C^*-algebra, it suffices for the exactness of the sequence (17.1) that the kernel of $B \otimes_{\min} X \to (B/\mathcal{I}) \otimes_{\min} X$ coincides with $\mathcal{I} \otimes_{\min} X$.

But in the operator space case, the exactness of (17.1) requires in addition that the map $B \otimes_{\min} X \to (B/\mathcal{I}) \otimes_{\min} X$ be surjective. (Since C^*-representations all have closed ranges, this is automatic in the C^*-case.) Moreover, in the C^*-case all morphisms have norm 1, which of course is no longer true for general c.b. maps, so we need to introduce the "constant of exactness" as follows: We have a complete contraction $B \otimes_{\min} X \to (B/\mathcal{I}) \otimes_{\min} X$ associated to the quotient map $q \colon B \to B/\mathcal{I}$. Since this map vanishes on $\mathcal{I} \otimes_{\min} X$, it defines a (completely contractive) map

$$T_X \colon (B \otimes_{\min} X)/(\mathcal{I} \otimes_{\min} X) \to (B/\mathcal{I}) \otimes_{\min} X.$$

By definition, X is exact iff T_X is an isomorphism for any B and $\mathcal{I}$. Note that, assuming T_X injective, $B \otimes_{\min} X \to (B/\mathcal{I}) \otimes_{\min} X$ is surjective iff T_X^{-1} is bounded so the norm $\|T_X^{-1}\|$ measures the degree of exactness of (17.1).

We then denote

$$\mathrm{ex}(X) = \sup \|T_X^{-1}\|, \tag{17.3}$$

where the supremum runs over all C^*-algebras B and all ideals $\mathcal{I} \subset B$.

It is easy to see that $ex(M_N) = 1$ or, more generally, that $ex(A) = 1$ for any nuclear C^*-algebra A.

Kirchberg discovered that exactness is closely connected to the existence of good embeddings into nuclear C^*-algebras and discussed various notions of "subnuclearity." For operator spaces, it is advantageous to first restrict the discussion to the finite-dimensional case: Let $E \subset B(H)$ be a finite-dimensional operator space. Recall the notation

$$d_{S\mathcal{K}}(E) = \inf\{d_{cb}(E, F) \mid F \subset \mathcal{K}\}. \tag{17.4}$$

By a simple perturbation argument (see Lemma 2.13.4), it can be shown that

$$d_{S\mathcal{K}}(E) = \inf\{d_{cb}(E, F) \mid F \subset M_N, N \geq \dim E\}. \tag{17.4}'$$

Theorem 17.1. *Let $X \subset B(H)$ be an operator space. For any fixed $\lambda \geq 1$, the following assertions are equivalent:*

(i) *X is exact and $ex(X) \leq \lambda$.*

(ii) *Let $u\colon X \to B(H)$ be the inclusion map. For any C^*-algebra C, the mapping $I_C \otimes u$ is bounded from $C \otimes_{\min} X$ to $C \otimes_{\max} B(H)$ with norm $\leq \lambda$.*

(iii) *For any finite-dimensional subspace $E \subset X$ we have $d_{S\mathcal{K}}(E) \leq \lambda$. In particular, for any n-dimensional subspace $E \subset B(H)$ we have*

$$d_{S\mathcal{K}}(E) = \sup\{\|E \otimes_{\min} C \to B(H) \otimes_{\max} C\|\}, \tag{17.5}$$

where the supremum runs over all possible C^*-algebras C.

First Part of the Proof of Theorem 17.1. We will show (i) $\Rightarrow$ (ii) $\Rightarrow$ (iii). Assume (i). Let C be any C^*-algebra. By Exercise 8.1, if G is a sufficiently large free group, then C is a quotient of $C^*(G)$. So if we denote $B = C^*(G)$, we have $C = B/\mathcal{I}$ for some ideal $\mathcal{I} \subset B$. Let us denote $Q(X) = (B \otimes_{\min} X)/(\mathcal{I} \otimes_{\min} X)$. By Theorem 13.1 (and Corollary 11.9), we have $Q(B(H)) = (B \otimes_{\max} B(H))/\mathcal{I} \otimes_{\max} B(H)$. Hence, by the exactness of (17.2),

$$Q(B(H)) = (B/\mathcal{I}) \otimes_{\max} B(H) = C \otimes_{\max} B(H).$$

Then $I_C \otimes u$ coincides with the following composition:

$$C \otimes_{\min} X = (B/\mathcal{I}) \otimes_{\min} X \to Q(X) \to Q(B(H)) = C \otimes_{\max} B(H),$$

which shows that

$$\|I_C \otimes u\colon C \otimes_{\min} X \to C \otimes_{\max} B(H)\| \leq \|T_X^{-1}\colon (B/\mathcal{I}) \otimes_{\min} X \to Q(X)\|$$
$$\leq ex(X).$$

Thus we obtain (ii). Assume (ii). Let $E \subset X$ be a finite-dimensional subspace. Then, by Corollary 12.6, for any $\varepsilon > 0$, there is a factorization of $u_{|E}$ of the form $E \xrightarrow{v} M_n \xrightarrow{w} B(H)$ with $\|v\|_{cb}\|w\|_{cb} < \lambda + \varepsilon$. Let $\widetilde{E} = v(E) \subset M_n$, and let $\theta\colon E \to \widetilde{E}$ be the map v but with range $\widetilde{E}$. Then $\theta^{-1}\colon \widetilde{E} \to E$ is the restriction of w; hence $\|\theta^{-1}\|_{cb} \leq \|w\|_{cb}$ and obviously $\|\theta\|_{cb} \leq \|v\|_{cb}$. Thus we obtain $d_{cb}(E, \widetilde{E}) \leq \|\theta\|_{cb}\|\theta^{-1}\|_{cb} \leq \|v\|_{cb}\|w\|_{cb} < \lambda + \varepsilon$. Since $\varepsilon > 0$ is arbitrary, this implies $d_{SK}(E) \leq \lambda$. Thus we have shown that (ii) $\Rightarrow$ (iii). ∎

To complete the proof, we need the following useful lemma from §2.4, and we first recall the specific notation used there. Let $\mathcal{I} \subset B$ be a (closed two-sided) ideal in a C^*-algebra B. Let E be an operator space. As before, we denote for simplicity

$$Q(E) = \frac{B \otimes_{\min} E}{\mathcal{I} \otimes_{\min} E}.$$

Then, if F is another operator space and if $u\colon E \to F$ is a c.b. map, we clearly have a c.b. map

$$u_Q\colon Q(E) \to Q(F)$$

naturally associated to $I_B \otimes u$ such that

$$\|u_Q\|_{cb} \leq \|u\|_{cb}.$$

Lemma 17.2. *If u is a complete isometry, then u_Q is also one.*

This was already proved above as Lemma 2.4.8.

End of the Proof of Theorem 17.1. It remains to show (iii) $\Rightarrow$ (i). Assume (iii). We claim that, for any t in $(B/\mathcal{I}) \otimes X$, we have $\|t\|_{Q(X)} \leq \lambda\|t\|_{\min}$. Indeed, we can assume $t \in (B/\mathcal{I}) \otimes E$ with $E \subset X$ finite-dimensional. By (iii) for any $\varepsilon > 0$ there is an n and a subspace $\widetilde{E} \subset M_n$ such that $d_{cb}(E, \widetilde{E}) < \lambda + \varepsilon$. Consider first an element $\widetilde{t}$ in $(B/\mathcal{I}) \otimes \widetilde{E}$. We have by Lemma 17.2

$$\|\widetilde{t}\|_{Q(\widetilde{E})} = \|\widetilde{t}\|_{Q(M_n)},$$

but

$$Q(M_n) = \frac{B \otimes_{\min} M_n}{\mathcal{I} \otimes_{\min} M_n} = \frac{M_n(B)}{M_n(\mathcal{I})} = M_n(B/\mathcal{I}) = (B/\mathcal{I}) \otimes_{\min} M_n;$$

hence

$$\|\widetilde{t}\|_{Q(\widetilde{E})} = \|\widetilde{t}\|_{(B/\mathcal{I}) \otimes_{\min} \widetilde{E}}.$$

But then using $d_{cb}(E, \widetilde{E}) < \lambda + \varepsilon$, it is easy to see that this implies

$$\|t\|_{Q(E)} \leq (\lambda + \varepsilon)\|t\|_{(B/\mathcal{I}) \otimes_{\min} E},$$

and hence we obtain the announced claim. From this claim it follows that T_X is invertible and $\|T_X^{-1}\| \leq \lambda$; hence $\mathrm{ex}(X) \leq \lambda$, which concludes the proof that (iii) $\Rightarrow$ (i).

Finally, (17.5) follows immediately from the equivalence of (ii) and (iii). ∎

Warning. Let $\mathcal{I} \subset B$ and $q\colon B \to B/\mathcal{I}$ be as before. For any C^*-algebra A, the morphism $q \otimes I_A\colon B \otimes_{\min} A \to (B/\mathcal{I}) \otimes_{\min} A$ is a metric surjection. Hence, for any x in $(B/\mathcal{I}) \otimes A$ with $\|x\|_{\min} < 1$, there is an $\hat{x}$ in $B \otimes_{\min} A$ with $\|\hat{x}\|_{\min} < 1$ such that $q \otimes I_A(\hat{x}) = x$. However, in general such an $\hat{x}$ *cannot be chosen* in the *algebraic* tensor product $B \otimes A$. Only when A is exact can we choose $\hat{x}$ in $A \otimes B$! We emphasize this point because it is a potential source of confusion and is at the heart of "exactness."

By Lemma 17.2, if $E \subset F$, we have

$$\mathrm{ex}(E) \leq \mathrm{ex}(F).$$

On the other hand, it is easy to see that $\mathrm{ex}(X) \leq \sup\{\mathrm{ex}(E) \mid E \subset X, \dim E < \infty\}$, and consequently

$$\mathrm{ex}(X) = \sup\{\mathrm{ex}(E) \mid E \subset X, \dim E < \infty\}. \tag{17.6}$$

A careful look at the preceding proof shows that we have actually proved

Theorem 17.3. *An operator space X is exact iff*

$$d_{SK}(X) \overset{\mathrm{def}}{=} \sup\{d_{SK}(E) \mid E \subset X, \ \dim E < \infty\} < \infty.$$

Moreover, for any finite-dimensional operator space E we have $\mathrm{ex}(E) = d_{SK}(E)$ *and*

$$\mathrm{ex}(X) = \sup\{d_{SK}(E) \mid E \subset X, \ \dim E < \infty\}.$$

Remark 17.4. As observed by Kirchberg, if X is a C^*-algebra and is exact as an operator space in the sense of Theorem 17.1, then necessarily $\mathrm{ex}(X) = 1$. Indeed, the mapping T_X is a C^*-representation; hence, as soon as it is injective it is isometric. Therefore we have

Corollary 17.5. *A C^*-algebra A is exact iff $d_{SK}(A) = 1$, or, equivalently, iff $d_{SK}(A) < \infty$.*

We will say that an operator space is λ-exact if $\mathrm{ex}(X)$(i.e. $d_{SK}(X)$) $\leq \lambda$. With this terminology, an exact C^*-algebra is automatically 1-exact.

Remark 17.6. In (17.1), let us make the specific choice $\mathcal{I} = K(\ell_2)$ and $B = B(\ell_2)$. It can be shown that if (17.1) is exact with this specific choice of

B and $\mathcal{I}$, then it is exact for any choice (see Lemma 16.11), and $\|T_X^{-1}\|$ for this choice is equal to $\mathrm{ex}(X)$.

Another interesting choice of B and $\mathcal{I}$ is given by the consideration of ultraproducts. Let

$$B_\infty = \bigoplus_{n \in \mathbb{N}} M_n$$

and let $\mathcal{I}_0 \subset B_\infty$ be the ideal formed of all sequences $x = (x_n)$ in B_∞ such that $\|x_n\|_{M_n} \to 0$ when $n \to \infty$. Again, for the exactness of X, it suffices to consider $B = B_\infty$ and $\mathcal{I} = \mathcal{I}_0$ in the above definition, and $\|T_X^{-1}\|$ also coincides in this case with the exactness constant $\mathrm{ex}(X)$. Now let $\mathcal{U}$ be a nontrivial ultrafilter on $\mathbf{N}$, and let $\mathcal{I}_\mathcal{U}$ be ideal of all sequences $x = (x_n)$ in B_∞ such that $\lim_\mathcal{U} \|x_n\|_{M_n} = 0$. The quotient C^*-algebra $B_\infty / \mathcal{I}_\mathcal{U}$ is nothing but the ultraproduct that we denoted earlier in §2.8 by $\prod_{n \in \mathbb{N}} M_n / \mathcal{U}$. As we will show in the next statement, this class of quotients also suffices to control the exactness of X. For this result, we advise the reader to read first Remark 2.8.3.

Theorem 17.7. *Let X be an operator space. The following properties are equivalent.*

(i) *X is exact.*

(ii) *For any family $(E_i)_{i \in I}$ of operator spaces and any ultrafilter $\mathcal{U}$ on I, we have a completely isomorphic embedding:*

$$\left(\prod_{i \in I} E_i / \mathcal{U} \right) \otimes_{\min} X \subset \prod_{i \in I} (E_i \otimes_{\min} X) / \mathcal{U}.$$

(iii) *There is a constant λ such that, for any $(E_i)_{i \in I}$ as in (ii), the embedding considered in (ii) has norm $\leq \lambda$.*

(iv) *Same as (iii) but restricted to $I = \mathbb{N}$ and to families $(E_n)_{n \in \mathbb{N}}$ of finite-dimensional spaces of the same arbitrary dimension.*

(v) *There is a constant λ such that, for all finite-dimensional subspaces $E \subset X$, we have*
$$d_{SK}(E) \leq \lambda.$$

Proof. (i) $\Rightarrow$ (ii): Assume (i). We can assume $E_i \subset A_i$ for some C^*-algebra A_i. By the injectivity of ultraproducts (cf. (2.8.2)) and of the minimal tensor product, it suffices to prove (ii) when $E_i = A_i$. But, then we have $\prod_{i \in I} A_i / \mathcal{U} = B / \mathcal{I}$ with $B = \bigoplus_{i \in I} A_i$ and $\mathcal{I} = \{x \mid \lim_\mathcal{U} \|x_i\| = 0\}$, so that (ii) follows from the definition of exactness. (ii) $\Rightarrow$ (iii) is routine, and (iii) $\Rightarrow$ (iv) is trivial. We now prove the key implication (iv) $\Rightarrow$ (v).

Assume (iv). Let $E \subset X$ be a fixed d-dimensional subspace. Let $j \colon E \to B(\ell_2)$ be a complete isometry. Let $P_n \colon B(\ell_2) \to M_n$ be the standard projection that leaves e_{ij} invariant if $i, j \leq n$ and takes it to zero otherwise. Let $E_n = P_n j(E)$ and let $u_n \colon E \to E_n$ be the restriction of $P_n j$ to E. Clearly

$$\sup_n \|u_n\|_{cb} \leq 1,$$

and u_n is invertible for all n large enough. Moreover, $x \to (u_n(x))_{n \geq 1}$ obviously defines a complete isometry $u \colon E \to \Pi E_n / \mathcal{U}$. Therefore, $\|u^{-1}\|_{CB(\Pi E_n / \mathcal{U}, E)} = 1$. But u^{-1} can be identified with a norm 1 element of $(\Pi E_n / \mathcal{U})^* \otimes_{\min} E$. By Lemma 2.8.1, the latter space is the same as $\left(\prod_{n \in \mathbb{N}} E_n^* / \mathcal{U}\right) \otimes_{\min} E$. Hence, since we assume (iv), we can associate to u^{-1} a sequence v_n with $v_n \in E_n^* \otimes_{\min} E = CB(E_n, E)$ with $\lim_{\mathcal{U}} \|v_n\|_{CB(E_n, E)} \leq \lambda$. Since (v_n) comes from u^{-1}, we have necessarily

$$\forall x \in E \qquad \lim_{\mathcal{U}} v_n u_n(x) = x.$$

But since E is d-dimensional, this also implies that $\lim_{\mathcal{U}} v_n u_n = I_E$ in the c.b. norm, or equivalently, that $\lim_{\mathcal{U}} \|v_n - u_n^{-1}\|_{cb} = 0$. Finally, we have clearly $d_{SK}(E) \leq \|u_n\|_{cb} \|u_n^{-1}\|_{cb}$ for all n, and hence

$$d_{SK}(E) \leq \lim_{\mathcal{U}} \|u_n\|_{cb} \|v_n\|_{cb} \leq \lambda,$$

so we obtain (v). Finally, (v) $\Rightarrow$ (i) is already included in Theorem 17.1. ■

We will also use the following refinement:

Lemma 17.8. *Let Γ be any set and let $H = \ell_2(\Gamma)$. For any subset $\mathcal{S} \subset \Gamma$ we view $\ell_2(\mathcal{S})$ as a subspace of $\ell_2(\Gamma)$. The following properties of an operator space $X \subset B(\ell_2(\Gamma))$ are equivalent:*

(i) *X is exact with $\mathrm{ex}(X) = 1$.*

(vi) *For any $\varepsilon > 0$ and any finite-dimensional subspace $E \subset X$, there is a finite subset $\mathcal{S} \subset \Gamma$ such that the natural completely contractive "restriction (or compression) map"*

$$P_{\mathcal{S}} \colon x \in X \to P_{\ell_2(\mathcal{S})} x_{|\ell_2(\mathcal{S})}$$

defines a complete isomorphism from E to $P_{\mathcal{S}}(E) \subset B(\ell_2(\mathcal{S}))$ such that

$$\|(P_{\mathcal{S}})^{-1} \colon P_{\mathcal{S}}(E) \to E\|_{cb} < 1 + \varepsilon.$$

Proof. Clearly (vi) $\Rightarrow d_{SK}(E) = 1$, so (vi) $\Rightarrow$ (i) follows from Theorem 17.1.

Conversely, assume (i). Consider E as in (vi). Since $d_{SK}(E) = 1$ (by Theorem 17.1), we can find n and $\widetilde{E} \subset M_n$ such that $d_{cb}(E, \widetilde{E}) < 1 + \varepsilon$. We have then by Proposition 1.12

$$\|(P_S)^{-1}\colon P_S(E) \to E\|_{cb} < (1 + \varepsilon)\|(P_S)^{-1}\colon P_S(E) \to E\|_n.$$

But if n remains fixed and if S tends to Γ along the net of finite subsets of Γ, it is easy to see (see Exercise 2.13.1) that $\|(P_S)^{-1}\colon P_S(E) \to E\|_n \to 1$. $\blacksquare$

We will now apply the ideas in Chapter 13 to exactness for C^*-algebras. Let E be a finite-dimensional operator space, and let

$$u_E\colon B/I \otimes_{\min} E \to \frac{B \otimes_{\min} E}{I \otimes_{\min} E}$$

be the canonical isomorphism.

As shown in (17.6), we know that for any exact operator space E

$$d_{SK}(E) = \sup\{\|u_E\|\} = \sup\{\|u_E\|_{cb}\}, \tag{17.7}$$

where the supremum runs over all possible pairs (I, B) with $I \subset B$. (Actually, it suffices to consider $I = K(\ell_2)$ and $B = B(\ell_2)$.) The point of the next result is that it suffices for the exactness of A to be able to embed (almost completely isometrically) the linear span of the unitary generators of A and the unit into $K(\ell_2)$ (or into a nuclear C^*-algebra).

Theorem 17.9. *Let $E \subset A$ be a closed subspace of a unital C^*-algebra A. We assume that $1_A \in E$ and that E is the closed linear span of a family of unitary elements of A. Moreover, we assume that E generates A (i.e., that the smallest C^*-subalgebra of A containing E is A itself). We denote by $[E^*E]_d$ the subspace spanned by all the products of the form $x_1^* y_1 x_2^* y_2 ... x_d^* y_d$ with $x_1, y_1, ..., x_d, y_d \in E$. Then, the following are equivalent.*

(i) *A is exact (i.e., $d_{SK}(A) = 1$, or, equivalently, $d_{SK}(A) < \infty$).*

(ii) *$d_{SK}(E) = 1$.*

(iii) *$\limsup_{d \to \infty}(d_{SK}([E^*E]_d))^{1/d} = 1$.*

Proof. Assume (ii). Let $(\mathcal{I}, B)$ be as above with B unital. By (17.7), if $d_{SK}(E) = 1$, the unital $*$-homomorphism

$$\pi\colon B/\mathcal{I} \otimes A \longrightarrow \frac{B \otimes_{\min} A}{\mathcal{I} \otimes_{\min} A}$$

becomes completely contractive when restricted to $(B/\mathcal{I}) \otimes_{\min} E$. By Proposition 13.6, π extends to a continuous (contractive) $*$-homomorphism on

$(B/\mathcal{I}) \otimes_{\min} A$. Hence A is exact. More generally, assume (iii). Let $T \in B/\mathcal{I} \otimes E$. To compute $\|\pi(T)\|$ we use $\|\pi(T)\|^{2d} = \|\pi((T^*T)^d)\|$. Note that $(T^*T)^d \in B/\mathcal{I} \otimes [E^*E]_d$, and hence by (17.7) $\|\pi((T^*T)^d)\| \leq d_{SK}([E^*E]_d)\|(T^*T)^d\|$, which yields $\|\pi(T)\| \leq (d_{SK}([E^*E]_d))^{1/2d}\|T\|$, and by (17.7) again

$$d_{SK}(E) \leq (d_{SK}([E^*E]_d))^{1/2d}.$$

Thus (iii) implies (ii). The converses are clear by Theorem 17.1. ∎

Corollary 17.10. *For any free group G, the reduced C^*-algebra $C^*_\lambda(G)$ is exact.*

Proof. Note that, for any $D \geq 1$, $[E^*E]_D$ is the closed span of $\{\lambda(t) \mid |t| \leq 2D\})$. Hence, by Remark 9.7.5,

$$d_{SK}([E^*E]_D) \leq 4(2D)(2D+1),$$

so that (iii) holds. ∎

We now turn to the connection between exactness and the approximation property. It will be convenient to introduce several variants of the approximation property for a mapping.

Definitions 17.11. *Let $u \colon X \to Y$ be a c.b. mapping between operator spaces.*

(i) *We will say that u is strongly approximable if there is a net $u_i \colon X \to Y$ of finite rank maps such that, for any C^*-algebra B, the maps $I_B \otimes u_i \colon B \otimes_{\min} X \to B \otimes_{\min} Y$ converge pointwise to $I_B \otimes u$. Note that it clearly suffices to consider $B = B(\ell_2)$.*

(ii) *We will say that u is exact if, for any ideal $\mathcal{I} \subset B$ in a C^*-algebra B, the mapping*

$$\tilde{u} \colon (B/\mathcal{I}) \otimes_{\min} X \longrightarrow \frac{B \otimes_{\min} Y}{\mathcal{I} \otimes_{\min} Y} \tag{17.8}$$

that takes $(b + \mathcal{I}) \otimes x$ to $b \otimes u(x) + \mathcal{I} \otimes_{\min} Y$ is bounded. We denote

$$\mathrm{ex}(u) = \sup\{\|\tilde{u}\|\}, \tag{17.9}$$

where the supremum runs over all $\mathcal{I}$ and B.

(iii) *Let E be a finite-dimensional operator space and let $u \colon E \to Y$. We denote*

$$\gamma_{SK}(u) = \inf\{\|v\|_{cb}\|w\|_{cb} d_{SK}(F)\},$$

where the infimum runs over all factorizations of u of the form $E \xrightarrow{w} F \xrightarrow{v} Y$. Equivalently, we can write $\gamma_{SK}(u) = \inf\{\|v\|_{cb}\|w\|_{cb}\}$ and restrict F to be a subspace of $\mathcal{K} = K(\ell_2)$. Now consider again X arbitrary and $u\colon X \to Y$. We then define

$$\gamma_{S\mathcal{K}}(u) = \sup\{\gamma_{S\mathcal{K}}(u_{|E}) \mid E \subset X, \dim E < \infty\}.$$

Note that if u is the identity on X, we recover the previous definition, that is, we have

$$\gamma_{S\mathcal{K}}(I_X) = d_{S\mathcal{K}}(X).$$

Then the proof of Theorem 17.1 can be adapted to this setting with very little changes (this was observed by Marius Junge). In particular, we have

Theorem 17.12. *Let $u\colon X \to Y$ be an operator between operator spaces. Then u is exact iff $\gamma_{S\mathcal{K}}(u) < \infty$ and we have*

$$\mathrm{ex}(u) = \gamma_{S\mathcal{K}}(u). \tag{17.10}$$

Definition. *If the identity map on a C^*-algebra (or an operator space) A is strongly approximable, we will say that A has the "strong operator space approximation property" (the strong OAP, in short).*

If A has the strong OAP, it is rather easy to verify that, for any ideal $\mathcal{J} \subset B$ in a C^*-algebra B, any element in the kernel of the mapping $B \otimes_{\min} A \to (B/\mathcal{J}) \otimes_{\min} A$ must be approximable by elements in $\mathcal{J} \otimes_{\min} A$, and hence it actually lies in $\mathcal{J} \otimes_{\min} A$. Thus the strong OAP implies exactness for a C^*-algebra.

More generally, we have

Theorem 17.13. *Let A be a C^*-algebra. Let X, Y be operator spaces. Then, for any c.b. map $w\colon X \to A$ and any strongly approximable map $v\colon A \to Y$, the composition vw is exact and satisfies*

$$\gamma_{S\mathcal{K}}(vw) \leq \|v\|_{cb}\|w\|_{cb}. \tag{17.11}$$

Proof. Let $\mathcal{J} \subset B$ be as before, and let $q\colon B \otimes_{\min} A \to (B/\mathcal{J}) \otimes_{\min} A$ be the morphism associated to the quotient map from B onto $B/\mathcal{J}$. We denote for any operator space A

$$R(A) = \frac{B \otimes_{\min} A}{\ker(q)}.$$

Note that, by definition, A is exact iff we have $R(A) = Q(A)$ for any $\mathcal{J}$ and B (here $Q(A) = (B \otimes_{\min} A)/\mathcal{J} \otimes_{\min} A$, as defined in Lemma 17.2).

Now assume first that $v_i\colon A \to Y$ is a finite rank map. Consider $\widehat{v}_i = I_B \otimes v_i\colon B \otimes_{\min} A \to B \otimes_{\min} Y$. Then, for any θ in $\ker(q)$, $\widehat{v}_i(\theta) \in \mathcal{J} \otimes_{\min} Y$.

(Indeed, it is easy to check that, for any ξ in A^*, we have $(I_B \otimes \xi)(\ker(q)) \subset \mathcal{J}$.) Now assume that $\widehat{v}_i$ tends pointwise to $\widehat{v}$: $B \otimes_{\min} A \to B \otimes_{\min} Y$. This certainly implies $\widehat{v}(\theta) \in \mathcal{J} \otimes_{\min} Y$. Therefore any strongly approximable map v: $A \to Y$ defines a mapping $\widehat{v}$: $B \otimes_{\min} A \to B \otimes_{\min} Y$ such that $\widehat{v}(\ker(q)) \subset \mathcal{J} \otimes_{\min} Y$. Passing to the quotient spaces, we obtain a map $\widetilde{v}$: $R(A) \to Q(Y)$ that (since $\|\widehat{v}\| \leq \|v\|_{cb}$) satisfies $\|\widetilde{v}\| \leq \|v\|_{cb}$. But now, since A is a C^*-algebra, the morphism q is onto and we have necessarily

$$(B/\mathcal{J}) \otimes_{\min} A = R(A).$$

On the other hand, for any c.b. map w: $X \to A$ we clearly have

$$\|I_{B/\mathcal{J}} \otimes w: B/\mathcal{J} \otimes_{\min} X \to B/\mathcal{J} \otimes_{\min} A\|_{cb} \leq \|w\|_{cb}.$$

Therefore, we conclude that, if $u = vw$, the map (17.8) can be factorized as $\widetilde{v} \circ (I_{B/\mathcal{J}} \otimes w)$, and hence it has norm $\leq \|v\|_{cb}\|w\|_{cb}$. By (17.9) and (17.10), this completes the proof. ∎

Applying this to $u = v = w = I_A$, we obtain:

Corollary 17.14. *If a C^*-algebra A has the strong OAP, then A is exact.*

More generally, as observed in [J1], we have

Corollary 17.15. *A C^*-algebra $A \subset B(H)$ is exact iff the inclusion $A \to B(H)$ can be approximated pointwise by a net of finite rank maps u_i: $A \to B(H)$ with $\sup_i \|u_i\|_{cb} < \infty$.*

Proof. The "only if" part follows e.g. from Theorem 17.1 and the factorization of maximizing maps described in Corollary 12.6. Conversely, assume that there is a net (u_i) as in Corollary 17.15. Of course, each u_i: $A \to B(H)$ is strongly approximable (since u_i has finite rank); hence, by Theorem 17.13, we have $\gamma_{S\mathcal{K}}(u_i) \leq \|u_i\|_{cb}$, so that $\sup_i \sup \gamma_{S\mathcal{K}}(u_i) < \infty$. Clearly this implies that A is exact by a perturbation argument (see Lemma 2.13.2). ∎

Corollary 17.16. *Let E be a finite-dimensional subspace of an arbitrary C^*-algebra A. Let $\lambda(E, A) = \inf\{\|P\|_{cb}\}$, where the infimum runs over all possible linear projections P from A onto E. Then*

$$d_{S\mathcal{K}}(E) \leq \lambda(E, A). \tag{17.12}$$

Proof. Let P: $A \to E$ be a projection onto E and let j: $E \to A$ be the inclusion map. We then have $I_E = Pj$. Hence, by (17.11),

$$d_{S\mathcal{K}}(E) = \gamma_{S\mathcal{K}}(E) \leq \|P\|_{cb}\|j\|_{cb} = \|P\|_{cb},$$

whence (17.12). ∎

Remark 17.17. (i) By [DCH], for any noncommutative free group $\mathbb{F}$ the reduced C^*-algebra $C^*_\lambda(\mathbb{F})$ (as defined in §9.7) has the strong OAP. Hence, this is another proof that it is exact.

(ii) In sharp contrast, the full C^*-algebra $C^*(\mathbb{F})$ (considered in Chapters 8 and 13) is not exact. This was first proved in [Wa4]. From the (more recent) viewpoint of operator spaces, this can be seen as a consequence of the following estimate from [P10]. Let $\mathbb{F}_n$ (resp. $\mathbb{F}_\infty$) be the free group with n (resp. countably many) generators ($n = 1, 2, \ldots$), and let $U_1, \ldots, U_n$ be the unitary operators in $C^*(\mathbb{F}_n)$ associated to the free generators. Let $E_U^n = \mathrm{span}(U_1, \ldots, U_n)$. Then we have $d_{SK}(E_U^n) \geq n(2\sqrt{n-1})^{-1}$, so that, if $n \geq 3$ this number is > 1; hence the containing C^*-algebra $C^*(\mathbb{F}_n)$ cannot be exact (a variant shows that $n \geq 2$ suffices for this). Then the fact that $C^*(\mathbb{F})$ is not exact follows from the following two well-known facts: First, any noncommutative free group $\mathbb{F}$ contains $\mathbb{F}_n$ and $\mathbb{F}_\infty$ as subgroups (cf., e.g., [FTP, p. 15]). Second, for any discrete group Γ and any subgroup $\Gamma' \subset \Gamma$, the full C^*-algebra $C^*(\Gamma')$ appears naturally embedded as a C^*-subalgebra of $C^*(\Gamma)$ (see Proposition 8.8).

Remark. As mentioned after (17.2), the exactness of a C^*-algebra can be summarized by saying that for any $\mathcal{I}, B$ the kernel of the map $q\colon B \otimes_{\min} A \to (B/\mathcal{I}) \otimes_{\min} A$ coincides with $\mathcal{I} \otimes_{\min} A$. This is what is called a "slice map property." The slice map property for C^*-algebras was introduced in [To3] and further studied in [Wa4, Kr, Ki7]. More generally, let X, Y be operator spaces and let $E \subset X$, $F \subset Y$ be closed subspaces. The Fubini-product of E, F in $X \otimes_{\min} Y$ is the subspace

$$\mathcal{F}(E, F) \subset X \otimes_{\min} Y$$

formed of all the elements t in $X \otimes_{\min} Y$ such that $\forall x^* \in X^*$ and $\forall y^* \in Y^*$,

$$(x^* \otimes I_Y)(t) \in F \quad \text{and} \quad (I_X \otimes x^*)(t) \in E.$$

Then, the slice map problem (see [Wa4, Kr, Ki7]) is to decide when we have

$$\mathcal{F}(E, F) = E \otimes_{\min} F.$$

For instance, for the map $q\colon B \otimes_{\min} A \to (B/\mathcal{I}) \otimes_{\min} A$, we have $\ker(q) = \mathcal{F}(\mathcal{I}, A)$. Thus, by definition, A is exact iff $\mathcal{F}(\mathcal{I}, A) = \mathcal{I} \otimes_{\min} A$ for any B and any ideal $\mathcal{I} \subset B$, or iff this holds for $\mathcal{I} = \mathcal{K}$ and $B = B(\ell_2)$.

We will say that an operator space X has the slice map property (SMP in short) with respect to $F \subset Y$ if $\mathcal{F}(X, F) = X \otimes_{\min} F$. We will say that X has the Y-SMP if this holds for all (closed) subspaces $F \subset Y$.

The SMP is closely related to the OAP. Here is a summary of the main known facts:

(i) The c_0-SMP is equivalent to the AP (this essentially goes back to Grothendieck).

(ii) The $\mathcal{K}$-SMP is equivalent to the OAP ([Kr]).

(iii) The $B(\ell_2)$-SMP is equivalent to the strong OAP ([Kr]).

Remark. Haagerup and Kraus ([HK]) and Kirchberg ([Ki7]) independently proved that a C^*-algebra A has the strong OAP if (and only if) it is exact and has the OAP (or actually iff it is locally reflexive – in the sense of the next chapter – and has the OAP). Moreover ([Ki7]) the OAP is stable under extensions; that is, if $\mathcal{I}$ and $B/\mathcal{I}$ both have the OAP, then so does B (this can be proved by using (ii) and the exactness of $\mathcal{K}$). However, by [Ki2] there is an example of a B such that $\mathcal{I} \simeq \mathcal{K}$, $B/\mathcal{I}$ has the strong OAP (actually the metric CBAP), but B is not exact. This shows that neither exactness nor the strong OAP is stable under extensions.

Haagerup and Kraus ([HK]) say that a discrete group G has the AP if $C^*_\lambda(G)$ has the OAP and the approximating net can be chosen formed of finitely supported multipliers (see Remark 8.4 for background on multipliers). They show that G has the AP iff $C^*_\lambda(G)$ has the OAP. Moreover, for $C^*_\lambda(G)$ (discrete case), the OAP implies automatically the strong OAP. They also study the AP for general locally compact groups.

Up to now, the Gromov group $\mathcal{G}$ that will be discussed in Remark 17.22 is the only known example of a discrete group failing the AP.

It is unknown whether there is an exact C^*-algebra failing the OAP (or even the AP!), but there is a prime suspect, namely, $C^*_\lambda(SL(3,\mathbb{Z}))$. Indeed, since $SL(3,\mathbb{Z})$ is a lattice in a (connected) Lie group, by a result due to Connes [Co1], $C^*_\lambda(SL(3,\mathbb{Z}))$ embeds in a nuclear C^*-algebra; hence it is exact. However, by [H8], it fails the CBAP. This motivates the conjecture (formulated in [HK]) that it fails the OAP.

Note that if G is the semi-direct product of $SL(2,\mathbb{Z})$ with $\mathbb{Z}^2$ (for the standard action), then $C^*_\lambda(G)$ has the OAP ([HK]) but fails the CBAP ([H8]).

We refer to [CoH] for important results on approximation properties using c.b. multipliers.

Corollary 17.5 means that exactness is characterized by a "local embeddability" into the nuclear C^*-algebra $\mathcal{K}$. Actually, Kirchberg [Ki5, KiP] (see also [An1]) recently obtained a considerably deeper result, which yields a global embedding:

Theorem 17.18. *Any separable exact C^*-algebra A embeds (as a C^*-subalgebra) in a separable nuclear C^*-algebra B. Actually, we can take for B the Cuntz algebra O_2. Moreover, if A is nuclear, there is an embedding $A \subset O_2$ for which there is a completely contractive (and completely positive) projection $P\colon O_2 \to A$.*

Curiously, it remains open whether a nonseparable exact C^*-algebra embeds in a nuclear one.

Remark. It is clear by Theorem 17.3 that exactness passes to subspaces and is stable under the minimal tensor product. It is also true [Ki1] that if a C^*-algebra A is exact, then all its quotients A/I are also exact. Surprisingly, however, the only known proofs are *extremely difficult*. One way or the other, they are all related to Kirchberg's embedding Theorem 17.18 (or previous versions of it in [Ki1]). The derivation is as follows: To show that A/I is exact if A itself is exact, it is easy to reduce to the case when A is separable. But then, by Theorem 17.18, A/I is a quotient of a subalgebra of a nuclear C^*-algebra, and such algebras are exact by Proposition 18.19 in the next chapter.

In the case $A = C^*_\lambda(G)$, Kirchberg's embedding theorem can be recovered as a corollary of the following striking result due to Ozawa [Oz2] (inspired by the previous paper [GK]).

Let $D(G) \subset B(\ell_2(G))$ be the (commutative) C^*-algebra of all diagonal operators. Note that obviously $D(G) \simeq \ell_\infty(G)$. We will denote by $UC^*(G)$ the C^*-subalgebra of $B(\ell_2(G))$ generated by $C^*_\lambda(G)$ and $D(G)$.

Theorem 17.19. *([Oz2]) Let G be any discrete group. The following are equivalent.*

  (i) $C^*_\lambda(G)$ *is exact.*

  (ii) $UC^*(G)$ *is nuclear.*

For the proof we will need several results. The first statement provides us with a description of the "typical" operators in $UC^*(G)$.

Lemma 17.20. *Let*

$$V = \mathrm{span}\{\lambda(t)D \mid t \in G, D \in D(G)\}.$$

  (i) Then V is a dense $*$-subalgebra in $UC^*(G)$.

  (ii) Any operator v in V can be uniquely written as a finite sum of the form

$$v = \sum_{\theta \in T} \lambda(\theta)D_\theta$$

   with $D_\theta \in D(G)$, where $T \subset G$ is a finite subset.

  (iii) An operator v in $B(\ell_2(G))$ is in V iff there is a finite subset $T \subset G$ such that the associated matrix $\{v(s,t) \mid s,t \in G\}$ satisfies

$$v(s,t) = 0 \qquad \forall\, s,t \quad \text{such that} \quad st^{-1} \notin T.$$

Proof. (i) follows from the elementary fact that $\lambda(t)D(G)\lambda(t)^{-1} \subset D(G)$ for any t in G. (ii) Clearly any v in V can be written as a finite sum $v = \sum_{\theta \in T} \lambda(\theta)D_\theta$ with $D_\theta \in D(G)$. Then the associated matrix $v(s,t)$ defined by $v(s,t) = \langle v\delta_t, \delta_s \rangle$ satisfies

$$\forall\, \theta \in T \qquad v(\theta s, s) = D_\theta(s,s),$$

which shows that D_θ and T are uniquely determined by v.

(iii) Assume $v(s,t) = 0$ if $st^{-1} \notin T$ with $T \subset G$ finite. Then, if $D_\theta(s,s) = v(\theta s, s)$ for all θ in T, we have $v = \sum_{\theta \in T} \lambda(\theta)D_\theta$, so that $v \in V$. The converse is clear by (ii). ∎

Definition. Let G be any set. We will say that a function $\varphi \colon G \times G \to \mathbb{C}$ is a positive definite kernel if, for any n and any $t_1, \ldots, t_n \in G$, the matrix $(\varphi(t_i, t_j))$ is positive definite, that is,

$$\sum \alpha_i \overline{\alpha_j} \varphi(t_i, t_j) \geq 0 \qquad \forall (\alpha_i) \in \mathbb{C}^n.$$

Remark. It is easy to see that this holds iff there is a Hilbert space H and a function $x \colon G \to H$ such that $\varphi(s,t) = \langle x(s), x(t) \rangle$. Indeed, if φ is a positive definite kernel, we may equip the space of all finitely supported functions $\alpha \colon G \to \mathbb{C}$ with the scalar product $\langle \alpha, \beta \rangle = \sum \alpha(s)\overline{\beta(t)}\varphi(s,t)$; after passing to the quotient by the kernel $\{\alpha \mid \langle \alpha, \alpha \rangle = 0\}$ and completing, we obtain a Hilbert space H, and we have $\langle \delta_s, \delta_t \rangle = \varphi(s,t)$. The converse is obvious.

We now come to the key point:

Lemma 17.21. *([Oz2]) If $C_\lambda^*(G)$ is exact, then for any $\varepsilon > 0$ and any finite subset $S \subset G$, there is a finite subset $T \subset G$ and a positive definite kernel $\varphi \colon G \times G \to \mathbb{C}$ such that*

$$|\varphi(s,t) - 1| \leq \varepsilon \quad \text{if} \quad st^{-1} \in S$$

and

$$\varphi(s,t) = 0 \quad \text{if} \quad st^{-1} \notin T.$$

Proof. Let $E = \operatorname{span}\{\lambda(s) \mid s \in S\}$. Let $j \colon E \to B(\ell_2(G))$ be the inclusion map. By Lemma 17.8, there is a finite subset $\mathcal{S} \subset G$ such that the compression map $P_{\mathcal{S}} \colon E \to \widetilde{E} = P_{\mathcal{S}}(E)$ is an isomorphism such that $\|P_{\mathcal{S}}^{-1} \colon \widetilde{E} \to E\|_{cb} < 1 + \varepsilon$.

Let $u = j(P_{\mathcal{S}})^{-1} \colon \widetilde{E} \to B(\ell_2(G))$. We have $\|u\|_{cb} < 1 + \varepsilon$. By the extension property (Corollary 1.7), u admits an extension $w \colon B(\ell_2(\mathcal{S})) \to B(\ell_2(G))$ with $\|w\|_{cb} < 1 + \varepsilon$. Recall (see (11.5)) that $\|w\|_{cb} = \|w\|_{dec}$. Hence, by Corollary 12.5, the tensor $\widetilde{u} \in (\widetilde{E})^* \otimes B(\ell_2(G))$ associated to u satisfies

$\delta(\widetilde{u}) < 1 + \varepsilon$. We will now apply Lemma 12.9 to $u \colon \widetilde{E} \to B(\ell_2(G))$. By enlarging S if necessary we may clearly assume that S is symmetric and $e \in S$. Then E, and hence also $\widetilde{E}$, is unital and self-adjoint and u is a unital self-adjoint map. By Lemma 12.9, there is a c.p. map $\Phi \colon B(\ell_2(\mathcal{S})) \to B(\ell_2(G))$ such that

$$\forall\, x \in \widetilde{E} \qquad \|\Phi(x) - u(x)\| \le \varepsilon \|x\|. \tag{17.13}$$

We then define

$$\varphi(s,t) = \langle \Phi(P_{\mathcal{S}}(\lambda(st^{-1})))\delta_t, \delta_s \rangle.$$

Using the complete positivity of the composition $\Phi P_{\mathcal{S}}$ it is easy to check that φ is a positive definite kernel: Given $t_1, \ldots, t_n$ in G we have $(\lambda(t_i t_j^{-1})) \ge 0$; hence $\Phi(P_{\mathcal{S}}(\lambda(t_i t_j^{-1}))) \ge 0$, which implies $(\varphi(t_i, t_j)) \ge 0$.

Moreover, if $st^{-1} \in S$, we have $\lambda(st^{-1}) \in E$, and hence $P_{\mathcal{S}}(\lambda(st^{-1})) \in \widetilde{E}$ and $u P_{\mathcal{S}}(\lambda(st^{-1})) = \lambda(st^{-1})$. Hence, by (17.13),

$$\|\Phi P_{\mathcal{S}}(\lambda(st^{-1})) - \lambda(st^{-1})\| \le \varepsilon,$$

which implies (since $\lambda(st^{-1})\delta_t = \delta_s$)

$$|\varphi(s,t) - 1| \le \varepsilon.$$

Finally, let $T = \mathcal{S}\mathcal{S}^{-1}$, that is, $T = \{ab^{-1} \mid a, b \in \mathcal{S}\}$. Then $P_{\mathcal{S}}(\lambda(st^{-1})) \ne 0$ iff there are a, b in $\mathcal{S}$ such that $\langle \lambda(st^{-1})\delta_b, \delta_a \rangle \ne 0$, or, equivalently, iff $st^{-1} \in T$. Thus we obtain a finite set T such that $\varphi(s,t) = 0$ if $st^{-1} \notin T$. ∎

Proof of Theorem 17.19. It suffices to show that the conclusion of Lemma 17.21 implies that $UC^*(G)$ is nuclear. Indeed this will show (i) $\Rightarrow$ (ii), and the converse is clear. Let φ be as in Lemma 17.21. Assuming $e \in S$, we have $tt^{-1} \in S$; hence

$$\forall\, t \in G \qquad |\varphi(t,t) - 1| < \varepsilon.$$

Then the Schur multiplier $M_\varphi \colon B(\ell_2(G)) \to B(\ell_2(G))$ that takes $(a(s,t))$ to $(\varphi(s,t)a(s,t))$ is completely positive (see Exercise 1.5) with

$$\|M_\varphi\| = \|M_\varphi(1)\| = \sup_{t \in G} |\varphi(t,t)| \le 1 + \varepsilon.$$

By Lemma 17.20,

$$M_\varphi(B(\ell_2(G)) \subset UC^*(G).$$

Let I be the set of pairs $\alpha = (S, \varepsilon)$ with S finite, $e \in S$, and $\varepsilon > 0$, directed so that $\alpha \to \infty$ corresponds to "S tends to G and ε tends to 0." Let φ_α be the kernel associated to $\alpha = (S, \varepsilon)$ as in Lemma 17.21, and let

$$u_\alpha \colon UC^*(G) \to UC^*(G)$$

be the restriction of M_{φ_α} to $UC^*(G)$. Note that u_α is c.p. and $\|u_\alpha\| \le \|M_{\varphi_\alpha}\| \le 1 + \varepsilon$.

We claim that

$$\forall \, x \in UC^*(G) \qquad \|u_\alpha x - x\| \to 0.$$

Indeed, since $\|u_\alpha\| \to 1$, it suffices to check this (by Lemma 17.20) for all x of the form $x = \lambda(t)D$ with $D \in D(G)$. Then we find

$$u_\alpha x - x = \sum_s D(s,s)(\varphi_\alpha(ts,s) - 1)e_{ts,s}.$$

Hence

$$\|u_\alpha x - x\| \le \sup_s |D(s,s)||\varphi_\alpha(ts,s) - 1|,$$

and if α is chosen large enough, we have $(ts)(s)^{-1} = t \in S$ for all s; hence $\sup_s |\varphi_\alpha(ts,s) - 1| \le \varepsilon$ so that $\|u_\alpha x - x\| \le \varepsilon \|D\|$, which proves our claim.

To show that $A = UC^*(G)$ is nuclear it suffices to show that, for any C^*-algebra B, $u_\alpha \otimes I_B$ is bounded from $A \otimes_{\min} B$ to $A \otimes_{\max} B$. Indeed, by Proposition 11.7, we then have

$$\|u_\alpha \otimes I_B \colon A \otimes_{\min} B \to A \otimes_{\max} B\| \le \|u_\alpha\|_{dec},$$

and since u_α is c.p., $\|u_\alpha\|_{dec} = \|u_\alpha\| \le 1 + \varepsilon$, so we conclude that $A \otimes_{\min} B = A \otimes_{\max} B$.

Finally, to check that $u_\alpha \otimes I_B$ is bounded from $A \otimes_{\min} B$ to $A \otimes_{\max} B$ it suffices to rewrite u_α as a finite sum of maps of the form $A \xrightarrow{v} C \xrightarrow{w} A$ with C nuclear, v c.b., and with w such that $w \otimes I_B$ is bounded from $C \otimes_{\max} B$ to $C \otimes_{\max} A$ (which holds if w is decomposable). We will show this with $C = D(G)$. Being commutative, this is a nuclear C^*-algebra (see Exercise 11.7). Let $P \colon B(\ell_2(G)) \to D(G)$ be the standard contractive projection taking $(a(s,t))_{s,t}$ to $(a(s,t)1_{\{s=t\}})$. Note that P is c.p. by Exercise 1.5 (since P can be viewed as Schur multiplication by $\langle \delta_s, \delta_t \rangle$). Then, if $\alpha = (S, \varepsilon)$ and if T is a finite set for which $\varphi_\alpha(s,t) = 0$ if $st^{-1} \notin T$, we can write for any x in V

$$u_\alpha x = \sum_{\theta \in T} \lambda(\theta) P(\lambda(\theta)^{-1} u_\alpha x).$$

Thus u_α can be written as announced, since $v \colon x \to P(\lambda(\theta)^{-1} u_\alpha x)$ is a c.b. map into $D(G)$ and $w \colon y \to \lambda(\theta)y$ is decomposable from $D(G)$ into $UC^*(G)$. (More directly, assuming B unital, $w \otimes I_B \colon C \otimes_{\max} B \to A \otimes_{\max} B$ is clearly contractive since it is the restriction to $C \otimes_{\max} B$ of the left multiplication by $\lambda(\theta) \otimes 1$ in $A \otimes_{\max} B$.)

This completes the proof that $UC^*(G)$ is nuclear. ∎

Remark 17.22. (i) It was a long-standing open problem whether $C^*_\lambda(G)$ is exact for any discrete group G. (The groups for which it is true are called exact in [KiW].) However, this was disproved very recently by N. Ozawa [Oz2] (see also [An2]): His result (the above Lemma 17.21) shows that, for a special group $\mathcal{G}$ constructed previously by Gromov [Gro] using very delicate arguments (not yet fully understood at the time of this writing), $C_\lambda^*(\mathcal{G})$ cannot be exact and hence is a counterexample to the above problem. A fortiori, by [HK], $C_\lambda^*(\mathcal{G})$ fails the OAP; in other words, $\mathcal{G}$ fails the AP.

(ii) On the other hand, it remains an outstanding open problem whether the full C^*-algebra $C^*(G)$ is *not* exact for any nonamenable discrete group G.

The proof of Kirchberg's embedding (Theorem 17.18) is quite difficult and beyond the scope of this book. However, there is a much simpler operator space version, which we can fully prove, following [EOR]. Fix $\lambda \geq 1$. We will say that an operator space X is λ-nuclear if the identity on X can be approximated pointwise by a net of finite rank maps of the form

$$X \xrightarrow{v_\alpha} M_{n_\alpha} \xrightarrow{w_\alpha} X$$

with $\sup_\alpha \|v_\alpha\|_{cb} \|w_\alpha\|_{cb} \leq 1$ and $n_\alpha < \infty$. When $\lambda = 1$, we may as well assume that v_α, w_α are complete contractions.

Note that, if X is a C^*-algebra, it is nuclear iff it is 1-nuclear in the preceding sense (see Theorem 11.5) and actually ([P1, p. 35]) iff it is λ-nuclear for some λ.

Recall we say that X is λ-exact if

$$\sup\{d_{SK}(E) \mid E \subset X, \dim(E) < \infty\} \leq \lambda.$$

We then have

Theorem 17.23. *([EOR]) Every 1-exact separable operator space X embeds completely isometrically in some 1-nuclear separable operator space.*

Proof. ([EOR]) Let $\{E_n\}$ be an increasing sequence of finite-dimensional subspaces of X with dense union. Let $\varepsilon(n) > 0$ be chosen so that $\sum \varepsilon(n) < \infty$. We claim that there is an increasing sequence $N(1) < N(2) < \ldots$ of natural numbers and maps as in the following diagram:

$$
\begin{array}{ccccccccc}
E_1 & \subseteq & \cdots & \subseteq & E_n & \subseteq & E_{n+1} & \subseteq & \cdots \\
i_1 \downarrow & & & & i_n \downarrow & & i_{n+1} \downarrow & & \\
M_{N(1)} & \overset{j_1}{\hookrightarrow} & \cdots & \overset{j_{n-1}}{\hookrightarrow} & M_{N(n)} & \overset{j_n}{\hookrightarrow} & M_{N(n+1)} & \overset{j_{n+1}}{\hookrightarrow} & \cdots
\end{array}
$$

such that

(i) j_n is a complete isometry,

(ii) $\|i_n\|_{cb} \leq 1$ and $\|i_{n|i_n(E_n)}^{-1}\|_{cb} \leq 1 + \varepsilon(n)$,

(iii) $\|i_{n+1|E_n} - j_n i_n\|_{cb} \leq \varepsilon(n)$.

We will prove this by induction on n. Since $d_{SK}(E_1) = 1$, we can find $N(1)$ and an embedding $i_1 \colon E_1 \to M_{N(1)}$ such that $\|i_1\|_{cb} \leq 1$ and

$$\|i_{1|i_1(E_1)}^{-1}\|_{cb} \leq 1 + \varepsilon(1).$$

Now suppose that $i_n \colon E_n \to M_{N(n)}$ is given satisfying (ii) above. We will prolong the above diagram to the right by constructing $N(n+1)$, $i_{n+1} \colon E_{n+1} \to M_{N(n+1)}$, and $j_n \colon M_{N(n)} \to M_{N(n+1)}$ satisfying the required conditions.

Since $d_{SK}(E_{n+1}) = 1$, we can find $N \geq 1$ and a complete contraction $u \colon E_{n+1} \to M_N$ with $\|u_{|u(E_{n+1})}^{-1}\|_{cb} < 1 + \varepsilon(n+1)$. By the injectivity of $M_{N(n)}$, the map $i_n \colon E_n \to M_{N(n)}$ admits an extension $\widetilde{i_n} \colon E_{n+1} \to M_{N(n)}$ with $\|\widetilde{i_n}\|_{cb} \leq 1$. By the injectivity of M_N, the map $u_{|E_n} i_n^{-1} \colon i_n(E_n) \to M_N$ admits an extension $v \colon M_{N(n)} \to M_N$ with $\|v\|_{cb} \leq 1 + \varepsilon(n)$. Note that $v i_n = u_{|E_n}$.

Then let $N(n+1) = N(n) + N$, and let $i_{n+1} \colon E_{n+1} \to M_{N(n+1)}$ and $j_n \colon M_{N(n)} \to M_{N(n+1)}$ be defined by the following block diagonal sums:

$$i_{n+1}(e) = i_n(e) \oplus u(e) \quad (e \in E_{n+1})$$
$$j_n(x) = x \oplus (1 + \varepsilon(n))^{-1} v(x) \quad (x \in M_{N(n)}).$$

Then it is easy to check the desired conditions. This establishes our claim.

Clearly we may now view the maps $j_n \colon M_{N(n)} \to M_{N(n+1)}$ as if they were inclusions (of operator spaces, but not of algebras!) and form the direct limit $Y = \overline{\cup M_{N(n)}}$ (i.e., the inductive limit of this system). Obviously the resulting o.s. Y is 1-nuclear by construction. Moreover, the mapping $i \colon \cup E_n \to Y$ defined by $i(x) = \lim_{n \to \infty} i_n(x)$ is well defined and completely isometric. By density, it extends to a completely isometric embedding of X into Y. ∎

Exercise

Exercise 17.1. Prove that a C^*-algebra is nuclear iff it is both exact and WEP.

Chapter 18. Local Reflexivity

Basic properties. In Banach space theory, the *principle of local reflexivity* ([LiR]) says that every Banach space X has the following property, called "local reflexivity": $B(E,X)^{**} = B(E,X^{**})$ (isometrically) for any finite-dimensional Banach space E.

In sharp contrast, the o.s. analog is not universally true, and local reflexivity has turned out to be a very important property.

Definition 18.1. Let $\lambda \geq 1$. An operator space X is called λ-locally reflexive if, for any finite-dimensional o.s. E, the natural linear isomorphism

$$CB(E, X^{**}) \longrightarrow CB(E,X)^{**}$$

has norm $\leq \lambda$.

It is easy to check that the inverse map $CB(E,X)^{**} \longrightarrow CB(E,X^{**})$ always has norm 1. Hence X is 1-locally reflexive iff, for any finite-dimensional E, we have an isometric identity

$$CB(E,X)^{**} = CB(E, X^{**}),$$

or equivalently, for any finite-dimensional o.s. F we have

$$(F \otimes_{\min} X)^{**} = F \otimes_{\min} X^{**}.$$

Replacing E^* and F by $M_n(E^*)$ and $M_n(F)$, it is easy to see that the preceding identities are actually completely isometric, but we will not need this.

If $Y \subset X$ (isometrically), we have $Y^{**} \subset X^{**}$ (isometrically). Hence, since $CB(E,Y) \subset CB(E,X)$ (isometrically) if $Y \subset X$ (completely isometrically), it is easy to check (just like for reflexivity) that $Y \subset X$ is λ-locally reflexive if X is. On the other hand, in general, the quotient X/Y is not λ-locally reflexive: Indeed, any separable o.s. is a quotient of S_1 that is locally reflexive by Theorem 18.7. Nevertheless, the quotient of a locally reflexive C^*-algebra by an ideal inherits local reflexivity (see Exercise 18.4).

Remark 18.2. By (2.5.1), we know that, for any operator space X:

$$(M_n \otimes_{\min} X)^{**} = M_n \otimes_{\min} X^{**} \quad \text{(isometrically)}.$$

Hence, for any subspace $S \subset M_n$ we have

$$(S \otimes_{\min} X)^{**} = S \otimes_{\min} X^{**} \quad \text{(isometrically)},$$

or equivalently

$$CB(S^*, X)^{**} = CB(S^*, X^{**}) \quad \text{(isometrically)}.$$

The λ-local reflexivity of X is used in the following equivalent reformulation:

Proposition 18.3. *An operator space X is λ-locally reflexive iff, for any finite-dimensional operator space E and any mapping $u: E \to X^{**}$ with $\|u\|_{cb} \leq 1$, there is a net u_i in $CB(E, X)$ with $\|u_i\|_{cb} \leq \lambda$ such that, for any x in E, $u_i(x) \to u(x)$ for the $\sigma(X^{**}, X^*)$-topology.*

Proof. That this is indeed equivalent follows easily from the classical density of the unit ball of a Banach space B in the unit ball of its bidual B^{**} for the topology $\sigma(B^{**}, B^*)$ applied to $B = CB(E, X) = E^* \otimes_{\min} X$.   ∎

In this formulation, it is easy to see (replacing E by $u(E)$) that it suffices to consider the case when $E \subset X^{**}$ and the map u to be approximated is the inclusion map $E \to X^{**}$. In the Banach space case, when $E \subset X^{**}$, a stronger principle holds: The maps $u_\alpha: E \to X$ approximating the inclusion $E \subset X^{**}$ can be chosen so that $\|u_\alpha\| \leq 1$ and $\|u_{\alpha|u_\alpha(E)}^{-1}\| \longrightarrow 1$ when $\alpha \to \infty$ (see Exercise 18.2).

The following consequence for operator spaces was observed in [GH]. Here, for any map $u: E \to F$, we denote $\|u\|_n = \|I \otimes u: M_n(E) \to M_n(F)\|$.

Proposition 18.4. *Let X be an arbitrary operator space. Let $E \subset X^{**}$ be a finite-dimensional subspace. There is a net of maps $u_\alpha: E \to X$ tending to the inclusion map $E \subset X^{**}$ in the point-$\sigma(X^{**}, X^*)$ topology and, moreover, such that, for each fixed $n \geq 1$, $\|u_\alpha\|_n \leq 1$ and $\|u_{\alpha|u_\alpha(E)}^{-1}\|_n \to 1$.*

Proof. Consider the inclusion $M_n(E) \subset M_n(X^{**})$. Recall that, by (2.1.4), $M_n(X^{**}) = M_n(X)^{**}$. By the local reflexivity principle for Banach spaces, there is a net of contractive maps $V_\alpha: M_n(E) \to M_n(X)$ such that $V_\alpha(e)$ tends $\sigma(M_n(X)^{**}, M_n(X)^*)$ to e for any e in $M_n(E)$. Let $\tilde{V}_\alpha$ be defined as in Exercise 18.3. Then, by Exercise 18.3, we have $\tilde{V}_\alpha = I \otimes u_\alpha$ and $\|u_\alpha\|_n \leq 1$. Clearly $\tilde{V}_\alpha(e)$ still tends to e for $\sigma(M_n(X)^{**}, M_n(X)^*)$. Therefore, by Exercise 18.2 applied to $\tilde{V}_\alpha$, we find that $\tilde{V}_\alpha$ is almost isometric when α is large enough, and hence $\|u_{\alpha|u_\alpha(E)}^{-1}\|_n \to 1$.   ∎

Proposition 18.5. *([EOR]) An operator space X is λ-locally reflexive iff the same is true for every separable subspace.*

Proof. Since local reflexivity passes to subspaces, we only need to prove the "if" part. Suppose that each separable subspace of X is λ-locally reflexive. Let $E \subset X^{**}$ be a finite-dimensional subspace. Fix a finite subset $\xi_1, \ldots, \xi_k$ in X^*. By Proposition 18.2 there is a sequence of maps $u_n: E \to X$ such that $\|u_n\|_n \leq 1$ such that

$$\lim_n \langle \xi_j, u_n(e) - e \rangle = 0 \quad \text{for each} \quad j = 1, \ldots, k.$$

Let X_1 be a separable subspace of X containing $u_n(E)$ for all $n \geq 1$. Let $\mathcal{U}$ be a free ultrafilter on $\mathbb{N}$. We define a mapping $v\colon E \to X_1^{**}$ by setting

$$\forall e \in E \qquad v(e) = \lim_{\mathcal{U}} u_n(e) \quad \text{(limit in } \sigma(X_1^{**}, X_1^*) \text{ sense)}.$$

Note that $\langle \xi_j, v(e) \rangle = \langle \xi_j, e \rangle$ for each $j = 1, \ldots, k$. Clearly $\|v\|_{cb} \leq 1$ (since $\|u_n\|_n \leq 1$ for each n). Then, by the local reflexivity of the separable subspace X_1, there is a net of maps $v_\alpha\colon E \to X_1$ such that $\|v_\alpha\|_{cb} \leq 1$ and

$$\forall e \in E \qquad \langle \xi_j, v_\alpha(e) \rangle \to \langle \xi_j, v(e) \rangle = \langle \xi_j, e \rangle.$$

This shows that X is locally reflexive. ∎

Remark 18.6. As pointed out to the author by Kirchberg, it is not hard to show that any L_1-space (in the usual, commutative sense) over a measure space (Ω, μ) is locally reflexive. However, the question whether the predual of a von Neumann algebra (for instance, the space S_1 of all trace class operators on ℓ_2 (i.e., the predual of $B(\ell_2)$)) is locally reflexive remained open for a while, but this was recently proved in [EJR] (see also [J1] and [JLM]). We include the simplified proof from [JLM].

Theorem 18.7. *The predual M_* of any von Neumann algebra M is locally reflexive.*

Proof. Let $X = M_*$ so that $X^{**} = M^*$. We will use the criterion in Proposition 18.3. Let E be a finite-dimensional operator space and let $u\colon E \to M^*$ be a complete contraction. We may assume $E^* \subset B(H)$ (completely isometrically). Let $i\colon E^* \subset B(H)$ be the inclusion map. Let $T\colon M \to E^*$ be the restriction of u^* to M. Note that $iT\colon M \to B(H)$ is a finite rank map with $\|iT\|_{cb} \leq 1$. Hence, by Theorem 12.7, for any $\varepsilon > 0$ there is, for some n, a factorization of iT of the form $M \xrightarrow{v} M_n \xrightarrow{w} B(H)$ with $\|v\|_{cb} = 1$, $\|w\|_{cb} < 1 + \varepsilon$. Let $S = v(M) \subset M_n$. Since $w(S) \subset E^*$, we have, by restriction, a factorization of T of the form

$$T\colon M \xrightarrow{v_1} S \xrightarrow{w_1} E^*$$

with $\|v_1\|_{cb} = 1$, $\|w_1\|_{cb} < 1 + \varepsilon$. Taking adjoints, since $u = T^*$, we find a factorization of u of the form

$$u\colon E \xrightarrow{w_1^*} S^* \xrightarrow{v_1^*} M^*.$$

But now, by Remark 18.2, v_1^* can be approximated in the point-$\sigma(M^*, M)$ sense by a net of complete contractions $\psi_\alpha\colon S^* \to M_*$. Then, letting $u_\alpha = \psi_\alpha w_1^*$, we find an approximating net as in Proposition 18.3, but with $\sup \|u_\alpha\|_{cb} \leq 1 + \varepsilon$. Since ε is arbitrarily small, this implies that M_* is 1-locally reflexive. ∎

A conjecture on local reflexivity and OLLP. We now return to the OLLP, which we encountered in Chapter 16.

Proposition 18.8. *Let $\lambda \geq 1$. Any operator space X for which X^{**} has the λ-OLLP is λ-locally reflexive and has the λ-OLLP.*

Proof. Fix $\varepsilon > 0$. Let $E \subset X^{**}$ be a finite-dimensional subspace. By Theorem 16.6, the inclusion $E \to X^{**}$ admits a factorization $E \xrightarrow{v} G^* \xrightarrow{w} X^{**}$ with $\|w\|_{cb} = 1$, $\|v\|_{cb} < \lambda + \varepsilon$, and $G \subset M_N$ for some N. Since $M_N(X^{**}) = M_N(X)^{**}$ (isometrically) (see (2.5.1)), we also have $G \otimes_{\min} X^{**} = (G \otimes_{\min} X)^{**}$; hence w can be approximated in the point-$\sigma(X^{**}, X^*)$ sense by maps $w_\alpha \colon G^* \to X$ with $\|w_\alpha\|_{cb} \leq 1$. Then the maps $u_\alpha = w_\alpha v \colon E \to X$ approximate in the same sense the inclusion $E \to X^{**}$ and $\sup \|u_\alpha\|_{cb} \leq \lambda + \varepsilon$. Thus X is $(\lambda + \varepsilon)$-locally reflexive for any $\varepsilon > 0$, and hence (by an elementary argument) it is λ-locally reflexive. To prove that X has the λ-OLLP, assume that $E \subset X$. Then we have, for any e in E, $u_\alpha(e) - e \to 0$ weakly in X; hence, passing to convex hulls, we can obtain a net such that $u_\alpha(e) - e \to 0$ strongly for any e in E. By a simple perturbation (see §2.13) argument, this gives us that X satisfies (iii) in Theorem 16.6, and hence X has the λ-OLLP. ∎

We will now show the equivalence of a number of interesting conjectures formulated by Ozawa [Oz3, Oz6] and closely related to previous work by Oikhberg [O3].

Definition 18.9. *We will say that an operator space X is submaximal if it embeds completely isometrically into a maximal o.s. Y.*

If X is separable and submaximal, we may as well take $Y = \max(B(\ell_2))$ (or, in the nonseparable case, $Y = \max(B(H))$ for some H). Indeed, if X is separable, we can assume $X \subset B(\ell_2)$. Then let $\widetilde{X}$ be the o.s. obtained by inducing on X the o.s. structure of $\max(B(\ell_2))$. Now assume $X \subset Y$ with Y maximal. By the injectivity of $B(\ell_2)$, there is a complete contraction $T \colon Y \to B(\ell_2)$ extending the inclusion $X \to B(\ell_2)$. Since Y is maximal, $\|T \colon Y \to \max(B(\ell_2))\|_{cb} \leq 1$; hence, restricting to X, we find $\|T \colon X \to \widetilde{X}\|_{cb} \leq 1$. On the other hand, since $\max(B(\ell_2)) \to B(\ell_2)$ is trivially completely contractive, we also have $\|T^{-1} \colon \widetilde{X} \to X\|_{cb} \leq 1$. Hence, we conclude that:

$$X \text{ submaximal} \Longleftrightarrow X = \widetilde{X} \text{ (completely isometrically)}.$$

Any maximal space X is (essentially by definition; see Proposition 3.3) completely isometric to a quotient of $\ell_1(\Gamma)$ for some set Γ. If X is finite-dimensional, it is easy by a compactness argument to see that we can achieve this with a finite set Γ, provided we replace "completely isometric" by $(1 + \varepsilon)$-completely isometric (see Exercise 2.13.2).

Similarly, submaximal spaces can be identified with subquotients of $\ell_1(\Gamma)$ with Γ an arbitrary set. However, in sharp contrast with the preceding case, it turns out to be apparently a quite delicate question of whether the corresponding finite-dimensional assertion is valid. This motivates the following

Definition 18.10. Let $\lambda \geq 1$. An operator space X will be called λ-hypermaximal if, for any finite-dimensional subspace $E \subset X$, there is a finite-dimensional normed space F and a factorization $E \xrightarrow{v} \max(F) \xrightarrow{w} X$ of the inclusion map $E \subset X$ such that $\|v\|_{cb}\|w\|_{cb} < \lambda + \varepsilon$.

We will say that X is λ-maximal if it is completely isomorphic to a maximal space and the c.b. norm of the identity $X \to \max(X)$ is $\leq \lambda$.

We claim that X λ-hypermaximal implies X, λ-maximal. Indeed, if X is as in Definition 18.10, any bounded $u\colon X \to B(H)$ satisfies, for any finite-dimensional subspace $E \subset X$, $\|u_{|E}\|_{cb} = \|uwv\|_{cb} \leq \|uw\|_{cb}\|v\|_{cb}$, and hence (since $\|uw\|_{cb} = \|uw\|$) a fortiori $\|u_{|E}\|_{cb} \leq \|uw\|\|v\|_{cb} \leq \|u\|\|w\|\|v\|_{cb} \leq (\lambda + \varepsilon)\|u\|$. Thus we must have $\|u\|_{cb} \leq \lambda\|u\|$, and we conclude that X is λ-maximal.

Remark 18.11. By a perturbation argument it is easy to verify that, if a maximal o.s. X has the metric approximation property (resp. the λ-BAP), then X is 1-hypermaximal (resp. λ-hypermaximal).

Remark 18.12. The reader should observe the analogy between the preceding definition and that of spaces with λ-OLLP (see (i)$\Rightarrow$ (iii) in Theorem 16.6): If $G = \max(F)$ with $\dim F < \infty$, then, for each $\varepsilon > 0$, $G^*(1 + \varepsilon)$-embeds into ℓ_∞^N for some N. Thus, in some sense, the λ-OLLP is the noncommutative analog of the preceding definition.

Theorem 18.13. ([Oz3]) Fix $\lambda \geq 1$. The following conjectures are equivalent:

 (i) Every maximal o.s. has the λ-OLLP.

 (ii) Every maximal o.s. is λ-locally reflexive.

 (ii)' Every submaximal o.s. is λ-locally reflexive.

 (iii) $\max(B(\ell_2))$ is λ-locally reflexive.

 (iv) Any maximal o.s. is λ-hypermaximal.

 (v) For any separable Banach space X, any bounded map $u\colon X \to \mathcal{B}/\mathcal{K}$ admits a bounded lifting $\widetilde{u}$ with $\|\widetilde{u}\| \leq \lambda\|u\|$ (recall the notation $\mathcal{B} = B(\ell_2)$, $\mathcal{K} = K(\ell_2)$).

Proof. (i) $\Rightarrow$ (ii). If X is maximal, then (by Exercise 3.3) X^{**} is maximal; so, by Proposition 18.8, (i) $\Rightarrow$ (ii).

(ii) $\Leftrightarrow$ (ii)' is obvious since local reflexivity passes to subspaces.

(ii) $\Rightarrow$ (iii) is trivial.

(iii) $\Rightarrow$ (ii). By the preceding remarks, if X is separable and submaximal, it embeds into $\max(B(\ell_2))$; hence it is λ-locally reflexive if (iii) holds. Now assume X possibly nonseparable. Let $X_1 \subset X$ be an arbitrary separable subspace. By Exercise 3.8, if X is maximal, there is *separable* subspace X_2

with $X_1 \subset X_2 \subset X$ that is also maximal and hence λ-locally reflexive by the separable case. A fortiori, its subspace X_1 is λ-locally reflexive. Then since local reflexivity is separably determined (see Proposition 18.5), we conclude that X itself must be λ-locally reflexive.

(ii) $\Rightarrow$ (iv). This uses a construction due to W. B. Johnson [Jo]. Consider a finite-dimensional subspace $E \subset \max(X)$. Fix $\varepsilon > 0$. Assume X separable for simplicity. Let $\{E_n\}$ be an increasing sequence of finite-dimensional subspaces of X such that $E \subset E_1$ and $\cup E_n$ is dense in X. Let $Y = \ell_1(\{E_n\})$. We claim that the identity on X^{**} admits a factorization through Y^{**} of the form

$$I_{X^{**}}\colon X^{**} \xrightarrow{\ J\ } Y^{**} \xrightarrow{\ P^{**}\ } X^{**},$$

where J and $P\colon Y \to X$ are complete contractions.

There is a natural metric surjection $P\colon Y \to X$ defined by $P((x_n)) = \sum x_n$. Note that $Y^* = \ell_\infty(\{E_n^*\})$. Let $\mathcal{U}$ be a free ultrafilter on $\mathbb{N}$. Let us denote by $J\colon X^{**} \to Y^{**}$ the mapping defined by

$$\forall\, x \in X^{**} \quad \forall\, \xi = (\xi_n) \in Y^* \qquad \langle J(x), \xi \rangle = \langle x, [\lim_{\mathcal{U}} \xi_n]\rangle,$$

where $[\lim_{\mathcal{U}} \xi_n] \in X^*$ is first defined as a pointwise limit on the union of the spaces $\{E_n\}$ and then extended by density to the whole of X.

Clearly $\|J\| \leq 1$. Note that $P^{**}J = I_{X^{**}}$. Let $j = J_{|E}\colon E \to Y^{**}$, so that, denoting by $i_E\colon E \to X^{**}$ the inclusion map, we have $P^{**}j = i_E$. Since $\|J\| \leq 1$, we have $\|J\colon \max(X^{**}) \to \max(Y)^{**}\|_{cb} \leq 1$, and hence a fortiori

$$\|j\colon E \to \max(Y)^{**}\|_{cb} \leq 1.$$

If we assume (ii), $\max(Y)$ is λ-locally reflexive; hence there is a net of maps $u_\alpha\colon E \to \max(Y)$ with $\|u_\alpha\|_{cb} \leq \lambda$ tending point-$\sigma(Y^{**}, Y)$ to j. A fortiori, $P^{**}u_\alpha \to P^{**}j = i_E$ in the point-$\sigma(X^{**}, X^*)$ topology. Hence, since $P^{**}u_\alpha = Pu_\alpha$, we have

$$\forall\, e \in E \qquad Pu_\alpha(e) \to e \tag{18.1}$$

with respect to $\sigma(X^{**}, X^*)$. But now, since both $Pu_\alpha(e)$ and e are in X, the convergence in (18.1) holds in the weak topology of X. Passing to convex hulls, we can modify the u_α and ensure that, actually, the convergence (18.1) holds in norm. This gives us a (strong) approximate factorization of i_E as follows:

$$E \xrightarrow{\ u_\alpha\ } \max(Y) \xrightarrow{\ P\ } \max(X)$$

with $\|u_\alpha\|_{cb} \leq \lambda$. Now, by Remark 18.11, $\max(Y)$ is 1-hypermaximal, so we immediately deduce from this factorization that $\max(X)$ is also λ-hypermaximal. This completes the proof that (ii) $\Rightarrow$ (iv).

Since it is clear that a λ-hypermaximal space has the λ-OLLP (see Remark 18.12), we have (iv) $\Rightarrow$ (i) and (i)–(iv) are equivalent.

By Theorem 16.10, (v) is equivalent to the assertion that every separable maximal o.s. has the λ-OLLP. Now let X be a nonseparable maximal o.s. and let $E \subset X$ be a finite-dimensional subspace. By Exercise 3.8 there is X_1 separable and maximal such that $E \subset X_1 \subset X$. Thus, if we know that X_1 always has the λ-OLLP, it immediately follows (by definition of the OLLP) that X has the λ-OLLP. This shows that (i) $\Leftrightarrow$ (v). ∎

It is a long-standing open question whether any ideal $I \subset B$ in a *separable C^*-algebra* is automatically complemented, that is, whether there is a *bounded* linear projection $P: B \to I$. Thus the next statement constitutes a strong motivation to disprove the above conjectures.

Proposition 18.14. *([Oz6]) If, for all values of λ, the equivalent conjectures in Theorem 18.13 fail to be true, there is an embedding $\mathcal{K} \subset B$ as an ideal in a separable C^*-algebra without any bounded linear projection from B onto $\mathcal{K}$.*

Proof. Indeed, for any λ, we can find X_λ, and $u_\lambda: X_\lambda \to \mathcal{B}/\mathcal{K}$ with $\|u_\lambda\| \leq 1$ such that any lifting $\widetilde{u}_\lambda$ must have $\|\widetilde{u}_\lambda\| > \lambda$. Let $B_\lambda \subset \mathcal{B}$ be a separable C^*-algebra such that $\mathcal{K} \subset B_\lambda$ and $u_\lambda(X_\lambda) \subset B_\lambda/\mathcal{K}$. If $P_\lambda: B_\lambda \to \mathcal{K}$ is any projection, we must have $\|P_\lambda\| \geq \lambda - 1$ (otherwise we would get a lifting $\widetilde{u}_\lambda$ associated to $I - P_\lambda$ with norm $\leq \lambda$). We will take $\lambda = 1, 2, \ldots$, and consider $I = c_0(\mathcal{K}) \subset c_0(\{B_n\})$. Clearly, I is an ideal in a separable C^*-algebra, and any projection $P: c_0(\{B_n\}) \to I$ must induce (after a well known averaging) a projection P_n from B_n to $\mathcal{K}$ with $\|P\| \geq \|P_n\| \geq n - 1$; hence P cannot be bounded.

Finally, if we let $B \subset \mathcal{K} \otimes_{\min} \mathcal{B}$ be the C^*-algebra generated by $\mathcal{K} \otimes_{\min} I$ and by $c_0(\{B_n\})$, then $\mathcal{K} \otimes_{\min} \mathcal{K}$ is an ideal (isomorphic to $\mathcal{K}$) in B, but there is no bounded projection $P: B \to \mathcal{K} \otimes_{\min} \mathcal{K}$ (because otherwise, composing P with the natural contractive projection $Q: \mathcal{K} \otimes_{\min} \mathcal{K} \to c_0(\mathcal{K})$, we would obtain a bounded projection from B onto I and a fortiori from $c_0(\{B_n\})$ onto I). ∎

Properties C, C', and C''. Exactness versus local reflexivity. In the C^*-algebra case, the ideas developed in this chapter go back to Archbold and Batty [AB], who introduced the properties C and C' defined below. Their work was continued and extended to the operator space setting by Effros and Haagerup [EH], who also added C'' to the list.

The reader should beware: This subject is full of traps; see Exercise 18.9 for an illustration!

Before we give the formal definitions, we need to introduce, given a pair of C^*-algebras A, B, the natural inclusion of $A^{**} \otimes B^{**}$ into $(A \otimes_{\min} B)^{**}$, as follows. (See also Exercise 11.6(iii).) Assuming A, B unital for simplicity, we obviously have embeddings

$$A \to (A \otimes_{\min} B)^{**} \quad \text{and} \quad B \to (A \otimes_{\min} B)^{**}$$

(defined by $a \to a \otimes 1$ and $b \to 1 \otimes b$) that extend to a pair of normal representations

$$A^{**} \to (A \otimes_{\min} B)^{**} \quad \text{and} \quad B^{**} \to (A \otimes_{\min} B)^{**}$$

with commuting ranges. The "product" of these representations gives us an $*$-homomorphism

$$J \colon A^{**} \otimes B^{**} \to (A \otimes_{\min} B)^{**}$$

that is clearly injective (since $A^* \otimes B^* \subset (A \otimes_{\min} B)^*$). Perhaps a slightly more concrete description of J emerges from the following observation: For any (a, b) in $A^{**} \times B^{**}$ whenever (a_α) and (b_β) are nets in A, B with $\|a_\alpha\| \leq \|a\|$, $\|b_\alpha\| \leq \|b\|$, tending respectively to a and b with respect to the Mackey topologies $\tau(A^{**}, A^*)$ and $\tau(B^{**}, B^*)$, then $J(a \otimes b)$ is the weak-$*$ limit of $J(a_\alpha \otimes b_\beta)$ in $(A \otimes_{\min} B)^{**}$.

We claim that for any t in $A^{**} \otimes B^{**}$ we have

$$\|t\|_{A^{**} \otimes_{\min} B^{**}} \leq \|J(t)\|_{(A \otimes_{\min} B)^{**}}. \tag{18.2}$$

Indeed, if we assume $\|J(t)\|_{(A \otimes_{\min} B)^{**}} \leq 1$, there is a net t_α in the unit ball of $A \otimes_{\min} B$ tending weak-$*$ to $J(t)$. Let $u_\alpha \colon A^* \to B$ and $u \colon A^* \to B^{**}$ be the linear maps associated, respectively, to t_α and t. Then, for any ξ in A^*, $u_\alpha(\xi) \to u(\xi)$ in the $\sigma(B^{**}, B^*)$ sense. But, by (2.3.2), we have $\|u_\alpha\|_{cb} = \|t_\alpha\|_{\min}$ and $\|u\|_{cb} = \|t\|_{\min}$; hence, we obtain

$$\|t\|_{\min} = \|u\|_{cb} \leq \sup_\alpha \|u_\alpha\|_{cb} \leq \sup_\alpha \|t_\alpha\|_{\min} \leq 1,$$

which establishes our claim (18.2).

Let A be a C^*-algebra. We will say that

(1) A has property C if, for any C^*-algebra B, the norm induced on the algebraic tensor product $A^{**} \otimes B^{**}$ by $(A \otimes_{\min} B)^{**}$ coincides with the minimal norm. In other words, we have an isometric embedding

$$A^{**} \otimes_{\min} B^{**} \subset (A \otimes_{\min} B)^{**}.$$

(2) A has property C' if, for any C^*-algebra B, the norm induced by $(A \otimes_{\min} B)^{**}$ on $A \otimes B^{**}$ coincides with the minimal norm. In other words, we have an isometric embedding

$$A \otimes_{\min} B^{**} \subset (A \otimes_{\min} B)^{**}.$$

(3) A has property C'' if, for any C^*-algebra B, the norm induced by $(A \otimes_{\min} B)^{**}$ on $A^{**} \otimes B$ coincides with the minimal norm. In other words, we have an isometric embedding

$$A^{**} \otimes_{\min} B \subset (A \otimes_{\min} B)^{**}.$$

The implications $C \Rightarrow C'$ and $C \Rightarrow C''$ are obvious by restriction. Conversely, it is not too difficult to verify (see Exercise 18.6) that A has property C iff it has C' and C''.

Clearly these notions (and the preceding observations) remain valid if A is merely an operator space, but the embeddings are now completely isomorphic embeddings and we must introduce the relevant constants. Let $\lambda \geq 1$ be a constant. Then we will say that an operator space X has property C (resp. C', resp. C'') with constant λ if, for any C^*-algebra B, we have a completely isomorphic embedding

$$X^{**} \otimes_{\min} B^{**} \to (X \otimes B)^{**}$$

(resp. $X \otimes_{\min} B^{**} \to (X \otimes_{\min} B)^{**}$, resp. $X^{**} \otimes_{\min} B \to (X \otimes_{\min} B)^{**}$) with c.b. norm majorized by λ. Note that it suffices to majorize the norms by λ, since by changing B to $M_n(B)$ ($n = 1, 2, \ldots$) we recover the c.b. norm.

Moreover, by Gelfand's embedding theorem for C^*-algebras, it clearly suffices in all these properties to consider the case $B = B(H)$ with H Hilbert. We will denote by $C(X)$ (resp. $C'(X)$, rresp. $C''(X)$) the smallest number λ for which X has property C (resp. C', rresp C'') with constant λ. We have clearly

$$C'(X) \leq C(X) \quad \text{and} \quad C''(X) \leq C(X).$$

It is useful to observe that we have trivially

$$C'(X) = \sup\{C'(E) \mid E \subset X \quad \dim E < \infty\}. \tag{18.3}$$

The most interesting constant seems to be the constant $C''(X)$, which is just the local reflexivity constant.

Proposition 18.15. *Let $\lambda \geq 1$ be a constant and let X be an operator space. The following are equivalent.*

(i) X has property C'' with $C''(X) \leq \lambda$.

(ii) X is λ-locally reflexive.

Proof. To show (i) $\Rightarrow$ (ii), we may assume $E \subset B$ for some C^*-algebra B. Note that $X^{**} \otimes_{\min} E \subset X^{**} \otimes_{\min} B$ and $(X \otimes_{\min} E)^{**} \subset (X \otimes_{\min} B)^{**}$ are isometric embeddings, so that, by restricting to $X \otimes_{\min} E$, (i) implies

$\|X^{**} \otimes_{\min} E \to (X \otimes_{\min} E)^{**}\| \leq \lambda$, which is the same as (ii). The proof of the converse is similarly easy. ∎

Theorem 18.16. *Let A be a C^*-algebra.*

(i) If A has property C, then A is exact.

(ii) A is exact iff it has property C'.

(iii) If A is exact, then A is locally reflexive, that is, it has property C''.

(iv) $C' \Leftrightarrow C \Leftrightarrow$ exactness.

Part (i) is due to Archbold and Batty [AB], and (ii) and (iii) are due to Kirchberg.

Part (iv) is then immediate from (ii) and (iii): Indeed, by Exercise 18.6 we know $C \Leftrightarrow (C' \& C'')$, but (ii) and (iii) together show that $C' \Rightarrow C''$; hence C and C' must actually be equivalent and by (ii) C' is equivalent to exactness.

Apparently, no simple direct proof of (iii) is known. It is derived from Theorem 17.18 (or from a previous result of Kirchberg representing separable exact C^*-algebras as quotients of sub- C^*-algebras of nuclear ones; cf. [Ki1, Wa3]). For part (ii), however, a simple argument is available: Marius Junge observed an extension to operator spaces of this result as follows.

Theorem 18.17. *Let X be an operator space. Then X has property C' with constant λ iff*

$$\sup\{d_{S\mathcal{K}}(E) \mid E \subset X, \dim E < \infty\} \leq \lambda. \tag{18.4}$$

In particular, $d_{S\mathcal{K}}(E) = C'(E)$ for any finite-dimensional operator space.

Proof. Assume (18.4). Note that the space M_N and a fortiori any of its subspaces obviously satisfy C' with constant 1. Therefore, any finite-dimensional operator space E satisfies C' with constant $d_{S\mathcal{K}}(E)$. Thus, if (18.4) holds, X satisfies C' with constant λ, because of (18.3).

Conversely, assume that X satisfies C' with constant λ. We will show that X is exact with constant λ. By (17.6) and (18.3) we can assume that X is finite-dimensional. Then let $\mathcal{I}$ be an ideal in a C^*-algebra B. Note that

$$B^{**} \simeq \mathcal{I}^{**} \oplus (B/\mathcal{I})^{**},$$

so that we have an isometric identification

$$(B/\mathcal{I})^{**} \otimes_{\min} X \simeq (B^{**} \otimes_{\min} X)/(\mathcal{I}^{**} \otimes_{\min} X).$$

Now, if X satisfies C' with constant λ, the inclusion $B^{**} \otimes_{\min} X \to (B \otimes_{\min} X)^{**}$ has norm $\leq \lambda$; hence, after passing to the quotient, the inclusion

$$(B^{**} \otimes_{\min} X)/(\mathcal{I}^{**} \otimes_{\min} X) \longmapsto \frac{(B \otimes_{\min} X)^{**}}{(\mathcal{I} \otimes_{\min} X)^{**}}$$

has norm $\leq \lambda$. By the preceding identification, this means that the inclusion

$$(B/\mathcal{I})^{**} \otimes_{\min} X \longmapsto \frac{(B \otimes_{\min} X)^{**}}{(\mathcal{I} \otimes_{\min} X)^{**}} = \left(\frac{B \otimes_{\min} X}{\mathcal{I} \otimes_{\min} X} \right)^{**}$$

has norm $\leq \lambda$. By restriction to $B/\mathcal{I} \otimes_{\min} X$, the inclusion

$$B/\mathcal{I} \otimes_{\min} X \longrightarrow \left(\frac{B \otimes_{\min} X}{\mathcal{I} \otimes_{\min} X} \right)$$

also has norm $\leq \lambda$, but this says that $\mathrm{ex}(X) \leq \lambda$. Whence the desired conclusion. ∎

The following obvious reformulation of C' is useful.

Proposition 18.18. *Let E be a finite-dimensional operator space. Let $\lambda = C'(E)$. Then, for any C^*-algebra B and any $u \colon E^* \to B^{**}$, there is a net $u_i \colon E^* \to B$ with $\sup_{i \in I} \|u_i\|_{cb} \leq \lambda \|u\|_{cb}$ that tends to u pointwise with B^{**} equipped with the $\sigma(B^{**}, B^*)$-topology.*

Proof. Indeed, $CB(E^*, B^{**}) = E \otimes_{\min} B^{**}$ isometrically; hence, if $\lambda = C'(E)$, u defines an element of norm $\leq \lambda \|u\|_{cb}$ in $CB(E^*, B)^{**}$. Therefore there is a net (u_i) as above tending to u in the topology $\sigma(CB(E^*, B)^{**}, CB(E^*, B)^*)$. But it is easy to check that this equivalently means that $u_i \to u$ in the point-$\sigma(B^{**}, B^*)$ topology. ∎

Proposition 18.19. *([AB]) Any nuclear C^*-algebra A is locally reflexive. More generally, any quotient C^*-algebra of a C^*-subalgebra of A is exact (and locally reflexive).*

Proof. By Exercise 18.7 (since A nuclear implies A^{**} injective) A must be locally reflexive (equivalently, it must have (C'')). Moreover, it is obvious that nuclear $\Rightarrow$ exact. Hence (by Theorem 18.17) nuclear implies $(C''$ and $C')$. By Exercise 18.6, this means that nuclearity implies (C). But now (C) passes to subalgebras (obviously) and to quotient C^*-algebras (by Exercise 18.8); hence any quotient of a subalgebra of A has (C) and therefore (Theorems 18.16 and 18.17) is exact. ∎

Remark 18.20. It seems to be a delicate open problem to decide whether the converse of Theorem 18.16(iii) holds. In other words: *Is it true that local reflexivity implies exactness for a C^*-algebra?* Note that for operator spaces this is clearly false, since any reflexive operator space is obviously locally reflexive, and of course it is not exact (take, e.g., OH; see Theorem 21.5).

Nevertheless, we have the following o.s. analog of Theorem 18.16(iii):

Theorem 18.21. *([EOR]) Any 1-exact operator space is 1-locally reflexive.*

Proof. The proof relies on a deep result of Kirchberg [Ki1] that is beyond the scope of this book. The latter result says that any separable 1-nuclear operator space X can be realized (completely isometrically) as a quotient $B/(L + R)$, where B is the CAR C^*-algebra (infinite tensor product of M_2 in the C^*-sense) and L, R are respectively a left and right closed ideal in B. By classical results on one-sided ideals in C^*-algebras, the bidual of $B/(L+R)$ embeds canonically completely isometrically into B^{**}. Using this, it is not hard to show (see Exercise 18.4) that the 1-local reflexivity of B (note that B is a nuclear C^*-algebra) passes to $B/(L+R)$. Thus any 1-nuclear operator space X is 1-locally reflexive. Since we have seen (Theorem 17.23) that any 1-exact separable o.s. embeds in a 1-nuclear one, we conclude (invoking Proposition 18.5) that 1-exact implies 1-locally reflexive. ∎

Note that, as a consequence, X is 1-exact iff it satisfies either (C) or (C′) with constant 1.

Remark. Apparently, all of the known proofs that a nuclear C^*-algebra A or an exact one is locally reflexive (see Exercise 18.7) use rather delicate (and somewhat indirect) arguments. If one could find a direct reasonably simple proof for the local reflexivity of A, one would have a much simpler demonstration that nuclearity and exactness pass to quotients. There is very recent work by S. Wassermann along this line.

Exercises

Exercise 18.1. Let A be a C^*-algebra. Show that A^{**} is injective if A is both WEP and locally reflexive.

Exercise 18.2. Let X be a Banach space. Let $E \subset X^{**}$ be a finite-dimensional subspace, and let $u_\alpha \colon E \to X$ be a net of mappings with $\|u_\alpha\| \le \lambda$ such that, for any e in E, $u_\alpha(e) \to e$ with respect to $\sigma(X^{**}, X^*)$. Show that u_α is injective when α is larger enough and

$$\overline{\lim_\alpha} \, \|u_\alpha^{-1}|_{u_\alpha(E)}\| \le 1.$$

Exercise 18.3. Fix $n \ge 1$. Let E, X be operator spaces. Let $V \colon M_n(E) \to M_n(X)$ be a bounded map. Let G be the (finite) group of all unitary matrices such that each row and each column has exactly one nonzero entry equal to ± 1. We define

$$\forall\, x \in M_n(E) \qquad \widetilde{V}(x) = \frac{1}{|G|^2} \sum_{g,h \in G} g^{-1} \cdot V(gxh) \cdot h^{-1}.$$

Show that we can write

$$\widetilde{V} = I \otimes u$$

for some u: $E \to X$ with

$$\|u\|_n \leq \|V\|.$$

(Note: Actually the same result holds if G is the whole unitary group.)

Exercise 18.4. Let X be a λ-locally reflexive operator space with a closed subspace $Y \subset X$. Let q: $X \to X/Y$ be the quotient map. Assume that there is a completely contractive map

$$r: (X/Y)^{**} \to X^{**}$$

such that $q^{**}r$ is the identity on $(X/Y)^{**}$. Show that X/Y is λ-locally reflexive. In particular, any quotient of a C^*-algebra by a (two-sided, closed) ideal is locally reflexive if it is the case for the C^*-algebra ([AB]).

Exercise 18.5. ([AB]) Show that $C^*(\mathbb{F}_2)$ and $C^*(\mathbb{F}_\infty)$ are not locally reflexive.

Exercise 18.6. ([EH]) Show that for a C^*-algebra

$$(C) \Leftrightarrow (C' \ \& \ C'').$$

More generally, if an operator space X satisfies C' with constant λ' and C'' with constant λ'', then it has property C with constant $\lambda'\lambda''$.

Exercise 18.7. Let A be a C^*-algebra. Show that if A^{**} is injective, then A is locally reflexive (hence nuclear implies locally reflexive).

Hint: Use the fact that A^{**} is semi-discrete, that is, there are nets of complete contractions v_α: $A^{**} \to M_{n(\alpha)}$ and w_α: $M_{n(\alpha)} \to A^{**}$ such that $w_\alpha v_\alpha \to I_{A^{**}}$ pointwise in the $\sigma(A^{**}, A^*)$-sense.

Exercise 18.8. Show that if a C^*-algebra A has property (C), then any quotient C^*-algebra A/I also has (C).

Exercise 18.9. Find what is wrong in the following FALSE ARGUMENT: Let $\mathcal{K} = K(\ell_2)$ and $\mathcal{B} = B(\ell_2)$, and let A be any C^*-algebra. Since $\mathcal{K} \subset \mathcal{B}$, we have $(\mathcal{K} \otimes_{\min} A)^{**} \subset (\mathcal{B} \otimes_{\min} A)^{**}$ (isometrically), but, on the other hand (see Exercise 5.6), $(\mathcal{K} \otimes_{\min} A)^{**} \simeq \mathcal{B} \overline{\otimes} A^{**}$; hence we have $\|\mathcal{B}\overline{\otimes}A^{**} \to (\mathcal{B} \otimes_{\min} A)^{**}\| \leq 1$ and restricting to $\mathcal{B} \otimes A^{**}$ we obtain $\|\mathcal{B} \otimes_{\min} A^{**} \to (\mathcal{B} \otimes_{\min} A)^{**}\| \leq 1$, which "shows" that A is locally reflexive.

Chapter 19. Grothendieck's Theorem for Operator Spaces

We first recall the noncommutative version of Grothendieck's theorem (in short, GT) due to U. Haagerup and the author (see [P4] for details). Let A, B be C^*-algebras. Then any bounded linear map $u \colon A \to B^*$ satisfies the following: For any finite sequences (a_i) in A and (b_i) in B

$$\left| \sum \langle u(a_i), b_i \rangle \right| \leq$$

$$K \|u\| \max \left\{ \left\| \sum a_i^* a_i \right\|^{1/2}, \left\| \sum a_i a_i^* \right\|^{1/2} \right\} \max \left\{ \left\| \sum b_i^* b_i \right\|^{1/2}, \left\| \sum b_i b_i^* \right\|^{1/2} \right\},$$

where K is a numerical constant independent of u.

When A, B are commutative C^*-algebras, this is a classical result due to Grothendieck. The best constant K in this case is called the (complex) Grothendieck constant and is denoted K_G. It is known that $1.338 \leq K_G \leq 1.405$. See [Kö] for the latest information on this. The operator space version of GT, which we prove below, applies to completely bounded maps; hence we assume more than in the classical version of GT, but on the other hand it applies to all mappings $u \colon E \to F^*$ where E and F are *exact* operator spaces. Thus we seem to be requiring much less structure on the domain and range of u.

For convenience, we will use the following.

Notation. Let E be an operator space. Then, for any finite sequence $(x_i)_{i \leq n}$ in E, we set

$$\|(x_i)\|_{RC} = \max \left\{ \left\| \sum e_{1i} \otimes x_i \right\|_{M_n(E)}, \left\| \sum e_{i1} \otimes x_i \right\|_{M_n(E)} \right\}.$$

Equivalently (see Remark 1.13), if $E \subset B(H)$, then

$$\|(x_i)\|_{RC} = \max \left\{ \left\| \sum x_i^* x_i \right\|^{1/2}, \left\| \sum x_i x_i^* \right\|^{1/2} \right\}.$$

Theorem 19.1. *Let E, F be exact operator spaces. Let $C = d_{SK}(E) d_{SK}(F)$. Then any c.b. map $u \colon E \to F^*$ satisfies the following inequality. For any finite sequences (a_i) in E and (b_i) in F we have*

$$\left| \sum \langle u(a_i), b_i \rangle \right| \leq 4C \|u\|_{cb} \|(a_i)\|_{RC} \|(b_i)\|_{RC}. \tag{19.1}$$

More precisely, if we denote for simplicity $X = C_\lambda^*(\mathbb{F}_\infty)$, we have

$$\left| \sum \langle u(a_i), b_i \rangle \right| \leq C \|u\|_{cb} \left\| \sum a_i \otimes \lambda(g_i) \right\|_{E \otimes_{\min} X} \left\| \sum b_i \otimes \lambda(g_i) \right\|_{F \otimes_{\min} X}. \tag{19.1$'$}$$

Recall that, given a map $v \colon Y \to Z$ between Banach spaces, we denote by $\gamma_2(v)$ its norm of factorization through a Hilbert space, that is,

$$\gamma_2(v) = \inf\{\|v_1\| \, \|v_2\|\},$$

where the infimum runs over all possible Hilbert spaces H and all factorizations of v of the form $Y \xrightarrow{v_1} H \xrightarrow{v_2} Z$.

Corollary 19.2. *In the situation of the preceding theorem, let A, B be C^*-algebras with completely isometric embeddings $E \subset A$ and $F \subset B$. Then there are states f_1, g_1 on A, f_2, g_2 on B, and $0 \le \theta_1, \theta_2 \le 1$ such that for any (a, b) in $E \times F$*

$$|\langle u(a), b \rangle| \le 4C\|u\|_{cb}[\theta_1 f_1(a^* a) + (1 - \theta_1)g_1(aa^*)]^{1/2}$$

$$\cdot [\theta_2 f_2(b^* b) + (1 - \theta_2)g_2(bb^*)]^{1/2}. \tag{19.2}$$

Consequently, there is a bounded linear map $\widetilde{u} \colon A \to B^*$ with $\|\widetilde{u}\| \le \gamma_2(\widetilde{u}) \le 4C\|u\|_{cb}$ that extends u in the sense that, if we view u and $\widetilde{u}$ as bilinear forms on $E \times F$ and $A \times B$, respectively, then $\widetilde{u}$ extends u. A fortiori, we have $\gamma_2(u) \le 4C\|u\|_{cb}$.

Proof. The proof of the first assertion is entirely analogous to the solution to Exercise 2.2.2, to which we refer the reader. One should simply note that

$$\|(a_i)\|_{RC} = \sup \left\{ \theta f\left(\sum a_i^* a_i\right) + (1 - \theta)g\left(\sum a_i a_i^*\right) \right\}^{1/2},$$

where the supremum runs over all states f, g on A and all $0 \le \theta \le 1$.

The second assertion is proved as follows. Equation (19.2) allows us to write

$$|\langle u(a), b \rangle| \le 4C\|u\|_{cb}(\langle a, a \rangle_1 \langle b, b \rangle_2)^{1/2}, \tag{19.3}$$

where $\langle \ , \ \rangle_1$ and $\langle \ , \ \rangle_2$ are scalar products on A and B respectively such that

$$\forall (a, b) \in A \times B \qquad \langle a, a \rangle_1 \le \|a\|^2 \quad \text{and} \quad \langle b, b \rangle_2 \le \|b\|^2.$$

If we denote by H_1 and H_2 the Hilbert spaces obtained after passing to the quotient and completing, we have contractive inclusions

$$J_1 \colon A \to H_1 \quad \text{and} \quad J_2 \colon B \to H_2,$$

so that we deduce from (19.3) a factorization of the form

$$\forall (a, b) \in E \times F \qquad \langle u(a), b \rangle = \langle J_2^* T J_1(a), b \rangle,$$

where $T \colon H_1 \to H_2^*$ is an operator such that $\|T\| \le 4C\|u\|_{cb}$. Thus, the mapping

$$\widetilde{u} = J_2^* T J_1$$

is an "extension" of u (in the sense of Corollary 19.2) satisfying

$$\gamma_2(\widetilde{u}) \le \|J_2^*\| \, \|T J_1\| \le 4C\|u\|_{cb}. \qquad \blacksquare$$

The preceding statement explains why GT is often described as a factorization theorem.

Actually, we will prove the following slightly more "abstract" result:

Generalized Theorem 19.1. *Let E, F be exact operator spaces and let $C = d_{SK}(E)d_{SK}(F)$. Let A_1, A_2 be C^*-algebras. Assume that either A_1 or A_2 is QWEP (i.e., is a quotient of a C^*-algebra with the WEP). Then any c.b. map $u\colon E \to F^*$ satisfies, for any finite sequences (a_i) in E, (b_j) in F, (x_i) in A_1, and (y_j) in A_2, the following inequality:*

$$\left\|\sum \langle u(a_i), b_j \rangle x_i \otimes y_j \right\|_{\max} \le C\|u\|_{cb} \left\|\sum a_i \otimes x_i \right\|_{\min} \left\|\sum b_j \otimes y_j \right\|_{\min}. \tag{19.1''}$$

To prove Theorem 19.1, the key ingredient will be the embedding of the von Neumann algebra of the free group into an ultraproduct described in §9.10.

In addition, we will use the following fact.

Lemma 19.3. *Let $C = d_{SK}(E)d_{SK}(F)$ as before. Let A_1, A_2 be C^*-algebras and assume $A_i = B_i/I_i$, where B_i are C^*-algebras and $I_i \subset B_i$ closed two-sided ideals. Let $q_i\colon B_i \to A_i$ be the quotient map. Let $\varphi \in (A_1 \otimes A_2)^*$ be a linear form such that*

$$\|\varphi(q_1 \otimes q_2)\|_{(B_1 \otimes_{\min} B_2)^*} \le 1.$$

Then, for any finite sequences $(a_i), (b_i), (x_i), (y_j)$ with $(a_i, b_j) \in E \times F$ and $(x_i, y_j) \in A_1 \times A_2$ and any linear map $u\colon E \to F^*$, we have

$$\left|\sum \langle u(a_i), b_j \rangle \varphi(x_i \otimes y_j)\right| \le C\|u\|_{cb} \left\|\sum a_i \otimes x_i \right\|_{\min} \left\|\sum b_j \otimes y_j \right\|_{\min}. \tag{19.4}$$

Remark. This result will be applied to a form φ that is unbounded on $A_1 \otimes_{\min} A_2$, so we really need to consider $\varphi(q_1 \otimes q_2)$ instead of φ.

Remark 19.4. If $q_1 \otimes q_2$ maps (contractively) $B_1 \otimes_{\min} B_2$ into $A_1 \otimes_{\max} A_2$, then (19.4) holds for all φ such that $\|\varphi\|_{(A_1 \otimes_{\max} A_2)^*} \le 1$, since this implies a fortiori

$$\|\varphi(q_1 \otimes q_2)\|_{(B_1 \otimes_{\min} B_2)^*} \le 1.$$

We will use the following simple fact (for a proof, see the solution to Exercise 19.1).

Lemma 19.5. *Let E, F, G be operator spaces. Consider a linear map $u\colon E \to CB(F, G)$. Then, for any C^*-algebras B_1, B_2, u defines a bilinear form*

$$\widehat{u}\colon E \otimes_{\min} B_1 \times F \otimes_{\min} B_2 \to G \otimes_{\min} B_1 \otimes_{\min} B_2$$

satisfying $\|\widehat{u}\| \le \|u\|_{cb}$ and

$$\widehat{u}(a \otimes x, b \otimes y) = u(a)(b) \otimes x \otimes y.$$

Proof of Lemma 19.3. We may clearly assume, without loss of generality, that E and F are finite-dimensional. We first assume $E \subset M_n$ and $F \subset M_m$. Let $x = \sum a_i \otimes x_i$ and $y = \sum b_j \otimes y_j$. Assume $\|x\|_{\min} < 1$ and $\|y\|_{\min} < 1$. Then, by Lemma 17.2, there are $\widehat{x}$ in $E \otimes_{\min} B_1$ and $\widehat{y}$ in $F \otimes_{\min} B_2$ such that $(I \otimes q_1)(\widehat{x}) = x$ and $(I \otimes q_2)(\widehat{y}) = y$, with $\|\widehat{x}\|_{\min} < 1$, $\|\widehat{y}\|_{\min} < 1$. Indeed, this is clear when $E = M_n$ and $F = M_m$, but Lemma 17.2 ensures that it remains automatically true for subspaces. We may clearly assume $\widehat{x} = \sum a_i \otimes \widehat{x}_i$, $\widehat{y} = \sum b_j \otimes \widehat{y}_j$ with $q_1(\widehat{x}_i) = x_i$ and $q_2(\widehat{y}_j) = y_j$. Then, with the notation of Lemma 19.5 (taking $G = \mathbb{C}$), we have

$$\widehat{u}(\widehat{x}, \widehat{y}) = \sum_{i,j} \langle u(a_i), b_j \rangle \widehat{x}_i \otimes \widehat{y}_j.$$

Hence, applying $\varphi \circ (q_1 \otimes q_2)$ to this, we obtain

$$\left| \sum_{i,j} \langle u(a_i), b_j \rangle \varphi(x_i \otimes y_j) \right| = |\varphi \circ (q_1 \otimes q_2)(\widehat{u}(\widehat{x}, \widehat{y}))|$$
$$\leq \|\varphi \circ (q_1 \otimes q_2)\|_{(B_1 \otimes_{\min} B_2)^*} \|\widehat{u}\| \, \|\widehat{x}\| \, \|\widehat{y}\|$$
$$\leq \|u\|_{cb}.$$

Thus we obtain the desired inequality when E, F are matricial spaces. The general case is then easy to prove using isomorphisms $v \colon E \to \widehat{E} \subset M_n$ and $w \colon F \to \widehat{F} \subset M_m$. Replacing $\widehat{E}$ by E and $\widehat{F}$ by F produces an extra factor equal to $\|v\|_{cb} \|v^{-1}\|_{cb} \|w\|_{cb} \|w^{-1}\|_{cb}$, whence the presence of the constant $C = d_{SK}(E) d_{SK}(F)$ in (19.4). $\blacksquare$

Proof of Theorem 19.1. We use the construction described in §9.10. We can take for (x_i) either $(\lambda(g_i))$ (with (g_i) the generators of F_∞) or a free semi-circular (or circular) sequence and we let $y_j = \overline{x_j}$. In both cases, we have (see §9.7 and §9.9)

$$\left\| \sum a_i \otimes x_i \right\| \leq 2 \|(a_i)\|_{RC} \quad \text{and} \quad \left\| \sum b_j \otimes y_j \right\| \leq 2 \|(b_j)\|_{RC} \qquad (19.5)$$

when a_i, b_j are elements of an arbitrary operator space.

Let M be the von Neumann algebra generated by (x_i). By §9.10, there is a family of (finite-dimensional) matrix algebras $M(n)$, a free ultrafilter $\mathcal{U}$, and an ideal $I_\mathcal{U}$ such that M embeds into $B/I_\mathcal{U}$ where $B = \ell_\infty(\{M(n) \mid n \geq 1\})$; and moreover, if $q \colon B \to B/I_\mathcal{U}$ denotes the quotient map and τ (resp. τ_n) the normalized trace on M (resp. $M(n)$), we have for any $t = (t_n)$ in B

$$\tau(q(t)) = \lim_\mathcal{U} \tau_n(t_n). \qquad (19.6)$$

We identify M with a subalgebra of $B/I_{\mathcal{U}}$ as described in §9.10. We will apply Lemma 19.3 with $y_i = \overline{x_i}$, (x_i) being as above with $B_1 = B$, $B_2 = \overline{B}$, $q_1 = q, q_2 = \overline{q}$, and with $\varphi \colon B/I_{\mathcal{U}} \otimes \overline{B/I_{\mathcal{U}}} \to \mathbb{C}$ given by

$$\varphi(s \otimes \overline{t}) = \lim_{\mathcal{U}} \varphi_n(s_n, \overline{t_n}),$$

where

$$\varphi_n(s_n, \overline{t_n}) = \tau_n(s_n t_n^*).$$

Note that, since the $M(n)$ are finite-dimensional matrix algebras, it is clear (see, e.g., Proposition 2.9.1) that $\|\varphi_n\|_{(M(n) \otimes_{\min} \overline{M(n)})^*} \leq 1$ for any n, and hence $\|\varphi\|_{(B \otimes_{\min} \overline{B})^*} \leq 1$. Then, recalling (19.5) and noting that (19.6) implies $\varphi(x_i \otimes \overline{x_j}) = \tau(x_i x_j^*) = \delta_{ij}$, we can finally deduce (19.1) and (19.1)' from (19.4) and (19.5). ∎

Proof of Generalized Theorem 19.1. We will use Lemma 19.3. Assume, say, that A_1 is QWEP. Let B_1 be WEP such that $B_1/I_1 \simeq A_1$, and let $B_2 = C^*(G)$ with G a suitable free group so that $A_2 \simeq B_2/I_2$ (see Exercise 8.1). By Kirchberg's Theorem (see Theorem 15.5 and the remark after it), we know that $B_1 \otimes_{\min} B_2 = B_1 \otimes_{\max} B_2$, and hence a fortiori

$$\|q_1 \otimes q_2 \colon B_1 \otimes_{\min} B_2 \to A_1 \otimes_{\max} A_2\| \leq 1.$$

Thus Remark 19.4 applies in this case and gives us (19.1)''. ∎

Lemma 19.6. *In the situation of Lemma 19.3, assume that $A_1 = A_2 = A$, where A is a quotient of a WEP (in short QWEP) C^*-algebra. Then, for any tracial state $\psi \colon A \to \mathbb{C}$, $\forall\, a_i \in E$, $\forall\, b_j \in F$, $\forall\, x_i \in A$, $\forall\, y_j \in A$, we have*

$$\left| \sum \langle u(a_i), b_j \rangle \psi(x_i y_j) \right|$$

$$\leq d_{SK}(E) d_{SK}(F) \|u\|_{cb} \left\| \sum a_i \otimes x_i \right\|_{\min} \left\| \sum b_j \otimes y_j \right\|_{\min}.$$

Proof. We may apply (19.1)'' with $A_1 = A$ and $A_2 = A^{op}$.

Now, since ψ is a *tracial* state, we can write $\psi(xy) = \langle \pi_1(x)\pi_2(y)\xi, \xi \rangle$, where π_1, π_2 are *commuting* representations of A and A^{op} associated to the GNS construction relative to ψ and $\|\xi\| = 1$. Thus we find that $x \otimes y \to \psi(xy)$ defines an element in the unit ball of $(A \otimes_{\max} A^{op})^*$, so that (19.1)'' gives us the conclusion. ∎

Remark. Note that the von Neumann algebra M appearing in the proof of Theorem 19.1 is QWEP by Exercise 15.4. This explains our terminology "Generalized Theorem 19.1."

As we observed earlier, the spaces R, C and their direct sum $R \oplus C$ are the only known examples of infinite-dimensional separable operator spaces E that are exact as well as their duals. It is a natural question to ask whether they are indeed the only ones. An affirmative answer was very recently given in [PiS]. The next statement is a first step in this direction.

Corollary 19.7. *If an operator space E is exact as well as its dual E^*, then E must be isomorphic to a Hilbert space. More precisely, we have*

$$\gamma_2(I_E) \leq 4d_{SK}(E)d_{SK}(E^*).$$

Proof. This is a immediate consequence of Corollary 19.2. ∎

Remark 19.8. In Lemma 19.3, the exactness of E (or F) is used only to lift elements of $E \otimes_{\min} (B_1/I_1)$ up into $E \otimes_{\min} B_1$. Therefore, if $I_1 = \{0\}$ (resp. if both $I_1 = I_2 = \{0\}$), then Lemma 19.3 remains valid when E (resp. when each space E or F) is an arbitrary operator space, and (19.4) is valid with $C = d_{SK}(F)$ (resp. with $C = 1$). In particular, we find:

Corollary 19.9. *Let E, F be arbitrary operator spaces. Let $\{c_n \mid n \geq 1\}$ be a system satisfying the CAR (see §9.3). Then any c.b. map $u: E \to F^*$ satisfies for any finite sequences (a_i) in E and (b_i) in F*

$$\left| \sum_1^n \langle u(a_i), b_i \rangle \right| \leq 2\|u\|_{cb} \left\| \sum_1^n a_i \otimes c_i \right\|_{\min} \left\| \sum_1^n b_i \otimes c_i \right\|_{\min}. \qquad (19.7)$$

Proof. Recall that $\{c_1, \ldots, c_n\}$ can be realized in a finite-dimensional C^*-algebra $M(n)$, with normalized trace τ_n. Since $c_i c_j^* + c_j^* c_i = \delta_{ij} I$, the system $\{c_i/2^{1/2} \mid i \leq n\}$ is orthonormal with respect to ψ_n. Hence we may apply the preceding Remark 19.8 with $A_1 = B_1 = M(n)$, $A_2 = B_2 = \overline{M(n)}$, $\varphi(x \otimes \overline{y}) = \tau_n(xy^*)$, and $x_i = c_i$, $y_j = \overline{c_j}$. Then (19.4) is valid with $C = 1$, so that we obtain (19.7). ∎

Corollary 19.10. *Let E and F be minimal operator spaces. Then any c.b. map $u: E \to F^*$ satisfies for all finite sequences (a_i) in E and (b_i) in F*

$$\left| \sum \langle u(a_i), b_i \rangle \right| \leq 2\|u\|_{cb} \sup_{\xi \in B_{E^*}} \left(\sum |\xi(a_i)|^2 \right)^{1/2} \sup_{\eta \in B_{F^*}} \left(\sum |\eta(b_i)|^2 \right)^{1/2}.$$

Proof. This is an immediate consequence of the preceding corollary and the fact that if $E = \min(E)$, we have by (9.3.1)

$$\left\| \sum a_i \otimes c_i \right\|_{\min} = \sup_{\xi \in B_{E^*}} \left\| \sum \xi(a_i)c_i \right\| \leq \sup_{\xi \in B_{E^*}} \left(\sum |\xi(a_i)|^2 \right)^{1/2},$$

and similarly for $\sum b_i \otimes c_i$ if $F = \min(F)$. ∎

Remark. The preceding proof works equally well if we use a spin system instead of the system (c_n).

Let E, F be Banach spaces. Consider a linear map $u: E \to F^*$. We denote by $\gamma_2^*(u)$ the smallest constant C such that, for any finite sequences (a_i) in E and (b_i) in F, we have

$$\left| \sum \langle u(a_i), b_i \rangle \right| \leq C \sup_{\xi \in B_{E^*}} \left(\sum |\xi(a_i)|^2 \right)^{1/2} \sup_{\eta \in B_{F^*}} \left(\sum |\eta(b_i)|^2 \right)^{1/2}.$$

As the notation indicates, this norm is dual to the γ_2-norm (introduced before Corollary 19.2) in the following sense. For any $v \in E \otimes F$, let us denote by $\gamma_2(v)$ the γ_2-norm of the linear map $\widetilde{v}: E^* \to F$ determined by v. It is easy to see that $\gamma_2(v) < 1$ iff v can be written as $v = \sum a_i \otimes b_i$ with

$$\sup_{\xi \in B_{E^*}} \left(\sum |\xi(a_i)|^2 \right)^{1/2} \sup_{\eta \in B_{F^*}} \left(\sum |\eta(b_i)|^2 \right)^{1/2} < 1.$$

Therefore we have a duality formula:

$$\gamma_2^*(u) = \sup\{ |\langle u, v \rangle| \mid v \in E \otimes F, \ \gamma_2(v) \leq 1 \}. \tag{19.8}$$

Note that, by Exercise 2.2.2, $\gamma_2^*(u) \leq 1$ iff there are probability measures λ and μ on $(B_{E^*}, \sigma(E^*, E))$ and $(B_{F^*}, \sigma(F^*, F))$ such that

$$\forall \, (a, b) \in E \times F \qquad |\langle u(a), b \rangle| \leq \left(\int |\xi(a)|^2 d\lambda(\xi) \right)^{1/2} \cdot \left(\int |\eta(b)|^2 d\mu(\eta) \right)^{1/2}.$$

$$\tag{19.9}$$

The duality between γ_2 and γ_2^* originally goes back to Grothendieck [Gr]. The norm γ_2^* is now fairly well understood in Banach space theory, and (19.9) can be reinterpreted in terms of 2-absolutely summing operators (see [P4] for more on this).

Thus the next result gives (at least in a special situation) a meaningful equivalent of the c.b. norm (due to V. Paulsen and the author on one hand, see [Pa5], and also independently to M. Junge).

Theorem 19.11. *Let E, F be minimal operator spaces. Then, for any c.b. map $u: E \to F^*$, we have*

$$2^{-1} \gamma_2^*(u) \leq \|u\|_{cb} \leq \gamma_2^*(u). \tag{19.10}$$

Proof. By (19.8) (and the remarks preceding it), Corollary 19.10 implies

$$\gamma_2^*(u) \leq 2\|u\|_{cb},$$

whence the left side of (19.10)). As for the right side, it simply follows from the observation that, for any v in $E \otimes F$ (here E, F can be arbitrary operator spaces), we have

$$\gamma_2(v) \leq \|v\|_{E \otimes^\wedge F}. \tag{19.11}$$

Indeed, assume $\|v\|_{E\otimes^\wedge F} < 1$. Then the map $\tilde{v}\colon E^* \to F$ associated to v admits a factorization of the form:

$$E^* \xrightarrow{\alpha} M_n \xrightarrow{L_a} S_2^n \xrightarrow{R_b} S_1^n \xrightarrow{\beta} F,$$

where $\|\alpha\|_{cb} < 1$, $\|\beta\|_{cb} < 1$ and where $L_a(x) = ax$, $R_b(x) = xb$ with $\|a\|_2 \cdot \|b\|_2 < 1$. Then, if we let $v_2 = L_a\alpha$ and $v_1 = \beta R_b$, we obtain $v = v_1 v_2$ and $\|v_1\| \cdot \|v_2\| < 1$; hence $\gamma_2(v) < 1$, which proves (19.11). By duality, using (19.8) and Chapter 4, (19.11) implies $\|u\|_{cb} \le \gamma_2^*(u)$. ∎

Remark. Let E be again minimal, let G be a maximal operator space, let $u\colon E \to G$ be a linear map, and let $i_G\colon G \to G^{**}$ be the canonical inclusion. We then have

$$2^{-1}\gamma_2^*(i_G u) \le \|u\|_{cb} \le \gamma_2^*(i_G u).$$

Indeed, by Exercise 3.2, G^* is a minimal operator space, so this follows from the preceding result.

Remark. Note that the preceding statement gives a satisfactory description of $CB(E, F^*)$ as a *Banach space*, but its operator space structure remains unclear, in particular, the following is open.

Problem. Let E, F be minimal operator spaces (take for instance $E = F = c_0$). Is $E^* \otimes_{\min} F^*$ completely isomorphic to the symmetrized Haagerup tensor product $E^* \otimes_\mu F^*$? Actually, this might even be true whenever E, F are exact. The same question arises with two maximal operator spaces instead of (E^*, F^*).

In addition, there is evidence that the preceding question may have a positive answer when E, F is any pair of C^*-algebras. A quite similar question is already raised in [B3] and at the end of [ER3].

Some very recent progress was made in [PiS]: It is proved there that if E, F are exact operator spaces, or if E, F are both C^*-algebras, it is indeed true that $E^* \otimes_{\min} F^* \simeq E^* \otimes_\mu F^*$ with equivalent norms. In particular, in the case $E = F = M_n$, it is proved in [PiS] that the norms of the identity maps

$$M_n^* \otimes_{\min} M_n^* \to M_n^* \otimes_\mu M_n^*$$

are bounded uniformly over n. Unfortunately it remains unclear at the time of this writing whether their c.b. norms are also uniformly bounded.

Exercises

Exercise 19.1. Prove Lemma 19.5.

Exercise 19.2. Show that

$$d_{SK}(\max(\ell_2^n)) \ge \sqrt{n}/4.$$

(Actually, this holds with any n-dimensional normed space in place of ℓ_2^n; see [JP].)

Chapter 20. Estimating the Norms of Sums of Unitaries: Ramanujan Graphs, Property T, Random Matrices

In this chapter, we estimate the growth of a specific sequence of numbers $C(n)$ that are closely related to the analysis of "asymptotic freeness" for sequences of n-tuples of $(N \times N)$ matrices with a common size N tending to infinity.

More precisely, for each $n \geq 1$, we define $C(n)$ as the infimum of the numbers C for which there exist integers $\{N_m \mid m \in \mathbb{N}\}$ and an *infinite* sequence of n-tuples of $N_m \times N_m$ unitary matrices $\{(u_i(m))_{1 \leq i \leq n} \mid m \in \mathbb{N}\}$ such that

$$\sup_{m \neq m'} \left\{ \left\| \sum_{i=1}^n u_i(m) \otimes \overline{u_i(m')} \right\|_{\min} \right\} \leq C.$$

The last norm is meant in $M_{N_m} \otimes_{\min} M_{N_{m'}}$, or, equivalently, in the (operator) norm of the space $M_{N_m \times N_{m'}}$ of all matrices of size $N_m N_{m'} \times N_m N_{m'}$. By the triangle inequality, we have $\left\| \sum_1^n u_i \otimes \overline{v_i} \right\|_{\min} \leq n$ whenever u_i, v_i are all unitary; hence we have the trivial bound:

$$C(n) \leq n,$$

which, as we will see, is far from the true value of $C(n)$ when n is large. However, we have $C(2) = 2$. This follows from Corollary 20.2 (the reader is invited to find a direct proof, as an exercise).

We will first estimate $C(n)$ from below. We start with a result from [P14]. The alternate proof that we give here is due to Szarek.

Theorem 20.1. *Let $u_1, \ldots, u_n$ be arbitrary unitary operators in $B(H)$ (H any Hilbert space). Then*

$$2\sqrt{n-1} \leq \left\| \sum_{i=1}^n u_i \otimes \overline{u_i} \right\|_{\min}.$$

Remark. If $\dim H < \infty$, the triangle inequality $\left\| \sum_1^n u_i \otimes \overline{u_i} \right\|_{\min} \leq n$ cannot be improved, and we have

$$\left\| \sum u_i \otimes \overline{u_i} \right\| = n. \tag{20.1}$$

Indeed, $t = I_H$ is an eigenvector for $t \to \sum u_i t u_i^*$ associated to the eigenvalue n.

More generally, it is easy to see that (20.1) still holds when $\dim H = \infty$ if $u_1, \ldots, u_n$ all belong to a finite injective von Neumann subalgebra $M \subset B(H)$. However, (20.1) is not true if we drop the injectivity assumption, as shown when M is the von Neumann algebra (factor actually) associated to the free group F_n on n generators. We first recall that $\lambda: G \to B(\ell_2(G))$ denotes the left regular representation that takes an element g in G to the unitary operator of left translation by g. Now, in the particular case $G = F_n$, let $g_1, \ldots, g_n$ be the generators of F_n. Then it is known that

$$\left\| \sum_1^n \lambda(g_i) \otimes \overline{\lambda(g_i)} \right\| = 2\sqrt{n-1} = \left\| \sum_1^n \lambda(g_i) \right\|. \qquad (20.2)$$

Indeed, by Fell's absorption principle (see Proposition 8.1), the left-hand side is the same as $\left\| \sum_1^n \lambda(g_i) \right\|$, and the latter norm was computed in [AO] and found to be equal to the middle term of (20.2) (see the following remark). The results of [AO] were partly motivated by Kesten's thesis [K], where it is proved that

$$\left\| \sum_1^n \lambda(g_i) + \lambda(g_i)^* \right\| = 2\sqrt{2n-1} \qquad (20.3)$$

and also that (20.3) realizes the minimum of all norms $\left\| \sum_{t \in S} \lambda(t) \right\|$ when S runs over all possible symmetric subsets of cardinality $2n$ of any discrete group G. Theorem 20.1 can be viewed as an abstract version of Kesten's lower bound.

Remark. We will not include the proof that $\left\| \sum_1^n \lambda(g_i) \right\| = 2\sqrt{n-1}$. Note, however, that $\left\| \sum_1^n \lambda(g_i) \right\| \geq \sqrt{n}$ is obvious and $\left\| \sum_1^n \lambda(g_i) \right\| \leq 2\sqrt{n}$ follows from (9.7.1). Thus, we have at least proved that

$$\sqrt{n} \leq \left\| \sum_1^n \lambda(g_i) \right\| \leq 2\sqrt{n}.$$

Proof of Theorem 20.1. Let $C = \{t \in S_2 \mid t \geq 0 \quad \|t\|_2 = 1\}$. We claim that, for any u_i in $B(H)$,

$$\left\| \sum u_i \otimes \overline{u_i} \right\| = \sup \left\{ \mathrm{tr}\left(\sum u_i t u_i^* s \right) \mid t, s \in C \right\}. \qquad (20.4)$$

To prove this, first note that, for any t, s in C,

$$\mathrm{tr}(u_i t u_i^* s) = \mathrm{tr}(s^{1/2} u_i t u_i^* s^{1/2}) \geq 0. \qquad (20.5)$$

By definition, if $T = \sum_1^n u_i \otimes \overline{u_i}$, we have

$$\|T\|_{\min} = \sup \left\{ \left| \mathrm{tr}\left(\sum u_i z u_i^* y \right) \right| \mid y, z \in S_2, \|y\|_2 \leq 1, \|z\|_2 \leq 1 \right\}.$$

Moreover, every z in S_2 can be written as $z_1 - z_2 + i(z_3 - z_4)$ with $z_1, \ldots, z_4$ all ≥ 0 such that $\sum_1^4 \|z_j\|_2^2 = \|z\|_2^2$. From this fact and the preceding observation it is easy to check our claim (20.4).

Let $T = \sum_1^n u_i \otimes \overline{u_i}$ and let $S = \sum_{i=1}^n \lambda(g_i)$. The idea of the proof is to show that, for any integer $m \geq 1$ and any t in C, we have

$$\langle (T^*T)^m t, t \rangle \geq \langle (S^*S)^m \delta_e, \delta_e \rangle, \qquad (20.6)$$

where δ_e denotes the basis vector in $\ell_2(\mathbb{F}_n)$ indexed by the unit element of $\mathbb{F}_n$. Note that the normalized trace τ in $VN(\mathbb{F}_n)$ is given by the formula

$$\forall x \in VN(F_n) \qquad \tau(x) = \langle x\delta_e, \delta_e \rangle.$$

To verify (20.6), note that we can expand $(T^*T)^m$ as a sum of the form $\sum_{\alpha \in I} u^\alpha \otimes \overline{u^\alpha}$, where the u^α are unitaries of the form $u_{i_1}^* u_{j_1} u_{i_2}^* u_{j_2} \cdots$.

Now, for certain α, we have $u_\alpha = I$ by formal cancellation (no matter what the u_i are). Let us denote by $I' \subset I$ the set of all such α. Then by (20.5) we have for all t in C

$$\langle (T^*T)^m t, t \rangle = \sum_{\alpha \in I} \text{tr}(u^\alpha t u^{\alpha*} t) \geq \sum_{\alpha \in I'} 1 = \text{card}(I'),$$

but by an elementary counting argument we have

$$\text{card}(I') = \langle (S^*S)^m \delta_e, \delta_e \rangle = \tau((S^*S)^m).$$

Hence we obtain (20.6). Therefore

$$\|T^*T\| \geq \lim_{m \to \infty} \langle (T^*T)^m t, t \rangle^{1/m} \geq \lim_{m \to \infty} (\tau((S^*S)^m))^{1/m} = \|S^*S\|,$$

so that we obtain $\|T\| \geq \|S\|$, whence Theorem 20.1 by (20.2). See Exercise 20.1 for another proof. ∎

Corollary 20.2. $2\sqrt{n-1} \leq C(n)$ *for all* $n \geq 1$.

Proof. Let $(u_i^m)_{i \leq n}$ be a sequence of n-tuples with $(u_1^m, \ldots, u_n^m)$ unitary in the space M_{N_m} of all $N_m \times N_m$ complex matrices. Let A be the space formed of all families $x = (x_m)_{m \in \mathbb{N}}$ with $x_m \in M_{N_m}$ and $\sup_m \|x_m\|_{M_{N_m}} < \infty$. Equipped with the norm $\|x\| = \sup \|x_m\|_{M_{N_m}}$, A becomes a C^*-algebra. Let $\mathcal{U}$ be a nontrivial ultrafilter and let $I_\mathcal{U} \subset A$ be the (closed two-sided self-adjoint) ideal formed of all sequences $x = (x_m)_{m \in \mathbb{N}}$ such that $\lim_\mathcal{U} \|x_m\| = 0$. Then the quotient space $A/I_\mathcal{U}$ is a C^*-algebra called the ultraproduct of $\{M_{N_m} \mid m \in \mathbb{N}\}$ with respect to $\mathcal{U}$. See also §2.8. By Gelfand theory we can view $A/I_\mathcal{U}$ as embedded into $B(\mathcal{H})$ for some Hilbert space $\mathcal{H}$. Let us denote by $\widehat{u}_1, \ldots, \widehat{u}_n$ the unitary elements in $A/I_\mathcal{U}$ associated to the families $(u_1^m)_{m \in \mathbb{N}}, \ldots, (u_n^m)_{m \in \mathbb{N}}$. We claim that, for any $a_1, \ldots, a_n$ in $B(H)$ (with H arbitrary), we have

$$\left\| \sum \widehat{u}_i \otimes a_i \right\| \leq \lim_{m,\mathcal{U}} \left\| \sum u_i^m \otimes a_i \right\|. \qquad (20.7)$$

Indeed, the quotient mapping $q\colon A \to A/I_\mathcal{U}$ being a C^*-representation is completely contractive, and hence (recalling (2.6.2)) $\|\sum \widehat{u}_i \otimes a_i\| \le \sup_m \|\sum u_i^m \otimes a_i\|$, but since the left side of (20.7) depends only on the equivalence class modulo $\mathcal{U}$, this last sup can be replaced by the limit along $\mathcal{U}$, and (20.7) follows.

Now, if we apply (20.7) with $a_i = \overline{\widehat{u}_i} \in B(\overline{\mathcal{H}})$, we obtain by Theorem 20.1

$$2\sqrt{n-1} \le \left\|\sum \widehat{u}_i \otimes \overline{\widehat{u}_i}\right\| \le \lim_{m,\mathcal{U}} \left\|\sum u_i^m \otimes \overline{\widehat{u}_i}\right\| = \lim_{m,\mathcal{U}} \left\|\sum \overline{u_i^m} \otimes \widehat{u}_i\right\|$$

$$= \lim_{m,\mathcal{U}} \left\|\sum \widehat{u}_i \otimes \overline{u_i^m}\right\|;$$

hence by (20.7) again

$$\le \lim_{m,\mathcal{U}} \lim_{m',\mathcal{U}} \left\|\sum u_i^{m'} \otimes \overline{u_i^m}\right\|,$$

and the last term is of course

$$\le \sup_{m \ne m'} \left\|\sum_{i=1}^n u_i^m \otimes \overline{u_i^{m'}}\right\|.$$

Thus we conclude that $2\sqrt{n-1} \le C(n)$. ∎

We now turn to the much more delicate task of majorizing $C(n)$. Until very recently, the best known estimates (see [Va]) used some deep number theoretic results ("Ramanujan graphs") due to Lubotzky, Phillips, and Sarnak ([LPS]), themselves based on André Weil's celebrated proof of "the Riemann hypothesis for curves over a finite field." Actually, it is somewhat easier for our exposition to derive our estimates from an application of these results (by the same authors [LPS]) to a "packing problem" on the sphere of $\mathbb{R}^3$, as follows. We will denote by S the Euclidean unit sphere in $\mathbb{R}^3$, equipped with its normalized surface measure σ. We let

$$L_2^0 = \left\{f \in L_2(S,\sigma) \mid \int f\, d\sigma = 0\right\}.$$

We consider the representation $\rho\colon SO(3) \to B(L_2^0)$ defined by:

$$\forall x \in S \qquad [\rho(\omega)f](x) = f(\omega^{-1}(x)).$$

This is usually called the "quasi-regular" representation of $SO(3)$, except that we restrict it to the orthogonal of the constant functions, namely, L_2^0.

Theorem 20.3. *([LPS]) For any n of the form $n = p+1$ with p prime ≥ 3, there are elements $t_1, \ldots, t_n$ in $SO(3)$ such that*

$$\left\|\sum_{i=1}^n \rho(t_i)\right\|_{B(L_2^0)} \le 2\sqrt{n-1}.$$

Corollary 20.4. *For any $n \geq 4$ of the form $n = p + 1$ with p prime we have*

$$C(n) \leq 2\sqrt{n-1}.$$

Proof. Since $SO(3)$ is a compact group, the unitary representation ρ: $SO(3) \to B(L_2^0)$ decomposes as a direct sum of a sequence $(\pi_m)_{m \geq 1}$ of irreducible *finite-dimensional* unitary representations of $SO(3)$, that is, we have

$$\rho \simeq \bigoplus_{m \geq 1} \pi_m.$$

The representation π_m is just the restriction of ρ to the spherical harmonics of degree m (for details see, e.g., p. 161 in [Fol]). In particular, for any $t_1, \ldots, t_n \in SO(3)$ we have

$$\left\| \sum_1^n \rho(t_i) \right\| = \sup_{m \geq 1} \left\| \sum_{i=1}^n \pi_m(t_i) \right\|.$$

These are distinct (i.e., mutually inequivalent) representations. Moreover, it is classical that the collection $\{\pi_m \mid m \geq 1\}$ exhausts all the nontrivial irreducible representations of $SO(3)$. (Note that this is special to the dimension 3.)

Therefore, if $m \neq m'$, then $\pi_m \otimes \overline{\pi_{m'}}$ decomposes again as a direct sum of a certain subset (depending of course on m and m') of $\{\pi_m \mid m \geq 1\}$. Note that by Schur's lemma, $\pi_m \not\simeq \pi_{m'}$ guarantees that the trivial representation is not contained in $\pi_m \otimes \overline{\pi_{m'}}$. Consequently, we have

$$\left\| \sum_{i=1}^n \pi_m(t_i) \otimes \overline{\pi_{m'}(t_i)} \right\| \leq \sup_{m \geq 1} \left\| \sum_1^n \pi_m(t_i) \right\| = \left\| \sum_1^n \rho(t_i) \right\|.$$

Therefore, if we choose the points $t_1, \ldots, t_n$ as in Theorem 20.3 and let $u_i(m) = \pi_m(t_i)$ with $N_m = \dim(\pi_m)$, then we obtain $C(n) \leq 2\sqrt{n-1}$. ∎

For the applications in the next chapter, the crucial point is that $C(n) < n$ for suitably large n. The ideas revolving around Kazhdan's property T (cf. [DHV, Vo1]) provide us with a different route to this fact:

Theorem 20.6. $C(n) < n$ for any $n > 2$.

Definition. *A finitely generated discrete group G, with generators g_1, $g_2, \ldots, g_n$, is said to have property T if the trivial representation is isolated in the set of all unitary representations of G. More precisely, this means that there is a number $\varepsilon > 0$ such that, for any unitary representation ρ, the condition*

$$(\exists \xi \in H_\rho, \ \|\xi\| = 1, \ \sup_{i \leq n} \|\rho(g_i)\xi - \xi\| < \varepsilon)$$

suffices to conclude that ρ admits a nonzero invariant vector, or, in other words, that ρ contains the trivial representation (as a subrepresentation).

It is easy to see that this property actually does not depend on the choice of the set of generators.

Remark. Let π, σ be two unitary irreducible representations of a discrete group G. Then Schur's classical lemma implies that, if $\pi \otimes \overline{\sigma}$ admits a nonzero invariant vector (in $H_\pi \otimes_2 \overline{H_\sigma}$), then π is unitarily equivalent to σ and both are finite-dimensional representations. Indeed, such a vector can be identified with a nonzero Hilbert-Schmidt operator $T \colon H_\sigma \to H_\pi$ such that $\pi(t) T \sigma(t)^* = T$ for all t in G. Equivalently, T intertwines π and σ (i.e. $\pi(t) T = T \sigma(t)$ for all t in G). By Schur's lemma, T must be an isomorphism, but since it is Hilbert-Schmidt, both H_π and H_σ must be finite-dimensional. Moreover, we have $T^* \pi(t)^* = \sigma(t)^* T^*$. Hence, multiplying this on the right with $\pi(t) T = T \sigma(t)$, we find $T^* T = \sigma(t)^* T^* T \sigma(t)$, which shows that $T^* T$ commutes with σ and therefore is a multiple of the identity. Thus we may assume that T is unitary, and we conclude that π and σ are unitarily equivalent. ∎

We will use the following well-known fact.

Lemma 20.7. *Let G be a discrete group with property T generated by $g_1, g_2, \ldots, g_n$ with g_1 equal to the unit. Then*

$$\sup \left\{ \left\| \sum_1^n \pi(g_i) \otimes \overline{\pi'(g_i)} \right\| \right\} < n, \tag{20.8}$$

where the supremum runs over all pairs of distinct (i,e., not unitarily equivalent) irreducible representations of G.

Proof. Let $\varepsilon > 0$ be as in the preceding definition of property T. Actually, we will show more generally that (20.8) holds with the sup running over all pairs of *disjoint* representations, that is, pairs (π, π'), such that $\pi \otimes \overline{\pi'}$ does not contain the trivial representation (equivalently has no invariant nontrivial vector). By the preceding remark, this happens whenever π, π' are two distinct *irreducible* representations. We now complete the argument: If (20.8) fails, then, for each m, we can find a disjoint pair (π_m, π'_m) and ξ_m with $\|\xi_m\| = 1$ such that

$$\lim_{m \to \infty} \left\| \left[\sum_1^n \pi_m(g_i) \otimes \overline{\pi'_m(g_i)} \right] (\xi_m) \right\| = n.$$

By the uniform convexity of Hilbert space (see Exercise 20.2), this forces the n-tuples of points $[\pi_m(g_i) \otimes \overline{\pi'_m(g_i)}](\xi_m)$ $(i = 1, \ldots, n)$ to all collapse to a single point, (i.e., the diameter of this n-tuple must tend to zero when $m \to \infty$), and since g_1 is the unit element, we have $\pi_m(g_1) \otimes \overline{\pi'_m(g_1)}(\xi_m) = \xi_m$; therefore we must have for each $i = 2, \ldots, n$

$$\lim_{m \to \infty} \|[\pi_m(g_i) \otimes \overline{\pi'_m(g_i)}](\xi_m) - \xi_m\| = 0.$$

In particular, when m is large enough we find that $\rho_m = \pi_m \otimes \overline{\pi'_m}$ satisfies $\sup_{i \le n} \|\rho_m(g_i)\xi_m - \xi_m\| < \varepsilon$, and hence by Property T that ρ_m contains the trivial representation, but this is impossible since each pair (π_m, π'_m) is disjoint. This contradiction completes the proof. ∎

Proof of Theorem 20.6. We use the fact that $G = SL_3(\mathbb{Z})$ has property T (cf. [DHV], see also [Sh] for a more recent simple proof) and that it can be generated by two elements g_2, g_3 (cf. [Tr]) as follows

$$g_2 = \begin{pmatrix} 1 & 0 & 0 \\ 1 & 1 & 0 \\ 0 & 0 & 1 \end{pmatrix} \quad \text{and} \quad g_3 = \begin{pmatrix} 0 & 1 & 0 \\ 0 & 0 & 1 \\ 1 & 0 & 0 \end{pmatrix}.$$

Let $\{\pi_m \mid m \ge 1\}$ be a sequence of distinct finite-dimensional irreducible representations of $SL_3(\mathbb{Z})$. (Note that the existence of such a sequence is immediate by considering the morphisms $SL_3(\mathbb{Z}) \to SL_3(\mathbb{Z}/p\mathbb{Z})$. Indeed, the irreducible representations of $SL_3(\mathbb{Z}/p\mathbb{Z})$ are finite-dimensional and give rise to irreducible representations of $SL_3(\mathbb{Z})$, with finite range, which separate points. Since $SL_3(\mathbb{Z})$ is infinite, there has to be infinitely many inequivalent representations appearing in this process.) If we set $u_i(m) = \pi_m(g_i)$ as before, then (20.8) implies that $C(n) < n$. ∎

We now turn to yet another proof that $C(n)$ is small, this time based on Gaussian random matrices. In a preliminary version of this book, we used the difficult estimates from [HT1] to show that

$$C(n) \le (3\pi/8)\sqrt{n}/2. \tag{20.9}$$

But, just as our manuscript was about to go to the printer, we learned of the improvements of [HT2], which prompted us to replace the proof of (20.9) by a brief outline of the results of [HT2] showing that $C(n) = 2\sqrt{n-1}$ for all $n \ge 2$, as follows.

Theorem 20.8. *([HT2]) For any $n \ge 2$, $C^*_\lambda(\mathbb{F}_n)$ embeds, as a unital C^*-subalgebra, into the C^*-algebraic ultraproduct $\Pi_\alpha M(\alpha)/\mathcal{U}$ of a family of matrix algebras $\{M(\alpha) \mid \alpha \ge 1\}$ (that is, there are integers $N(\alpha)$ such that $M(\alpha) = M_{N(\alpha)}$ and $\mathcal{U}$ is an ultrafilter on $\mathbb{N}$).*

Note that the difficulty lies in the fact that the ultraproduct $\Pi_\alpha M(\alpha)/\mathcal{U}$ is meant here *in the norm sense* as in §2.8 (and not in the von Neumann sense as in §9.10). The proof of Haagerup and Thorbjørnsen [HT2] rests on a delicate computation of the norms of Gaussian random matrices with matrix coefficients. Then, using this, they can embed the C^*-algebra $C^*(c_1, \ldots, c_n)$ generated by an n-tuple of free circular elements (see §9.9) into an ultraproduct of the form $\Pi_\alpha M(\alpha)/\mathcal{U}$. Since $C^*_\lambda(\mathbb{F}_n)$ embeds into $C^*(c_1, \ldots, c_n)$, the result follows a fortiori for $C^*_\lambda(\mathbb{F}_n)$.

As an immediate consequence, we have

Corollary 20.9. *([HT2]) For any $n \geq 1$ and any $\alpha \geq 1$, we can find an n-tuple of unitary $\alpha \times \alpha$ matrices $(u_i^\alpha)_{1 \leq i \leq n}$ such that, for any N and any $a_1, \ldots, a_n$ in M_N, we have*

$$\lim_{\alpha, \mathcal{U}} \left\| \sum_1^n a_i \otimes u_i^\alpha \right\|_{\min} = \left\| \sum_1^n a_i \otimes \lambda(g_i) \right\|_{\min}. \tag{20.10}$$

Proof. According to Theorem 20.8, we may view $C^*_\lambda(\mathbb{F}_n)$ as embedded into $\Pi_\alpha M(\alpha)/\mathcal{U}$. Then, for each $1 \leq i \leq n$, let $(u_i^\alpha)_{\alpha \geq 1}$ be a representative of $\lambda(g_i)$ in $\bigoplus_\alpha M(\alpha)$. Since $\lambda(g_i)$ is unitary, we must have $\lim_\mathcal{U} \|1 - (u_i^\alpha)^*(u_i^\alpha)\| = 0$; but an elementary argument (based on the polar decomposition of u_i^α) shows that, actually, we may as well assume that u_i^α is unitary for each α. We have then, for any fixed N, an isometric embedding

$$M_N(C^*_\lambda(\mathbb{F}_n)) \subset M_N(\Pi_\alpha M(\alpha)/\mathcal{U}) = \Pi_\alpha M_N(M(\alpha))/\mathcal{U}.$$

In particular, for any $a_1, \ldots, a_n$ in M_N we have

$$\left\| \sum_1^n a_i \otimes \lambda(g_i) \right\|_{M_N(C^*_\lambda(\mathbb{F}_n))} = \lim_{\alpha, \mathcal{U}} \left\| \sum_1^n a_i \otimes u_i^\alpha \right\|_{M_N(M(\alpha))}. \qquad \blacksquare$$

Remark. Actually, using Proposition 13.6, one can deduce conversely Theorem 20.8 from Corollary 20.9 (but, at the time of this writing, no direct proof of Corollary 20.9 is known).

By Corollaries 20.2 and 20.4, we have $C(p+1) = 2\sqrt{p}$ for all prime numbers $p \geq 3$. By [Mor], this remains true if p is a power of a prime number, but the general case was settled only very recently by Haagerup and Thorbjørnsen.

Theorem 20.10. *([HT2]) $C(n) \leq 2\sqrt{n-1}$ (and hence $C(n) = 2\sqrt{n-1}$) for any $n \geq 2$.*

Proof. Fix $\varepsilon > 0$. Obviously it suffices to construct a sequence of n-tuples $\{(u_i(m))_{1 \leq i \leq n} \mid m \geq 1\}$ of unitary matrices (note: $u_i(m)_{1 \leq i \leq m}$ is assumed to be, say, of size $N_m \times N_m$) such that, for any integer $p \geq 1$, we have

$$\sup_{1 \leq m \neq m' \leq p} \left\| \sum u_i(m) \otimes \overline{u_i(m')} \right\|_{\min} < 2\sqrt{n-1} + \varepsilon. \tag{20.11}$$

We will construct this sequence by induction on p. Assume that we already know the result up to p. That is, we already know a family $\{(u_i(m))_{1\leq i\leq m} \mid 1 \leq m \leq p\}$ formed of p n-tuples satisfying (20.11). We need to produce an additional n-tuple $(u_i(p+1))_{1\leq i\leq n}$ of unitary matrices (possibly of some larger size $N_{p+1} \times N_{p+1}$) such that (20.11) still holds for the enlarged family $\{(u_i(m))_{1\leq i\leq m} \mid 1 \leq m \leq p+1\}$ formed of one more n-tuple. By (20.10), for any $1 \leq m \leq p$, we have

$$\lim_{\alpha,\mathcal{U}} \left\|\sum\nolimits_1^n \overline{u_i(m)} \otimes u_i^{\alpha}\right\|_{\min} = \left\|\sum \overline{u_i(m)} \otimes \lambda(g_i)\right\|_{\min}.$$

On the other hand, by the absorption principle (Proposition 8.1) and by (20.2), we have

$$\left\|\sum\nolimits_1^n \overline{u_i(m)} \otimes \lambda(g_i)\right\|_{\min} = \left\|\sum\nolimits_1^n \lambda(g_i)\right\| = 2\sqrt{n-1}.$$

Hence, if α is chosen large enough, we can ensure that, for all $1 \leq m \leq p$ simultaneously, we have

$$\left\|\sum\nolimits_1^n \overline{u_i(m)} \otimes u_i^{\alpha}\right\|_{\min} < 2\sqrt{n-1} + \varepsilon.$$

But then, if we set $N_{p+1} = N(\alpha)$ and $u_i(p+1) = u_i^{\alpha}$, the extended family $\{(u_i(m))_{1\leq i\leq n} \mid 1 \leq m \leq p+1\}$ clearly still satisfies (20.11). ∎

Remark. Let $U(\alpha)$ denote the group of all $\alpha \times \alpha$ unitary matrices ($\alpha \geq 1$). Let $U_1^{(\alpha)},\ldots,U_n^{(\alpha)}$ be a sequence of independent matrix valued random variables, each having as its distribution the normalized Haar measure on $U(\alpha)$. It is very likely to be true that, for all N and for all $a_1,\ldots,a_n$ in M_N, we have, for almost all ω,

$$\limsup_{\alpha\to\infty} \left\|\sum\nolimits_1^n a_i \otimes U_i^{\alpha}(\omega)\right\|_{\min} = \left\|\sum\nolimits_1^n a_i \otimes \lambda(g_i)\right\|_{\min} \tag{20.12}$$

and in particular, if $a_1,\ldots,a_n$ are all unitary, for almost all ω,

$$\limsup_{\alpha\to\infty} \left\|\sum\nolimits_1^n a_i \otimes U_i^{\alpha}(\omega)\right\|_{\min} = 2\sqrt{n-1}.$$

If true, this would yield a more direct proof that $C(n) \leq 2\sqrt{n-1}$. The paper [HT2] contains an analog of (20.12) with Gaussian random matrices instead of $\{U_i^{\alpha}(\omega)\}$.

Remark 20.11. It is sometimes useful to replace $C(n)$ by another constant, $\beta(n)$. We define $\beta(n)$ as the supremum of all the numbers $\beta \geq 0$ for which there are two infinite sequences $\{(x_i(m))_{1\leq i\leq n} \mid m \in \mathbb{N}\}$ and

$\{(y_i(m))_{1 \le i \le n} \mid m \in \mathbb{N}\}$ of n-tuples of operators in $B(H)$ (H arbitrary) satisfying $\sup_{m,i} \|x_i(m)\| + \|y_i(m)\| < \infty$ together with:

$$\sup_{m \ne m'} \left\| \sum_{i=1}^{n} x_i(m) \otimes y_i(m') \right\|_{\min} \le 1 \text{ and } \beta \le \left\| \sum_{i=1}^{n} x_i(m) \otimes y_i(m) \right\|_{\min} \quad \forall m \in \mathbb{N}.$$

It would be more natural to assume $x_i(m), y_i(m) \in B(H_m)$ with H_m Hilbert, but the resulting $\beta(n)$ would be the same (since we can replace each H_m by the direct sum $H = \oplus_m H_m$). This definition is closely related to the notion of "coding sequence" introduced by Voiculescu in [Vo1].

Note that we have obviously

$$n/C(n) \le \beta(n).$$

Exercises

Exercise 20.1. (Alternate proof of Theorem 20.1.) Let $a(n) = \|\sum_1^n \lambda(g_i)\|$. Let $u_1, \ldots, u_n$ be unitary operators. Using (7.2) and Proposition 8.1 (Fell's absorption), show that

$$\left\| \sum u_i \otimes \lambda(g_i) \right\| \le \left\| \sum u_i \otimes \overline{u_i} \right\|^{1/2} a(n)^{1/2}$$

and conclude that

$$a(n) \le \left\| \sum_1^n u_i \otimes \overline{u_i} \right\|.$$

Exercise 20.2. Let $x = (x_1, \ldots, x_n)$ be an n-tuple in the unit ball of a Hilbert space H. Let $M(x) = n^{-1} \sum_1^n x_k$ and $\Delta(x) = \max_{1 \le i \ne j \le n} \|x_i - x_j\|$. Show that

$$\|M(x)\|^2 + n^{-1} \sum_1^n \|x_k - M(x)\|^2 \le 1.$$

Deduce from this that, if a sequence of n-tuples $\{x(m) \mid m \in \mathbb{N}\}$ (in the unit ball) is such that $\|M(x(m))\| \to 1$ when $m \to \infty$, then $\Delta(x(m)) \to 0$.

Chapter 21. Local Theory of Operator Spaces. Nonseparability of OS_n

One of the most interesting benefits of operator space theory is the possibility to employ *finite-dimensional* methods in C^*-algebra theory, and some ideas from the "local theory" of Banach spaces have already proved very fruitful in the study of operator spaces.

The "local theory" is the part of the theory that studies infinite-dimensional Banach spaces through the asymptotic properties of the collection of their finite-dimensional subspaces. If E and F are Banach spaces of the same dimension, they clearly are isomorphic, but one can measure their "degree of isomorphism" using their "Banach-Mazur distance," which is defined as follows:

$$d(E, F) = \inf\{\|u\|\|u^{-1}\|\},$$

where the infimum runs over all the isomorphisms u between E and F. We have clearly for any G of the same dimension $d(E, F) \leq d(E, G)d(G, F)$. Moreover, $d(E, F) = 1$ iff E and F are isometric (hint: use the compactness of the unit ball of $B(E, F)$ to check this). Thus, if we identify E and F whenever they are isometric, we can equip the set B_n of all n-dimensional Banach spaces with the metric

$$\delta(E, F) = \operatorname{Log} d(E, F).$$

By a classical theorem due to Fritz John (1948) we have $d(E, \ell_2^n) \leq \sqrt{n}$ for every n-dimensional normed space (see the remark after Corollary 7.7), and consequently $d(E, F) \leq n$ for every pair E, F in B_n. Moreover:

Theorem 21.1. *For each $n \geq 1$, (B_n, δ) is a compact metric space.*

Proof. Let $\{E_m\}$ be a sequence in B_n. Choosing an appropriate basis and using Fritz John's aforementioned theorem, we may assume that $E_m = (\mathbb{C}^n, \|\ \|_m)$ and that

$$\forall\, x \in \mathbb{C}^n \qquad |x| \leq \|x\|_m \leq \sqrt{n}\,|x|,$$

where we set

$$|x| = \left(\sum_1^n |x_i|^2\right)^{1/2}.$$

Let $\Omega = \{x \in \mathbb{C}^n \mid |x| \leq 1\}$, and let $f_m\colon \Omega \to \mathbb{R}$ be defined by $f_m(x) = \|x\|_m$. Note that $|f_m(x) - f_m(y)| \leq f_m(x - y) \leq \sqrt{n}|x - y|$ and $f_m(0) = 0$; hence, by Ascoli's Theorem, the sequence $\{f_m\}$ is relatively compact in $C(\Omega)$. Thus, after passing to a subsequence, we may assume that f_m converges uniformly on Ω to a limit; therefore, by homogeneity, for any x in $\mathbb{C}^n$, $\|x\|_m$ converges to a limit $\|x\|_\infty$ such that

$$\forall\, x \in \mathbb{C}^n \qquad |x| \le \|x\|_\infty \le \sqrt{n}\,|x|.$$

Let $E_\infty = (\mathbb{C}^n, \| \ \|_\infty)$. We claim that $\delta(E_m, E_\infty) \to 0$. Indeed, if we let $\varepsilon(m) = \sup\{|\|x\|_\infty - \|x\|_m| \mid x \in \Omega\}$, we have for any x in $\mathbb{C}^n$

$$|\|x\|_\infty - \|x\|_m| \le \varepsilon(m)|x| \le \varepsilon(m)\|x\|_m;$$

hence

$$(1 - \varepsilon(m))\|x\|_m \le \|x\|_\infty \le (1 + \varepsilon(m))\|x\|_m,$$

which implies $d(E_\infty, E_m) \le \frac{1+\varepsilon(m)}{1-\varepsilon(m)}$. Since $\varepsilon(m) \to 0$, we obtain $d(E_\infty, E_m) \to 1$ and $\delta(E_\infty, E_m) \to 0$.

Thus we have shown that any sequence $\{E_m\}$ in (B_n, δ) contains a convergent subsequence, which means that (B_n, δ) is compact. ∎

Remark 21.2. Fix an integer N. Assume that all the spaces E_m are given equipped with an isometric embedding $J_m \colon E_m \to B(H_m)$. For $a_1, \cdots, a_n \in M_N$ we set

$$F_m(a_1, \cdots, a_n) = \Big\| \sum a_i \otimes J_m(e_i) \Big\|_{M_N(B(H_m))}.$$

Then it is easy to show by the same reasoning as above (see also Exercise 21.1) that there is a subsequence of (E_m) such that the functions F_m converge uniformly on the unit ball of $(M_N)^n$. If we repeat this successively for $N = 1, 2, \ldots$ and pass to a further subsequence each time, we obtain a subsequence for which this holds for all N. By Ruan's Theorem (or by the stability of C^*-algebras under ultraproducts), there is (for some H_∞) an isometric embedding $J_\infty \colon E_\infty \to B(H_\infty)$, such that, for each $N = 1, 2, \ldots$,

$$\lim_{m \to \infty} \Big\| \sum a_i \otimes J_m(e_i) \Big\|_{M_N(B(H_m))} = \Big\| \sum a_i \otimes J_\infty(e_i) \Big\|_{M_N(B(H_\infty))}$$

uniformly on the unit ball of $(M_N)^n$. In particular, we conclude that $M_N(J_m(E_m))$ tends to $M_N(J_\infty(E_\infty))$ in the metric space $B_{N^2 n}$.

We now describe what remains of all this in the category of operator spaces. The main difference is that the "quantized" scalars are the elements of $\mathcal{K}$, which is an infinite-dimensional space; hence the compactness arguments are no longer available.

Fix an integer $n \ge 1$. Let us denote by OS_n the set of all operator spaces of dimension n. We consider that two spaces are *the same* if they are completely isometric. Then OS_n can be equipped with the metric

$$\delta_{cb}(E, F) = \mathrm{Log}\, d_{cb}(E, F),$$

which is analogous to the Banach-Mazur distance.

As in the Banach space case, the space OH_n is a center of this metric space (see the above Corollary 7.7):

Theorem 21.3.
$$\forall E \in OS_n \quad d_{cb}(E, OH_n) \leq \sqrt{n}.$$

Therefore
$$\forall E, F \in OS_n \quad d_{cb}(E, F) \leq n.$$

We recall that, since $d_{cb}(R_n, C_n) = n$, this cannot be improved. The analogous result (i.e., the optimality of the estimate) for Banach spaces is much more delicate (see [Gl]).

In another direction, it is natural to wonder whether a given finite-dimensional operator space can be realized (at least approximately) as a subspace of the compact operators. For that purpose, we have introduced above the constant $d_{SK}(E)$ (see (17.4) and (17.4)').

By Corollary 7.7, we have

Theorem 21.4. *For all E in OS_n, $d_{SK}(E) \leq \sqrt{n}$.*

Quite surprisingly it turns out that this bound cannot be improved asymptotically. Indeed, let us denote

$$E_U^n = (\ell_\infty^n)^*$$

equipped with its dual operator space structure, or, equivalently,

$$E_U^n = \max(\ell_1^n).$$

By Theorem 9.6.1 (or Corollary 8.13), E_U^n can also be realized as the span of the n unitary generators in the full C^*-algebra of the free group with n generators. Note that when $n = 2$ we have $d_{SK}(E_U^n) = 1$ by Proposition 3.10. However, this no longer holds when $n > 2$:

Theorem 21.5. *([P6]) Let $a_n = n/(2\sqrt{n-1})$. We have*

$$a_n \leq d_{SK}(E_U^n) \leq \sqrt{n}$$
$$(a_n)^{1/2} \leq d_{SK}(OH_n) \leq n^{1/4}.$$

Note $a_n \sim \sqrt{n}/2$ when $n \to \infty$ and $a_n > 1$ as soon as $n > 2$.

Proof. We may assume $E_U^n = \text{span}[U(g_i) \mid 1 \leq i \leq n]$ as in §9.6. Note that $(E_U^n)^* \simeq \ell_\infty^n$, so that $d_{SK}((E_U^n)^*) = 1$. Moreover, by the absorption principle (Proposition 8.1) and by (20.2), we have

$$\left\| \sum_1^n U(g_i) \otimes \lambda(g_i) \right\|_{\min} = 2\sqrt{n-1}.$$

Applying (19.1)' with u equal to the identity on E_U^n, we obtain $n \leq d_{SK}(E_U^n)2\sqrt{n-1}$, which proves the first line.

We now turn to OH_n. Let (T_i) be an orthonormal basis of OH_n. Applying (19.1)' with u equal to the identity on OH_n and using $OH_n^* \simeq \overline{OH_n}$, we obtain

$$n \leq d_{SK}(OH_n)^2 \left\| \sum T_i \otimes \lambda(g_i) \right\|_{\min} \left\| \sum \overline{T}_i \otimes \lambda(g_i) \right\|_{\min}$$

$$\leq d_{SK}(OH_n)^2 \left\| \sum T_i \otimes \lambda(g_i) \right\|_{\min}^2$$

$$= d_{SK}(OH_n)^2 \left\| \sum \lambda(g_i) \otimes \overline{\lambda(g_i)} \right\|_{\min} ;$$

hence, by (20.2), $n \leq d_{SK}(OH_n)^2 a_n$, and thus $(n/a_n)^{1/2} \leq d_{SK}(OH_n)$.

Conversely, by (10.8), since $d_{SK}(R_n \cap C_n) = 1$, we have $d_{SK}(OH_n) \leq n^{1/4}$.

■

This phenomenon is in sharp contrast with the Banach space case. Indeed, for every n-dimensional Banach space E and any $\varepsilon > 0$, there is an integer $N = N(\varepsilon, n) > 0$ and a subspace $F \subset \ell_\infty^N$ such that $d(E, F) < 1 + \varepsilon$ (see Exercise 2.13.2). Thus the analog of $d_{SK}(E)$ in this case is always 1!

Also, every separable Banach space (a fortiori every finite-dimensional one) embeds isometrically into a single separable space, namely, the space $C([0, 1])$. Thus it is natural to ask:

Problem: *Is there a single separable operator space $\mathcal{X}$ such that every finite-dimensional operator space embeds completely isometrically into $\mathcal{X}$?*

We have just seen that $\mathcal{X} = \mathcal{K}$ does not work. We will give the answer in Corollary 21.12. But before that, we observe that, if $\mathcal{X}$ is separable, a simple perturbation argument (see Corollary 2.13.3) shows that, for any n, the subset of OS_n formed of all the n-dimensional subspaces of $\mathcal{X}$ is a δ_{cb}-separable metric space. This brings us to the following question raised by E. Kirchberg:

Problem: *Is (OS_n, δ_{cb}) separable?*

A negative answer will guarantee that there does not exist any "universal" separable operator space $\mathcal{X}$ as above.

Contrary to the Banach space case, the space (OS_n, δ_{cb}), which is a complete metric space, is *not compact* if $n > 2$. Even the subset of all isometrically Hilbertian operator spaces is not compact! There is however a weaker metric structure that one can consider on OS_n. For any $N \geq 1$ and for $u \colon E \to F$, we denote

$$\|u\|_N = \|I_{M_N} \otimes u\|_{M_N(E) \to M_N(F)}.$$

Then, $\forall E, F \in OS_n$ we define

$$d_N(E, F) = \inf\{\|u\|_N \|u^{-1}\|_N \mid u \colon E \to F \text{ isomorphism}\}.$$

It is easy to check (by a compactness argument) that for all E, F in OS_n we have

$$d_{cb}(E, F) = \sup_N d_N(E, F). \tag{21.1}$$

Hence $d_{cb}(E_i, E) \to 1$ iff $d_N(E_i, E) \to 1$ uniformly in N.

But we are interested in the topology for which

$$E_i \to E \Leftrightarrow (\forall N \geq 1 \quad d_N(E_i, E) \to 1).$$

We call it the weak topology. It is associated to the metric

$$\delta_w(E, F) = \sum_{N \geq 1} 2^{-N} \operatorname{Log} d_N(E, F).$$

For the weak topology, the space OS_n is compact by Lemma 21.7) and a fortiori separable. But the strong topology (the one associated to d_{cb}) is strictly stronger than the weak one, so that the identity

$$f \colon (OS_n, \delta_w) \longrightarrow (OS_n, \delta_{cb})$$

is discontinuous (at least if $n > 2$). But if OS_n is assumed δ_{cb}-separable, the function f must be in *the first Baire class*. Indeed, by (21.1), if B is any closed ball in (OS_n, δ_{cb}), then $f^{-1}(B)$ is closed in (OS_n, δ_w); therefore, for any (strongly) open set U, $f^{-1}(U)$ must be (weakly) an F_σ-set, and the latter property characterizes functions of Baire class 1. Quite curiously, Kirchberg's question was first answered by applying the classical *Baire* theory to this map f.

Theorem 21.6. *([JP]) For $n > 2$, the metric space (OS_n, δ_{cb}) is non-separable.*

We will use the following elementary facts.

Lemma 21.7. *The metric space (OS_n, δ_w) is compact.*

Proof. This is an easy consequence of Remark 21.2.   ∎

Lemma 21.8. *For any E, F in OS_n we have*

$$d_{cb}(E^*, F^*) = d_{cb}(E, F),$$

and for any $N \geq 1$

$$d_N(E^*, F^*) = d_N(E, F).$$

Proof. The first equality is an immediate consequence of (2.3.3). Similarly, the second one follows from (2.3.4).   ∎

Lemma 21.9. *For any E in OS_n, there is a sequence $\{E_m\}$ in OS_n tending weakly to E and such that E_m is "matricial" (i.e., $E_m \subset M_N$ for a suitable $N = N(m)$) for any m.*

Proof. Since E is separable, we may clearly assume that $E \subset B(H)$ with H separable. Then let $\{H_m\}$ be an increasing sequence of finite-dimensional subspaces of H with $\cup H_m$ dense in H. Let $v_m \colon E \to B(H_m)$ be defined by $v_m(e) = P_{H_m} e_{|H_m}$, and let $E_m \subset B(H_m)$ be the range of v_m. Since (see Exercise 2.1.1) $\|v_m(e)\| \uparrow \|e\|$, we have

$$\big| \|v_m(e)\| - \|v_m(e')\| \big| \le \|v_m(e - e')\| \le \|e - e'\|$$

for any e, e' in E, and since the unit ball of E is compact, we must have $\|v_m(e)\| \to \|e\|$ uniformly on B_E. In particular, when m is large enough, v_m is an isomorphism from E to E_m, and moreover $d(E, E_m) \to 1$ when $m \to \infty$.

More generally, for any fixed $N \ge 1$, the mappings $(v_m)_N = I_{M_N} \otimes v_m \colon M_N(E) \to M_N(E_m)$ satisfy

$$\|(v_m)_N(x)\|_{M_N(E_m)} \uparrow \|x\|_{M_N(E)}$$

for any x in $M_N(E)$. Repeating the same argument, we find that the convergence is actually uniform on the unit ball of $M_N(E)$ and hence $d_N(E, E_m) \to 1$ when $m \to \infty$. Thus we conclude that E_m tends weakly to E. ∎

We will also need to characterize (following ([P6]) the points of continuity of f.

Lemma 21.10. *An element $E \in OS_n$ is a point of continuity of the map f iff*

$$d_{SK}(E) = d_{SK}(E^*) = 1.$$

Proof. Let $\{E_m\}$ be as in the preceding lemma. Then, if E is a point of continuity, E_m must tend strongly to E, and hence $d_{cb}(E_m, E) \to 1$. But since E_m is matricial, we have $d_{SK}(E_m) = 1$; hence we also obtain $d_{SK}(E) = 1$. Applying Lemma 21.9 to E^*, we find a sequence of matricial spaces $\{E_m\}$ that tends weakly to E^*. Then, by Lemma 21.8, $\{E_m^*\}$ tends weakly to E; but if E is a point of continuity, the convergence must be strong, so that, by Lemma 21.8 again, $\{E_m\}$ also tends strongly to E^*, whence $d_{SK}(E^*) = 1$.

Conversely, assume $d_{SK}(E) = 1$. Fix $\varepsilon > 0$. Then there is N and $\widetilde{E} \subset M_N$ such that $d_{cb}(E, \widetilde{E}) < 1 + \varepsilon$. By Proposition 1.12, this implies that for any F and any $u \colon F \to E$ we have $\|u\|_{cb} \le (1 + \varepsilon)\|u\|_N$. Now, if $d_{SK}(E^*) = 1$, we may use (2.3.4) and we find by the same argument an integer N' such that, for any F and any $v \colon E \to F$, we have $\|v\|_{cb} \le (1 + \varepsilon)\|v\|_{N'}$. Replacing

N by the largest of N, N', we obtain that for any F and any isomorphism $u: F \to E$ we have

$$\|u\|_{cb} \leq (1+\varepsilon)\|u\|_N \quad \text{and} \quad \|u^{-1}\|_{cb} \leq (1+\varepsilon)\|u^{-1}\|_N.$$

Thus we obtain

$$d_{cb}(E, F) \leq (1+\varepsilon)d_N(E, F).$$

From this last result, it is clear that if F tends to E weakly, that is, $d_N(E, F) \to 1$ for any N), then F to E strongly (i.e., $d_{cb}(E, F) \to 1$). ∎

Then we can use Chapter 19 to obtain:

Lemma 21.11. *Let $E \in OS_n$ be the weak limit of a sequence (E_α) in OS_n such that $d_{SK}(E_\alpha) = d_{SK}(E_\alpha^*) = 1$, for all α. Then $d(E, \ell_2^n) \leq 4$. Moreover, for any biorthogonal system (e_i, ξ_i) $(1 \leq i \leq n)$ in $E \times E^*$, we have*

$$n \leq \left\| \sum_1^n \lambda(g_i) \otimes e_i \right\|_{C_\lambda^*(\mathbb{F}_\infty) \otimes_{\min} E} \left\| \sum_1^n \lambda(g_i) \otimes \xi_i \right\|_{C_\lambda^*(\mathbb{F}_\infty) \otimes_{\min} E^*}. \tag{21.2}$$

Proof. By the last assertion of Corollary 19.2 (applied to the identity on E_α), we have $d(E_\alpha, \ell_2^n) = \gamma_2(I_{E_\alpha}) \leq 4$. Since weak convergence implies a fortiori $d(E_\alpha, E) \to 1$, we obtain the limit $d(E, \ell_2^n) \leq 4$.

Moreover, to show the second part, let $X = C_\lambda^*(\mathbb{F}_\infty)$. Recall (Corollary 17.10) that X is exact with $d_{SK}(X) = 1$. By Exercise 21.5 applied with $X = C_\lambda^*(\mathbb{F}_\infty)$, we have for each α a basis $\{e_i(\alpha) \mid 1 \leq i \leq n\}$ of E_α with dual basis $\{\xi_i(\alpha) \mid 1 \leq i \leq n\}$ in E_α^* such that

$$\lim_\alpha \left\| \sum \lambda(g_i) \otimes e_i(\alpha) \right\|_{\min} = \left\| \sum \lambda(g_i) \otimes e_i \right\|_{\min}$$

$$\lim_\alpha \left\| \sum \lambda(g_i) \otimes \xi_i(\alpha) \right\|_{\min} = \left\| \sum \lambda(g_i) \otimes \xi_i \right\|_{\min}.$$

By Theorem 19.1 applied to $u = I_{E_\alpha}$, we have

$$n \leq \left\| \sum \lambda(g_i) \otimes e_i(\alpha) \right\|_{\min} \left\| \sum \lambda(g_i) \otimes \xi_i(\alpha) \right\|_{\min}.$$

Hence passing to the limit we obtain

$$n \leq \left\| \sum \lambda(g_i) \otimes e_i \right\|_{\min} \left\| \sum \lambda(g_i) \otimes \xi_i \right\|_{\min},$$

as announced. ∎

First Proof of Theorem 21.6. Assume OS_n strongly separable. Then f is of the first Baire class and OS_n is weakly compact and hence weakly Baire. By Baire's famous theorem (see [Ku, 31, X, Theorem 1, p. 394]), f must have a (weakly) dense set of points of continuity, but this is impossible, at least if $n > 2$. Indeed, by Lemma 21.11, the spaces appearing in Lemma 21.10 are too few to be weakly dense in OS_n. To verify this, it suffices to produce a single element of OS_n failing the conclusion of Lemma 21.11. Obviously, we cannot have $d(E, \ell_2^n) \leq 4$ for any E in OS_n, at least if $n > 16$: Indeed it is easy to check that $d(\ell_1^n, \ell_2^n) = \sqrt{n}$, and this is > 4 when $n > 16$!

More precisely, for any $n > 2$, one can produce an E in OS_n failing (21.2): We set $E = E_\lambda^n$, that is, $E = \mathrm{span}[\lambda(g_i) \mid i = 1, ..., n]$, and $x_i = \lambda(g_i)$. Then, by Remark 8.10 and (20.2), we have $\| \sum_1^n \lambda(g_i) \otimes x_i \| = \| \sum_1^n \lambda(g_i) \| = 2\sqrt{n-1}$, and since $\sum \lambda(g_i) \otimes \xi_i$ represents the identity on E, we must have (by (2.3.2)) $\| \sum_1^n \lambda(g_i) \otimes \xi_i \| = \|I_E\|_{cb} = 1$. Hence, (21.2) implies $n \leq 2\sqrt{n-1}$, or, equivalently, $n \leq 2$. ∎

We will give two additional proofs of Theorem 21.6; see Theorem 21.14 and Corollary 21.15.

Remark. The only known examples of spaces E such that $d_{SK}(E) = d_{SK}(E^*) = 1$ are (in dimension > 1) the spaces ℓ_∞^2, ℓ_1^2 (say with their minimal-actually unique-o.s.s.), the spaces R_n and C_n ($n > 1$) or the spaces R and C, and also the spaces $\mathbb{C} \oplus R_n, \mathbb{C} \oplus_1 C_n$ or $\mathbb{C} \oplus R, \mathbb{C} \oplus_1 C$. For the row and column spaces this is clear since $R_n^* \simeq C_n$ (see Exercise 2.3.5). For $E = \ell_\infty^2$ or $E = \ell_1^2$, we have $\alpha(E) = 1$ in the sense of Chapter 3; that is, $\min(E) = \max(E)$ (see Proposition 3.10), hence $\min(E)^* = \min(E^*)$, and it is easy to check that $d_{SK}(\min(G)) = 1$ for any normed space G (for instance, because commutative C^*-algebras are nuclear by Exercise 11.7). Finally, for the spaces $\mathbb{C} \oplus R_n, \mathbb{C} \oplus_1 C_n$ (or $\mathbb{C} \oplus R, \mathbb{C} \oplus_1 C$), it follows from Corollary 9.4.2 (and Theorem 9.4.1) that the spaces $\mathbb{C} \oplus_1 R_n, \mathbb{C} \oplus_1 C_n$ embed completely isometrically into the Cuntz algebra, which is nuclear, and hence (by Theorem 11.5) they satisfy $d_{SK}(E) = 1$, and the same equality for their duals $\mathbb{C} \oplus_\infty C_n, \mathbb{C} \oplus_\infty R_n$ is obvious.

In [JP], it is shown that, if $d_{SK}(E) = d_{SK}(E^*) = 1$, then the Banach-Mazur distance $d(E, \ell_2^n)$ is at most 4 (see [PiS] for a more recent result). Consideration of ℓ_∞^2, ℓ_1^2 (or $\mathbb{C} \oplus R_n, \mathbb{C} \oplus_1 C_n$) shows that 4 cannot be replaced by something $< \sqrt{2}$, but we do not know what is the best possible number here.

The preceding remarks naturally lead to the following open question.

Problem. *Are the spaces* ℓ_∞^2, ℓ_1^2, R_n, C_n, $\mathbb{C} \oplus R_n, \mathbb{C} \oplus C_n, \mathbb{C} \oplus_1 R_n, \mathbb{C} \oplus_1 C_n$ *(resp.* $R, C, \mathbb{C} \oplus R, \mathbb{C} \oplus C, \mathbb{C} \oplus_1 R, \mathbb{C} \oplus_1 C$*) the only finite-dimensional (resp. separable infinite-dimensional) operator spaces* E *such that* $d_{SK}(E) = d_{SK}(E^*) = 1$?

Actually, the original proof in [JP] also yields the nonseparability of the subset of OS_n formed of all the (isometrically) Hilbertian operator spaces. Even the further subset of all the "homogeneous" Hilbertian operator spaces, in the sense of §9.2, fails to be separable if $n > 2$. When $n = 2$, the nonseparability remains an open question. Note, however, that if we restrict ourselves to operator spaces E spanned by n linearly independent unitaries, then $n > 2$ is necessary, since in the case $n = 2$ it suffices to consider the spaces E spanned by 1 and a unitary, and these lie in a commutative C^*-algebra; consequently they are minimal operator spaces, and hence they satisfy $d_{SA}(E) = 1$ with $A = c_0$ (see Exercise 2.13.2) and therefore (Corollary 2.13.3) form a separable collection.

Let E_0 be an n-dimensional Banach space. Let $OS_n(E_0)$ be the subset of OS_n formed of all the spaces isometric to E_0. (Equivalently, $OS_n(E_0)$ represents the set of all possible operator space structures on E_0 compatible with its norm.) This set may be a singleton (for instance, if $E_0 = \ell^2_\infty$; see Chapter 3), in which case it is compact and separable. Curiously, the converses are open:

Problem. *Consider the following properties of an n-dimensional normed space E_0.*

(i) $OS(E_0)$ is separable.

(ii) $OS(E_0)$ is compact.

(iii) $OS(E_0)$ is a singleton.

Are these properties equivalent?

We now generalize the numbers $d_{S\mathcal{K}}(.)$: For any operator space $\mathcal{X}$ (actually $\mathcal{X}$ will often be a C^*-algebra) and any finite-dimensional operator space E, we introduce

$$d_{S\mathcal{X}}(E) = \inf\{d_{cb}(E, F) \mid F \subset \mathcal{X}\}. \tag{21.3}$$

Of course, if $\mathcal{X} = B(H)$, $d_{S\mathcal{X}}(E) = 1$ for all E.

By Corollary 17.5, if A is an exact C^*-algebra, we have

$$d_{S\mathcal{K}}(E) \leq d_{SA}(E) \tag{21.4}$$

for all E. Moreover if A is roughly "large enough" (precisely, if A contains, for each n and $\varepsilon > 0$, a subspace X_n with $d_{cb}(X_n, M_n) < 1 + \varepsilon$), then the converse also holds. Thus the number $d_{SA}(E)$ is the same for all "large enough" nuclear or exact C^*-algebras.

In [Ki2], Kirchberg's candidate for a "universal" separable $\mathcal{X}$ as above was the full C^*-algebra $C^*(F_\infty)$ of the free group F_∞ with infinitely many generators, which the next result invalidates.

Corollary 21.12. *If $n > 2$, there does not exist any separable operator space $\mathcal{X}$ such that $d_{S\mathcal{X}}(E) = 1$ for any E in OS_n.*

Proof. If $\mathcal{X}$ is separable, the subset $\{E \in OS_n \mid d_{S\mathcal{X}}(E) = 1\}$ also is separable (by Corollary 2.13.3); hence this follows from Theorem 21.6. ∎

In the next chapter, we will concentrate on the special case when $\mathcal{X} = C^*(F_\infty)$, and to simplify the notation we will set

$$d_f(E) = d_{SC^*(F_\infty)}(E).$$

We now wish to measure the "degree of nonseparability" of OS_n. For that purpose, we introduce, for any $n \geq 1$, the number $\delta(n)$, which is the infimum of the numbers $\delta > 0$ for which OS_n admits a countable δ-net. More precisely, we prefer to use d_{cb} instead of δ_{cb}, so we define for any $E \in OS_n$ and any subset $\mathcal{D} \subset OS_n$

$$d_{cb}(E, \mathcal{D}) = \inf\{d_{cb}(E, F) \mid F \in \mathcal{D}\}.$$

Then we have by definition

$$\delta(n) = \inf_{\mathcal{D} \text{ countable}} \sup_{E \in OS_n} d_{cb}(E, \mathcal{D}). \tag{21.5}$$

Of course, we may replace "countable" by "separable" in this definition. Thus, for any *separable* operator space $\mathcal{X}$, considering $\mathcal{D} = \{E \subset \mathcal{X} \mid \dim E = n\}$, we obtain (by Corollary 2.13.3)

$$\delta(n) \leq \sup_{E \in OS_n} d_{S\mathcal{X}}(E). \tag{21.6}$$

The second and third proofs of Theorem 21.6 make use of the constant $C(n)$, introduced in the preceding chapter.

Recall that we have trivially $C(1) = 1$ and (exercise) $C(2) = 2$, but for $n > 2$ we have $C(n) < n$, and this is crucial to estimate $\delta(n)$, as the next result shows:

Theorem 21.13. *For all* $n \geq 2$, *we have* $n/C(n) \leq \delta(n)$.

Proof. Let C be any number with $C > C(n)$. By definition of $C(n)$, there is a sequence of n-tuples of unitary matrices $\{u_i(m)\}$ such that

$$\sup_{m \neq m'} \left\{ \left\| \sum_{i=1}^n u_i(m) \otimes \overline{u_i(m')} \right\|_{\min} \right\} \leq C. \tag{21.7}$$

Let e_i $(1 \leq i \leq n)$ be the canonical basis of ℓ_∞^n. We define for any $\Omega \subset \mathbb{N}$

$$x_i(\Omega) = Ce_i \oplus \left(\bigoplus_{m \in \Omega} u_i(m) \right),$$

and we let
$$E_\Omega = \mathrm{span}[x_1(\Omega), \dots, x_n(\Omega)].$$

Note that the presence of e_i ensures that E_Ω has dimension n. Now fix $\delta > \delta(n)$. We will show that $\delta \geq n/C$. By definition of $\delta(n)$, there is a sequence $\{E_m\}$ in OS_n such that, for any E in OS_n, there is an integer m such that $d_{cb}(E, E_m) < \delta$, and in particular for any Ω there is an m with $d_{cb}(E_\Omega, E_m) < \delta$. Since the set of all Ω is continuous (i.e., has the cardinality of the continuum), there is a continuous collection $\mathcal{C}$ of subsets of $\mathbb{N}$ for which the same m (fixed from now on) must be used, that is, such that for each Ω in $\mathcal{C}$ there is a map $v_\Omega \colon E_\Omega \to E_m$ satisfying:

$$\|v_\Omega\|_{cb} < \delta \quad \text{and} \quad \|v_\Omega^{-1}\|_{cb} = 1. \tag{21.8}$$

Now consider the continuous family $(v_\Omega(x_i(\Omega)))_{i \leq n}$ of n-tuples of elements of E_m. Since $(E_m)^n$ is norm-separable, for any $\eta > 0$ there exists a continuous subcollection $\mathcal{C}_1 \subset \mathcal{C}$ such that, for all Ω, Ω' in $\mathcal{C}_1$, we have

$$\sum_{i=1}^n \|v_\Omega(x_i(\Omega)) - v_{\Omega'}(x_i(\Omega'))\| < \eta. \tag{21.9}$$

A fortiori, $\mathcal{C}_1$ has cardinality > 1; hence we can find Ω, Ω' in $\mathcal{C}_1$ such that $\Omega' \not\subset \Omega$, so that there is an integer $m' \in \Omega'$ with $m' \notin \Omega$. This implies, by the definition of $x_i(\Omega)$

$$\left\| \sum x_i(\Omega) \otimes \overline{u_i(m')} \right\|_{\min} = \max\left\{ C, \sup_{m \in \Omega} \left\| \sum_{i=1}^n u_i(m) \otimes \overline{u_i(m')} \right\|_{\min} \right\} \leq C$$

and, on the other hand, by (20.1), since $m' \in \Omega'$, we have

$$n \leq \left\| \sum x_i(\Omega') \otimes \overline{u_i(m')} \right\|.$$

Thus, by (21.8), we have on one hand

$$\left\| \sum v_\Omega(x_i(\Omega)) \otimes \overline{u_i(m')} \right\|_{\min} \leq \|v_\Omega\|_{cb} \left\| \sum x_i(\Omega) \otimes \overline{u_i(m')} \right\|_{\min} \leq \delta C,$$

and on the other hand

$$n \leq \|v_{\Omega'}^{-1}\|_{cb} \left\| \sum v_{\Omega'}(x_i(\Omega')) \otimes \overline{u_i(m')} \right\|_{\min}$$
$$\leq \left\| \sum v_{\Omega'}(x_i(\Omega')) \otimes \overline{u_i(m')} \right\|,$$

which gives by (21.9)

$$n \leq \left\| \sum v_\Omega(x_i(\Omega)) \otimes \overline{u_i(m')} \right\| + \eta;$$

hence $n \le \delta C + \eta$. Finally, since $\eta > 0$ is arbitrary, we obtain $n \le \delta C$, and therefore $n/C(n) \le \delta(n)$.   ∎

Remark. A simple modification of the preceding proof shows that the constant $\beta(n) \ge n/C(n)$, introduced in Remark 20.11, satisfies $\beta(n) \le \delta(n)$.

Thus, if we want to show that, for n large, OS_n is "very" nonseparable, that is, that $\delta(n)$ is very big, we are reduced to showing that $C(n)$ is "much smaller" than n. Here is what we have seen in the preceding chapter.

Theorem 21.14.

   (i) For all $n \ge 3$, $C(n) < n$; hence $\delta(n) > 1$.

   (ii) More precisely, for all $n \ge 2$, $C(n) = 2\sqrt{n-1}$; hence $\delta(n) \ge n/(2\sqrt{n-1})(\ge \sqrt{n}/2)$.

Corollary 21.15. *Fix $n \ge 1$. Then, for any $\delta < \delta(n)$, there is an uncountable family (E_i) in OS_n such that*

$$\delta < d_{cb}(E_i, E_j). \qquad\qquad (21.10) \qquad \forall i \ne j$$

In particular, this holds for some $\delta > 1$ for all $n \ge 3$.

Proof. This follows by a standard argument: Let (E_i) be a maximal collection in OS_n satisfying (21.10). Let $D = \{E_i\}$. By maximality we have $d_{cb}(E, D) \le \delta$ for all E in OS_n, and since $\delta < \delta(n)$, the set D cannot be countable.   ∎

Remark. Actually, "countable" can be replaced by "with cardinality less than the continuum" in the proofs, so we effectively obtain a continuous collection satisfying (21.10).

Remark. By Theorem 21.3 (or Theorem 21.4) we have $\delta(n) \le \sqrt{n}$; hence the above lower bound for $\delta(n)$ is sharp, at least asymptotically. However, it might be that Corollary 21.15 itself can be considerably improved: Perhaps its conclusion is even true for all $\delta < n$. A result very close to this conjecture has been recently established in [ORi].

We will now apply Theorem 21.13 and (21.6) to the special case $X = C^*(F_\infty)$. To simplify the notation, we set

$$d_f(E) = d_{SC^*(F_\infty)}(E).$$

Then, recalling Theorem 21.14, we find:

Corollary 21.16. *For any $n \ge 2$ we have*

$$n/C(n) \le \sup\{d_f(E) \mid E \in OS_n\}.$$

In particular, for any $n > 2$ there is an E such that $d_f(E) > 1$.

Problem. *Find an explicit example of an E such that $d_f(E) > 1$.*

Remark. The class of all n-dimensional operator spaces E such that $d_f(E) = 1$ has surprisingly nice stability properties. This class is studied in detail in [Har1]. We will see in (22.4) that it is stable by duality. In addition, it is proved in [Har1] that this class is stable under the Haagerup tensor product and under complex interpolation. More precisely, we have (see [Har1] for details):

(i) For any operator spaces E_1, E_2 we have
$$d_f(E_1 \otimes_h E_2) \le d_f(E_1)d_f(E_2)$$
and
$$d_f(E_1 \otimes_{\min} E_2) \le d_f(E_1)d_{S\mathcal{K}}(E_2).$$

(ii) For any compatible pair (E_0, E_1) of n-dimensional operator spaces we have
$$d_f((E_0, E_1)_\theta) \le d_f(E_0)^{1-\theta}d_f(E_1)^\theta.$$
In particular, $d_f((E_0, E_1)_\theta) = 1$ if $d_f(E_0) = d_f(E_1) = 1$.

These stability properties (in particular the last one) explain why all the natural examples belong to the class of operator spaces E such that $d_f(E) = 1$. Indeed, first observe that all exact operator spaces (for instance, $\mathcal{K}$, $\min(\ell_2)$, $C[0, 1], \ldots$) are in this class by (22.2) below. Then the stability under duality shows that the class also contains, for example, S_1, $\max(\ell_2)$, $L_1[0, 1]$. Finally, using interpolation, we find that $S_p, L_p, \ldots$ also belong to the class for all $1 < p < \infty$. In particular, the operator Hilbert space OH is in this class, that is, we have
$$d_f(OH) = 1.$$

Exercises

Exercise 21.1. Let (p_m) be a sequence of norms (or seminorms) on a finite-dimensional vector space E. If (p_m) converges pointwise, then it automatically converges uniformly over any bounded subset of E.

Exercise 21.2. Let E_m $(m \ge 1)$ and E be n-dimensional Banach spaces. Show that the following assertions are equivalent:

(i) $d(E_m, E) \to 1$ when $m \to \infty$.

(ii) For any basis $\{e_i \mid 1 \le i \le n\}$ in E there is, for each m, a basis $\{e_i(m) \mid 1 \le i \le n\}$ in E_m such that
$$\forall\, x = (x_i)_{i \le n} \in \mathbb{C}^n \qquad \lim_{m \to \infty} \left\| \sum_1^n x_i e_i(m) \right\|_{E_m} = \left\| \sum_1^m x_i e_i \right\|_E.$$

(ii)′ For some basis, the same as (ii) holds.

(iii) For any free ultrafilter $\mathcal{U}$ on $\mathbb{N}$, we have

$$\Pi E_m / \mathcal{U} = E \text{ (isometrically)}.$$

Exercise 21.3. Let E_m $(m \geq 1)$ and E be n-dimensional operator spaces. Show that the following are equivalent.

(i) $d_N(E_m, E) \to 1$ for any $N \geq 1$ (i.e., E_m tends weakly to E in OS_n).

(ii) For any basis (e_i) in E, there is, for each N and m, a basis $\{e_i^N(m) \mid 1 \leq i \leq m\}$ in E_m such that, for any n-tuple $(a_i)_{i \leq n}$ in M_N, we have

$$\lim_{m \to \infty} \left\| \sum a_i \otimes e_i^N(m) \right\|_{M_N(E_m)} = \left\| \sum a_i \otimes e_i \right\|_{M_N(E)}.$$

(ii)′ For some basis (e_i) the same as (ii) holds.

(iii) For any free ultrafilter $\mathcal{U}$ on $\mathbb{N}$, we have $\Pi E_m / \mathcal{U} = E$ (completely isometrically).

Exercise 21.4. Let $\{e_i(m) \mid 1 \leq i \leq n\}$ and $\{e_i \mid 1 \leq i \leq n\}$ be bases in E_m and E, respectively, such that, for any N and any n-tuple $(a_i)_{i \leq n}$ in M_N, we have

$$\lim_{m \to \infty} \left\| \sum a_i \otimes e_i(m) \right\|_{M_N(E_m)} = \left\| \sum a_i \otimes e_i \right\|_{M_N(E)}.$$

Let $(\xi_i(m))$ and (ξ_i) denote the biorthogonal basis of E_m^* and E^*, respectively. Then show that

$$\forall\, N \geq 1\ \forall\, (a_i) \in (M_N)^n \qquad \lim_{m \to \infty} \left\| \sum a_i \otimes \xi_i(m) \right\|_{M_N(E_m^*)}$$
$$= \left\| \sum a_i \otimes \xi_i \right\|_{M_N(E^*)}.$$

Hint: Recall Proposition 1.12.

Exercise 21.5. Let $\{e_i(m) \mid 1 \leq i \leq n\}$ and $\{e_i \mid 1 \leq i \leq n\}$ be as in Exercise 21.4. Let X be any exact operator space with $C = d_{S\mathcal{K}}(X)$. We then have for any $(x_i)_{i \leq n}$ in X^n .

$$C^{-1} \left\| \sum x_i \otimes e_i \right\| \leq \liminf \left\| \sum x_i \otimes e_i(m) \right\| \leq \limsup \left\| \sum x_i \otimes e_i(m) \right\|$$
$$\leq C \left\| \sum x_i \otimes e_i \right\|.$$

In particular, if $d_{S\mathcal{K}}(X) = 1$, we have

$$\lim_{m \to \infty} \left\| \sum x_i \otimes e_i(m) \right\|_{\min} = \left\| \sum x_i \otimes e_i \right\|_{\min}.$$

Chapter 22. $B(H) \otimes B(H)$

Recently, E. Kirchberg [Ki2, Ki5] revived interest in the study of nuclear pairs of C^*-algebras, that is, pairs (A, B) such that $A \otimes_{\min} B = A \otimes_{\max} B$, or equivalently, such that there is only one C^*-norm on the algebraic tensor product $A \otimes B$. Recall that this happens for all B iff A is nuclear (see Chapters 11 and 12).

The C^*-algebras $C^*(\mathbb{F}_\infty)$ and $B(H)$ are typical examples of nonnuclear C^*-algebras. Their nonnuclearity was first proved respectively in [Ta2] and [Wa1]. (However, this is now quite clear since every separable C^*-algebra is a quotient of $C^*(\mathbb{F}_\infty)$ and a subalgebra of $B(H)$. Since nuclear C^*-algebras have nuclear quotients and exact subalgebras, it suffices to know the existence of a single nonexact C^*-algebra, and the above Theorem 21.5 clearly guarantees that $C^*(\mathbb{F}_\infty)$ is not exact.) Therefore, Kirchberg's Theorem (already proved above as Theorem 13.1) came as a surprise: For any free group $\mathbb{F}$ (and any H) we have

$$C^*(\mathbb{F}) \otimes_{\min} B(H) = C^*(\mathbb{F}) \otimes_{\max} B(H).$$

For a long time, the following question remained open: If a C^*-algebra A satisfies

$$A \otimes_{\min} A^{op} = A \otimes_{\max} A^{op}, \tag{22.1}$$

is A nuclear?

In [Ki2], Kirchberg finally gave a counterexample, using Theorem 13.1 as a starting point. Recall the following generalization of Theorem 13.1 proved above as Theorem 16.2:

Theorem 22.1. *Let A, B be C^*-algebras. If A has the LLP and B the WEP, then*

$$A \otimes_{\min} B = A \otimes_{\max} B.$$

In particular, if A has both the LLP and the WEP, then

$$A \otimes_{\min} A^{op} = A \otimes_{\max} A^{op}. \tag{22.1}$$

Kirchberg also showed that (22.1) implies that A has the WEP. This was clarified by Haagerup [H4] (see Theorem 15.6), who showed that a C^*-algebra A has the WEP iff for any finite sequence (a_i) in A we have

$$\left\| \sum a_i \otimes a_i^* \right\|_{A \otimes_{\min} A^{op}} = \left\| \sum a_i \otimes a_i^* \right\|_{A \otimes_{\max} A^{op}}.$$

Thus, to produce a nonnuclear example satisfying (22.1), it would suffice to solve positively the following.

Problem. Is there a nonnuclear C^*-algebra with both the WEP and the LLP?

Kirchberg's construction in [Ki2] (of a nonnuclear C^*-algebra satisfying (22.1)) comes very close, but unfortunately the preceding problem remains open.

Nevertheless, as Kirchberg [Ki2] pointed out, there might be much simpler examples, and in that direction he raised in [Ki2] the following problems.

Problem A: *Does (22.1) holds if $A = B(H)$?*

Problem B: *Does (22.1) holds if $A = C^*(\mathbb{F}_\infty)$?*

Note that in these two cases A and A^{op} are isomorphic C^*-algebras, so that Problem A (resp. B) can be reformulated simply as:

Is there a unique C^*-norm on $B(H) \otimes B(H)$ (resp. $C^*(\mathbb{F}_\infty) \otimes C^*(\mathbb{F}_\infty)$)?

As we will see in the following, the nonseparability of OS_n, discussed in the preceding chapter, gives a negative answer to Problem A. However, Problem B is still open.

We now wish to study the finite-dimensional operator subspaces of $C^*(\mathbb{F}_\infty)$, in analogy with our study of the subspaces of $\mathcal{K}$ in Chapter 17. Consider an n-dimensional subspace $E \subset C^*(\mathbb{F}_\infty)$ and a C^*-embedding $C^*(\mathbb{F}_\infty) \subset B(\mathcal{H})$. Then, for any u in $E \otimes B(H)$, we have by Theorem 22.1

$$\|u\|_{B(\mathcal{H}) \otimes_{\max} B(H)} \leq \|u\|_{C^*(\mathbb{F}_\infty) \otimes_{\max} B(H)} = \|u\|_{C^*(\mathbb{F}_\infty) \otimes_{\min} B(H)} = \|u\|_{E \otimes_{\min} B(H)}.$$

Thus the min-norm on $E \otimes B(H)$ coincides with the norm induced by $B(\mathcal{H}) \otimes_{\max} B(H)$. It turns out that this property characterizes subspaces of $C^*(\mathbb{F}_\infty)$.

Recall that we denote

$$d_f(E) = d_{SC^*(F_\infty)}(E) = \inf\{d_{cb}(E, \widehat{E}) \mid \widehat{E} \subset C^*(\mathbb{F}_\infty)\}.$$

By Corollary 21.12, we already know that, for any $n > 2$, there are n-dimensional operator spaces E such that $d_f(E) > 1$. Moreover, by (21.6) and Theorem 21.13, we know this number can actually be fairly large when n itself is large. Following [JP], we will now relate this to the ratio between the minimal and maximal norms on tensors of rank n in $B(H) \otimes B(H)$.

Theorem 22.2. *([JP]) Let $\alpha \geq 0$ be a constant and let $X \subset B(\mathcal{H})$ be an operator space. The following are equivalent.*

(i) *$d_f(E) \leq \alpha$ for all finite-dimensional subspaces $E \subset X$.*

(ii) *For any H and any operator space $F \subset B(H)$, we have*

$$\forall u \in E \otimes F \qquad \|u\|_{B(\mathcal{H}) \otimes_{\max} B(H)} \leq \alpha \|u\|_{\min}.$$

(iii) *The same as (ii) with $H = \ell_2$ and $F = B(\ell_2)$.*

Corollary 22.3. *([JP]) Let $E \subset B(\mathcal{H})$ be an n-dimensional operator space. Then*

$$d_f(E) = \sup\left\{\frac{\|u\|_{B(\mathcal{H}) \otimes_{\max} B(\ell_2)}}{\|u\|_{\min}} \mid u \in E \otimes B(\ell_2)\right\}.$$

Remark. Note that, using (17.5), this shows in particular that

$$d_f(E) \le d_{SK}(E) \tag{22.2}$$

for any finite-dimensional operator space E. In particular, by Theorem 21.4, this implies

$$\forall E \in OS_n \qquad d_f(E) \le \sqrt{n}. \tag{22.3}$$

Remark. It is observed in [JP] that, for a C^*-algebra A, the condition $d_f(A) = 1$ ensures that any finite-dimensional subspace $E \subset A$ is completely isometric to a subspace of $C^*(F_\infty)$. Note, however, that the algebra A need not embed as a C^*-algebra into $C^*(F_\infty)$. For instance, let A be the Cuntz algebra (see §9.4). It is nuclear and hence a fortiori exact, so that $d_f(A) = d_{SK}(A) = 1$; however, A does not embed into $C^*(F_\infty)$, because $C^*(F_\infty)$ embeds into a direct sum of matrix algebras (see Exercise 13.1); hence left invertible elements in it are right invertible, and the latter property obviously fails in the Cuntz algebra. Nevertheless, by the Choi and Effros lifting theorem (cf., e.g., [Wa2, p. 53]; see also Remark 16.3(v)) there is a unital completely positive (and completely contractive) factorization of the identity of the Cuntz algebra (or any separable nuclear C^*-algebra) through $C^*(F_\infty)$.

The preceding theorem leads naturally to the following.

Definition. Let $E_1 \subset B(H_1), E_2 \subset B(H_2)$ be arbitrary operator spaces. We will denote by $\| \ \|_M$ the norm induced on $E_1 \otimes E_2$ by $B(H_1) \otimes_{\max} B(H_2)$ and by $E_1 \otimes_M E_2$ its completion with respect to this norm. Clearly, $E_1 \otimes_M E_2$ can be viewed as an operator space embedded into $B(H_1) \otimes_{\max} B(H_2)$.

It can be checked easily, using the extension property of c.b. maps into $B(H)$ (Corollary 1.7), together with (11.5) and (11.7) (see [JP] for more details) that $\| \ \|_M$ and $E_1 \otimes_M E_2$ do not depend on the particular choices of complete embeddings $E_1 \subset B(H_1), E_2 \subset B(H_2)$. Moreover, we have

Lemma 22.4. Let F_1, F_2 be two operator spaces.
  (i) Consider c.b. maps $u_1 \colon E_1 \to F_1$ and $u_2 \colon E_2 \to F_2$. Then $u_1 \otimes u_2$ defines a c.b. map from $E_1 \otimes_M E_2$ to $F_1 \otimes_M F_2$ with $\|u_1 \otimes u_2\|_{CB(E_1 \otimes_M E_2, F_1 \otimes_M F_2)} \le \|u_1\|_{cb} \|u_2\|_{cb}$.
  (ii) If u_1 and u_2 are complete isometries, then $u_1 \otimes u_2 \colon E_1 \otimes_M E_2 \to F_1 \otimes_M F_2$ is also a complete isometry.

Remark. When E_1, E_2 are C^*-algebras, $E_1 \otimes_M E_2$ can be identified with a C^*-subalgebra of $B(H_1) \otimes_{\max} B(H_2)$, so that this tensor product $\otimes_M$ makes sense in both categories of operator spaces and C^*-algebras.

The next result analyzes more closely the significance of $\|u\|_M < 1$ for $u \in E \otimes F$. It turns out to be closely connected to the factorizations of the associated linear operator $U \colon F^* \to E$ through a subspace of $C^*(F_\infty)$. This can be considered as analogous to Theorem 17.1 or Theorem 17.2 but with $C^*(F_\infty)$ in the place of $\mathcal{K}$.

Proposition 22.5. *Let* E, F *be operator spaces; let* $u \in E \otimes F$, *and let* $U \colon F^* \to E$ *be the associated finite rank linear operator. Consider a finite-dimensional subspace* $S \subset C^*(\mathbb{F}_\infty)$ *and a factorization of* U *of the form* $U = ba$ *with bounded linear maps* $a \colon F^* \to S$ *and* $b \colon S \to E$, *where* $a \colon F^* \to S$ *is weak-∗ continuous. Then*

$$\|u\|_M = \inf\{\|a\|_{cb}\|b\|_{cb}\},$$

where the infimum runs over all such factorizations of U.

Proof. Assume $E \subset B(H)$ and $F \subset B(K)$ with H, K Hilbert. It clearly suffices to prove this in the case when E and F are both finite-dimensional. Assume U factorized as above with $\|a\|_{cb}\|b\|_{cb} < 1$. Then, by Kirchberg's Theorem 13.1, the min- and max-norms are equal on $C^*(\mathbb{F}_\infty) \otimes B(K)$. Hence, by Lemma 22.4(ii), we have isometrically $S \otimes_{\min} F = S \otimes_M F$, so that if $\widehat{a}$ is the element of $S \otimes_{\min} F$ associated to a, we have $\|\widehat{a}\|_M = \|a\|_{cb}$ and $u = (b \otimes I_F)(\widehat{a})$. Therefore, by Lemma 22.4(i), we have $\|u\|_M \leq \|b\|_{cb}\|\widehat{a}\|_M \leq \|a\|_{cb}\|b\|_{cb} < 1$. Thus we obtain $\|u\|_M \leq \inf\{\|a\|_{cb}\|b\|_{cb}\}$.

We now turn to the converse inequality. Assume $\|u\|_M < 1$. Let G be a large enough free group so that $B(H)$ is a quotient of $C^*(G)$, and let $q \colon C^*(G) \to B(H)$ be the quotient ∗-homomorphism. By the exactness of the maximal tensor product (see Exercise 11.1), if we view u as sitting in $B(H) \otimes B(K)$, there is a lifting $\widehat{u} \in C^*(G) \otimes B(K)$ of u with $\|\widehat{u}\|_{\max} < 1$. A fortiori $\|\widehat{u}\|_{\min} < 1$. By Lemma 17.2, since $u \in B(H) \otimes F$, there must actually exist a lifting $\widehat{u}$ in $C^*(G) \otimes F$ with $\|\widehat{u}\|_{\min} < 1$. Let $S \subset C^*(G)$ be a finite-dimensional subspace such that $\widehat{u} \in S \otimes F$. Note that since S is separable (say), there is a subgroup $G_1 \subset G$ isomorphic to $\mathbb{F}_\infty$ such that $S \subset C^*(G_1) \simeq C^*(\mathbb{F}_\infty)$. Let $a \colon F^* \to S$ be the linear map associated to $\widehat{u}$, and let b be the restriction of q to S. Since $\widehat{u}$ lifts u, we have $U = ba$, $\|b\|_{cb} \leq \|q\|_{cb} \leq 1$, and $\|a\|_{cb} = \|\widehat{u}\|_{\min} \leq \|\widehat{u}\|_{\max} < 1$. Thus we conclude $\inf\{\|a\|_{cb}\|b\|_{cb}\} \leq \|u\|_M$. ∎

Applying this result to $U = I_E$, we obtain

Corollary 22.6. *Let* E *be a finite-dimensional operator space. Let* $i_E \in E \otimes E^*$ *be the tensor associated to the identity on* E. *Then*

$$\|i_E\|_M = d_f(E).$$

In particular, we have

$$d_f(E) = d_f(E^*). \tag{22.4}$$

The next result was pointed out to me by N. Ozawa.

Corollary 22.7. For any finite-dimensional operator space E, there is a subspace of $\widehat{E} \subset C^*(\mathbb{F}_\infty)$ and an isomorphism $u \colon E \to \widehat{E}$ such that

$$\|u\|_{cb}\|u^{-1}\|_{cb} = d_f(E).$$

In particular, E satisfies $d_f(E) = 1$ iff E is completely isometric to a subspace of $C^*(\mathbb{F}_\infty)$.

Proof. Indeed, assume $E \subset B(H)$, $E^* \subset B(K)$ and let $t \in E^* \otimes B(H)$ be the tensor representing the inclusion $E \subset B(H)$. By the preceding, we have $d_f(E) = \|t\|_{B(K) \otimes_{\max} B(H)}$. By the proof of Proposition 22.5 (and with the same notation), $d_f(E)$ coincides with the infimum of $\|T\|_{E^* \otimes_{\min} C^*(G)}$ over all T in $E^* \otimes C^*(G)$ lifting t. By Lemma 2.4.5, this infimum is actually attained: There is a T with $\|T\|_{\min} = d_f(E)$. But now T defines a linear map $T_1 \colon E \to C^*(G)$ with $\|T_1\|_{cb} = d_f(E)$. Letting $\widehat{E} = T_1(E)$, and with $u = T_1$ viewed as acting from E to $\widehat{E}$, we find $\|u\|_{cb} = d_f(E)$ and $\|u^{-1}\|_{cb} \le \|q\|_{cb} \le 1$. ∎

Remark. (Due to N. Ozawa) Using Arveson's ideas in [Ar4], it is possible to show more generally that, if a separable operator space X is such that $d_f(E) = 1$ for every finite-dimensional subspace $E \subset X$ and if, in addition, X admits a net of completely contractive finite rank maps tending pointwise to the identity, then X embeds completely isometrically into $C^*(\mathbb{F}_\infty)$. This follows from Exercise 2.4.2. In particular, OH embeds completely isometrically into $C^*(\mathbb{F}_\infty)$. Unfortunately, however, we cannot describe the embedding more explicitly.

We can now relate the nonseparability of OS_n with the possible C^*-norms on $B(H) \otimes B(H)$, as follows (we denote by $\operatorname{rk}(u)$ the rank of u).

Proposition 22.8. Let $H = \ell_2$. For any $n \ge 1$ we define

$$\lambda(n) = \sup \left\{ \frac{\|u\|_{\max}}{\|u\|_{\min}} \;\middle|\; u \in B(H) \otimes B(H), \quad \operatorname{rk}(u) \le n \right\}.$$

Then

$$\lambda(n) = \sup\{d_f(E) \mid E \in OS_n\} \quad \text{and} \quad \lambda(n) \ge \delta(n). \tag{22.5}$$

Proof. Any u in $B(H) \otimes B(H)$ with rank at most n can be viewed as an element of $E \otimes B(H)$ for some n-dimensional subspace of $B(H)$. Then (22.5) follows immediately from Corollary 22.3 and (21.6). ∎

We can now exploit the results of the previous chapter to settle Problem A.

Theorem 22.9. *We have*

$$B(H) \otimes_{\min} B(H) \neq B(H) \otimes_{\max} B(H). \tag{22.6}$$

More precisely:

    (i) $\lambda(n) \leq \sqrt{n}$ for all n.

    (ii) $\lambda(n) > 1$ for all $n > 2$.

    (iii) $\lambda(n) \geq n/(2\sqrt{n-1}) \geq \sqrt{n}/2$ for all $n \geq 3$.

Proof. (i) follows from (22.3) and (22.5), and (ii) and (iii) follow from Theorem 21.14 and (22.5). Of course (ii) implies (22.6) since the equality can only be isometric (cf. Proposition 1.5). ∎

Remark. Kirchberg proved in [Ki2] that if his conjecture 16.12 is correct, then, for any C^*-algebra A, for any $\varepsilon > 0$, and for any finite-dimensional operator subspace $E \subset A^*$, there is a subspace $\widehat{E} \subset S_1$ with $d_{cb}(E, \widehat{E}) < 1 + \varepsilon$. In particular, it would follow (by Corollary 2.13.3) that the metric subspace $\mathcal{C}_n \subset OS_n$ formed by the n-dimensional subspaces of duals (or preduals) of C^*-algebras would be d_{cb}-separable for any n. Thus, to disprove conjecture 16.12, in analogy with Theorem 22.9, one could have hoped to show that $\mathcal{C}_n$ is not separable. Unfortunately, this totally fails; indeed, $\mathcal{C}_n$ is d_{cb}-separable ([JLM]), and actually it is even compact ([Oz1])!

Chapter 23. Completely Isomorphic C^*-Algebras

It is natural to wonder whether two C^*-algebras that are completely isomorphic must be isomorphic (as C^*-algebras). The answer is negative. For instance, Christensen and Sinclair (see [CS5]) (extending previous remarks of Haagerup and Lindenstrauss) proved that any injective (i.e. hyperfinite [Co1]) von Neumann algebra with separable predual is completely isomorphic either to ℓ_∞^n ($n \in \mathbb{N}$), ℓ_∞ or $B(H)$. So $B(H)$ and, for example, $\bigoplus_n M_n$ provide a negative answer to the above question. (See also [RW] and [Blo] for related results.)

In a different direction, A. Arias [A1] proved that the reduced C^*-algebras $C_\lambda^*(F_n)$ associated to the free groups with n generators are mutually completely isomorphic, that is, $C_\lambda^*(F_n) \simeq C_\lambda^*(F_k)$ as operator spaces for any n, k. Similarly, the von Neumann algebras are all mutually completely isomorphic. However, it is known from K-theory that $C_\lambda^*(F_n)$ is not C^*-isomorphic to $C_\lambda^*(F_k)$ when $n \neq k$ (see [PV1–2]). The analogous nonisomorphism in the von Neumann case remains a major outstanding open problem.

These negative results suggest that completely isomorphic C^*-algebras might be quite different. Nevertheless, several basic structural properties of C^*-algebras, such as nuclearity, injectivity (in the von Neumann case), the weak expectation property (in short, WEP), and exactness, are preserved under complete isomorphisms. The rest of this chapter is devoted to this. We introduced nuclearity in Chapter 11. It is known (see Theorem 11.5) that a C^*-algebra A is nuclear iff the identity on A is approximable by a net (u_i) of finite rank maps, of the form $A \xrightarrow{a_i} M_{n_i} \xrightarrow{b_i} A$ such that $\sup_{i \in I} \|a_i\|_{cb} \|b_i\|_{cb} \leq 1$. A priori, this property does not seem stable under complete isomorphism. However (cf. [P1]), it turns out that A is nuclear iff there is a constant C for which the identity of A is approximable by a net as above with $\sup_{i \in I} \|a_i\|_{cb} \|b_i\|_{cb} \leq C$. The latter is clearly invariant under complete isomorphisms.

A similar situation reappears for injective von Neumann algebras. A von Neumann algebra $M \subset B(H)$ is called injective if there is a completely contractive and completely positive projection $P \colon B(H) \to M$. (Actually, by a result of Tomiyama [To1], in this situation, any contractive P is automatically completely both contractive and positive.) It was proved in [P3, P1] and independently in [CS3] that M is injective as soon as there is a constant C and a projection $P \colon B(H) \to M$ with $\|P\|_{cb} \leq C$.

This was extended with more general von Neumann algebras in the place of $B(H)$ in [P7, CS4, H4].

A similar remark is valid for the class of C^*-algebras with WEP (i.e., those A for which the canonical inclusion $i_A \colon A \to A^{**}$ factors completely positively and (completely) contractively through $B(H)$; see Chapter 15).

Recapitulating, all these statements give us

Theorem 23.1. *Let A, B be two C^*-algebras that are completely isomorphic. Then:*

    (i) If A is nuclear, so is B.

    (ii) If A, B are von Neumann algebras, and if A is injective, so is B.

    (iii) If A has the WEP, so does B.

    (iv) If A is exact, so is B.

Proof. (i), (ii), and (iii) are immediate consequences of the preceding facts. Concerning (iv), note that A is exact iff there is a constant C such that, for all finite-dimensional subspaces $E \subset A$, we have $d_{SK}(E) \leq C$ (see, e.g., Corollary 17.5). From this (iv) follows, but it was already a consequence of Kirchberg's earlier work. ∎

Remark. Concerning the injectivity of a von Neumann algebra $M \subset B(H)$, it is unclear whether M must be injective if there merely exists a *bounded* projection from $B(H)$ onto M. For some results in this direction see [P13, HP2]. As observed in [P13] in the C^*-case, for any operator space $X \subset B(H)$ the projection constant

$$\lambda(X) = \inf\{\|P\| \mid P: \ B(H) \to X, \ \text{projection onto } X\}$$

is invariant under completely isometric isomorphism. (The proof uses the extension property of c.b. maps into $B(H)$; see Corollary 1.7.) In particular, $\lambda(X)$ does not depend on the particular embedding $X \subset B(H)$. Analogously, if Y is another operator space, we have $\lambda(X) \leq \lambda(Y)d_{cb}(X, Y)$. Let M_1, M_2 be two von Neumann algebras and let $M_1 \bar{\otimes} M_2$ denote their von Neumann algebraic tensor product. It is proved in [P13] that

$$\lambda(M_1 \bar{\otimes} M_2) \geq \lambda(M_1)\lambda(M_2).$$

Similar supermultiplicative estimates can be proved for certain constants related to the approximation property; see [SS1].

    Of course similar estimates hold for the *completely* bounded projection constant in $B(H)$ defined as $\lambda_{cb}(X) = \inf\{\|P\|_{cb}\}$, and $\lambda_{cb}(X) \geq \lambda(X)$, but not much is known about the relationship between these two constants even when X is a C^*-algebra.

Chapter 24. Injective and Projective Operator Spaces

This chapter is mainly a survey without proof.

In general, injective spaces are those that satisfy an extension property while projective spaces satisfy a lifting property. The extension property corresponds to the following diagram:

$$
\begin{array}{ccc}
Y & & \\
\cup & & \searrow \\
S & \xrightarrow{\;u\;} & X
\end{array}
$$

where S is a subspace of Y and u is a "morphism" that we wish to extend to the whole of Y. When this is possible for any S, Y, and u with values in X, we say that X has the extension property or is *injective*. This property is interesting in various categories. See, for example, [MN, p. 170] for Banach lattices, [CE3] for operator systems, [Li, Lac, Bou5] for Banach spaces, and [Co1] or [KR] for von Neumann algebras.

In all these cases, injective objects play a fundamental role. Thus it is not surprising that they should also be of interest in the operator space category. There, of course, X, Y are operator spaces, $S \subset Y$ is a closed subspace, and u is c.b. We will say that an operator space X has the extension property (or is injective) if any c.b. map u as above admits a c.b. extension $\widetilde{u} \colon Y \to X$ so that the resulting diagram commutes:

$$
\begin{array}{ccc}
Y & & \\
& \searrow \widetilde{u} & \\
\cup & & \\
S & \xrightarrow{\;u\;} & X
\end{array}
$$

It is easy to see that when this holds there is a constant λ such that $\widetilde{u}$ can always be found with $\|\widetilde{u}\|_{cb} \leq \lambda \|u\|_{cb}$. We then say that X is λ-injective. For example, by Corollary 1.7, we know that $B(H)$ is 1-injective for any H. For Banach spaces, $\ell_\infty(I)$ is 1-injective for any set I, and any Banach space embeds isometrically into $\ell_\infty(I)$ for some I; therefore, in the Banach category, λ-injective spaces are just the λ-complemented subspaces of $\ell_\infty(I)$. Since $B(H)$ is also "universal" for operator spaces (see §2.12), the analogous statement for operator spaces is immediate, as follows.

Proposition 24.1. *Let $\lambda \geq 1$. The following properties of an operator space X are equivalent:*

(i) *X is λ-injective.*

(ii) *For any completely isometric embedding $X \subset Y$ into an operator space Y, there is a c.b. projection $P \colon Y \to X$ with $\|P\|_{cb} \leq \lambda$.*

(iii) *There is a completely isometric embedding $X \subset B(H)$ and a c.b. projection $P \colon B(H) \to X$ with $\|P\|_{cb} \leq \lambda$.*

(iv) The identity on X admits, for a suitable H, a factorization through $B(H)$ of the form $X \xrightarrow{v} B(H) \xrightarrow{w} X$ with $\|v\|_{cb}\|w\|_{cb} \leq \lambda$.

Proof. (i) $\Rightarrow$ (ii): Let $u = I_X$. Any extension $\tilde{u}: Y \to X$ must be a projection onto X, whence (ii).

(ii) $\Rightarrow$ (iii) and (iii) $\Rightarrow$ (iv) are obvious.

(iv) $\Rightarrow$ (i): Consider $u: S \to X$ and let $T = vu: S \to B(H)$. Since T is $B(H)$-valued, by Corollary 1.7, T admits an extension $\tilde{T}: Y \to B(H)$ with $\|\tilde{T}\|_{cb} = \|T\|_{cb}$. Then the operator $\tilde{u} = w\tilde{T}: Y \to X$ extends u and satisfies $\|\tilde{u}\|_{cb} \leq \|w\|_{cb}\|\tilde{T}\|_{cb} = \|w\|_{cb}\|T\|_{cb} = \|w\|_{cb}\|vu\|_{cb} \leq \|w\|_{cb}\|v\|_{cb}\|u\|_{cb} \leq \lambda\|u\|_{cb}$. ∎

The particular case $\lambda = 1$ is really special. In the (real or complex) Banach category, a space X is 1-injective iff X is isometric to the space of (real- or complex-valued) continuous functions on an extremally disconnected compact (also called Stonean) space (cf., e.g., [Lac, p. 92]). If, moreover, X is a dual space, then this holds iff X is isometric to $L_\infty(\Omega, \mu)$ for some measure space (Ω, μ). In the operator space framework, the 1-injective objects were characterized as follows by Ruan ([Ru2]), using some important previous work by Hamana [Ham].

Theorem 24.2. *([Ru2]) An operator space X is 1-injective iff there is a 1-injective C^*-algebra A and projections p, q in A such that*

$$X \simeq pAq \text{ (completely isometrically)}.$$

Remark. Roger Smith observed (unpublished) that, if X is finite-dimensional, we can choose A finite-dimensional too.

Remark. Note that the row (resp. column) space R (resp. C) is clearly 1-injective. This corresponds to $A = B(\ell_2)$ with $p = e_{11}$ and $q = I$ (resp. $p = I$ and $q = e_{11}$) in Theorem 24.2. By [P13], a reflexive operator space can be λ-injective for some λ only if the underlying Banach space is isomorphic to a Hilbert space.

Theorem 24.2 reduces the classification of 1-injective operator spaces to that of 1-injective C^*-algebras. For dual spaces, the analogous result is as follows.

Theorem 24.3. *([EOR]) Let X be an operator space that is dual as a Banach space. Then X is 1-injective iff there is an injective von Neumann algebra M and a projection p in M such that*

$$X \simeq pM(1-p) \text{ (completely isometrically)}.$$

Moreover, when this holds we can make sure that the preceding isomorphism is also a weak-$*$ isomorphism, in such a way that $X_* = pM_*(1 - p)$ appears as the operator space predual of X.

The preceding result is closely connected to the theory of "triple operator systems," for which we refer the reader in particular to [NR1, NR2]. The interest of the preceding theorem is that injective von Neumann algebras seem much better understood ([Co1]) than injective C^*-algebras. In any case, the preceding two characterizations are valid only for $\lambda = 1$, and the classification of λ-injective operator spaces for $\lambda > 1$ seems to be much more delicate, just as it is for Banach spaces. For the latter spaces, the theory of the so-called $\mathcal{L}_p$-spaces, especially for $p = 1$ and $p = \infty$, plays a vital role; it is thus natural to develop the analogous theory for operator spaces. This was recently started in the papers [ER12, JNRX, JR, JOR].

In another direction, it is natural to investigate injectivity in the separable context, by restricting the extension property discussed at the beginning of this chapter to separable spaces Y and X. A (separable) Banach space X with this property is called *separably injective*. It is a classical result proved in 1941 by Sobczyk that $X = c_0$ is separably injective, and the problem whether this is the only (up to isomorphism) separable infinite-dimensional Banach space with this property remained open for a long time, until M. Zippin finally proved in the late 1970s that it is indeed so (see the references in [OR]).

Of course, the space $\mathcal{K}$ of all compact operators on ℓ_2 is the obvious operator space analog of c_0. Another possible analog is the direct sum in the c_0-sense of the spaces $\{M_n \mid n \geq 1\}$, which we will denote by $\mathcal{K}_{(0)}$. Thus, one would expect these to satisfy the analogous "separable injectivity" in the c.b. context. Unfortunately, however, Kirchberg [Ki2] showed that, in either case, it is not so: Thus, there is a separable operator space Y and a c.b. map $u\colon S \to \mathcal{K}$, defined on a subspace $S \subset Y$ that does not admit any c.b. extension. Nevertheless, Haskell Rosenthal [R] discovered that the separable (c.b.) extension property holds if the space $\mathcal{K}$ is replaced by either

$$\left(\oplus \sum_{n \geq 1} C_n \right)_{c_0} \quad \text{or} \quad \left(\oplus \sum_{n \geq 1} R_n \right)_{c_0}.$$

Moreover, he proved that, if Y is restricted to be locally reflexive (in addition to being separable), then $\mathcal{K}_{(0)}$ satisfies a closely related complementation property, which he called the CSCP: A separable locally reflexive operator space X has the CSCP if, whenever X is completely isomorphically embedded in a locally reflexive, separable superspace Y, there is a c.b. projection $P\colon Y \to X$ (this corresponds to the extension property when $u\colon S \to X$ is a complete isomorphism *onto* X). Subsequently, with several co-authors ([OR,

AR]) he proved that $\mathcal{K}$ itself has this property. Ozawa also found another proof of this.

Remark. It still remains an open problem (even if Y is locally reflexive) whether, when $S \subset Y$ with Y separable, any c.b. map $u\colon S \to \mathcal{K}$ admits a *bounded* extension to the whole of Y. Actually, it is an old open problem whether, for any embedding $\mathcal{K} \subset A$ of $\mathcal{K}$ as an ideal in a separable C^*-algebra A, there is a *bounded* projection $P\colon A \to \mathcal{K}$. See Proposition 18.14 for more on this.

We note here in passing that Ozawa [Oz3] exhibited an example of a C^*-algebra $A \subset B(H)$ and a *bounded* linear map $u\colon A \to B(\ell_2)$ that does not admit any *bounded* extension to the whole of $B(H)$.

Remark. We should mention also here the related paper [Blo], where it is proved that if A is a 1-injective operator system on a separable Hilbert space, and if P is a c.b. projection on A, then either the range of P or that of $1 - P$ is completely isomorphic to A.

We now turn to projective spaces, or equivalently, to spaces satisfying a lifting property. For instance, in the Banach setting, the basic example is the space ℓ_1: If $X = \ell_1$, then any contractive map $u\colon X \to Y/S$ into a quotient space admits, for any $\varepsilon > 0$, a lifting $\widetilde{u}\colon X \to Y$ with $\|\widetilde{u}\| < 1 + \varepsilon$. (In general, we cannot take $\varepsilon = 0$.) Contrary to what one would expect at first glance, the analog of this phenomenon for operator spaces is not the space $X = S_1$ of all trace class operators, but instead the direct sum $\ell_1(\{S_1^n \mid n \geq 1\})$ of the family of all the finite-dimensional versions of S_1. More generally:

Definition 24.4. *Let $\lambda \geq 1$. An operator space X is called λ-projective if, for any $\varepsilon > 0$, any c.b. map $u\colon X \to Y/S$ into a quotient operator space (here Y is any operator space and $S \subset Y$ any closed subspace) admits a lifting $\widetilde{u}\colon X \to Y$ with $\|\widetilde{u}\|_{cb} \leq (\lambda + \varepsilon)\|u\|_{cb}$.*

Examples. (i) For any (finite) integer n, S_1^n is 1-projective. Indeed, by (2.3.2) and (2.4.1)′, $CB(S_1^n, Y/S) = M_n(Y/S) = M_n(Y)/M_n(S)$ (isometrically) and $M_n(Y) = CB(S_1^n, Y)$; hence any u in the open unit ball of $CB(S_1^n, Y/S)$ admits a lifting in the open unit ball of $CB(S_1^n, Y)$, which means that S_1^n is 1-projective.

(ii) More generally, if p, q are (orthogonal) projections in M_n, then $X = \{pxq \mid x \in S_1^n\}$ is also 1-projective (since $x \to pxq$ is a completely contractive projection from S_1^n onto X).

(iii) Any direct sum in the ℓ_1-sense of a family $\{X_i \mid i \in I\}$ of λ-projective spaces is again λ-projective. This is easy to see using the defining property of ℓ_1-direct sums (see §2.6): Indeed we have isometrically

$$CB(\ell_1(\{X_i \mid i \in I\}), Y) = \ell_\infty\{CB(X_i, Y) \mid i \in I\},$$

from which our assertion follows.

(iv) In particular, let $\{n_i \mid i \in I\}$ be a family of integers with $n_i \geq 1$. Let p_i, q_i be projections in M_{n_i}. Let $T_i = \{p_i x q_i \mid x \in S_1^{n_i}\}$. Then the space $\ell_1(\{T_i \mid i \in I\})$ is 1-projective. Conversely, any finite-dimensional 1-projective space is of this form (this can be deduced from the remark after Theorem 24.2).

The projective counterpart to Proposition 24.1 is the following simple observation ([B2]).

Proposition 24.5. *Let $\lambda \geq 1$. The following properties of an operator space X are equivalent.*

(i) X is λ-projective.

(ii) For any $\varepsilon > 0$, there is a space Z of the form $Z = \ell_1(\{S_1^{n_i} \mid i \in I\})$ as above such that the identity on X factorizes through Z as follows:

$$I_X\colon X \xrightarrow{\ v\ } Z \xrightarrow{\ w\ } X$$

with $\|v\|_{cb}\|w\|_{cb} < \lambda + \varepsilon$.

Proof. (i) $\Rightarrow$ (ii): By (2.12.2), any space X is completely isometric to a quotient Z/S for a suitable $Z = \ell_1(\{S_1^{n_i} \mid i \in I\})$. Let $u\colon Z \to X$ be the quotient map. Fix $\varepsilon > 0$. If X is assumed λ-projective, there is a lifting $\tilde{u}\colon X \to Z$ with $\|\tilde{u}\|_{cb} < \lambda + \varepsilon$. Then $I_X = u\tilde{u}$ provides the factorization in (ii).

(ii) $\Rightarrow$ (i): Let Z be as in (ii). Since we already know that Z is 1-projective, it is easy to deduce from (ii) that X is λ-projective. $\blacksquare$

Remark. Note that (ii) above implies that X is completely isomorphic to a completely complemented subspace of Z. Since $Z^* = \oplus M_{n_i}$ is clearly injective, we immediately deduce:

Corollary 24.6. *If X is λ-projective, then X^* is λ-injective.*

While there are rather few projective Banach spaces, many more spaces satisfy the "local" version of projectivity (or, equivalently, a local form of the lifting property). The resulting class of Banach spaces is the class of $\mathcal{L}_1$ spaces (see [LiR]), which can be defined in many equivalent ways. One of these is X is $\mathcal{L}_1$ iff X^{**} is isomorphic to a complemented subspace of an L_1-space. In sharp contrast, the operator space versions of the various definitions of $\mathcal{L}_1$-spaces (or, more generally, $\mathcal{L}_p$-spaces) lead to possibly distinct classes of operator spaces; see [ER12]. This difficulty is of course related to the lack of local reflexivity in general.

One of the possible variants is studied in [KyR] under the name of "operator local lifting property," but since we have already used this terminology for a different notion in Chapter 16; so we will change it: An operator space will be called λ-locally projective if, for any map $u\colon X \to Y/S$, any $\varepsilon > 0$, and any finite-dimensional subspace $E \subset X$, the restriction of u to E admits a lifting $\tilde{u}\colon E \to Y$ with $\|\tilde{u}\|_{cb} \leq (\lambda + \varepsilon)\|u\|_{cb}$. It is proved in [KyR] that X has the λ-LLP iff X^* is λ-injective. More recently, in [EOR] the authors prove that this happens for $\lambda = 1$ iff there is an injective von Neumann algebra R and a (self-adjoint) projection p in R such that

$$X^* \simeq (1-p)Rp \quad \text{(completely isometrically)}.$$

It follows that X is 1-locally projective iff there is a net of finite rank maps of the form $X \xrightarrow{a_i} S_1^{n_i} \xrightarrow{b_i} X$ with $\|a_i\|_{cb}$, $\|b_i\|_{cb} \leq 1$ that tend pointwise to the identity on X.

In another direction, the results of [EOR] provide an extension to operator spaces of the classical work of Choi and Effros and Connes (see [CE1, CE2]) on nuclear C^*-algebras. Recall that an operator space X is λ-nuclear if there is a net of maps of the form $X \xrightarrow{a_i} M_{n_i} \xrightarrow{b_i} X$ with $\|a_i\|_{cb}\|b_i\|_{cb} \leq \lambda$ that tends pointwise to the identity on X. Moreover, (see Theorem 11.6 above) a C^*-algebra A is 1-nuclear iff A^{**} is injective (equivalently, is a 1-injective operator space).

The o.s. version of this result proved in [EOR] now reads like this:

Theorem 24.7. *An operator space X is 1-nuclear iff X is 1-locally reflexive and 1-WEP.*

Recall that X is 1-WEP if the canonical inclusion $X \to X^{**}$ factors completely contractively through $B(H)$.

PART III

OPERATOR SPACES AND NON-SELF-ADJOINT OPERATOR ALGEBRAS

Chapter 25. Maximal Tensor Products and Free Products of Operator Algebras

In the category of algebras (resp. unital algebras), the natural morphisms are of course algebra-homomorphisms (resp. unital ones). In this chapter, we work in the category of operator algebras (i.e., closed subalgebras of $B(H)$ for some H), and the natural morphisms there are the completely contractive homomorphisms. Actually, for the most part, we will work with *unital* operator algebras, and the morphisms will then be understood as *unital* completely contractive homomorphisms. When the context is sufficiently clear, we will use the term "morphism" and the reader will be supposed to know which category we are working in.

We start by recalling the definition of the free product of algebras: Let $(A_i)_{i \in I}$ be a family of algebras (resp. unital algebras). We will denote by $\dot{\mathcal{A}}$ (resp. $\mathcal{A}$) their free product in the category of algebras (unital algebras). This object is characterized as the unique algebra (resp. unital algebra) A containing each A_i as a (resp. unital) subalgebra and such that, if we are given another object B and morphisms $\varphi_i \colon A_i \to B$ $(i \in I)$, there is a unique morphism $\varphi \colon A \to B$ such that $\varphi_{|A_i} = \varphi_i$ for all i. If we now assume that $(A_i)_{i \in I}$ is a family of operator algebras (resp. unital ones), then we can equip $\dot{\mathcal{A}}$ (resp. $\mathcal{A}$) with a (resp. unital) operator algebra structure in the following way.

Let $\mathcal{F}$ be either $\mathcal{A}$ or $\dot{\mathcal{A}}$. Let C be the collection of all morphisms $u \colon \mathcal{F} \to B(H_u)$ such that $\|u_{|A_i}\|_{cb} \leq 1$ for all i in I. Let $j \colon \mathcal{F} \to \bigoplus_{u \in C} B(H_u)$ be the embedding defined by $j(x) = \bigoplus_{u \in C} u(x)$ for all x in $\mathcal{F}$. Clearly j is a morphism and (by standard algebraic facts) it is injective. This allows us to equip $\mathcal{F}$ with the noncomplete operator algebra structure associated to j, and, after completion, we obtain an operator algebra (resp. a unital one), admitting $\mathcal{F}$ as a dense subalgebra. We will denote by $\dot{*}_{i \in I} A_i$ (resp. $*_{i \in I} A_i$) the resulting (resp. unital) operator algebra, which we call the free product of the family of (resp. unital) operator algebras $(A_i)_{i \in I}$.

Let A be any of these two free products. Let $\sigma_i \colon A_i \to A$ be the natural embedding. Then, for any family of morphisms $u_i \colon A_i \to B(H)$, there is a unique morphism $u \colon A \to B(H)$ such that $u\sigma_i = u_i$ for all i.

We now turn to the maximal tensor product. This notion originally was introduced for C^*-algebras. We already discussed this in detail in Chapter 11. In the more general (unital) operator algebra case, it was first considered in [PaP]. For simplicity we will restrict ourselves to a pair of (resp. unital) operator algebras A_1, A_2. For any pair $\varphi = (\varphi_1, \varphi_2)$ of morphisms $\varphi_i \colon A_i \to B(H_\varphi)$ with values in a common space $B(H_\varphi)$ we denote by $\varphi_1 \cdot \varphi_2$ the linear mapping from $A_1 \otimes A_2$ to $B(H_\varphi)$ that takes $a_1 \otimes a_2$ to $\varphi_1(a_1)\varphi_2(a_2)$. If we assume, moreover, that φ_1 and φ_2 have commuting ranges (i.e., $\varphi_1(a_1)\varphi_2(a_2) = (\varphi_2(a_2)\varphi_1(a_1)$ for all a_1 in A_1, a_2 in A_2), then $\varphi_1 \cdot \varphi_2$ is a morphism from $A_1 \otimes A_2$ into $B(H_\varphi)$. Conversely, in the unital case, it is easy to see that any morphism $\psi\colon A_1 \otimes A_2 \to B(H)$ must be of the form $\psi = \psi_1 \cdot \psi_2$ for some pair (ψ_1, ψ_2) of morphisms with commuting ranges. For any $x = \sum_1^n a_i^1 \otimes a_i^2$ in $A_1 \otimes A_2$ we define

$$\|x\|_{\max} = \sup\{\|\varphi_1 \cdot \varphi_2(x)\|_{B(H_\varphi)}\} = \sup\left\{\left\|\sum_1^n \varphi_1(a_i^1)\varphi_2(a_i^2)\right\|_{B(H_\varphi)}\right\},$$
(25.1)

where the supremum runs over all pairs $\varphi = (\varphi_1, \varphi_2)$ of (resp. unital) completely contractive morphisms $\varphi_i\colon A_i \to B(H_\varphi)$ with commuting ranges.

We denote by $A_1 \otimes_{\max} A_2$ the completion of $A_1 \otimes A_2$ equipped with the norm $\|\ \|_{\max}$. Clearly this is an operator algebra (resp. a unital one).

Remark 25.1.

(i) It is easy to see that if $\varphi_i\colon A_i \to B_i$ are morphisms, then $\varphi_1 \otimes \varphi_2$ extends to a (contractive) morphism from $A_1 \otimes_{\max} A_2$ to $B_1 \otimes_{\max} B_2$.

(ii) Since the morphisms are different in each of the three categories of operator algebras, unital operator algebras, and unital C^*-algebras, the definition of the maximal tensor product leads to three different notions. Fortunately, however, at least two of these notions "match" each other; that is, given two unital C^*-algebras, their maximal tensor products as unital operator algebras and as C^*-algebras coincide. Thus the various notions are simply extensions of each other, when a comparison makes sense. Indeed, if A_1, A_2 are unital C^*-algebras, then any unital completely contractive morphism $\sigma_i\colon A_i \to B(H_i)$ is "automatically" a $*$-homomorphism. Thus the two associated "max-norms" are equal.

Lemma 25.2. *In the unital case, the mappings*

$$a_1 \to a_1 \otimes 1 \quad and \quad a_2 \to 1 \otimes a_2$$

from A_1 to $A_1 \otimes_{\max} A_2$ and A_2 to $A_1 \otimes_{\max} A_2$ extend to a morphism

$$q\colon A_1 * A_2 \longrightarrow A_1 \otimes_{\max} A_2,$$

which is a complete metric surjection, that is, it induces a completely isometric isomorphism between the unital operator algebras $A_1 * A_2 / \ker(q)$ and $A_1 \otimes_{\max} A_2$. Moreover, let $\mathcal{F}$ denote the algebraic free product of A_1 and A_2. Then the restriction of q to $\mathcal{F}$ defines a complete isometry between $\mathcal{F} / \ker(q) \cap \mathcal{F}$ and $A_1 \otimes A_2$ (where $\mathcal{F} \subset A_1 * A_2$ and $A_1 \otimes A_2 \subset A_1 \otimes_{\max} A_2$ are equipped with the induced operator space structures).

Proof. This is essentially routine. By the universal property of the free product, q is uniquely defined and is completely contractive. Let $Q = A_1 * A_2 / \ker(q)$ and let $v\colon Q \to A_1 \otimes_{\max} A_2$ be the associated injective complete contraction. Obviously $A_1 \otimes A_2$ is included is the range of v. Therefore, $\psi = v^{-1}_{|A_1 \otimes A_2}$ is a morphism into Q. Let $\sigma_i \colon A_i \to A_1 * A_2$ $(i = 1, 2)$ be the inclusion into the free product. It is easy to check that we have

$$\psi = (q\sigma_1) \cdot (q\sigma_2).$$

By definition of the maximal tensor product, since $\|q\sigma_i\|_{cb} = 1$, this implies that ψ extends to a complete contraction on $A_1 \otimes_{\max} A_2$. In other words, v is onto and completely isometric. ∎

We will see that, when one of the two algebras is the universal operator algebra $OA_u(E)$ associated to an operator space E, certain interesting identities appear, for instance, the following one.

Lemma 25.3. *Let E be an arbitrary operator space. Consider the linear mapping $T \colon A \otimes E \otimes A \longrightarrow OA_u(E) * A$ that takes $a \otimes e \otimes b$ $(a, b \in A, e \in E)$ to the product $a \cdot e \cdot b$, where E is identified with a subspace of $OA_u(E)$. Then T extends to a completely isometric embedding from $A \otimes_h E \otimes_h A$ into $OA_u(E) * A$.*

Proof. Let $x = \sum a_i \otimes e_i \otimes b_i$ be an element in $A \otimes E \otimes A$. With the notation of Chapter 5, we have

$$\|x\|_f = \sup \left\{ \left\| \sum \varphi_1(a_i)\sigma(e_i)\varphi_2(b_i) \right\| \right\},$$

where the supremum runs over all possible complete contractions

$$\varphi_i \colon A \to B(H) \qquad (i = 1, 2) \text{ and } \sigma \colon E \to B(H)$$

into the same $B(H)$.

By the factorization of c.b. maps (cf. Theorem 1.6) we can assume that $\varphi_i(\cdot) = V_i\pi_i(\cdot)W_i$, where $\pi_i \colon A \to B(H_i)$ is a representation restricted to A and V_i, W_i are contractions $(i = 1, 2)$. Replacing $\sigma(\cdot)$ by $W_1\sigma(\cdot)V_2$ and deleting V_1 and W_2 we find

$$\|x\|_f = \sup \left\{ \left\| \sum \pi_1(a_i)\sigma(e_i)\pi_2(b_i) \right\| \right\}, \tag{25.2}$$

where π_1, π_2 are representations restricted to A and $\|\sigma\|_{cb} \le 1$. Let

$$\pi(a) = \begin{pmatrix} \pi_1(a) & 0 \\ 0 & \pi_2(a) \end{pmatrix} \quad \text{and} \quad \widehat{\sigma}(e) = \begin{pmatrix} 0 & \sigma(e) \\ 0 & 0 \end{pmatrix}.$$

Denote by P_i: $H_1 \oplus H_2 \to H_i$ the projection onto the i-th coordinate. A simple calculation shows that

$$\sum \pi_1(a_i)\sigma(e_i)\pi_2(b_i) = P_1 \left(\sum \pi(a_i)\widehat{\sigma}(e_i)\pi(b_i) \right) P_2^*,$$

and hence we deduce from (25.2) that

$$\|x\|_f = \sup \left\{ \left\| \sum \pi(a_i)\widehat{\sigma}(e_i)\pi(b_i) \right\| \right\},$$

where the supremum runs over all π: $A \to B(H)$ completely contractive morphisms and all $\widehat{\sigma}$: $E \to B(H)$ with $\|\widehat{\sigma}\|_{cb} = 1$. Since any such $\widehat{\sigma}$ extends to a completely contractive morphism on $OA_u(E)$, we conclude that

$$\|x\|_f = \|T(x)\|_{OA_u(E)*A}.$$

This shows that T is isometric. We leave it to the reader to complete the proof. ∎

By a simple modification, we can prove:

Lemma 25.4. *Let $E_1, \ldots, E_{n-1}$ be arbitrary operator spaces ($n \geq 2$), and let A be a unital operator algebra. Consider the linear mapping*

$$T: A \otimes E_1 \otimes A \otimes E_2 \otimes \cdots \otimes E_{n-1} \otimes A \longrightarrow OA_u(E_1) * \cdots * OA_u(E_{n-1}) * A,$$

which takes $a_1 \otimes e_1 \otimes a_2 \otimes e_2 \otimes \cdots \otimes a_n$ to the product $a_1 e_1 a_2 e_2 \ldots a_n$ in the free product. Then T is a complete isometry from $A \otimes_h E_1 \otimes_h A \otimes_h E_2 \otimes \cdots \otimes_h A$ into the free product $OA_u(E_1) * \cdots * OA_u(E_{n-1}) * A$.

Proof. We only give a hint: Given morphisms π_i: $A \to B(H_i)$ ($i \leq n$) and complete contractions σ_i: $E_i \to B(H_{i+1}, H_i)$ we introduce

$$\pi = \begin{pmatrix} \pi_1 & & 0 \\ & \ddots & \\ 0 & & \pi_n \end{pmatrix} \quad \text{and} \quad \widehat{\sigma} = \begin{pmatrix} 0 & \sigma_1 & & & 0 \\ & \ddots & \ddots & & \\ & & & \ddots & \sigma_{n-1} \\ & & & & 0 \\ & & & & 0 \end{pmatrix}.$$

The rest is as before. ∎

Lemma 25.5. *Let E, A be respectively an operator space and a unital operator algebra. Let $\mathcal{F}$ be the algebraic free product of $OA_u(E)$ and A. For any*

z in $\mathbb{C}$ with $|z| \leq 1$, we denote by $\sigma_z \colon OA_u(E) \to OA_u(E)$ the unique morphism such that $\sigma_z(e) = ze$ for all e in E. Let $\widehat{\sigma}_z \colon \mathcal{F} \to \mathcal{F}$ be the morphism that is the free product of σ_z and the identity on A. Let $\mathcal{F}_j \subset \mathcal{F}$ ($j \geq 0$) be the linear subspace of $\mathcal{F}$ spanned by elements of the form

$$a_0 e_1 a_1 e_2 \ldots e_j a_j \qquad (e_1, \ldots, e_j \in E, \quad a_0, \ldots, a_j \in A).$$

Then any element x in $\mathcal{F}$ can be uniquely written as a finite sum $x_0 + x_1 + \cdots + x_N$, where each x_j is $\mathcal{F}_j$. Moreover, we have

$$\widehat{\sigma}_z(x) = x_0 + z x_1 + \cdots + z^N x_N. \tag{25.3}$$

Proof. Clearly any element x in $\mathcal{F}$ can be written as above and (25.3) is obvious. The unicity follows from (25.3). ∎

Consider an arbitrary operator space E and a unital operator algebra A. We will now introduce two (a priori distinct) o.s.s.'s on $E \otimes A$, and we will show that they actually coincide.

First, we note that $E \otimes A$ can be viewed as a linear subspace of $OA_u(E) \otimes A$; therefore we may equip $E \otimes A$ with the operator space structure induced by $OA_u(E) \otimes_{\max} A$. We denote by Δ the induced norm and by $E \otimes_\Delta A$ the operator space obtained by completing $E \otimes A$ equipped with this structure.

Let y be an element of $E \otimes A$. We have clearly

$$\Delta(y) = \sup \|\sigma \cdot \pi(y)\|, \tag{25.4}$$

where the supremum runs over all pairs (σ, π) where $\sigma \colon E \to B(H)$ is a complete contraction, $\pi \colon A \to B(H)$ a morphism, and, moreover, σ and π have commuting ranges. Indeed, by definition of $OA_u(E)$, any complete contraction $\sigma \colon E \to \pi(A)'$ is the restriction of a morphism $\widehat{\sigma} \colon OA_u(E) \to \pi(A)'$.

Remark. Note that, by Remark 25.1, if A is a unital C^*-algebra, the preceding definition of the norm Δ coincides with the one given in Theorem 12.1.

We now introduce an a priori different structure on $E \otimes A$, as follows. Note that any y in $E \otimes A$ can be written, for some integer N, as $y = \sum_{ij=1}^N x_{ij} \otimes a_i b_j$, where $x \in M_N(E)$, $a_i \in A$, $b_j \in A$. (Indeed, since A is assumed unital, this is clear). We define

$$\delta(y) = \inf \left\{ \|x\|_{M_N(E)} \left\| \sum a_i a_i^* \right\|^{1/2} \left\| \sum b_j^* b_j \right\|^{1/2} \right\}, \tag{25.5}$$

where the infimum runs over all possible N and all possible representations of y as above. Let $q\colon A \otimes E \otimes A \to E \otimes A$ be the linear mapping that takes $a \otimes e \otimes b$ to $e \otimes ab$. It is easy to check that

$$\delta(y) = \inf\{\|\widehat{y}\|_{A \otimes_h E \otimes_h A} \mid \widehat{y} \in A \otimes E \otimes A, \quad q(\widehat{y}) = y\}.$$

Note in particular that δ is a norm. We will denote by $E \otimes_\delta A$ the completion of $E \otimes A$ with respect to this norm δ. More generally, for any n and any y in $M_n(E \otimes A)$, we define

$$\|y\|_n = \inf\{\|\widehat{y}\|_{M_n(A \otimes_h E \otimes_h A)} \mid \widehat{y} \in M_n(A \otimes E \otimes A) \quad (I \otimes q)(\widehat{y}) = y\}.$$

After completion, we obtain (by Ruan's Theorem) an operator space structure on $E \otimes_\delta A$, and, from now on, we will consider $E \otimes_\delta A$ as an operator space. Note that q extends to a complete contraction from $A \otimes_h E \otimes_h A$ to $E \otimes_\delta A$ (which we still denote abusively by q), which is a complete metric surjection from $A \otimes_h E \otimes_h A$ onto $E \otimes_\delta A$. In other words, we have

$$A \otimes_h E \otimes_h A / \ker(q) \simeq E \otimes_\delta A \quad \text{(completely isometrically)}.$$

We will now essentially repeat Theorem 12.1, but in a more general setting and with a different proof.

Theorem 25.6. *Let E, A be respectively an operator space and a unital operator algebra. Then for any y in $E \otimes A$ we have*

$$\Delta(y) = \delta(y).$$

More precisely, we have

$$E \otimes_\Delta A \simeq E \otimes_\delta A \quad \text{(completely isometrically)}.$$

Proof. Let $y \in E \otimes A$. First recall that $\Delta(y) = \|y\|_{OA_u(E)*A}$. Let $\psi\colon OA_u(E) * A \to OA_u(E) \otimes_{\max} A$ be the complete surjection appearing in Lemma 25.2. We will use the notation in Lemma 25.5. By Lemma 25.2, we have $\Delta(y) < 1$ iff there is an element x in $\mathcal{F}$ such that $\|x\|_{OA_u(E)*A} < 1$ and $\psi(x) = y$.

We claim that x can be replaced by an element x_1 in $\mathcal{F}_1 = \mathrm{span}(A \cdot E \cdot A)$. Indeed, we have (cf. Lemma 25.5) $x = x_0 + x_1 + x_2 + \cdots + x_N$ and $\widehat{\sigma}_z(x) = x_0 + zx_1 + z^2 x_2 + \cdots + z^N x_N$. Moreover, it is easy to check that

$$\psi\widehat{\sigma}_z = (\sigma_z \otimes I_A)\psi.$$

Therefore we have $\psi\widehat{\sigma}_z(x) = zy$ for all z, which implies

$$y = \psi(x_1).$$

Moreover, denoting by m the normalized Lebesgue measure on the unit circle, we have by (25.3)

$$x_1 = \int \widehat{\sigma}_z(x)\bar{z}\, dm(z),$$

which ensures (by Jensen's inequality) that

$$\|x_1\| \le \|x\| < 1.$$

Finally, by Lemma 25.3, x_1 can be identified with an element $\widehat{x}$ in the unit ball of $A \otimes_h E \otimes_h A$. Thus, we conclude that $\Delta(y) < 1$ iff there is an element $\widehat{x}$ in $A \otimes E \otimes A$ such that $q(\widehat{x}) = y$ and $\|\widehat{x}\|_{A \otimes_h E \otimes_h A} < 1$. This proves that $\Delta(y) = \delta(y)$, so that $E \otimes_\Delta A \simeq E \otimes_\delta A$ isometrically. It is easy to modify the argument to prove that this is a complete isometry. We leave the details to the reader. ∎

Proposition 25.7. *Consider an operator space F and a unital operator algebra $A \subset B(\mathcal{H})$. Let $u\colon F \to A$ be a finite rank linear map, and let $\widetilde{u} \in F^* \otimes A$ denote the associated tensor. Then $\delta(\widetilde{u}) < 1$ iff u admits, for some N, a factorization of the form $F \xrightarrow{\alpha} M_N \xrightarrow{\beta} A$ with $\|\alpha\|_{cb} < 1$ and β of the form $\beta(e_{ij}) = a_i b_j$ with $a_i, b_j \in A$ such that*

$$\left\| \sum a_i a_i^* \right\| \left\| \sum b_j^* b_j \right\| < 1. \tag{25.6}$$

Moreover, if F is a "minimal" operator space, that is, if $F = \min(F)$ (completely isometrically), then $\delta(\widetilde{u}) < 1$ iff u admits, for some N, a factorization of the form $F \xrightarrow{\alpha} \ell_\infty^N \xrightarrow{\beta} A$ with $\|\alpha\|_{cb} < 1$ and β of the form $\beta(e_i) = a_i b_i$ with $a_i, b_i \in A$ such that (25.6) holds.

Proof. The first part is essentially obvious by (25.5) with $E = F^*$. Now, if $F = \min(F)$, then any element α in the unit ball of $M_N(\max(F^*))$ admits a special factorization as described in Theorem 3.1. Using this, one easily completes the proof. ∎

Definition 25.8. *Let F, A be an operator space and a unital operator algebra. A linear map $u\colon F \to A$ will be called δ-boundedly approximable if there is a constant C and a net $u_i\colon F \to A$ of finite rank maps tending pointwise to u and such that the associated elements $\widetilde{u}_i$ in $F^* \otimes A$ satisfy $\sup_{i \in I} \delta(\widetilde{u}_i) \le C$. We will denote by $D(u)$ the smallest constant C such that this holds.*

Theorem 25.9. *The following properties of a unital operator algebra A are equivalent:*

(i) *For any unital operator algebra B, $B \otimes_{\min} A = B \otimes_{\max} A$ (isomorphically).*

(ii) *The identity of A is δ-boundedly approximable.*

Moreover, if A is a C^*-algebra, this holds iff A is nuclear and, in the C^*-case, the identity in (i) *is automatically completely isometric.*

Proof. We first claim that, for any x in $B \otimes A$ and any $u \colon A \to A$ of finite rank with associated tensor $\tilde{u} \in A^* \otimes A$, we have

$$\|(I_B \otimes u)(x)\|_{B \otimes_{\max} A} \le \delta(\tilde{u}) \|x\|_{\min}. \tag{25.7}$$

Indeed, let $E = A^*$ and let $v \colon E \to B$ be the finite rank map associated to x. Assume $\|x\|_{\min} = \|v\|_{cb} \le 1$. Then v extends to a morphism $\hat{v} \colon OA_u(E) \to B$, so that, by Remark 25.1.(i) we have

$$\|(\hat{v} \otimes I_A)[\tilde{u}]\|_{\max} \le \|\tilde{u}\|_{OA_u(E) \otimes_{\max} A} = \Delta(\tilde{u}) = \delta(\tilde{u}).$$

Since $(\hat{v} \otimes I_A)(\tilde{u}) = (v \otimes I_A)(\tilde{u}) = (I_B \otimes u)(x)$, we obtain (25.7) by homogeneity. Assume (ii) and let u_i be as in Definition 25.8, with $u = I_A$. Note that, for any x in $B \otimes A$, $(I_B \otimes u_i)(x) \to x$ in the largest Banach tensor norm (say); hence it is in the norm of $B \otimes_{\max} A$. Then we have for any x in $B \otimes A$

$$\|(I_B \otimes u_i)(x)\|_{\max} \le C \|x\|_{\min},$$

which yields in the limit

$$\|x\|_{\max} \le C \|x\|_{\min}. \tag{25.8}$$

This shows that (ii) $\Rightarrow$ (i).

Conversely, assume (i). By an easy direct sum argument, there is a constant C such that for all B and all x in $B \otimes A$ we have (25.8). Consider then an arbitrary finite-dimensional subspace $F \subset A$ and let $i_F \colon F \to A$ denote the inclusion map with associated tensor $\tilde{i}_F \in F^* \otimes A$. Note that $\|\tilde{i}_F\|_{\min} = \|i_F\|_{cb} = 1$; hence (i) implies $\Delta(\tilde{i}_F) \le C$. By Theorem 25.6 and a simple extension argument, we obtain for any $\varepsilon > 0$ a finite rank map $u \colon A \to A$ extending $i_F \colon F \to A$ and such that $\|\tilde{u}\|_{A^* \otimes_\delta A} < C(1 + \varepsilon)$. Letting F run over the directed net of finite-dimensional subspaces $F \subset A$ we obtain a net as in Definition 25.8, whence (ii). Finally, if A is a C^*-algebra, then any unital completely contractive homomorphism $\pi \colon A \to B(H)$ is a $*$-homomorphism, and if A is nuclear, $\pi(A)$ is also and the min- and max-norm coincide on $\pi(A)' \otimes \pi(A)$. Therefore, for any unital operator algebra B and any morphism $\sigma \colon B \to \pi(A)'$, we have for any x in $B \otimes A$:

$$\begin{aligned}
\|\sigma \cdot \pi(x)\| &\le \|(\sigma \otimes \pi)(x)\|_{\pi(A)' \otimes_{\max} \pi(A)} = \|(\sigma \otimes \pi)(x)\|_{\min} \\
&\le \|x\|_{\min}.
\end{aligned}$$

Hence we obtain $\|x\|_{\max} \le \|x\|_{\min}$, which shows that (i) holds isometrically. Replacing A by $M_n(A)$ $(n \ge 1)$, we obtain a complete isometry. Conversely, if a C^*-algebra A satisfies (i), then a fortiori (i) holds whenever B is a C^*-algebra, which is the definition of nuclearity (cf. Chapter 11). ∎

Remark 25.10. Let A be a unital operator algebra. The preceding proof shows that the smallest constant C such that, for any unital operator algebra B and any x in $B \otimes A$, we have $\|x\|_{\max} \leq C\|x\|_{\min}$, is equal to $D(I_A)$.

We will denote by $D(A)$ this constant. In sharp contrast with the C^*-case, the constant $D(A)$ can obviously be both finite and > 1 in certain cases (for instance, when A is finite-dimensional; see Remark 25.14). If A is a C^*-algebra, $D(A)$ can only take the value 1 if it is finite. The converse is true (in a very strong sense), as shown by the next result.

Theorem 25.11. Let $A \subset B(\mathcal{H})$ be a unital closed subalgebra. Then $D(A) = 1$ implies that A is self-adjoint. Equivalently, $D(A) = 1$ iff A is a nuclear C^*-subalgebra of $B(\mathcal{H})$.

We will use the following elementary fact:

Lemma 25.12. Fix $\varepsilon_1 > 0$ and $\varepsilon_2 > 0$. Let $a_i, \beta_j \in B(H)$ be such that $1 = \sum_1^n a_i\beta_i$, $\sum_1^n a_i a_i^* \leq 1 + \varepsilon_1$, $\sum \beta_j^*\beta_j \leq 1 + \varepsilon_2$. Then we have

$$\sum (\beta_i - a_i^*)^*(\beta_i - a_i^*) \leq \varepsilon_1 + \varepsilon_2.$$

Proof. For any h in H with $\|h\| = 1$, we have $1 = \langle h, h \rangle = \sum \langle \beta_i h, a_i^* h \rangle$; hence

$$\sum \|\beta_i h - a_i^* h\|^2 = \left\langle \sum \beta_i^* \beta_i h, h \right\rangle + \left\langle \sum a_i a_i^* h_1 h \right\rangle - 2$$
$$\leq \varepsilon_1 + \varepsilon_2.$$

Proof of Theorem 25.11. We have seen in Theorem 25.9 that $D(A) = 1$ if A is a nuclear C^*-algebra. Conversely, assume $D(A) = 1$. Using the same argument and the same notation as in the proof of (i) ⇒ (ii) in Theorem 25.9, we find that, for any finite-dimensional subspace $F \subset A$ and any $\varepsilon > 0$, there are, for some n, an element (ξ_{ij}) in $M_n(F^*)$ and a_i, b_j in A such that

$$\sum_1^n a_i a_i^* < 1, \qquad \sum_1^n b_j^* b_j < 1, \qquad \|(\xi_{ij})\|_{M_n(F^*)} < (1 + \varepsilon)^{1/2},$$

and

$$\forall x \in F \qquad x = \sum_{i,j=1}^n \xi_{ij}(x) a_i b_j.$$

[Indeed, this simply makes explicit the fact that $\delta(\widetilde{i}_F) < (1 + \varepsilon)^{1/2}$.]

Let us now assume $I \in F$, so that we have

$$I = \sum a_i \beta_i = \sum \gamma_j b_j,$$

where $\beta_i = \sum_j \xi_{ij}(I) b_j$, $\gamma_j = \sum_i \xi_{ij}(I) a_i$.

Note that $\left\| \sum \beta_i^* \beta_i \right\|^{1/2} \leq \|(\xi_{ij}(I)\|_{M_n} \left\| \sum b_j^* b_j \right\|^{1/2} < (1+\varepsilon)^{1/2}$, and similarly $\left\| \sum \gamma_j \gamma_j^* \right\| < 1 + \varepsilon$. By Lemma 25.12, it follows that

$$\sum (\beta_i - a_i^*)^* (\beta_i - a_i^*) \leq \varepsilon \quad \text{and} \quad \sum (b_j - \gamma_j^*)^* (b_j - \gamma_j^*) \leq \varepsilon.$$

Note that $\beta_i, \gamma_j \in A$, so that the last two inequalities show that a_i^* and b_j^* are "close" to being also in A. More precisely, for any x in F, we have

$$x^* = \sum_{ij=1}^{n} \overline{\xi_{ij}(x)} b_j^* a_i^*.$$

Assume $\|x\| \leq 1$. Then, if we set

$$y = \sum_{ij=1}^{n} \overline{\xi_{ij}(x)} \gamma_j \beta_i,$$

we have by (1.12)

$$
\begin{aligned}
\|x^* - y\| &\leq \left\| \sum \overline{\xi_{ij}(x)} (b_j^* - \gamma_j) a_i^* \right\| + \left\| \sum \overline{\xi_{ij}(x)} \gamma_j (a_i^* - \beta_i) \right\| \\
&\leq \left\| \sum (b_j^* - \gamma_j)(b_j^* - \gamma_j)^* \right\|^{1/2} \left\| \sum a_i a_i^* \right\|^{1/2} (1+\varepsilon)^{1/2} \\
&\quad + \left\| \sum \gamma_j \gamma_j^* \right\|^{1/2} \left\| \sum (a_i^* - \beta_i)^* (a_i^* - \beta_i) \right\|^{1/2} (1+\varepsilon)^{1/2} \\
&\leq 2(\varepsilon(1+\varepsilon))^{1/2}.
\end{aligned}
$$

Thus we proved $\text{dist}(x^*, A) \leq 2(\varepsilon(1+\varepsilon))^{1/2}$, and since $\varepsilon > 0$ is arbitrary, we conclude that x^* belongs to A. Applying this to $F_x = \text{span}[I, x]$, where x runs over all elements of A, we conclude that A is self-adjoint. ∎

Remark. See [LeM4] for generalizations of the preceding statement.

Remark 25.13. Actually, we only use the fact that (ξ_{ij}) defines a map $\xi \colon F \to M_n$ with ordinary norm ≤ 1. In other words, we only use $\max(F)$ instead of F or, equivalently, $\min(F^*)$ instead of F^*. Moreover, we only consider two-dimensional subspaces F. Thus the conclusion of Theorem 25.11 holds if $B \otimes_{\min} A = B \otimes_{\max} A$ (isometrically) for any B of the form $B = OA_u(G)$, where G is any two-dimensional subspace of ℓ_∞.

Remark 25.14. It is easy to show (using an Auerbach basis; cf. [LiT, p. 16]) that, for any n-dimensional operator space E, any unital operator algebra A, and any x in $E \otimes A$, we have

$$\delta(x) \le n\|x\|_{\min}. \tag{25.9}$$

Therefore, if $\dim(A) = n$, then we have $D(A) \le n$.

Remark 25.15. The conclusion of the preceding statement can fail if $D(A) > 1$. Indeed, consider the two-dimensional algebra $A \subset B(H)$ ($\dim H = 2$) formed by all matrices of the form $\begin{pmatrix} a & b \\ 0 & a \end{pmatrix}$ ($a, b \in \mathbb{C}$). Note that the matrix $\begin{pmatrix} 0 & 1 \\ 0 & 0 \end{pmatrix}$ is a nonzero element in A with zero square. Since A is finite-dimensional, we have trivially $D(A) < \infty$; but on the other hand, A is not isomorphic to a C^*-algebra. (Indeed, if it were, since A is commutative, it would have to be isomorphic to $\ell_\infty^{(2)}$, but this is absurd since $\ell_\infty^{(2)}$ does not contain any nonzero element with zero square.)

In the remainder of this chapter, we wish to discuss several examples for which the following well-known fact will be useful. We first recall some notation. Let D be the open unit disc in $\mathbb{C}$ with boundary ∂D. For any integer $k \ge 1$, we denote by $A(D^k)$ the closure in $C((\partial D)^k)$ of the algebra of all (analytic) polynomials on $\mathbb{C}^k$ equipped with the induced operator algebra structure. Thus $A(D^k)$ is equipped with its minimal o.s. Note the completely isometric identities

$$A(D^k) = A(D) \otimes_{\min} \cdots \otimes_{\min} A(D) \text{ and } C((\partial D)^k) = C(\partial D) \otimes_{\min} \cdots \otimes_{\min} C(\partial D).$$

Proposition 25.16. Let $u\colon A(D) \otimes \cdots \otimes A(D) \to B(H)$ be a unital homomorphism. Let $T_1 = u(z \otimes 1 \otimes \cdots \otimes 1)$, $T_2 = u(1 \otimes z \otimes 1 \otimes \cdots \otimes 1), \ldots,$ and $T_k = u(1 \otimes \cdots \otimes 1 \otimes z)$. The following assertions are equivalent.

(i) u extends completely contractively to $A(D) \otimes_{\min} \cdots \otimes_{\min} A(D)$.

(ii) There is a Hilbert space $\widehat{H}$ with $H \subset \widehat{H}$ and a $*$-homomorphism $\pi\colon C(\partial D) \otimes_{\min} \cdots \otimes_{\min} C(\partial D) \to B(\widehat{H})$ such that

$$\forall f \in A(D) \otimes \cdots \otimes A(D) \quad u(f) = P_H \pi(f)_{|H}.$$

(iii) There is a Hilbert space $\widehat{H}$ with $H \subset \widehat{H}$ and a k-tuple of mutually commuting unitaries $(U_1, \ldots, U_k)$ on $\widehat{H}$ such that for any polynomial P we have

$$P(T_1, \ldots, T_k) = P_H P(U_1, \ldots, U_k)_{|H}.$$

Proof. (i) $\Rightarrow$ (ii) is an immediate consequence of Corollary 1.8.

(ii) $\Rightarrow$ (iii) is immediate: We simply set $U_k = \pi(z_k)$, where z_k denotes the k-th coordinate function on $(\partial D)^k$ viewed as an element of $C(\partial D) \otimes \cdots \otimes C(\partial D)$.

Finally, assume (iii). The mapping $P \to P(U_1, \ldots, U_k)$ obviously extends to a $*$-homomorphism π on $C(\partial D) \otimes_{\max} \cdots \otimes_{\max} C(\partial D) = C(\partial D) \otimes_{\min} \cdots \otimes_{\min} C(\partial D)$. Restricting to the subspace $A(D) \otimes_{\min} \cdots \otimes_{\min} A(D)$, we obtain (i). ∎

Example 25.17. We have completely isometrically

$$A(D) \otimes_{\min} A(D) = A(D) \otimes_{\max} A(D). \qquad (25.10)$$

As observed in [PaP], this is essentially a reformulation of a famous dilation theorem due to Ando (see e.g.[P10]): Any pair of commuting contractions T_1, T_2 in $B(H)$ admits a unitary dilation, that is, there is $\widehat{H}$ with $H \subset \widehat{H}$ and commuting unitaries U_1, U_2 on $\widehat{H}$ such that, for any polynomial P, we have $P(T_1, T_2) = P_H P(U_1, U_2)_{|H}$.

Proof of (25.10). It suffices to show that, for any pair of mutually commuting completely contractive morphism $\pi_j \colon A(D) \to B(H)$, $(j = 1, 2)$, the morphism $\pi_1 \cdot \pi_2 \colon A(D) \otimes A(D) \to B(H)$ extends completely contractively to $A(D) \otimes_{\min} A(D)$. Taking into account the preceding proposition, this follows from Ando's dilation theorem. ∎

By a well-known example of Varopoulos (see e.g.[P10]) Ando's dilation theorem does not extend to three mutually commuting contractions. The following related example due to S. Parrott [Par2] is very important.

Example 25.18. ([Par]) There is a contractive homomorphism $u \colon A(D^3) \to B(H)$ that is not completely contractive. Equivalently, there is a triple (T_1, T_2, T_3) of commuting contractions such that, for any polynomial P in three variables, we have

$$\|P(T_1, T_2, T_3)\| \leq \sup_{z \in D^3} |P(z_1, z_2, z_3)| \qquad (25.11)$$

but for which the morphism

$$P \to P(T_1, T_2, T_3)$$

is *not* completely contractive.

The operators T_1, T_2, T_3 will be of the form

$$T_j = \begin{pmatrix} 0 & 0 \\ a_j & 0 \end{pmatrix}$$

acting on $H = K \oplus K$ as in Exercise 25.1, and a_j, $(j = 1, 2, 3)$ will be suitably chosen contractions on K. Note that $T_j T_k = 0$ for all j, k, so (T_1, T_2, T_3) mutually commute.

Let $P(z_1, z_2, z_3)$ be a polynomial on D^3. Let $P = P_0 + P_1 + \cdots$ be its decomposition into a sum of homogeneous polynomials. Note that $P_d(T_1, T_2, T_3) = 0$ for all $d \geq 2$, so that we have

$$P(T_1, T_2, T_3) = \lambda_0 I + \sum_1^3 \lambda_j T_j,$$

where $P_0(z) = \lambda_0$ and $P_1(z) = \sum_1^3 \lambda_j z_j$. Fix (z_1, z_2, z_3) in $\overline{D}^3$. Let $T = \left(\sum_1^3 |\lambda_j| \right)^{-1} \sum_1^3 \lambda_j T_j$. By von Neumann's inequality (see (8.11) above) we have

$$\left\| \lambda_0 I + \left(\sum \lambda_j z_j \right) \cdot T \right\| = \| P(z_1 T, z_2 T, z_3 T) \|$$
$$\leq \sup_{z \in D} |P(z_1 z, z_2 z, z_3 z)| \leq \sup_{D^3} |P|.$$

Hence, choosing $z_j \in \partial D$ such that $\lambda_j z_j = |\lambda_j|$, we find

$$\left\| \lambda_0 I + \sum \lambda_j T_j \right\| = \left\| \lambda_0 I + \left(\sum \lambda_j z_j \right) T \right\| \leq \sup_{D^3} |P|,$$

which establishes (25.11).

We now claim that, for a suitable choice of unitary operators a_j, $(j = 1, 2, 3)$, the morphism $u \colon A(D^3) \to B(H)$ taking P to $P(T_1, T_2, T_3)$ is not completely contractive on $A(D^3) \simeq A(D) \otimes_{\min} A(D) \otimes_{\min} A(D)$. Indeed, by Proposition 25.16, if $\|u\|_{cb} = 1$, then there is $\widehat{H}$ with $H \subset \widehat{H}$ and a triple (U_1, U_2, U_3) of mutually commuting unitaries on $\widehat{H}$ dilating (T_1, T_2, T_3). In particular we have $T_j = P_H U_{j|H}$.

By Exercise 25.1, if a_j itself is unitary, we must have for any j, k

$$\forall\, h \in K \qquad U_j^{-1} U_k \begin{pmatrix} h \\ 0 \end{pmatrix} = \begin{pmatrix} a_j^{-1} a_k h \\ 0 \end{pmatrix}.$$

Therefore, if U_1, U_2, U_3 commute, then $U_3^{-1} U_1$ and $U_3^{-1} U_2$ commute, and the preceding identity shows that $a_3^{-1} a_1$ and $a_3^{-1} a_2$ must commute, but it is very easy to produce examples of unitaries (a_1, a_2, a_3) for which this fails! Just take, for instance, $a_3 = I$ and choose for a_1, a_2 any pair of 2×2 unitary matrices that do *not* commute.

Remark. The preceding example shows that three commuting contractions may fail to dilate to three commuting unitaries. Nevertheless, it is proved in [GaR] that any k-tuple of contractions (T_j), $(j = 1, \ldots, k)$ that cyclically commute (i.e., such that

$$T_1 T_2 \ldots T_k = T_2 \ldots T_k T_1 = T_3 T_4 \ldots T_k T_1 T_2 = \cdots = T_k T_1 \ldots T_{k-1})$$

admits unitary dilations that cyclically commute. See [OP] for a discussion of a tensor product, analogous to $\otimes_\mu$, but associated to cyclic commutation.

The following variant of Parrott's example was kindly pointed out to us by C. Foias.

Example 25.19. There is an example of four contractions (T_j), $(j = 1, \ldots, 4)$ such that

$$T_i T_j = T_j T_i \qquad 1 \le i \le 2, \quad 3 \le j \le 4$$

but which cannot be dilated to four unitaries (U_j), $(j = 1, \ldots, 4)$ satisfying

$$U_i U_j = U_j U_i \qquad 1 \le i \le 2, \quad 3 \le j \le 4.$$

Indeed, consider again $H = K \oplus K$ and $T_j \in B(H)$ of the form

$$T_j = \begin{pmatrix} 0 & 0 \\ a_j & 0 \end{pmatrix}.$$

Let (U_j) be unitaries on $\widehat{H} \supset H$ such that $T_j = P_H U_{j|H}$. If each of U_1, U_2 commutes with each of U_3, U_4, then a fortiori $U_1^{-1} U_2$ must commute with $U_3^{-1} U_4$. By Exercise 25.1, this forces $a_1^{-1} a_2$ to commute with $a_3^{-1} a_4$, but here again it is very easy to produce unitary matrices (a_j) for which this fails! Just take $a_1 = a_3 = I$ and a_2, a_4 noncommuting unitaries. ∎

Consider two operator spaces E, F. Recall (see Remark 8.12) that we have completely isometric embeddings

$$OA_u(E) \subset C_u^*\langle E \rangle \quad \text{and} \quad OA_u(F) \subset C_u^*\langle F \rangle.$$

Taking the tensor product of these (unital) morphisms we obtain a (unital) morphism

$$\Phi \colon OA_u(E) \otimes_{\max} OA_u(F) \to C_u^*\langle E \rangle \otimes_{\max} C_u^*\langle F \rangle. \tag{25.12}$$

such that

$$\|\Phi\|_{cb} = 1. \tag{25.13}$$

In the particular case $E = F = \mathbb{C}$, we have $OA_u(E) = OA_u(F) = A(D)$ (by Theorem 8.11), and (25.10) implies a fortiori that Φ is completely isometric (since (25.10) says that Φ composed with the completely contractive morphism $C_u^*\langle E \rangle \otimes_{\max} C_u^*\langle F \rangle \to C_u^*\langle E \rangle \otimes_{\min} C_u^*\langle F \rangle$ is completely isometric).

However, we will show in the following that Φ fails to be completely isometric in the case $E = F = \ell_1^2$ equipped with (say) its maximal o.s.s. (recall that, actually, by Proposition 3.10, all o.s.s.'s coincide on ℓ_1^2).

The following statement is immediate from the definition of the symmetrized Haagerup tensor product in Chapter 5.

Proposition 25.20. *The natural embedding $E \otimes F \subset OA_u(E) \otimes_{\max} OA_u(F)$ induces a completely isometric embedding*

$$E \otimes_\mu F \subset OA_u(E) \otimes_{\max} OA_u(F).$$

More generally, for any N-tuple of operator spaces $E_1, \ldots, E_N$ we have a completely isometric embedding

$$(E_1 \otimes \cdots \otimes E_N)_\mu \subset OA_u(E_1) \otimes_{\max} \cdots \otimes_{\max} OA_u(E_N).$$

By the preceding proposition, this implies:

Proposition 25.21. *If E, F are maximal operator spaces, then the norms induced on $E \otimes F$ by $C_u^*\langle E \rangle \otimes_{\max} C_u^*\langle F \rangle$ and $OA_u(F) \otimes_{\max} OA_u(F)$ are equivalent.*

Proof. Consider x in $E \otimes F$. Let $u \colon E^* \to F$ be the associated map. Note that E^*, F^* are minimal operator spaces by Exercise 3.2. By Theorem 19.11, we have

$$2^{-1} \gamma_2^*(u) \leq \|u\|_{cb} = \|x\|_{\min} = \|x\|_{C^*\langle E \rangle \otimes_{\min} C^*\langle F \rangle} \leq \|x\|_{C^*\langle E \rangle \otimes_{\max} C^*\langle F \rangle}.$$

Since it is not hard to check (see Exercise 5.7) to check that

$$\|x\|_\mu = \gamma_2^*(u),$$

we obtain

$$2^{-1} \|x\|_{OA_u(E) \otimes_{\max} OA_u(F)} = 2^{-1} \|x\|_\mu \leq \|x\|_{C^*\langle E \rangle \otimes_{\max} C^*\langle F \rangle}.$$

The converse follows from (25.13). ∎

Let I, I' be two arbitrary sets. Let $E = \max(\ell_1(I))$ and $F = \max(\ell_1(I'))$. Then, by Theorem 8.8, we have a (unital) morphism

$$\Psi \colon OA_u(E) \otimes_{\max} OA_u(F) \to C^*(F_I) \otimes_{\max} C^*(F_{I'})$$

such that

$$\|\Psi\|_{cb} = 1.$$

Let e_i be the canonical basis of $\ell_1(I)$. Note that, for any i in I, the inclusion

$$OA_u(\max(\ell_1(I))) \to C^*(F_I)$$

takes e_i to a unitary U_i in $C^*(F_I)$, so that, for any i' in I', $\Psi(e_i \otimes e_{i'}) = U_i \otimes U_{i'}$ is also unitary in $C^*(F_I) \otimes_{\max} C^*(F_{I'})$.

Example 25.22. Consider the case $I = I' = \{1, 2\}$, so that $E = F = \max(\ell_1^2)$. Then Ψ is *not* completely isometric. More precisely, the restriction of Ψ to the five-dimensional space $S = 1 \otimes 1 + E \otimes 1 + 1 \otimes F$ is not completely isometric.

Proof. Let (T_j), $(1 \le j \le 4)$ be 4 contractions as in Example 25.19. Then let $\varphi_1 \colon OA_u(\ell_1^2) \to B(H)$ (resp. $\varphi_2 \colon OA_u(\ell_1^2) \to B(H)$) be the unique completely contractive unital morphisms such that

$$\varphi_1(e_1) = T_1, \quad \varphi_1(e_2) = T_2$$

(resp. $\varphi_2(e_1) = T_3, \varphi_2(e_2) = T_4$). Our (partial) commutation assumption on (T_j) ensures that φ_1, φ_2 have commuting ranges; hence $\varphi_1 \cdot \varphi_2$ extends to a completely contractive morphism on $OA_u(E) \otimes_{\max} OA_u(F)$. Now, if Ψ were completely isometric, we would be able to extend $\varphi = \varphi_1 \cdot \varphi_2$ to a completely contractive unital map φ on $C^*(F_2) \otimes_{\max} C^*(F_2)$. By Corollary 1.8, this would imply that there is $\widehat{H}$ with $H \subset \widehat{H}$ and a representation $\pi \colon C^*(F_2) \otimes_{\max} C^*(F_2) \to B(\widehat{H})$ such that $\forall\, x \in OA_u(E) \otimes OA_u(F)$

$$\varphi_1 \cdot \varphi_2(x) = P_H \pi(x)_{|H}.$$

In particular, setting $U_j = \pi(e_j \otimes 1)$ and $U_{j+2} = \pi(1 \otimes e_j)$, we would obtain four unitaries contradicting Example 25.19.

This contradiction shows that Ψ is not completely isometric. By Proposition 13.6, this implies that the restriction $\Psi_{|\Psi(S)}^{-1}$ has c.b. norm > 1, or equivalently, that $\Psi_{|S}$ is not completely isometric. ∎

Example 25.23. Let A be the disc algebra $A(D)$, let φ be a finite Blaschke product of degree n, and let $I \subset A$ be the ideal generated by φ, that is, $I = \{\varphi f \mid f \in A\}$. By Corollary 6.4, we may consider the quotient A/I as an operator algebra. Let $q \colon A \to A/I$ be the quotient morphism. Let $\pi \colon A/I \to B(H)$ be any morphism, and consider the composition $\widehat{\pi} = \pi q \colon A \to B(H)$. Since A is generated by the single function z, $\widehat{\pi}$ and π are entirely determined by the single contraction $T = \widehat{\pi}(z) \in B(H)$. Note that $\varphi(T) = \widehat{\pi}(\varphi) = 0$. Conversely, any contraction T such that $\varphi(T) = 0$ determines uniquely a morphism π as above. Moreover, $\{\pi(A/I)\}' = \{T\}'$. Let us denote by $C(\varphi)$ the smallest constant C such that, for any T satisfying $\varphi(T) = 0$ and for any polynomial $P(z) = \sum_0^N a_k z^k$ with coefficients in $\{T\}'$, we have

$$\left\| \sum_0^N a_k T^k \right\|_{B(H)} \le C \sup_{|z|=1} \left\| \sum_0^N a_k z^k \right\|_{B(H)}. \tag{25.14}$$

The following fact is essentially due to Bourgain [Bou4] but was observed by Daher (see [P10, p. 90]): There is a numerical constant K such that, for any Blaschke product φ of degree n, we have

$$C(\varphi) \leq K \, \text{Log}(n+1). \tag{25.15}$$

Daher observed that Bourgain's proof of a related result in [Bou4] actually establishes (25.15). Let $\tilde{q} \in A^* \otimes A/I$ be the tensor associated to q. It is not difficult to see that $C(\varphi) = \Delta(\tilde{q})$; hence, by Theorem 25.6, we have

$$C(\varphi) = \delta(\tilde{q}).$$

Therefore, the second part of Proposition 25.7 implies a striking factorization of q (which is closely related to, but a bit different from, the one appearing in [Bou4]). We now consider the special case when $\varphi = \varphi_n$ where $\varphi_n(z) = z^n$, and we will show in this case that (25.15) is sharp, that is, there is a constant $K' > 0$ such that

$$\forall n \geq 1 \qquad K' \, \text{Log}(n+1) \leq C(\varphi_n). \tag{25.16}$$

To verify this we will use Hardy's classical inequality (cf., e.g., [Ka, p. 91]) concerning "analytic" measures on the unit circle $\mathbb{T}$. We denote by $M(\mathbb{T})$ the Banach space of all complex measures on $\mathbb{T}$. Recall that $M(\mathbb{T}) = C(\mathbb{T})^*$ isometrically. A measure μ in $M(\mathbb{T})$ will be called analytic if $\int \omega^j \mu(d\omega) = 0$ for all $j > 0$. With this terminology, Hardy's inequality (together with the F. and M. Riesz' Theorem) asserts that there is a constant $K_1 > 0$ such that any analytic measure μ satisfies

$$\sum_{k>0} \frac{|\hat{\mu}(k)|}{k} \leq K_1 \|\mu\|_{M(\mathbb{T})}, \tag{25.17}$$

where the Fourier coefficients are defined as

$$\forall k \in \mathbb{Z} \qquad \hat{\mu}(k) = \int \omega^{-k} \mu(d\omega).$$

Fix an integer $n > 1$. We set $\varphi(z) = z^n$, and we consider the operator $T \in M_n$ defined by $Te_1 = 0$ and $Te_i = e_{i-1}$ for $i = 1, 2, \ldots, n$. Clearly, $T^n = 0$, $\|T\| = 1$ and $\|T^{n-1}\| = 1$. Assume that (25.14) holds. We claim that this implies that there is an analytic measure μ such that $\|\mu\|_{M(\mathbb{T})} \leq C(\varphi_n)$ and $\hat{\mu}(k) = 1$ for all $k = 1, \ldots, n-1$. Using (25.17), we immediately deduce that $(1 + \cdots + 1/n - 1) \leq K_1 C(\varphi_n)$, whence (25.16). Therefore, to complete the proof of (25.16), it suffices to verify the above claim.

Now let f be a function in the linear span in $C(\mathbb{T})$ of the functions $\{z^j \mid j \in \mathbb{Z}, \ j \le n - 1\}$. For any fixed z in $\mathbb{T}$, we define

$$\forall \xi \in \mathbb{T} \qquad f_z(\xi) = \sum_{j \le n-1} \widehat{f}(j) z^j \xi^{n-1-j}.$$

By (25.14) applied with $a_k = \widehat{f}(k) T^{n-1-k}$ $(0 \le k \le n - 1)$ we have

$$\left| \sum_0^{n-1} \widehat{f}(k) \right| = \left\| \sum_0^{n-1} \widehat{f}(k) T^{n-1} \right\|$$

$$\le C \sup_{|z|=1} \left\| \sum_0^{n-1} \widehat{f}(k) T^{n-1-k} z^k \right\|.$$

Now observe that f_z is a polynomial such that $f_z(T) = \sum_0^{n-1} \widehat{f}(k) T^{n-1-k} z^k$, and by von Neumann's inequality we have

$$\| f_z(T) \| \le \sup_{\xi \in \mathbb{T}} |f_z(\xi)|.$$

Thus we obtain

$$\left| \sum_0^{n-1} \widehat{f}(k) \right| \le C \sup_{\xi \in \mathbb{T}} |f_z(\xi)|.$$

But it is easy to check by translation invariance on the circle that

$$\sup_{\xi \in \mathbb{T}} |f_z(\xi)| = \sup_{z \in \mathbb{T}} \left| \sum_{j \le n-1} \widehat{f}(j) z^{-j} \right|,$$

whence finally

$$\left| \sum_0^{n-1} \widehat{f}(k) \right| \le C \left\| \sum_{j \le n-1} \widehat{f}(j) z^{-j} \right\|_{C(\mathbb{T})}.$$

By the Hahn–Banach Theorem, there is a linear form $\mu \in C(\mathbb{T})^*$ with norm $\le C$ such that $\widehat{\mu}(j) = 1$ if $0 \le j \le n - 1$ and $\widehat{\mu}(j) = 0$ if $j < 0$. Identifying μ with a (complex) measure on $\mathbb{T}$, we obtain the above claim. ∎

Remark. The proof of the preceding estimate (25.16) establishes a conjecture formulated in the first edition of [P10]. Moreover, the equality $C(\varphi) = \delta(\widetilde{q})$ together with (25.16) yields a minor improvement over Bourgain's estimate in [Bou4]; namely, if $T \in M_n$ is polynomially bounded with constant C, then there is an invertible matrix $S \in M_n$ such that $\|S^{-1}TS\| \le 1$ and satisfying $\|S^{-1}\|\|S\| \le K \operatorname{Log}(n+1)C^2$.

Exercises

Exercise 25.1. Let K be a Hilbert space and let $H = K \oplus K$. Let $T \in B(K \oplus K)$ be the operator defined by the matrix

$$T = \begin{pmatrix} 0 & 0 \\ a & 0 \end{pmatrix}.$$

Let U be a unitary dilation of T on a Hilbert space $\widehat{H}$ with $\widehat{H} \supset H$. Show that, if a is isometric, then for all x in K we have necessarily (cf. [SNF, Par2]):

$$U \begin{pmatrix} x \\ 0 \end{pmatrix} = \begin{pmatrix} 0 \\ ax \end{pmatrix}.$$

Moreover, if a is unitary, we also have

$$U^{-1} \begin{pmatrix} 0 \\ x \end{pmatrix} = \begin{pmatrix} a^{-1}x \\ 0 \end{pmatrix}.$$

Exercise 25.2. Let E_1, E_2 be a pair of operator spaces. Consider the symmetrized Haagerup tensor product $E_1 \otimes_\mu E_2$ defined before Theorem 5.17. Show that we have a canonical completely isometric embedding

$$E_1 \otimes_\mu E_2 \subset OA(E_1) \otimes_{\max} OA(E_2).$$

Chapter 26. The Blecher-Paulsen Factorization. Infinite Haagerup Tensor Products

We begin this chapter by a striking factorization theorem due to Blecher and Paulsen ([BP2]).

Theorem 26.1. *Let $A \subset B(\mathcal{H})$ be a unital closed subalgebra that is generated by a unital subset S of the closed unit ball of A so that, if $\mathcal{A}$ denotes the algebra generated by S, then $\mathcal{A}$ is dense in A. Let $K \geq 0$ be a constant. The following are equivalent:*

(i) *Any unital homomorphism $u\colon \mathcal{A} \to B(H)$ such that $\sup_{a \in S} \|u(a)\| \leq 1$ is c.b. and satisfies $\|u\|_{cb} \leq K$.*

(ii) *For any n and any x in $M_n(\mathcal{A})$ with $\|x\|_{M_n(A)} < 1$, there is, for some m, a factorization of the form $x = \alpha_0 D_1 \alpha_1 \ldots D_m \alpha_m$, where $\alpha_0, \ldots, \alpha_m$ are (possibly rectangular) matrices with scalar entries, such that $\prod_{j=0}^m \|\alpha_j\| < K$ and $D_1, \ldots, D_m$ are diagonal (hence square) matrices with entries in S.*

(iii) *Same as (ii) but with each D_i having one diagonal entry in S and all others equal to the unit of A.*

Proof. The implication (ii) $\Rightarrow$ (i) is easy and left as an exercise for the reader. Conversely, assume (i). For any x in $M_n(\mathcal{A})$ let us denote

$$|||x|||_n = \inf \left\{ \prod_{j=0}^m \|\alpha_j\| \right\},$$

where the infimum runs over all possible factorizations of x as in (ii) above.

Note that $\|x\|_{M_n(A)} \leq |||x|||_n$. It is rather easy to verify that the sequence of the norms $(||| \ |||_n)$ satisfies Ruan's Axioms (R_1') and (R_2'); hence, after completion, we obtain an operator space E such that E is the completion of $\mathcal{A}$ and $|||x|||_n = \|x\|_{M_n(E)}$ for all x in $M_n(\mathcal{A})$. (See §2.2).

The very definition of the norms $||| \ |||_n$ shows that the product of $\mathcal{A}$ extends to a completely contractive mapping from $E \otimes_h E$ to E and $\|1_A\|_E = \|1_A\|_A = 1$, so that, by Theorem 6.1, E is an operator algebra for this product. Hence there is a completely isometric unital homomorphism $j\colon E \to B(H)$. Obviously (using a trivial factorization of the 1×1 matrix (x)) we have, by definition of $||| \ |||_n$,

$$\forall x \in S \qquad \|x\|_E = |||(x)|||_1 \leq 1.$$

Therefore, our assumption (i) implies

$$\|j\colon \mathcal{A} \to B(H)\|_{cb} \leq K.$$

In particular, this implies that, for any x in $M_n(\mathcal{A})$, we have

$$|||x|||_n = \|(I \otimes j)(x)\|_{M_n(B(H))} \leq K\|x\|_{M_n(\mathcal{A})},$$

whence (ii). Finally, the equivalence of (ii) and (iii) is obvious, since each $N \times N$ diagonal matrix D_i with entries in $\mathcal{S}$ can clearly be written as a product of N matrices of the form appearing in (iii). ∎

Definition. When the properties in Theorem 26.1 hold, we say that $\mathcal{S}$ K-completely generates A, and if $K = 1$, we simply say that $\mathcal{S}$ completely generates A.

The simplest example consists in taking $\mathcal{S}$ equal to the whole closed unit ball of A. Then we immediately obtain.

Corollary 26.2. Let $A \subset B(\mathcal{H})$ be a unital closed subalgebra and let $K \geq 0$ be a constant. The following are equivalent:

(i) Any contractive unital homomorphism $u\colon A \to B(H)$ is c.b. and satisfies $\|u\|_{cb} \leq K$.

(ii) For any n and any x in $M_n(A)$ with $\|x\|_{M_n(A)} < 1$, there is a factorization of the form $x = \alpha_0 D_1 \alpha_1 D_2 \ldots D_m \alpha_m$, where $\alpha_0, \alpha_1, \ldots, \alpha_m$ are (possibly rectangular) matrices with scalar entries and $D_1, \ldots, D_m$ are diagonal (hence square) matrices with entries in A satisfying

$$\prod \|\alpha_i\| \prod \|D_i\| < K.$$

(Note that m and the sizes of the matrices α_i, D_i are arbitrary.)

Definition 26.3. We will say that a unital operator algebra A is "full" (perhaps "completely full" would be better!) if every contractive unital homomorphism $u\colon A \to B(H)$ is completely contractive. In other words, A is full iff it satisfies (i) in Corollary 26.2 with $K = 1$, or, equivalently, if it is completely generated by its unit ball.

For example, every unital C^*-algebra A satisfies this. In this case, every contractive unital homomorphism $u\colon A \to B(H)$ is automatically a *-homomorphism (hence a fortiori a complete contraction). Indeed, first observe that an element x in A (or in $B(H)$) is unitary iff x is invertible and both x and x^{-1} are contractive. Thus, if we denote by $\mathcal{U}(A)$ the set of all unitary elements in A, we see that any unital homomorphism $u\colon A \to B(H)$ satisfying $\sup\{\|u(x)\| \mid x \in \mathcal{U}(A)\} \leq 1$ must take unitaries to unitaries and hence must be a *-homomorphism.

In addition to C^*-algebras, the disc algebra $A(D)$ and the bidisc algebra $A(D^2)$ are both full. More precisely, $A(D)$ (resp. $A(D^2)$) is completely generated by $\{1, z\}$ (resp. $\{1, z_1, z_2\}$). This follows, as explained in the following, from two classical dilation theorems due to Sz.-Nagy and Ando (see [Pa1] or [P10]). However, by Example 25.18, $A(D^n)$ is not full when $n \geq 3$. Note that $A(D^n)$ is a minimal operator space; hence, for $n = 1$ or $n = 2$, if a unital operator algebra is isometrically isomorphic to $A(D^n)$, it is automatically completely isometric to it. The classical Sz.-Nagy dilation theorem says that, given any contraction T on a Hilbert space H, there is a larger Hilbert space $\widehat{H}$ containing H and a unitary operator $U \colon \widehat{H} \to \widehat{H}$ such that

$$\forall n \geq 0 \quad T^n = P_H U^n_{|H}. \tag{26.1}$$

Let $\mathcal{A}$ be the algebra of (analytic) polynomials on the open unit disc $D \subset \mathbb{C}$, and let

$$u \colon \mathcal{A} \to B(H)$$

be the homomorphism defined by

$$u(P) = P(T).$$

Then (26.1) implies

$$u(T) = P_H \pi(P)_{|H}, \tag{26.2}$$

where $\pi \colon \mathcal{A} \to B(\widehat{H})$ is defined by $\pi(P) = P(U)$. Note that (by the spectral functional calculus), since U is unitary (hence a fortiori normal), π extends to a C^*-representation from $C(\partial D)$ to $B(\widehat{H})$, and in particular $\|\pi\|_{cb} = 1$.

Let $A(D) \subset C(\partial D)$ be the closure of $\mathcal{A}$ in $C(\partial D)$. Then, (26.2) implies

$$\|u\|_{CB(A(D), B(H))} = 1.$$

Thus, Sz.-Nagy's dilation theorem shows that, in the disc algebra, the set $\mathcal{S} = \{1, z\}$ satisfies (i) in Theorem 26.1 with $K = 1$. Whence:

Corollary 26.4. *([BP2]) Fix $n \geq 1$. Let $f \colon D \to M_n$ be an analytic matrix valued function such that $\sup_{z \in D} \|f(z)\|_{M_n} < 1$. Assume moreover that all the entries $f_{i,j}(z)$ are polynomials. Then there is, for some integer m, a factorization of the form*

$$f(z) = (a_1 + zb_1) \dots (a_m + zb_m),$$

where a_j, b_j are scalar (possibly rectangular) matrices such that $\sup_{z \in D} \|a_j + zb_j\| < 1$ for all $j \leq m$. More precisely, we can obtain $a_j + zb_j$ of the following form:

$$a_j + zb_j = \alpha_{j-1} D_j \alpha_j,$$

where α_{j-1}, α_j are scalar matrices such that $\|\alpha_{j-1}\|, \|\alpha_j\| < 1$ and where D_j is a diagonal matrix of size (say) $N_j \times N_j$ of the form

$$\begin{pmatrix} 1 & & & & & \\ & \ddots & & & \bigcirc & \\ & & 1 & & & \\ & & & z & & \\ & & & & \ddots & \\ & \bigcirc & & & & z \end{pmatrix}$$

with 1 appearing p_j times and z appearing q_j times ($p_j \geq 0, q_j \geq 0, p_j + q_j = N_j$).

In the case of the bidisc algebra $A(D^2)$, let $\mathcal{A}$ be the algebra of all polynomials $P(z_1, z_2)$ on D^2 and let $u\colon \mathcal{A} \to B(H)$ be the unital homomorphism that takes P to $P(T_1, T_2)$, where $T_1, T_2 \in B(H)$ are two commuting contractions. Then, Ando's dilation theorem (see e.g.[P10]) asserts that there is a Hilbert space $\widehat{H} \supset H$ and two commuting unitaries U_1, U_2 such that for any P we have

$$P(T_1, T_2) = P_H P(U_1, U_2)_{|H}.$$

As above, this shows that, in the bidisc algebra $A(D^2) \subset C(\partial D \times \partial D)$, the set $\mathcal{S} = \{1, z_1, z_2\}$ satisfies the property (i) in Theorem 26.1. Whence:

Corollary 26.5. Fix $n \geq 1$. Let $f\colon D^2 \to M_n$ be an analytic matrix valued function, with polynomial entries, such that $\sup_{z_1, z_2 \in D} \|f(z_1, z_2)\|_{M_n} < 1$. Then f admits for some $m \geq 1$ a factorization of the form

$$f = \alpha_0 D_1 \alpha_1 \ldots D_m \alpha_m,$$

where $\prod \|\alpha_j\| < 1$ and where each D_j is a diagonal matrix of size (say) $N_j \times N_j$ with diagonal of the form

$$[\underbrace{1, \ldots 1}_{p_j}, \underbrace{z_1, \ldots, z_1}_{q_j}, \underbrace{z_2, \ldots, z_2}_{r_j}]$$

($p_j \geq 0, q_j \geq 0, r_j \geq 0, p_j + q_j + r_j = N_j$).

Remark. If we wish, we can assume in the preceding statements that $N_1 = N_2 = \cdots = N_m = N$, so that only α_0 and α_m are rectangular matrices (of size $n \times N$ and $N \times n$) and all the other ones are $N \times N$ matrices.

Remark. The preceding two corollaries are already significant for $n = 1$, that is, for ordinary *complex-valued* analytic functions!

Problem. Surprisingly, the following seems to be still open: Let $K(k)$ be the supremum of $\|u\|_{cb}$ over all possible contractive unital morphisms $u\colon A(D^k) \to$

$B(H)$. By the Sz.-Nagy and Ando dilation theorems we have $K(1) = 1$ and $K(2) = 1$. By Example 25.18, we know that $K(3) > 1$ and a fortiori $K(k) > 1$ for any $k \geq 3$. We suspect that Corollary 26.2 might be useful to show that $K(3) = \infty$. However, although it is unlikely to be true, it is unknown whether $K(3)$ is finite!

Consider now a unital C^*-algebra A and let $\mathcal{S}$ be the group formed of all the unitaries in A. Then (see Remark 26.3) any unital homomorphism $u: A \to B(H)$ such that $\sup\{\|u(x)\| \mid x \in \mathcal{S}\} \leq 1$ is a $*$-homomorphism and therefore has $\|u\|_{cb} = 1$. Whence, by Theorem 26.1:

Corollary 26.6. *Fix $n \geq 1$. Let A be a unital C^*-algebra. Consider $x \in M_n(A)$ with $\|x\|_{M_n(A)} < 1$. Then x admits a factorization of the form*

$$x = \alpha_0 D_1 \alpha_1 D_2 \ldots D_m \alpha_m,$$

where α_j are complex matrices (possibly rectangular) such that $\prod \|\alpha_j\| < 1$ and where D_j are diagonal matrices with entries in the unitary group of A.

Remark 26.7. Let G be a discrete group, and let $\Gamma \subset G$ be a subset containing the unit and generating G (so that any element of G is a finite product of elements of Γ). Let $A = C^*(G)$, let $\mathcal{A} = \mathbb{C}[G] \subset C^*(G)$, and let $\mathcal{S}$ be the copy of Γ sitting naturally inside the unitary group of A. Then, it is easy to check that Theorem 26.1 applies in this situation so that, in the preceding statement, if $x \in M_n(\mathcal{A})$ and $\|x\|_{M_n(A)} < 1$, we can find a factorization as above but with the D_j having entries in $\mathcal{S}$ (i.e., in Γ with the obvious identification).

In Theorem 26.1, we consider a subset $\mathcal{S}$ of the unit ball of A. Actually, in operator space theory, it is more natural to consider the unit ball of $M_n(A)$ or of $\mathcal{K} \otimes_{\min} A$. This leads to the following easy generalization of Theorem 26.1.

Theorem 26.8. *Let A be a unital operator algebra and let $\mathcal{A} \subset A$ be a dense unital subalgebra. We will identify as usual $M_n(\mathcal{A})$ with a subset of $M_{n+1}(\mathcal{A})$ and with a subset of $\mathcal{K} \otimes_{\min} A$. Let $\mathcal{S}$ be a subset of the unit ball of $\mathcal{K} \otimes_{\min} A$. We assume that*

$$\mathcal{S} \subset \mathcal{K} \otimes \mathcal{A}$$

and, moreover, that the algebra generated by $\mathcal{S}$ contains $\bigcup_{n \geq 1} M_n(\mathcal{A})$. Then, for any fixed constant $K \geq 0$, the following are equivalent:

(i) *Any unital homomorphism $u: \mathcal{A} \to B(H)$ such that $\sup_{x \in \mathcal{S}} \|I_{\mathcal{K}} \otimes u(x)\| \leq 1$ is c.b. and satisfies $\|u\|_{cb} \leq K$.*

(ii) *For any n, any x in $M_n(\mathcal{A})$ with $\|x\|_{M_n(A)} < 1$ admits (for some m) a factorization of the form $x = \alpha_0 D_1 \alpha_1 \ldots D_m \alpha_m$, where $\alpha_0, \ldots, \alpha_m$ are in $\mathcal{K} \otimes 1$ with $\Pi\|\alpha_j\| < K$ and where $D_1, \ldots, D_m$ are elements of*

$\mathcal{K} \otimes_{\min} A$ represented by block diagonal matrices of the form

$$
D_j = \begin{pmatrix} \overline{y_1(j)} & & & & \bigcirc \\ & \overline{|y_2(j)|} & & & \\ & & \ddots & & \\ & & & \ddots & \\ \bigcirc & & & & \overline{|y_{N_j}(j)|} \end{pmatrix}
$$

with $y_k(j) \in \mathcal{S}$ for all k and j.

Proof. The proof is analogous to that of Theorem 26.1, so we merely outline it. Assume (i). Note that any x in the algebra generated by $\mathcal{S}$ admits a factorization as in (ii). Thus, for any x in $\bigcup_{n \geq 1} M_n(\mathcal{A})$, we can define the norm

$$
|||x||| = \inf \left\{ \prod_{j=0}^m \|\alpha_j\| \right\},
$$

where the infimum runs over all possible factorizations as in (ii). By Ruan's Theorem and the Blecher-Ruan-Sinclair Theorem, there is a unital homomorphism $u \colon \mathcal{A} \to B(H)$ such that $I_\mathcal{K} \otimes u$ defines an isometry from $\left(\bigcup_{n \geq 1} M_n(\mathcal{A}),\right.$ $|||\ |||)$ into $\mathcal{K} \otimes_{\min} B(H)$. By definition of $|||\ |||$, if we denote by $P_n \colon \mathcal{K} \to M_n$ the natural projection, for any x in $\mathcal{S}$ and for any n, we have $|||(P_n \otimes I)(x)||| \leq 1$; hence $\|P_n \otimes u(x)\|_{M_n(B(H))} \leq 1$. Therefore, $\|I_\mathcal{K} \otimes u(x)\| = \lim_{n \to \infty} \|P_n \otimes u(x)\|_{M_n(B(H))} \leq 1$. Thus, by (i), this implies that $\|u\|_{CB(A,B(H))} \leq K$, or, equivalently, for any x in $\bigcup_{n \geq 1} M_n(\mathcal{A})$ we have

$$
|||x||| = \|I_\mathcal{K} \otimes u(x)\| \leq K \|x\|_{\mathcal{K} \otimes_{\min} A},
$$

so that we conclude that (ii) holds.

This shows (i) $\Rightarrow$ (ii). The converse is easy and left to the reader. ∎

We will now show that the product factorization appearing in the preceding statements can be interpreted in terms of infinite Haagerup tensor products. We first define the latter.

Let I be a totally ordered set and let $(E_j)_{j \in I}$ be a collection of operator spaces, given together with a family $(\xi_j)_{j \in I}$ with $\xi_j \in E_j$ and $\|\xi_j\|_{E_j} = 1$.

For any finite subset $J \subset I$ we can form the Haagerup tensor product

$$
E_J = \bigotimes_{j \in J} E_j.
$$

Note that this definition depends on the ordering of J; that is, what we mean here is that, if $J = \{j_1, j_2, \ldots, j_m\}$ with $j_1 < j_2 < \cdots < j_m$ (in the order of I), then $E_J = E_{j_1} \otimes_h \cdots \otimes_h E_{j_m}$. For any finite set $J' = \{j'_1, \ldots, j'_{m'}\}$ with $j'_1 < \cdots < j'_{m'}$ containing J, we define the linear mapping

$$i_{J,J'} \colon E_J \to E_{J'},$$

which takes $x_{j_1} \otimes \cdots \otimes x_{j_m}$ to $y_{j'_1} \otimes \cdots \otimes y_{j'_{m'}}$, where $y_{j'} = x_{j'}$ if $j' \in J$ and $y_{j'} = \xi_{j'}$ if $j' \notin J$. Clearly (since we assume $\|\xi_j\| = 1$ for all j), $i_{J,J'}$ is a completely isometric embedding of E_J onto $E_{J'}$, and if J'' is another finite subset of I containing J', we clearly have

$$i_{J',J''} \circ i_{J,J'} = i_{J,J''}.$$

Therefore, we may unambiguously define the inductive limit of the system $\{E_J\}$ equipped with the above embeddings. We will denote by

$$\left(\bigotimes_{j \in I} E_j \right)_h$$

the resulting operator space. [More precisely, we introduce a vector space V containing each E_J in such a way that the diagrams

$$
\begin{array}{ccc}
E_J & \longrightarrow & V \\
\scriptstyle i_{J,J'} \searrow & & \nearrow \\
& E_{J'} &
\end{array}
$$

all commute and such that V is the union of E_J when J runs over all finite subsets of I. Then we obtain a norm on $M_n(V) = \bigcup_J M_n(E_J)$ and we apply Ruan's Theorem. After completion V becomes an operator space denoted by $\left(\bigotimes_{j \in I} E_j \right)_h$.]

In the above situation, let $\mathcal{H}$ be a Hilbert space and assume given, for each j in I, a complete contraction $i_j \colon E_j \to B(\mathcal{H})$ such that $i_j(\xi_j) = 1$. For each finite subset $J \subset I$, say, $J = \{j_1, \ldots, j_m\}$ with $j_1 < \cdots < j_m$, there is clearly a unique linear map from $\bigotimes_{j \in J} E_j$ into $B(\mathcal{H})$ that takes $x_{j_1} \otimes \cdots \otimes x_{j_m}$ to $i_{j_1}(x_{j_1}) \ldots i_{j_m}(x_{j_m})$. By Corollary 5.4, this mapping extends to a complete contraction $i_J \colon E_J \to B(\mathcal{H})$. Since $i_j(\xi_j) = 1$ for all j, for any finite subset J' containing J we have $i_{J'|E_J} = i_J$. Whence the next statement.

Lemma 26.9. *With the preceding notation, the system of maps* $\{i_J \mid J \subset I, |J| < \infty\}$ *extends unambiguously to a complete contraction from* $\left(\bigotimes_{j \in I} E_j \right)_h$ *to* $B(\mathcal{H})$.

To illustrate this notion, let us return to the situation discussed in Theorem 26.1. Let us first show that the subset $\mathcal{S}$ naturally defines an operator space structure on its linear span. Let V be the linear span of $\mathcal{S}$. For any x in $M_n(V)$ we define

$$\|x\|_n = \inf\{\|\alpha_0\| \, \|\alpha_1\|\},$$

where the infimum runs over all possible factorizations of x of the form $x = \alpha_0 D \alpha_1$, where α_0, α_1 are scalar rectangular matrices and where D is a diagonal matrix with entries in $\mathcal{S}$. By Ruan's Theorem, these norms correspond to an operator space structure on V for which the inclusion $V \subset A$ is completely contractive. Therefore, the inclusion map $V \subset A$ extends to a complete contraction $i\colon \widehat{V} \to A$ defined on the completion of V. Let us denote by E the resulting operator space, that is, E is $\widehat{V}$ equipped with the o.s.s. associated to the above norms $\|\ \|_n$. We will form the infinite Haagerup tensor product $\left(\bigotimes_{j \in I} E_j\right)_h$ in the case $J = \mathbb{N}$ (with its usual ordering), $E_j = E$, $i_j = i$, and $\xi_j = 1_A$ for all j. (Recall that we assume $1_A \in \mathcal{S}$.) We will denote simply by E_∞ the resulting operator space and by $E_n = E \otimes_h \cdots \otimes_h E$ (n times) the finite tensor product considered as a subset of E_∞, so that $\bigcup_{n \geq 1} E_n$ is a dense linear subspace of E_∞.

By Lemma 26.9, the product mapping of A unambiguously defines an inductive system of complete contractions

$$p_n\colon E_n \to A,$$

which extend to a single complete contraction

$$p_\infty\colon E_\infty \to A,$$

which plays the role of an "infinite product map." We can now reformulate Theorem 26.1 with this notion.

Proposition 26.10. *With the preceding notation, the two properties in Theorem 26.1 imply that the completely contractive mapping $p_\infty\colon E_\infty \to A$ is a complete surjection. More precisely, it is surjective and the resulting isomorphism $\sigma\colon E_\infty/\ker(p_\infty) \to A$ satisfies*

$$\|\sigma\|_{cb} \leq 1 \quad \text{and} \quad \|\sigma^{-1}\|_{cb} \leq K.$$

Remark 26.11. It is easy to check that the two properties in Theorem 26.1 are equivalent to the following one:

(iii) Let $\mathcal{E} = \bigcup_{n \geq 1} E_n \subset E_\infty$ be equipped with the induced operator space structure and similarly for $\mathcal{A} \subset A$. The restriction of p_∞ to $\mathcal{E}_\infty$ is a surjection from $\mathcal{E}$ onto $\mathcal{A}$ that defines a completely contractive isomorphism

$$\widehat{\sigma}\colon \mathcal{E}/\ker(p_\infty) \cap \mathcal{E} \to \mathcal{A}$$

such that $\|\widehat{\sigma}^{-1}\|_{cb} \leq K$.

Similarly we have:

Proposition 26.12. *In the situation of Corollary 26.2, let $I = \mathbb{N}$, $E = \max(A)$, let $i: E \to \mathcal{A} = A$ be the identity map, and let $i_j = i$ and $\xi_j = 1_A$ for all j. Then the infinite product map $p_\infty: E_\infty \to A$ defines a completely contractive isomorphism $\sigma: E_\infty / \ker p_\infty \to A$ such that $\|\sigma^{-1}\|_{cb} \leq K$. Thus any full operator algebra (in particular every unital C^*-algebra A) is completely isometric to a quotient of an infinite Haagerup tensor product of copies of the operator space $\max(A)$.*

Let us now return to the disc and bidisc algebras $A(D)$ and $A(D^2)$, as in Corollaries 26.4 and 26.5. Note that the linear span V of $\mathcal{S} = [1, z]$ (resp. $\mathcal{S} = [1, z_1, z_2]$) in $A(D)$ (resp. $A(D^2)$) is isometric to ℓ_1^2 (resp. ℓ_1^3). By Theorem 2.7.2, the operator space E introduced above on the vector space V can be identified respectively with $\max(\ell_1^2)$ and $\max(\ell_1^3)$. Hence we obtain the following application of Proposition 26.10.

Corollary 26.13. *The algebra $A(D)$ (resp. $A(D^2)$) is completely isometric to a quotient of an infinite Haagerup tensor product of copies of $\max(\ell_1^2)$ (resp. $\max(\ell_1^3)$).*

Remark. Actually, any full operator algebra A is completely isometric to an infinite Haagerup tensor product of copies of $\max(\ell_1^{(2)})$. Indeed, let $I = B_A$ equipped with any total order, and let $\mathcal{J}$ be a countable disjoint union of copies of I simply ordered consecutively (we could use the lexicographical order on $\mathbb{N} \times I$). Thus an element j in $\mathcal{J}$ is simply an element of $I = B_A$ placed in one of the successive copies of I inside $\mathcal{J}$. We denote by (e_0, e_1) the canonical basis of $\ell_1^{(2)}$. Let $E_j = \max(\ell_1^{(2)})$ and $\xi_j = e_0$ for all j in $\mathcal{J}$. Then we have a mapping

$$\widehat{p}: \left(\bigotimes_{j \in \mathcal{J}} E_j \right)_h \to A$$

that is a complete metric surjection onto A. The latter is defined as follows: For any j in $\mathcal{J}$ corresponding to some a in $I = B_A$ we define $u_j: \max(\ell_1^{(2)}) \to A$ by $u_j(e_0) = 1_A$ and $u_j(e_1) = a$. Clearly the construction described in Lemma 26.9 yields a completely contractive map

$$\widehat{p}: \left(\bigotimes_{j \in \mathcal{J}} E_j \right)_h \to A.$$

Note that any product $a_1 a_2 ... a_n$ of elements in B_A can be written as $u_{j_1}(e_1) u_{j_2}(e_1) ... u_{j_n}(e_1)$ by suitably choosing the placement $j_1 < j_2 < ... < j_n$ of the elements $a_1, a_2, ..., a_n$ in $\mathcal{J}$. The fact that $\widehat{p}$ is a complete metric surjection then follows immediately from part (iii) in Theorem 26.1.

Now let E be an *arbitrary* operator space. We can also apply Theorem 26.1 to the universal operator algebra $OA_u(E)$ associated to E, as defined in

Chapter 6. We will consider E as a subspace of of $OA_u(E)$ and will denote by e the unit element in $OA_u(E)$. Let $\widetilde{E} \subset OA_u(E)$ be the operator space spanned by e and E, that is,

$$\widetilde{E} = \mathbb{C}e + E.$$

This space $\widetilde{E}$ is the *unitization* of E, already considered in Proposition 8.19.

Then, we take again $I = \mathbb{N}$, $E_j = \widetilde{E}$, and $\xi_j = e$ for all j; we denote by $(\widetilde{E})_\infty$ the infinite Haagerup tensor product associated to this data; and we let $\mathcal{E} \subset (\widetilde{E})_\infty$ be the union of all the finite tensor products $\widetilde{E} \otimes_h \cdots \otimes_h \widetilde{E}$ (n times) $n \geq 1$.

We can then apply Theorem 26.1 as follows.

Proposition 26.14. *The product mapping from $\widetilde{E} \otimes \widetilde{E}$ to $OA_u(E)$ extends to a complete contraction p_∞ from $(\widetilde{E})_\infty$ onto $OA_u(E)$, which defines a completely isometric isomorphism from $(\widetilde{E})_\infty / \ker(p_\infty)$ onto $OA_u(E)$. More precisely, the restriction of p_∞ to $\mathcal{E}$ defines a completely isometric isomorphism between the (noncomplete) operator spaces*

$$\mathcal{E} / \ker(p_\infty) \cap \mathcal{E} \quad and \quad T_u(E) \subset OA_u(E).$$

Proof. Let $A = OA_u(E)$, $\mathcal{A} = T_u(E)$ and let $S \subset \mathcal{K} \otimes_{\min} A$ be the unit ball of $\mathcal{K} \otimes_{\min} \widetilde{E}$. Then the property (i) in Theorem 26.8 clearly holds with $K = 1$. Therefore, the proposition can be deduced from (ii) in Theorem 26.8 by a simple reformulation. ∎

We now return to the general setting of a family $(E_j)_{j \in I}$ of operator spaces, given together with a family of complete contractions $i_j: E_j \to B(\mathcal{H})$ and a family $(\xi_j)_{j \in I}$ such that $\xi_j \in E_j$ and $i_j(\xi_j) = 1$ for all j in I. We will denote by $\mathcal{A}$ the (unital) subalgebra of $B(\mathcal{H})$ generated by $\bigcup_{j \in I} i_j(E_j)$.

We will say that the triple $\tau = \{\{E_j \mid j \in I\}, \{\xi_j \mid j \in I\}, \{i_j \mid j \in I\}\}$ is a "generating family of operator spaces."

Given such a generating family, we can introduce the enveloping unital operator algebra $\widetilde{A}_\tau$ as follows: We consider the "collection" C_τ of all unital homomorphisms $u: \mathcal{A} \to B(H_u)$ such that $\|u i_j\|_{cb} = 1$ for all j in I, and we introduce the embedding

$$j_\tau: \mathcal{A} \to \bigoplus_{u \in C_\tau} B(H_u)$$

defined by $j_\tau(x) = \bigoplus_{u \in C_\tau} u(x)$. Note that, since the inclusion $\mathcal{A} \subset B(\mathcal{H})$ belongs to C_τ, this map j_τ is injective; hence it defines a new operator space (and operator algebra) structure on $\mathcal{A}$. We will denote by $\widetilde{A}_\tau$ the completion of $\mathcal{A}$ for the latter structure. Note that the mappings $i_j\colon E_j \to \mathcal{A} \subset \widetilde{A}_\tau$ are completely contractive as mappings into $\widetilde{A}_\tau$, so that, by Lemma 26.9, we have a natural complete contraction from $\big(\bigotimes_{j\in I} E_j\big)_h$ into $\widetilde{A}_\tau$. But actually, we will need to enlarge the ordered set I: We will denote by $\mathcal{I}$ the totally ordered set that is the disjoint union of countably many copies of I, ordered consecutively. Let $\{E_j \mid j \in \mathcal{I}\}$ be the extended family where each E_j $(j \in I)$ is repeated countably infinitely many times.

Theorem 26.15. *Let τ be a generating family of operator spaces as above and let $p\colon \big(\bigotimes_{j\in\mathcal{I}} E_j\big)_h \to \widetilde{A}_\tau$ be the natural completely contractive map, associated to the "duplicated" family $\{E_j \mid j \in \mathcal{I}\}$. Then p is a complete metric surjection, that is, it is surjective and it defines a completely isometric isomorphism from $\big(\bigotimes_{j\in\mathcal{I}} E_j\big)_h/\ker(p)$ to $\widetilde{A}_\tau$. Moreover, if we denote by $\mathcal{E}$ the subspace of $\big(\bigotimes_{j\in\mathcal{I}} E_j\big)_h$ that is the union of the (algebraic) tensor products $\bigotimes_{j\in J} E_j$ (with J running over all finite subsets of $\mathcal{I}$), then the restriction of p to $\mathcal{E}$ induces a completely isometric isomorphism between the (noncomplete) operator spaces $\mathcal{E}/\ker(p) \cap \mathcal{E}$ and $\mathcal{A}$ viewed as a subspace of $\widetilde{A}_\tau$.*

Proof. A proof can be given using Theorem 26.8 applied to $\mathcal{S} = \bigcup_{j\in I}(I_\mathcal{K} \otimes i_j)(B_{\mathcal{K}\otimes_{\min} E_j})$ and applied to $A = \widetilde{A}_\tau$.

For the convenience of the reader, here is a direct argument (which in essence is the same): One first observes that $p(\mathcal{E}) = \mathcal{A}$, so that the induced map $\sigma\colon \mathcal{E}/\ker(p) \cap \mathcal{E} \to \mathcal{A}$ is a linear isomorphism. Note that the quotient space $Q = \mathcal{E}/\ker(p) \cap \mathcal{E}$ is equipped with a natural o.s.s. as a (noncomplete) quotient of a subspace of $\big(\bigotimes_{j\in\mathcal{I}} E_j\big)_h$. Let B be the unital operator algebra obtained by equipping Q with the product $x \cdot y = \sigma^{-1}(\sigma(x)\sigma(y))$. Then, after completion, B is a unital operator algebra, that is, we may view B as embedded into some $B(\mathcal{H})$. Indeed, this follows from Theorem 6.1, because the repetition of I infinitely many times in $\mathcal{I}$ allows us to show that this product map is a complete contraction from $Q \otimes_h Q$ to Q. But then $\sigma^{-1}\colon \mathcal{A} \to B$ is a unital homomorphism such that $\sigma^{-1}i_j$ coincides with the composition of the inclusion $E_j \to \mathcal{E}$ followed by the quotient map $\mathcal{E} \to Q$. In particular, $\|\sigma^{-1}i_j\|_{cb} = 1$ for any j, which implies (by definition of $\widetilde{A}_\tau$) that σ^{-1} extends completely contractively to $\widetilde{A}_\tau$. Since σ is completely contractive (cf. Lemma 26.9), we obtain the second assertion in Theorem 26.15, and the first assertion follows from it. ∎

We will now illustrate the preceding statement in the case of free products. For simplicity of notation, we restrict ourselves to free products of pairs of algebras, but we could just as well consider arbitrary families. We have seen earlier (see Theorem 5.13) that the Haagerup tensor product $A_1 \otimes_h A_2$ of two operator algebras naturally embeds into their free product $A_1 * A_2$. We will now see that the free product itself can be viewed as the quotient of an infinite Haagerup tensor product.

Corollary 26.16. *Let A_1, A_2 be two unital operator algebras, and let $A_1 * A_2$ be their free product (as unital operator algebras). Let $(E_j)_{j \in \mathbb{N}}$ be the family of operator spaces defined by setting $E_j = A_1$ if j is odd and $E_j = A_2$ if j is even. Then $A_1 * A_2$ is naturally completely isometric to a quotient of $\left(\bigotimes_{j \in \mathbb{N}} E_j \right)_h$.*

Proof. We take $I = \{0, 1\}, E_1 = A_1, E_0 = A_2$ and take for i_j the natural embedding into $A_1 * A_2 \subset B(\mathcal{H})$. Then the algebra $\widetilde{A}_\tau$ associated to this generating family actually coincides with $A_1 * A_2$, by definition of the latter. Thus Corollary 26.16 follows from Theorem 26.15. ∎

Chapter 27. Similarity Problems

The following conjecture formulated in 1955 by R. Kadison is still open. It will serve as the motivation for all further developments in this chapter.

Kadison's conjecture. *Every bounded unital homomorphism*

$$u\colon A \to B(H)$$

on a unital C^*-algebra A is similar to a $*$-homomorphism, that is, there is an isomorphism $\xi\colon H \to H$ such that

$$x \to \xi u(x)\xi^{-1}$$

is a $*$-homomorphism.

Note that u is only assumed to be a bounded Banach algebra morphism, and we conclude that after conjugation by ξ it becomes a C^*-algebra morphism. More explicitly, our assumption are that $u(xy) = u(x)u(y) \; \forall \, x, y \in A$ and $u(1) = 1$, and the conclusion is that

$$\forall \, x \in A \qquad \xi u(x^*)\xi^{-1} = (\xi u(x)\xi^{-1})^*.$$

When this conclusion holds, Kadison says that u is "orthogonalizable." Many partial results are known on this, mainly due to E. Christensen and U. Haagerup. In particular, this conjecture is known whenever A is a nuclear C^*-algebra or when $A = B(H)$ (or, more generally, when A has no tracial states) and also when A is a II_1-factor with property Γ. We will return on these examples later. Moreover, Haagerup [H6] (see also [Chr3]) proved that the conjecture is correct for all *cyclic* homomorphisms $u\colon A \to B(H)$. Nevertheless, the problem remains open in full generality, and, as shown by the next statement, it is really a problem on c.b. maps.

Theorem 27.1. *([H6]) Let $u\colon A \to B(H)$ be a unital homomorphism on a unital C^*-algebra. Then u is similar to a $*$-homomorphism iff u is c.b. and moreover*

$$\|u\|_{cb} = \inf\{\|\xi^{-1}\| \, \|\xi\|\},$$

where the infimum (which is actually attained) runs over all isomorphisms ξ for which $\xi u(\cdot)\xi^{-1}$ is a $*$-homomorphism.

Thus, Kadison's conjecture boils down to the implication

$$\|u\| < \infty \overset{?}{\Rightarrow} \|u\|_{cb} < \infty$$

for unital homomorphisms into $B(H)$. Equivalently, we want to know whether complete boundedness is *automatic* for bounded homomorphisms.

The above conjecture is closely connected to the following well-known question.

Derivation problem. *Let A be a C^*-algebra. Given a $*$-homomorphism $\pi\colon A \to B(H_\pi)$, a bounded linear map $\delta\colon A \to B(H_\pi)$ is called a π-derivation if*

$$\forall\, a, b \in A \qquad \delta(ab) = \pi(a)\delta(b) + \delta(a)\pi(b).$$

We say that δ is π-inner, or simply "inner", if there is T in $B(H_\pi)$ such that

$$\forall\, a \in A \qquad \delta(a) = \pi(a)T - T\pi(a).$$

The problem is to show that all π-derivations are inner.

Once again this problem turns out to be a problem on c.b. maps, because of the following result due to E. Christensen.

Theorem 27.2. *([Chr5]) Let $\delta\colon A \to B(H_\pi)$ be a π-derivation (as above) on a C^*-algebra A. Then δ is inner iff it is c.b. Moreover, we have*

$$\|\delta\|_{cb} = \inf\{2\|T\| \mid T \in B(H_\pi), \quad \delta = \delta_T\}, \tag{27.1}$$

where we have set $\delta_T(a) = \pi(a)T - T\pi(a)$.

Proof. If δ is inner, say, if $\delta = \delta_T$, we clearly have by the triangle inequality $\|\delta\|_{cb} = \|\delta_T\|_{cb} \le 2\|T\|$; hence $\|\delta\|_{cb} \le \inf\{2\|T\| \mid \delta = \delta_T\}$. The nontrivial direction is the converse for which we will use Theorem 27.1. Indeed, given a π-derivation $\delta\colon A \to B(H)$, the formula

$$u(a) = \begin{pmatrix} \pi(a) & \delta(a) \\ 0 & \pi(a) \end{pmatrix}$$

obviously defines a homomorphism into $B(H \oplus H) \simeq M_2(B(H))$. Moreover, if δ is c.b., u must also be c.b.; hence, by Theorem 27.1, u is similar to a $*$-homomorphism $\rho\colon A \to B(H \oplus H)$, that is, for some invertible $\xi\colon H \oplus H \to H \oplus H$ we have $u(x) = \xi\rho(x)\xi^{-1}$. Then let $a = \xi\xi^*$ so that $u(x)a = \xi\rho(x)\xi^*$ for all x in A. Since $\rho(x) = \rho(x^*)^*$, we have $u(x)a = (u(x^*)a)^*$, and hence

$$\forall\, x \in A \qquad u(x)a = au(x^*)^*.$$

In matricial notation, this becomes

$$\begin{pmatrix} \pi(x) & \delta(x) \\ 0 & \pi(x) \end{pmatrix} \begin{pmatrix} a_{11} & a_{12} \\ a_{12}^* & a_{22} \end{pmatrix} = \begin{pmatrix} a_{11} & a_{12} \\ a_{12}^* & a_{22} \end{pmatrix} \begin{pmatrix} \pi(x) & 0 \\ \delta(x^*)^* & \pi(x) \end{pmatrix},$$

which implies (consider the $(2,2)$ and $(1,2)$ entries only)

$$\pi(x)a_{22} = a_{22}\pi(x) \quad \text{and} \quad \pi(x)a_{12} + \delta(x)a_{22} = a_{12}\pi(x). \tag{27.2}$$

Since $a \geq 0$ is invertible, $\exists \, \delta > 0$ such that $\langle ay, y \rangle \geq \delta \|y\|^2$ for all y in $H \oplus H$; hence the same holds for a_{22} on H and a_{22} must be invertible. Moreover, a_{22} commutes with $\pi(x)$ so that (27.2) implies

$$\delta(x) = (a_{12}\pi(x) - \pi(x)a_{12})a_{22}^{-1} = T\pi(x) - \pi(x)T$$

with $T = a_{12}a_{22}^{-1}$, which shows that δ is inner. Unfortunately, this method only seems to yield T with $\|T\| \leq 4\|\delta\|_{cb}$. In order to replace this factor 4 by the correct factor $1/2$, we will need more work.

First we will reduce to the von Neumann algebra case. Assume that $\delta\colon A \to B(H_\pi)$ is c.b. Then, by a routine argument, π and δ both extend to normal maps $\widetilde{\pi}\colon A^{**} \to B(H_\pi)$ and $\widetilde{\delta}\colon A^{**} \to B(H_\pi)$, so that $\widetilde{\pi}$ is a $*$-homomorphism and $\widetilde{\delta}$ is a $\widetilde{\pi}$-derivation with the same c.b. norm. Note that $\widetilde{\delta}$ must vanish on the kernel of $\widetilde{\pi}$ (since the latter ideal admits approximate units). Thus, replacing A^{**} by its image under $\widetilde{\pi}$, we can reduce to the case when A is a von Neumann algebra and π is its natural inclusion into $B(H)$. Moreover, by the first part of the proof, we can assume that δ is inner.

Thus it suffices to prove (27.1) when $\delta\colon M \to B(H)$ is an inner derivation, say, $\delta = \delta_T$, on a von Neumann subalgebra $M \subset B(H)$. Recall that δ_T is defined by $\delta_T(x) = Tx - xT$. Then we will prove

$$\|\delta\|_{cb} = 2\inf\{\|T\| \mid \delta = \delta_T\}$$

or, equivalently,

$$\|\delta_T\|_{cb} = 2\inf\{\|T - \widehat{T}\| \mid \widehat{T} \in M'\}, \tag{27.3}$$

where M' denotes the commutant of M. We will denote by $d(T, M')$ the right side of (27.3). Here is a quick proof of (27.3) taken from [Chr5] (but the argument is similar to Arveson's one in [Ar3, p. 12]). We will prove that

$$d(T, M') \leq (1/2)\|\delta_T\|_{cb}. \tag{27.4}$$

First note that $\|\delta_T\|_{cb}$ is equal to the norm of the operator $I_{B(\ell_2)} \otimes \delta_T$ acting on the von Neumann algebra $B(\ell_2)\overline{\otimes}M \subset B(\ell_2(H))$ (i.e., the one generated by the algebraic tensor product $B(\ell_2) \otimes M)$). Let $B(H)_*$ denote the predual of $B(H)$. By Hahn-Banach, we have

$$d(T, M') = \sup\{|f(T)| \mid f \in B(H)_*, f \perp M', \|f\| = 1\}. \tag{27.5}$$

Consider $f \in B(H)_*$ with $f \perp M'$ and $\|f\| = 1$. By the classical identification of $B(H)_*$ with the space of trace class operators on H, there are unit vectors ξ, η in $\ell_2(H)$ such that

$$\forall x \in B(H) \quad f(x) = \langle (I \otimes x)\xi, \eta \rangle. \tag{27.6}$$

Hence, for any x in M', $(I \otimes x)\xi$ is orthogonal to η. Let p be the orthogonal projection on $\ell_2(H)$ onto the closure of $\{(I \otimes x)\xi \mid x \in M'\} \subset \ell_2(H)$. Clearly $p\xi = \xi$ and $p\eta = 0$. Moreover, since p commutes with $(I \otimes M')$, we have $p \in B(\ell_2)\overline{\otimes}M$. By (27.6), we can write

$$
\begin{aligned}
|f(T)| &= |\langle (I \otimes T)p\xi, (1-p)\eta \rangle| \\
&\leq \|(1-p)(I \otimes T)p\| = \|[(I \otimes T)p - p(I \otimes T)]p\| \\
&\leq \|(I \otimes T)p - p(I \otimes T)\| \\
&= \tfrac{1}{2}\|(I \otimes T)(2p-1) - (2p-1)(I \otimes T)\| = \tfrac{1}{2}\|(I \otimes \delta_T)(2p-1)\|;
\end{aligned}
$$

hence note that $2p - 1 = p - (1-p)$ is unitary)

$$
|F(T)| \leq \tfrac{1}{2}\|I \otimes \delta_T\| = \tfrac{1}{2}\|\delta_T\|_{cb}.
$$

Taking the supremum over all possible f and using (27.5), we obtain (27.4). The converse is obvious. ∎

As we just saw, there is a very natural way to construct homomorphisms using derivations. Indeed, given a π-derivation $\delta\colon A \to B(H_\pi)$, the formula

$$
u(a) = \begin{pmatrix} \pi(a) & \delta(a) \\ 0 & \pi(a) \end{pmatrix}
$$

obviously defines a homomorphism into $B(H_\pi \oplus H_\pi) \simeq M_2(B(H_\pi))$. More generally, given two homomorphisms

$$
\pi_1\colon A \to B(H) \quad \text{and} \quad \pi_2\colon A \to B(H)
$$

on the same operator algebra, and given a (π_1, π_2)-derivation $\delta\colon A \to B(H)$ (by this we mean $\forall\, a, b \in A$, $\delta(ab) = \pi_1(a)\delta(b) + \delta(a)\pi_2(b)$), the formula

$$
u(a) = \begin{pmatrix} \pi_1(a) & \delta(a) \\ 0 & \pi_2(a) \end{pmatrix} \tag{27.7}
$$

defines a homomorphism from A to $B(H \oplus H) \simeq M_2(B(H))$.

In particular, this observation shows that, if a C^*-algebra A satisfies Kadison's conjecture, then the derivation problem has an affirmative solution for A. Recently, E. Kirchberg [Ki9] showed that the converse also holds, so that Kadison's similarity problem for a C^*-algebra A is actually equivalent to the derivation problem for A as formulated above.

Actually, Theorem 27.1 also holds for non-self-adjoint operator algebras. This is a very important result due to Paulsen.

Theorem 27.3. *([Pa4]) Let $A \subset B(\mathcal{H})$ be a unital operator algebra. Let $u \colon A \to B(H)$ be a unital homomorphism, and let C be a constant. The following properties are equivalent:*

(i) *u is c.b. and $\|u\|_{cb} \leq C$.*

(ii) *There is an isomorphism $\xi \colon H \to H$ with $\|\xi\| \, \|\xi^{-1}\| \leq C$ such that $x \to \xi u(x)\xi^{-1}$ is completely contractive.*

(iii) *There is a Hilbert space $\widehat{H}$ with $H \subset \widehat{H}$, a $*$-homomorphism $\pi \colon B(\mathcal{H}) \to B(\widehat{H})$ and an isomorphism $\xi \colon H \to H$ with $\|\xi\| \, \|\xi^{-1}\| \leq C$ such that*

$$\forall \, a \in A \qquad u(a) = \xi^{-1} P_H \pi(a)_{|H} \xi.$$

Proof. The implications (iii) $\Rightarrow$ (ii) $\Rightarrow$ (i) are obvious. We will show (i) $\Rightarrow$ (iii). Assume (i). By the fundamental factorization Theorem 1.6, we can write

$$u(x) = V \pi(x) W,$$

where $\pi \colon B(\mathcal{H}) \to B(\widehat{H})$ is a representation (C^*-sense) and where $V \colon \widehat{H} \to H$ and $W \colon H \to \widehat{H}$ satisfy $\|V\| \, \|W\| \leq C$. Let $E_1 = \overline{\operatorname{span}}[\pi(A)W(H)]$. Note that E_1 is invariant under $\pi(A)$ and contains $W(H)$ (and actually is minimal with these properties). We may rewrite the above factorization trivially as

$$u(x) = (V_{|E_1}) \pi(x)_{|E_1} P_{E_1} W. \tag{27.8}$$

Let $E_2 = \ker(V_{|E_1})$. We claim that $E_2 \subset E_1$ is also $\pi(A)$-invariant. The simplest way to see this is to observe that

$$\forall \, a \in A \qquad V \pi(a)_{|E_1} = u(a) V_{|E_1}. \tag{27.9}$$

Indeed, it suffices to check (27.8) on a vector k of the form $k = \pi(x)Wh$ with $h \in H$, $x \in A$. We then have $V\pi(a)k = V\pi(ax)Wh = u(ax)h = u(a)u(x)h = u(a)V\pi(x)Wh = u(a)Vk$, which establishes (27.9).

From (27.9) it is clear that $\pi(a)E_2 \subset E_2$ for all a in A. We will work with the (so-called semi-invariant) subspace $E_1 \ominus E_2 = E_1 \cap E_2^{\perp}$ (as in the proof of Theorem 6.3). We start by the observation that, since $u(1) = 1$, we have $VW = I$, and hence $V(E_1) = H$, so V is surjective. Passing to the quotient by $\ker(V)$ we obtain from (27.8)

$$\forall \, x \in A \qquad u(x) = S P_{E_1 \ominus E_2} \pi(x) P_{E_1}(W),$$

where $S \colon E_1 \ominus E_2 \to H$ is an isomorphism with $\|S\| \leq \|V\|$. Moreover, since $P_{E_1 \ominus E_2} \pi(x)$ vanishes on E_2 (since E_2 is invariant), we actually have

$$\forall \, x \in A \qquad u(x) = S P_{E_1 \ominus E_2} \pi(x) P_{E_1 \ominus E_2} W.$$

Taking $x = 1$, this gives

$$1 = SP_{E_1 \ominus E_2} W,$$

which implies that $P_{E_1 \ominus E_2} W$ is the inverse of the isomorphism S, or, equivalently, $S^{-1} = P_{E_1 \ominus E_2} W$. Hence

$$\forall \, x \in A \qquad u(x) = S\pi(x)S^{-1},$$

which shows that $S^{-1}u(\cdot)S$ is a $*$-homomorphism and we have $\|S\| \leq \|V\|$, $\|S^{-1}\| \leq \|W\|$; hence $\|S\| \, \|S^{-1}\| \leq \|u\|_{cb}$. This gives us (iii), except that S is an isomorphism from $E_1 \ominus E_2$ to H. But since $E_1 \ominus E_2$ and H have the same (Hilbertian) dimension, they are isometric; hence there is a unitary $U \colon E_1 \ominus E_2 \to H$ such that S can be written as $S = \xi U$. This gives us an isomorphism $\xi \colon H \to H$ with $\|\xi\| \, \|\xi^{-1}\| \leq \|u\|_{cb}$ such that $\xi^{-1}u(\cdot)\xi = U(S^{-1}u(\cdot)S)U^*$ is a $*$-homomorphism. This completes the proof that (i) $\Rightarrow$ (iii). ∎

Definition 27.4. *We say that a unital operator algebra $A \subset B(\mathcal{H})$ has the similarity property, in short SP, if any bounded unital homomorphism $u \colon A \to B(H)$ is automatically completely bounded.*

As we will see, this property is closely connected with a certain notion of "length" for A, as follows.

Definition 27.5. *An operator algebra $A \subset B(\mathcal{H})$ is said to be of length $\leq d$ if there is a constant K such that, for any n and any x in $M_n(A)$, there is an integer $N = N(n, x)$ and scalar matrices $\alpha_0 \in M_{n,N}(\mathbb{C})$, $\alpha_1 \in M_N(\mathbb{C}), \ldots, \alpha_{d-1} \in M_N(\mathbb{C})$, $\alpha_d \in M_{N,n}(\mathbb{C})$ together with diagonal matrices $D_1, \ldots, D_d$ in $M_N(A)$ satisfying*

$$\begin{cases} x = \alpha_0 D_1 \alpha_1 D_2 \ldots D_d \alpha_d \\ \prod_0^d \|\alpha_i\| \prod_1^d \|D_i\| \leq K\|x\|. \end{cases} \tag{27.10}$$

We denote by $\ell(A)$ the smallest d for which this holds, and we call it the length of A (so that A has length $\leq d$ is indeed the same as $\ell(A) \leq d$).

Remarks. (i) Fix an integer n and consider x in $M_n(A)$. We denote

$$\|x\|_{(d)} = \inf\{\prod_{i=0}^d \|\alpha_i\| \prod_{i=1}^d \|D_i\|\},$$

where the infimum runs over all possible representations $x = \alpha_0 D_1 \alpha_1 \ldots D_d \alpha_d$ of the form appearing in (27.10) above. By (27.11) below, this is a norm on $M_n(A)$. Clearly

$$\|x\|_{M_n(A)} \leq \|x\|_{(d)}.$$

(ii) When A is unital (or has a contractive approximate unit, which is the case whenever A is a C^*-algebra) we have

$$\forall x \in M_n(A) \qquad \|x\|_{(d+1)} \le \|x\|_{(d)}.$$

Indeed we can always insert a redundant diagonal matrix with coefficients all equal to the unit. Moreover, if A is a C^*-algebra, the results of Chapter 26 (see Corollary 26.6) show that

$$\forall x \in M_n(A) \qquad \|x\|_{M_n(A)} = \lim_{d \to \infty} \downarrow \|x\|_{(d)}.$$

(iii) Let $E = \max(A)$. Let $p_d\colon E \otimes_h \cdots \otimes_h E \to A$ be the product map. Clearly p_d is completely contractive (since it is completely contractive on $A \otimes_h \cdots \otimes_h A$). We claim that the norm $\|\cdot\|_{(d)}$ on $M_n(A)$ is equal to the quotient norm $M_n(E \otimes_h \cdots \otimes_h E)/M_n(\ker(p_d))$.

Indeed, consider $x \in M_n(A)$ and let $y \in M_n(E \otimes \cdots \otimes E)$ be such that $(I_{M_n} \otimes p_d)(y) = x$. Then we have

$$\|x\|_{(d)} = \inf\{\|y\|_{M_n(E \otimes_h \cdots \otimes_h E)}\}, \tag{27.11}$$

where the infimum runs over all possible such y. This follows immediately from (5.6) combined with Theorem 3.1. But, since p_d is completely contractive on the completion $E \otimes_h \cdots \otimes_h E$, the equality (27.11) must remain (a fortiori) true when the infimum runs over all y in $M_n(E \otimes_h \cdots \otimes_h E)$ such that $(I_{M_n} \otimes p_d)(y) = x$.

Proposition 27.6. *If $\ell(A) \le d$ as in (27.10) above, then A has the similarity property (SP), and for any bounded homomorphism $u\colon A \to B(H)$ we have $\|u\|_{cb} \le K\|u\|^d$.*

Proof. Let $u\colon A \to B(\mathcal{H})$ be any bounded homomorphism. Consider $x \in M_n(A)$ factorized as in (27.10). We then have

$$(I_{M_n} \otimes u)(x) = \alpha_0(I_{M_N} \otimes u)(D_1) \ldots (I_{M_N} \otimes u)(D_d)\alpha_d,$$

and clearly for any $i = 1, \ldots, d$

$$\|I_{M_N} \otimes u(D_i)\| \le \|u\| \, \|D_i\|;$$

hence

$$\|(I_{M_n} \otimes u)(x)\| \le \|u\|^d \prod_0^d \|\alpha_i\| \prod_1^d \|D_i\| \le K\|u\|^d\|x\|.$$

Therefore $\|u\|_{cb} \le K\|u\|^d$. ∎

Curiously, the preceding *sufficient* condition, which a priori seems rather strong, is actually also *necessary* for A to satisfy the SP. This is the main result of [P17] (see also [P18]).

Theorem 27.7. *Let A be a unital operator algebra. The following assertions are equivalent.*

(i) *There is a number $c > 1$ such that any unital homomorphism $u\colon A \to B(H)$ with $\|u\| \le c$ is completely bounded.*

(ii) *A satisfies the similarity property.*

(iii) *There are $\alpha \ge 0$ and a constant K such that any bounded unital homomorphism $u\colon A \to B(H)$ satisfies $\|u\|_{cb} \le K\|u\|^\alpha$.*

(iv) *There is an integer d such that $\ell(A) \le d$. Moreover, if we define*

$$d(A) = \inf\{\alpha \ge 0 \mid \exists K \quad \forall\, u \text{ unital homomorphism } \quad \|u\|_{cb} \le K\|u\|^\alpha\},$$

then we have

$$d(A) = \ell(A), \tag{27.12}$$

and the infimum defining $d(A)$ is attained when $\alpha = \ell(A)$.

Remark. Let A be a unital C^*-algebra. Assume that there is a constant $\alpha \ge 0$ such that, given an arbitrary $*$-homomorphism $\pi\colon A \to B(H)$, any π-derivation $\delta\colon A \to B(H)$ satisfies

$$\|\delta\|_{cb} \le \alpha\|\delta\|. \tag{27.13}$$

Then, for any bounded unital homomorphism $u\colon A \to B(H)$, we have $\|u\|_{cb} \le \|u\|^\alpha$. This is proved by a modification of Kirchberg's argument in [Ki9]; see [P17] for details. By the preceding statement, this implies that $d(A) \le [\alpha]$ (integral part of α), which leads us to conjecture that (in the C^*-case) the best possible α in (27.13) is "automatically" an integer.

Proof of Theorem 27.7. We have already shown that (iv) $\Rightarrow$ (iii) in the preceding statement. Moreover, (iii) $\Rightarrow$ (ii) $\Rightarrow$ (i) are obvious. Thus for the equivalence it suffices to show (i) $\Rightarrow$ (iv). Assume (i). By an easy direct sum argument, we can show that there is a constant β such that any u as in (i) with $\|u\| \le c$ must satisfy $\|u\|_{cb} \le \beta$. Then we select d so that

$$\beta \sum_{k>d} c^{-k} < 1/2. \tag{27.14}$$

We will show that $\ell(A) \le d$. Equivalently, it suffices to show that there is a constant K' (independent of n) such that for all x in $M_n(A)$ we have

$$\|x\|_{(d)} \le K'\|x\|_{M_n(A)}. \tag{27.15}$$

Consider x with $\|x\|_{M_n(A)} < 1$. We will show that x can be written as

$$x = x' + x'' \quad \text{with } \|x'\|_{(d)} \le (d+1)\beta \text{ and } \|x''\|_{M_n(A)} < 1/2.$$

By a well-known iteration argument, this implies (27.15) for some K'.

Let $\mathcal{S} = \{c^{-1}B_A\} \cup \{1_A\}$. We will apply Theorem 24.1 to this set. We have for any unital homomorphism $u\colon A \to B(H)$

$$\sup_{x \in \mathcal{S}} \|u(x)\| = 1 \Rightarrow \|u\|_{cb} \leq \beta.$$

Hence any x in $M_n(A)$ with $\|x\| < 1$ can be written in the form

$$x = \alpha_0 D_1 \alpha_1 \dots D_m \alpha_m,$$

where $\prod \|\alpha_i\| < \beta$ and where each matrix D_i is of the form

$$D_i = \begin{pmatrix} c^{-1}x_i & & & \\ & 1 & & \bigcirc \\ & & \ddots & \\ \bigcirc & & & 1 \end{pmatrix}$$

for suitable elements x_i in the unit ball of A. (Indeed, by rearranging the lines and columns of the α_i we can assume for simplicity that D_i has this form.)

Then we denote for z in $\mathbb{C}$

$$D_i(z) = \begin{pmatrix} zx_i & & & \\ & 1 & & \bigcirc \\ & & \ddots & \\ \bigcirc & & & 1 \end{pmatrix}.$$

Let

$$x(z) = \alpha_0 D_1(z)\alpha_1 \dots D_m(z)\alpha_m.$$

Developing the product we find

$$x(z) = \sum_0^m z^k \Delta_k.$$

Note that

$$\sup_{|z| \leq 1} \|x(z)\|_{M_n(A)} \leq \beta.$$

Hence, by Cauchy's formula (since $\Delta_k = \int e^{-ikt} x(e^{it}) dt / 2\pi$), we have

$$\sup_k \|\Delta_k\|_{M_n(A)} \leq \beta. \tag{27.16}$$

Then we decompose x as

$$x = x' + x''$$

with $x' = \sum_0^d c^{-k} \Delta_k$ and

$$x'' = \sum_{k>d} c^{-k} \Delta_k.$$

Note that our original choice of d in (27.14) guarantees with (27.16) that

$$\|x''\|_{M_n(A)} < 1/2.$$

Thus it remains only to majorize $\|x'\|_{(d)}$ as announced earlier. In order to do this we will "lift" $x(z)$ up into the algebra $OA_u(E)$ studied in Chapters 6 and 8, with $E = \max(A)$.

Let e be the unit in $OA_u(E)$. Let $\widetilde{E}$ be the unitization of E, that is, $\widetilde{E}$ is the span of e and E in $OA_u(E)$. For any i we denote

$$\widehat{D}_i(z) = \begin{pmatrix} z\widehat{x}_i & & & \\ & e & & \bigcirc \\ & & \ddots & \\ \bigcirc & & & e \end{pmatrix},$$

where $\widehat{x}_i$ denotes $x_i \in E$ viewed as an element of $OA_u(E)$. Thus $\widehat{D}_i(z) \in M_n(\widetilde{E})$. We let

$$\widehat{x}(z) = \alpha_0 \widehat{D}_1(z) \alpha_1 \dots \widehat{D}_m(z) \alpha_m.$$

Then $\widehat{x}(z) \in M_n(OA_u(E))$, and if $|z| \le 1$,

$$\|\widehat{x}(z)\|_{M_n(OA_n(E))} \le \prod \|\alpha_i\| < \beta.$$

Moreover, if we denote by $\widehat{p}\colon OA_u(E) \to A$ the product map, then $I_{M_n} \otimes \widehat{p}$ applied to $\widehat{x}(z)$ gives us back $x(z)$. Expanding $\widehat{x}(z)$ as a polynomial in z we find

$$\widehat{x}(z) = \sum_0^m z^k \widehat{\Delta}_k,$$

where $\widehat{\Delta}_k \in M_n(E \otimes \dots \otimes E)$ (k times). By Cauchy's formula again, we have

$$\|\widehat{\Delta}_k\|_{M_n(OA_u(E))} < \beta;$$

but now, since $E \otimes_h \dots \otimes_h E$ is completely isometrically embedded into $OA_u(E)$ (see Proposition 6.6), $\widehat{\Delta}_k$ can be identified with an element of $M_n(E \otimes \dots \otimes E)$ (k times) such that $\|\widehat{\Delta}_k\|_{M_n(E \otimes_h \dots \otimes_h E)} < \beta$.

Since $(I_{M_n} \otimes \widehat{p})(\widehat{x}(z)) = x(z)$, we have $(I_{M_n} \otimes p_k)(\widehat{\Delta}_k) = \Delta_k$. Hence the preceding remarks give us for any $k \le d$

$$\|\Delta_k\|_{(d)} \le \|\Delta_k\|_{(k)} < \beta.$$

Finally, we conclude as announced

$$\|x'\|_{(d)} \leq \sum_0^d \|\Delta_k\|_{(d)} \leq (d+1)\beta. \qquad \blacksquare$$

Corollary 27.8. *If $d(A) \leq d$, then any linear map $u\colon A \to B(H)$ such that there are bounded linear maps $\xi_i\colon A \to B(H_{i+1}, H_i)$ $(i = 1, \ldots d)$ with $H_{d+1} = H_1 = H$ satisfying*

$$\forall\, x_i \in A \qquad u(x_1 x_2 \ldots x_d) = \xi(x_2)\xi_2(x_2) \ldots \xi_d(x_d)$$

must be completely bounded and we have

$$\|u\|_{cb} \leq K \prod_{i=1}^d \|\xi_i\|,$$

where K is a constant independent of u.

Proof. By the preceding result we have $\ell(A) \leq d$. Then this is easy to prove, just like Proposition 27.6 above. $\qquad \blacksquare$

Remark. The analog of Theorem 27.7 for dual operator algebras (and normal homomorphisms) is proved in [LeM7].

Examples. Obviously any finite-dimensional A satisfies $d(A) = 1$. Assume A infinite-dimensional. Then, if A is a nuclear C^*-algebra, we have $d(A) = 2$ (due to Bunce and Christensen; see [Chr 1–2]). Moreover, if A is a C^*-algebra without tracial states, such as $B(H)$, then $d(A) \leq 3$ (see [H6]), and we proved in [P17] that $d(B(H)) = 3$ (see Exercise 5.10 for a direct proof that $\ell(B(H)) \leq 3$). Moreover, if $\mathcal{K}_1$ denotes the unitization of $\mathcal{K}$, then for *any* C^*-algebra B we have $d(\mathcal{K}_1 \otimes_{\min} B) \leq 3$. In [P19] it is observed that $d(A) \geq 3$ if A is a II_1-factor and that $d(A) = 3$ for the hyperfinite one. More recently, in [Chr6] it is proved that $d(A) = 3$ for any II_1-factor with property Γ, improving on the previous successive estimates $d(A) \leq 44$ ([Chr4]) and $d(A) \leq 5$ ([P19]). Finally, in [P18] we prove that, for any integer d, there is a unital (non-self-adjoint) operator algebra A_d with length equal to d (i.e., we have $d(A_d) = d$). However, it remains open whether there exist C^*-algebras, or uniform algebras, or Q-algebras with arbitrarily large *finite* length.

Exercise

Exercise 27.1. Let $I \subset A$ be a closed two-sided ideal in an operator algebra. Show that $\ell(A/I) \leq \ell(A)$. Then prove that

$$\ell(A) \leq \max\{\ell(I), \ell(A/I)\},$$

and that the equality holds (i.e., $\ell(I) \leq \ell(A)$) when A is a C^*-algebra (use the fact that I admits a quasi-central approximate unit) or when $A \simeq I \times (A/I)$.

Chapter 28. The Sz.-Nagy-Halmos Similarity Problem

This chapter concentrates on the following question: How can we recognize when an operator T in $B(H)$ is similar to a contraction?

Of course T similar to a contraction means that there is an isomorphism $\xi\colon H \to H$ such that $\|\xi^{-1}T\xi\| \le 1$. This problem originates from a question of Sz.-Nagy [SN], who observed that an obvious necessary condition is that T should be *power bounded*, that is, we should have

$$\sup_{n\ge 1} \|T^n\| < \infty.$$

He asked whether the converse holds (and proved it for compact T), but Foguel [Fo] quickly found a counterexample. Soon after that, Paul Halmos (and perhaps others, too) noticed that there was a stronger necessary condition, namely, T should be *polynomially bounded*, which means that there is a constant C such that for all polynomials P we have

$$\|P(T)\| \le C\|P\|_\infty, \tag{28.1}$$

where

$$\|P\|_\infty = \sup_{z\in D} |P(z)|.$$

Here, of course, if $P(z) = \sum_0^n a_k z^k$, we set $P(T) = a_0 I + \sum_1^n a_k T^k$. In that case, we will say that T is *polynomially C-bounded*. This is indeed necessary: By von Neumann's classical inequality (see (8.11)) any contraction T must satisfy (28.1) with $C = 1$. Hence, if $T = \xi^{-1}\theta\xi$ with $\|\theta\| \le 1$, we must have (28.1) with $C \le \|\xi^{-1}\|\,\|\xi\|$.

Foguel's example was shown to fail (28.1) ([Leb]). Halmos then asked in 1970 (see [Hal]) whether this condition is sufficient, that is, whether, conversely, every polynomially bounded operator is similar to a contraction. This question was recently answered negatively in [P20]. The original proof of the polynomial boundedness in [P20], which was rather complicated, was simplified in [Kis1] and in [DP]. In [DP], the polynomial boundedness is deduced very simply from the "vectorial Nehari theorem," which we discussed in §9.1. We will follow the same approach. See [P10] for more background on these questions.

We start with a very useful criterion for similarity to a contraction due to Paulsen [Pa4]. This is nothing but Theorem 25.2 applied when A is the disc algebra.

Note. Throughout this chapter, unless specified otherwise, we always equip the disc algebra $A(D)$ with its minimal operator space structure, associated, for instance, to the embedding $A(D) \subset C(\partial D)$.

Theorem 28.1. *([Pa4]) Consider an operator T in $B(H)$ and fix a constant $C \geq 0$. The following are equivalent:*

(i) *There is an isomorphism $\xi\colon H \to H$ with $\|\xi\| \, \|\xi^{-1}\| \leq C$ such that $\|\xi^{-1}T\xi\| \leq 1$.*

(ii) *The homomorphism $P \to P(T)$ defined on polynomials extends to a c.b. homomorphism $u_T\colon A(D) \to B(H)$ with $\|u_T\|_{cb} \leq C$.*

(iii) *For any n and any $n \times n$ matrix (P_{ij}) with polynomial entries, we have*

$$\|(P_{ij}(T))\|_{M_n(B(H))} \leq C \sup_{|z| \leq 1} \|(P_{ij}(z))\|_{M_n}.$$

(iv) *There is a Hilbert space $\widehat{H}$ with $\widehat{H} \supset H$, a unitary operator U on $\widehat{H}$, and an isomorphism $\xi\colon H \to H$ with $\|\xi\| \, \|\xi^{-1}\| \leq C$ such that*

$$\forall\, n \geq 1 \qquad T^n = \xi^{-1}P_H U^n{}_{|H}\xi.$$

Proof. The equivalence between (ii) and (iii) and (iv) $\Rightarrow$ (i) are obvious. (ii) $\Rightarrow$ (iv) follows from Theorem 27.3 and (i) $\Rightarrow$ (ii) from Theorem 8.7. ∎

Note. When $C = 1$ in the preceding statement, (i) just means that T is a contraction. In that case, the equivalence between (i) and (iv) is known as Sz.-Nagy's dilation theorem, saying that any contraction admits a unitary dilation.

The next statement gives a counterexample to Halmos's question.

Theorem 28.2. *For any $\varepsilon > 0$, there is a polynomially $(1+\varepsilon)$-bounded operator on ℓ_2 that is not similar to a contraction. More precisely, for any n there is an $n \times n$ matrix $T_{\varepsilon,n} \in M_n$ that is polynomially $(1+\varepsilon)$-bounded but such that*

$$\delta\varepsilon(\mathrm{Log}(n+1))^{1/2} \leq \inf\{\|\xi\| \, \|\xi^{-1}\| \mid \|\xi^{-1}T_{\varepsilon,n}\xi\| \leq 1\}, \qquad (28.2)$$

where $\delta > 0$ is a number independent of (ε, n).

Notation. For any T in $B(H)$, assumed similar to a contraction, we let

$$\mathrm{Sim}(T) = \inf\{\|\xi^{-1}\| \, \|\xi\|\},$$

where the infimum runs over all the isomorphisms $\xi\colon H \to H$ such that $\|\xi^{-1}T\xi\| \leq 1$.

Remarks. (i) If $T_{\varepsilon,n}$ is as above, then $T = \bigoplus_n T_{\varepsilon,n}$ is polynomially $(1+\varepsilon)$-bounded but is not similar to a contraction. Indeed, it is easy to check directly that the restriction of an arbitrary operator T in $B(H)$ to an invariant subspace $E \subset H$ satisfies

$$\mathrm{Sim}(T_{|E}) \leq \mathrm{Sim}(T).$$

(This can also be seen from Theorem 28.1.)

(ii) The lower estimate (28.2) should be compared with Bourgain's upper bound [Bou4], slighly improved in Example 25.23: For any $\varepsilon > 0$, $n \geq 1$ and any polynomially $(1 + \varepsilon)$-bounded T in M_n we have

$$\text{Sim}(T) \leq K(1 + \varepsilon)^2 \text{Log}(n + 1),$$

where K is a numerical constant. It would be nice to close the gap between these two estimates. In particular, in (28.2), can $(\text{Log}(n + 1))^{1/2}$ be replaced by $\text{Log}(n + 1)$?

The obstruction to "similar to a contraction" will come from the following.

Lemma 28.3. *Let (B_n) be a sequence in $B(H)$ such that, for some $\beta > 0$, we have*

$$\forall \, (\alpha_n) \in \ell_2 \quad \left\| \sum \alpha_n B_n \right\| \leq \beta \left(\sum |\alpha_n|^2 \right)^{1/2}. \tag{28.3}$$

Let T, V, W in $B(H)$ be such that

$$\forall \, n \geq 0 \qquad B_n = VT^{2^n}W.$$

If

$$\sup_N N^{-1/2} \left\| \sum_1^N B_n \otimes \overline{B}_n \right\| = \infty$$

(in particular, this holds if B_n is a spin system or a sequence satisfying the CAR), then T is not similar to a contraction. More precisely, we always have

$$\sup_N N^{-1/2} \left\| \sum_1^N B_n \otimes \overline{B}_n \right\| \leq \beta \, \text{Sim}(T) \|V\| \|W\|. \tag{28.4}$$

Proof. We use the implication (i) $\Rightarrow$ (iv) in Paulsen's criterion. (This is actually Sz.-Nagy's dilation theorem.) Let (B_n) be as in Lemma 28.3. Assume (iv) in Theorem 28.1. Then we have

$$B_n = V\xi^{-1}P_H U^{2^n}{}_{|H}\xi W;$$

hence there are operators V_1, W_1 such that

$$\forall \, n \geq 1 \qquad B_n = V_1 U^{2^n} W_1.$$

Then we can write

$$\left\| \sum_1^N B_n \otimes \overline{B}_n \right\| \leq \|V_1\| \, \|W_1\| \left\| \sum_1^N U^{2^n} \otimes \overline{B}_n \right\|.$$

But, by the classical spectral theorem, the C^*-algebra generated by the unitary (hence normal) operator U is commutative and can be identified with

$C(\sigma)$, where σ is the spectrum of U and of course $\sigma \subset \{z \in \mathbb{C} \mid |z| = 1\}$. Hence, by (28.3),

$$\left\| \sum_1^N U^{2^n} \otimes \overline{B}_n \right\| = \sup_{z \in \sigma} \left\| \sum_1^N z^{2^n} \overline{B}_n \right\| \leq \beta \sqrt{N},$$

which, by our assumption on (B_n), is impossible when $N \to \infty$. A more precise recapitulation yields (28.4). ∎

Remark. Let $E \subset A(D) \subset C(\partial D)$ be the closed linear subspace spanned by $\{z^{2^n} \mid n \geq 1\}$. The operator space interpretation of the preceding lemma is simply that the linear mapping $v \colon E \to B(H)$ defined by $v(z^{2^n}) = B_n$ is *not* completely bounded. Indeed, the preceding argument shows that

$$\left\| \sum_1^N B_n \otimes \overline{B}_n \right\| \leq \left\| \sum_1^N v(z^{2^n}) \otimes \overline{B}_n \right\| \quad \text{while} \quad \left\| \sum_1^N z^{2^n} \otimes \overline{B}_n \right\| \leq \beta \sqrt{N}.$$

The crucial step to prove Theorem 28.2 is the following.

Lemma 28.4. *Let $H = \ell_2$. Let (B_n) be any sequence in $B(H)$ satisfying (28.3). Then, for any $0 < \varepsilon \leq 1$, there is a polynomially $(1 + \varepsilon)$-bounded operator T in $B(\ell_2)$ and operators V, W with $\|V\| \, \|W\| \leq 4\beta \varepsilon^{-1}$ such that*

$$\forall \, n \geq 0 \qquad B_n = VT^{2^n}W$$

and actually also $\forall \, k \notin \{2^n \mid n \geq 0\} \quad 0 = VT^k W$.

To prove Lemma 28.4, we need to construct a special class of polynomially bounded operators. We first discuss operators of the form:

$$T = \begin{pmatrix} S^* & \Gamma \\ 0 & S \end{pmatrix}, \tag{28.5}$$

where $S \colon \ell_2 \to \ell_2$ is the shift and where $\Gamma \colon \ell_2 \to \ell_2$ is a Hankel operator, which means that $\Gamma S = S^* \Gamma$.

Operators of the form (28.5) were first considered by Peller [Pe1], who explicitly proposed them as possible counterexamples for the Halmos similarity problem. Independently, these examples also appeared, together with analogous considerations, in unpublished work by C. Foias and the late J. P. Williams (see also [CCFW] and [CC1–2]). But the hope to find a counterexample among these operators was much reduced by Bourgain [Bou4] in 1985, and finally, in the summer of 1995, Alexandrov and Peller [APe] finished it off by showing that such a T is polynomially bounded iff it is similar to a contraction, and, moreover, this happens iff the "symbol" of Γ, the function

$\varphi(t) = \sum_{n \geq 0} \Gamma_{0n} e^{int}$, has its derivative φ' in BMO. However, all this applies only to Hankel matrices with scalar entries, that is, with $\Gamma_{ij} \in \mathbb{C}$, and as we will soon see, the situation changes dramatically when we turn to matrices $\Gamma = (\Gamma_{ij})$ with *operator* coefficients, that is, with $\Gamma_{ij} \in B(H)$. To describe this, we need to modify our notation.

We denote $\ell_2(H) = \ell_2 \otimes_2 H \simeq \bigoplus_{n \in \mathbb{N}} H_n$ with $H_n = H \ \forall \ n \in \mathbb{N}$. Let

$$\widehat{S} \colon \ell_2(H) \to \ell_2(H)$$

be the "multivariate" shift operator defined by $\widehat{S} = S \otimes I_H$, or, equivalently,

$$\widehat{S} \colon (h_0, h_1, \ldots) \to (0, h_0, h_1, \ldots).$$

Let $\Gamma \colon \ell_2(H) \to \ell_2(H)$ be a Hankelian operator with associated matrix (Γ_{ij}), where $\Gamma_{ij} \in B(H)$. The Hankelian character of Γ means that

$$\widehat{S}^* \Gamma = \Gamma \widehat{S},$$

or, equivalently, that Γ_{ij} depends only on $i + j$, that is, there is a sequence $\{a(n) \mid n \in \mathbb{N}\}$ in $B(H)$ such that

$$\Gamma_{ij} = a(i + j).$$

Our examples will be of the form

$$T_\Gamma = \begin{pmatrix} \widehat{S}^* & \Gamma \\ 0 & \widehat{S} \end{pmatrix}, \tag{28.6}$$

where $\Gamma \colon \ell_2(H) \to \ell_2(H)$ is a Hankel matrix with entries in $B(H)$, so that T_Γ acts on $\ell_2(H) \oplus \ell_2(H)$.

A simple calculation shows that, for any $n \geq 1$,

$$T_\Gamma^n = \begin{pmatrix} \widehat{S}^{*n} & n\Gamma\widehat{S}^{n-1} \\ 0 & \widehat{S}^n \end{pmatrix};$$

hence for any polynomial P we have

$$P(T_\Gamma) = \begin{pmatrix} P(\widehat{S}^*) & \Gamma P'(\widehat{S}) \\ 0 & P(\widehat{S}) \end{pmatrix}. \tag{28.7}$$

Thus (as Peller observed [Pe2]) T_Γ is polynomially bounded iff there is a constant C such that for any polynomial P

$$\|\Gamma P'(\widehat{S})\| \leq C\|P\|_\infty.$$

Remark. In passing, note that the homomorphism $u \colon P \to P(T_\Gamma)$ is associated to a derivation just like in (27.7): Indeed, if we set $\delta(P) = \Gamma P'(\widehat{S})$, then $\delta(PQ) = P(\widehat{S}^*)\delta(Q) + \delta(P)Q(\widehat{S})$ for all polynomial P, Q.

Let $\gamma_n \colon \ell_2 \to \ell_2$ be the basic Hankel operator defined (for any $n \geq 0$) by

$$\gamma_n = \sum_{i+j=n} e_{ij}.$$

We will use the following elementary fact.

Sublemma 28.5. *Let D be the unbounded diagonal (derivation-like) operator defined on the canonical basis of ℓ_2 by*

$$De_j = je_j.$$

Let

$$a_n = 2^{-n}\gamma_{2^n}D = \sum_{i+j=2^n} e_{ij}(j/2^n).$$

(Note in passing that $\gamma_{2^n} = a_n + a_n^*$.) We then have:

(i) *For any polynomial P and any $n \geq 0$*

$$2^{-n}\gamma_{2^n}SP'(S) = a_nP(S) - P(S^*)a_n.$$

(ii)

$$\left\| \sum_{n\geq 0} a_n^*a_n \right\| \leq 4/3.$$

Proof. (i) By the Hankelian property of γ_{2^n}, we have $\gamma_{2^n}P(S) = P(S^*)\gamma_{2^n}$; hence

$$a_nP(S) - P(S^*)a_n = 2^{-n}\gamma_{2^n}(DP(S) - P(S)D).$$

Thus it suffices to check that

$$DP(S) - P(S)D = SP'(S),$$

which is entirely elementary. (Indeed, we may assume $P(S) = S^m$; then $SP'(S) = mS^m$ and $(DS^m - S^mD)(e_j) = me_{m+j} = SP'(S)e_j$ for all $j \geq 0$.)

(ii) It is easy to check that

$$(a_n^*a_n)_{ij} = 0 \quad \text{if} \quad i \neq j$$

and

$$(a_n^*a_n)_{ii} = 2^{-2n}i^2 1_{\{i\leq 2^n\}},$$

so that $\sum a_n^*a_n$ is a diagonal operator on ℓ_2 with norm

$$\left\| \sum a_n^*a_n \right\| \leq \sup_i \sum_{2^n\geq i} 2^{-2n}i^2 \leq \sum_{n\geq 0} 2^{-2n} = 4/3. \qquad \blacksquare$$

Theorem 28.6. *Fix $\beta \geq 0$. Let B_n in $B(H)$ be such that*

$$\forall\, (\alpha_n) \in \ell_2 \quad \left\| \sum \alpha_nB_n \right\| \leq \beta \left(\sum |\alpha_n|^2 \right)^{1/2}.$$

Let

$$\Gamma = \sum_{n \geq 0} 2^{-n} \gamma_{2^n} S \otimes B_n \in B(\ell_2(H)). \tag{28.8}$$

Then, for any polynomial P, we have

$$\|\Gamma P'(\widehat{S})\| = \| \sum 2^{-n} \gamma_{2^n} S P'(S) \otimes B_n\| \leq 4\beta\|P\|_\infty. \tag{28.9}$$

Consequently, the operator T_Γ given by (28.6) is polynomially bounded. However, if

$$\sup_N N^{-1/2} \left\| \sum_1^N B_n \otimes \overline{B}_n \right\| = \infty$$

(in particular, if B_n is a spin system or a sequence satisfying the CAR), then T_Γ is not similar to a contraction.

Proof. The idea of the proof is to start with the case $B_n = e_{n1}$, which will turn out to be very easy using Sublemma 28.5. Then we observe that we have a bounded linear mapping $v\colon B(\ell_2) \to B(H)$ with $\|v\| \leq \beta$ such that

$$v(e_{n1}) = B_n.$$

(Indeed, we simply take $v = wP$, where $P\colon B(\ell_2) \to B(\ell_2)$ is the natural contractive projection onto $\overline{\mathrm{span}}(e_{n1})$ and where $w(e_{n1}) = B_n$.) Note that we have

$$\left[\sum 2^{-n} \gamma_{2^n} S \otimes B_n \right] P'(\widehat{S}) = \sum 2^{-n} \gamma_{2^n} S P'(S) \otimes B_n, \tag{28.10}$$

and this is a Hankel operator on $\ell_2(H)$. Hence, if we let

$$B = \sum 2^{-n} \gamma_{2^n} S P'(S) \otimes B_n$$

and

$$A = \sum 2^{-n} \gamma_{2^n} S P'(S) \otimes e_{n1},$$

then we have

$$B = (Id \otimes v)(A);$$

hence by Corollary 9.1.6

$$\|B\| \leq \|v\| \, \|A\| \leq \beta\|A\|. \tag{28.11}$$

But now $\|A\|$ is very easy to estimate! Indeed, by Sublemma 28.5 we have

$$\|A\| = \left\| \sum [a_n P(S) - P(S^*)a_n] \otimes e_{n1} \right\|$$

$$\leq \left\| \sum a_n \otimes e_{n1} \right\| \|P(S)\| + \|P(S^*)\| \left\| \sum a_n \otimes e_{n1} \right\|$$

$$\leq 2\|P\|_\infty \left\| \sum a_n^* a_n \right\|^{1/2}$$

$$\leq 4/\sqrt{3}\|P\|_\infty \leq 4\|P\|_\infty.$$

Thus, recalling (28.10) and (28.11), we obtain (28.9). This yields the polynomial boundedness of T_Γ as announced.

Finally note that the 2×2 matrix of $P(T_\Gamma)$ admits $\Gamma P'(\widehat{S})$ as its $(1\,2)$-entry and, in turn, if $P(z) = \sum p_k z^k$, the $(0\,0)$-entry of $\Gamma P'(\widehat{S})$ is given by:

$$[\Gamma P'(\widehat{S})]_{0\,0} = \left[\sum 2^{-n} \gamma_{2^n} S P'(S) \otimes B_n \right]_{0\,0} = \sum p_{2^n} B_n.$$

Thus we obviously have operators V, W with $\|V\|, \|W\| \leq 1$ such that for all P

$$\sum p_{2^n} B_n = V P(T_\Gamma) W. \tag{28.12}$$

In particular, for all $n \geq 0$

$$B_n = V T_\Gamma^{2^n} W;$$

hence we deduce the second part of Theorem 28.6 from Lemma 28.3. ∎

Remark. Let Γ be a Hankel operator with entries Γ_{ij} in $B(H)$. Assume that there is a *finite-dimensional* subspace $K \subset H$ such that either all the operators Γ_{ij} have range in K, or all their adjoints have range in K (so that we can think of $B(H)$ as replaced by either $B(H, K)$ or $B(K, H)$). Then (see [DP]) if an operator T_Γ of the form (28.6) is polynomially bounded, it is similar to a contraction (i.e., there is no counterexample in this class). This extends the case $\dim(H) = 1$ proved in [APe].

Proof of Lemma 28.4. We use Γ given by (28.7) and replace T_Γ by $T_{\alpha\Gamma}$ with $\alpha = \varepsilon/(4\beta)$. Then by (28.9) and (28.7) we have

$$\|P(T_{\alpha\Gamma})\| \leq (1 + \varepsilon)\|P\|_\infty,$$

and using (28.12) (with $T_{\alpha\Gamma}$ instead of T_Γ hence with αB_n instead of B_n) we obtain Lemma 28.4 with $\|V\| \, \|W\| \leq 1/\alpha$. ∎

Proof of Theorem 28.2. Let $(B_1, ..., B_N)$ be a system satisfying the CAR (see Theorem 9.3.1) on a Hilbert space H_N. Recall that we may assume $\dim(H_N) = 2^N$. Moreover, by Exercise 3.7, we know that

$$N/2 \leq \left\| \sum_{k=1}^N B_k \otimes \overline{B}_k \right\|.$$

Let $\Gamma = \sum_{k=0}^N 2^{-k} \gamma_{2^k} S \otimes B_k$, as above, acting on $\ell_2(H_N)$, and let T_Γ be the associated operator. Let $K_N \subset \ell_2(H_N)$ be the subspace spanned by $\{e_{ij} \otimes H_N \mid 0 \leq i, j \leq 2^N\}$. Note that K_N is a reducing subspace for Γ. Moreover, since K_N (resp. $K_N^\perp$) is invariant for $\widehat{S}^*$ (resp. $\widehat{S}$), the compressions

$P \to P_{K_N} P(\widehat{S}^*)_{|K_N}$ and $P \to P_{K_N} P(\widehat{S})_{|K_N}$ are homomorphisms. Therefore, if we let $s = P_{K_N} \widehat{S}_{|K_N}$ and $\gamma = P_{K_N} \Gamma_{|K_N}$, and if we define (for $\alpha > 0$)

$$T = \begin{pmatrix} s^* & \alpha\gamma \\ 0 & s \end{pmatrix} \in B(K_N \oplus K_N),$$

then T is polynomially $(1 + 4\alpha\beta)$-bounded, and by (28.4) we have

$$(\alpha/2)N^{1/2} \le \beta \mathrm{Sim}(T).$$

On the other hand, T acts on a Hilbert space of dimension $2\dim(K_N) = 2(2^N + 1)\dim(H_N) = 2(2^N + 1)2^N$. Hence, taking $\alpha = \varepsilon(4\beta)^{-1}$ and (say) $n = 2(2^N + 1)2^N$, we find a polynomially $(1 + \varepsilon)$-bounded $n \times n$ matrix T with

$$\varepsilon/(4\beta^2)N^{1/2} \le \mathrm{Sim}(T),$$

and since $N \simeq \mathrm{Log}(n)$, this yields (28.2). ∎

Let $T \in B(H)$ be polynomially bounded, and let $u \colon A(D) \to B(H)$ be the homomorphism taking P to $P(T)$. Let $u_n \colon M_n(A(D)) \to M_n(B(H))$ be the homomorphism $I_{M_n} \otimes u$. By Theorem 28.1, we know that T is similar to a contraction iff $\sup_{n \ge 1} \|u_n\| < \infty$, and we are thus lead to try to compare $\|u\|$ and $\|u_n\|$; but it turns out that only a very crude estimate can hold, as shown by the next statement.

Theorem 28.7. *There is a numerical constant K such that, for any unital homomorphism $u \colon A(D) \to B(H)$, we have*

$$\|u_n\| \le K\sqrt{n}\|u\|. \tag{28.13}$$

Conversely, there is a $\delta > 0$ such that for any $0 < \varepsilon \le 1$ and any $n \ge 1$ there is a unital homomorphism $u \colon A(D) \to B(H)$ with $\|u\| \le (1 + \varepsilon)$ such that

$$\delta\varepsilon\sqrt{n}\|u\| \le \|u_n\|. \tag{28.14}$$

The upper estimate (28.13) is a rather simple consequence of the following result due to Bourgain [Bou3] (see [P21] for a simpler proof).

Theorem 28.8. *There is a constant K such that, for any bounded linear map*

$$v \colon A(D) \to H \quad (H \text{ Hilbert}),$$

we have

$$\forall\, f_i \in A(D) \ (i \le n) \qquad \left(\sum \|v(f_i)\|^2\right)^{1/2} \le K\|v\| \sup_{|z|=1} \left(\sum |f_i(z)|^2\right)^{1/2}.$$

$$\tag{28.15}$$

Furthermore, for any linear map $u: A(D) \to B(H)$ we have

$$\left\| \sum u(f_i)^* u(f_i) \right\|^{1/2} \le K\|u\| \sup_{|z|=1} \left(\sum |f_i(z)|^2 \right)^{1/2}. \qquad (28.16)$$

Note that (28.16) follows immediately from (28.15) applied to $v(f) = u(f)\xi$ with ξ fixed in the unit ball of H.

Proof of Theorem 28.7. Let (P_{ij}) be an $n \times n$ matrix in the unit ball of $M_n(A(D))$. Let

$$Y_j = \sum_{i=1}^{n} e_{ij} \otimes u(P_{ij}).$$

By (28.16) we have for each j

$$\|Y_j\| \le K\|u\| \sup_{|z|=1} \left(\sum_{i=1}^{n} |P_{ij}(z)|^2 \right)^{1/2} \le K\|u\|.$$

On the other hand,

$$\left\| \sum_{i,j} e_{ij} \otimes u(P_{ij}) \right\| = \left\| \sum_{j=1}^{n} Y_j \right\| = \left\| \sum Y_j Y_j^* \right\|^{1/2}$$

$$\le \left(\sum \|Y_j\|^2 \right)^{1/2} \le \sqrt{n} \sup_{j} \|Y_j\| \le K\sqrt{n}\|u\|.$$

Hence we obtain $\|u_n\| \le K\sqrt{n}\|u\|$, which proves (28.13). We now turn to (28.14).

We will use the following fact, which is easy to prove using random matrices and a concentration of measure argument (see Exercise 28.1). For some numerical constant β, we can find for each n an n-tuple $(B_1, ..., B_n)$ in M_n satisfying (28.3) but such that

$$n \le \left\| \sum_{k=1}^{n} B_k \otimes \overline{B}_k \right\|. \qquad (28.17)$$

Again we set $\alpha = \varepsilon(4\beta)^{-1}$. Now, applying again the preceding construction with

$$\Gamma = \sum_{k=1}^{n} \gamma_{2^k} S \otimes [\alpha B_k],$$

we then find for the associated map $u: P \to P(T_\Gamma)$

$$\|u_n\| = \|u \otimes I_{M_n}\| \ge \left\| \sum_{k=1}^{n} u(z^{2^k}) \otimes \overline{B}_k \right\| \times \left\| \sum_{k=1}^{n} z^{2^k} \otimes \overline{B}_k \right\|^{-1};$$

hence, by (28.12) (applied with B_k replaced by αB_k) and by (28.17),

$$\geq \left\| \sum_{k=1}^n \alpha B_k \otimes \overline{B}_k \right\| \times \left\| \sum_{k=1}^n z^{2^k} \otimes \overline{B}_k \right\|^{-1} \geq \alpha\beta^{-1}\sqrt{n} \geq \varepsilon(4\beta^2)^{-1}\sqrt{n},$$

and T_Γ is polynomially $(1+\varepsilon)$-bounded, so $\|u\| \leq 1 + \varepsilon$. ∎

Exercise

Exercise 28.1. Show that there is $\beta > 0$ such that, for each n, there is an n-tuple $(B_1, \ldots, B_n)$ in M_n such that

$$\forall\, (\alpha_k) \in \mathbb{C}^n \qquad \left\| \sum_1^n \alpha_k B_k \right\|_{M_n} \leq \beta \left(\sum |\alpha_k|^2 \right)^{1/2} \tag{28.18}$$

and

$$n \leq \left\| \sum_1^n B_k \otimes \overline{B}_k \right\|. \tag{28.19}$$

(Actually, one can even show this with an n-tuple of unitary matrices in M_n.)

SOLUTIONS TO THE EXERCISES

Exercise 1.1. Consider $u\colon R \to C$. Fix $n \geq 1$, and let $a \in M_n(R) \simeq M_n \otimes R$ be the column matrix with entries in R defined by $a = \sum_1^n e_{j1} \otimes e_{1j}$. We have $(I_{M_n} \otimes u)(a) = \sum_1^n e_{j1} \otimes u(e_{1j})$. Hence, by Remark 1.13, we have

$$\left\| \sum_1^n u(e_{1j})^* u(e_{1j}) \right\|^{1/2} \leq \|u\|_{cb} \|a\| = \|u\|_{cb}.$$

But now let (u_{ij}) be the matrix associated to u, so that

$$u(e_{1j}) = \sum_i u_{ij} e_{i1}.$$

We then have $u(e_{1j})^* u(e_{1j}) = \sum_i |u_{ij}|^2 e_{11} = \|u(e_{1j})\|^2 e_{11}$; hence we find

$$\left(\sum_{ij \leq n} |u_{ij}|^2 \right)^{1/2} = \left(\sum_1^n \|u(e_{1j})\|^2 \right)^{1/2} = \left\| \sum_1^n u(e_{1j})^* u(e_{1j}) \right\|^{1/2} \leq \|u\|_{cb},$$

whence $\|u\|_{HS} \leq \|u\|_{cb}$.

To prove the converse, consider a finite sequence x_j in M_n. We can write:

$$\left\| \sum_j x_j \otimes u(e_{1j}) \right\|_{M_n(C)} = \left\| \sum_i \sum_j u_{ij} x_j \otimes e_{i1} \right\|_{M_n(C)}$$

$$= \left\| \sum_i \left(\sum_j u_{ij} x_j \right)^* \left(\sum_j u_{ij} x_j \right) \right\|_{M_n}^{1/2}$$

$$\leq \left(\sum_i \left\| \sum_j u_{ij} x_j \right\|_{M_n}^2 \right)^{1/2}$$

$$\leq \left(\sum_i \sum_j |u_{ij}|^2 \left\| \sum_j x_j x_j^* \right\|_{M_n} \right)^{1/2}$$

$$\leq \|u\|_{HS} \left\| \sum x_j \otimes e_{1j} \right\|_{M_n(R)},$$

and we conclude $\|u\|_{cb} \leq \|u\|_{HS}$.

Exercise 1.2. We first recall that $\|\tau_{n|R_n}\colon R_n \to C_n\|_{cb} = \|\tau_{n|C_n}\colon C_n \to R_n\|_{cb} = \sqrt{n}$. Let H be any Hilbert space. Consider now $x = (x_{ij})$ in $M_n(B(H))$. We claim that the transposed matrix $^t x$ satisfies $\|^t x\| \leq n \|x\|$.

From this claim (taking $B(H) = M_m$, $m \geq 1$ and using $M_m(M_n) \simeq M_n(M_m)$) it follows that $\|\tau_n\|_{cb} \leq n$.

To check this claim, we identify $M_n(B(H))$ with $M_n \otimes B(H)$, and we write

$$\|{}^t x\| = \left\| \sum e_{ij} \otimes x_{ji} \right\| \leq \left(\sum_{i=1}^{n} \left\| \sum_{j=1}^{n} e_{ij} \otimes x_{ji} \right\|^2 \right)^{1/2}$$

$$= \left(\sum_{i=1}^{n} \left\| \sum_{j=1}^{n} x_{ji} x_{ji}^* \right\| \right)^{1/2} \leq \sqrt{n} \sup_i \left\| \sum_j x_{ji} x_{ji}^* \right\|^{1/2}.$$

Hence, since $\|\tau_{n|C_n} \colon C_n \to R_n\|_{cb} \leq \sqrt{n}$, we find

$$\|{}^t x\| \leq \sqrt{n} \cdot \sqrt{n} \sup_i \left\| \sum_j x_{ji}^* x_{ji} \right\|^{1/2} \leq n\|x\|,$$

which proves our claim, and hence $\|\tau_n\|_{cb} \leq n$. To verify the equality, consider the element $T = \sum_{ij} e_{ij} \otimes e_{ji} \in M_n \otimes M_n \simeq M_n(M_n)$. We have $(I_{M_n} \otimes \tau_n)(T) = \sum_{ij} e_{ij} \otimes e_{ij}$. Viewing T as an element of $B(\ell_2^n \otimes_2 \ell_2^n)$, it is easy to verify that $T(e_j \otimes e_i) = e_i \otimes e_j$, hence $\|T\|_{M_n(M_n)} = \|T\|_{B(\ell_2^n \otimes_2 \ell_2^n)} = 1$. On the other hand, a similar analysis shows that $n^{-1} \sum e_{ij} \otimes e_{ij}$ is the orthogonal projection onto $n^{-1/2} \sum e_j \otimes e_j$ on the space $\ell_2^n \otimes_2 \ell_2^n$; hence

$$\left\| \sum_{ij=1}^{n} e_{ij} \otimes e_{ij} \right\|_{M_n(M_n)} = n.$$

Thus we conclude that $\|\tau_n\|_{cb} \geq n$.

Exercise 1.3. This follows immediately from the first part of Remark 1.13.

Exercise 1.4. Assume $F \subset B(K)$. Let $(K_\alpha)_{\alpha \in I}$ be a directed net of finite-dimensional subspaces of K such that $\cup K_\alpha$ is dense in K. Then, for any y in F, we have obviously

$$\|y\| = \sup_{\alpha \in I} \|P_{K_\alpha} y_{|K_\alpha}\|.$$

Let $d(\alpha) = \dim(K_\alpha)$. Using an orthonormal basis in K_α, we may identify $B(K_\alpha)$ with $M_{d(\alpha)}$. Let $v_\alpha \colon F \to M_{d(\alpha)}$ be the map defined by $v_\alpha(y) = P_{K_\alpha} y_{|K_\alpha}$. Clearly $\|v_\alpha\|_{cb} \leq 1$. Moreover, we have

$$\|y\| = \sup_{\alpha \in I} \|v_\alpha(y)\|.$$

More generally (since $\bigcup_\alpha \ell_2^n(K_\alpha)$ is dense in $\ell_2^n(K)$), for any (y_{ij}) in $M_n(F)$ we have

$$\|(y_{ij})\|_{M_n(F)} = \sup_{\alpha \in I} \|(v_\alpha(y_{ij}))\|_{M_n(M_{d(\alpha)})}.$$

Hence, for any (x_{ij}) in the unit ball of $M_n(E)$, we can write

$$\|(u(x_{ij}))\|_{M_n(F)} = \sup_{\alpha \in I} \|(v_\alpha u(x_{ij}))\|_{M_n(M_{d(\alpha)})} \leq \sup_{\alpha \in I} \|v_\alpha u\|_{cb},$$

which implies $\|u\|_{cb} \leq \sup_{\alpha \in I} \|v_\alpha u\|_{cb}$.

The announced equality is now obvious (recalling $\|vu\|_{cb} \leq \|v\|_{cb}\|u\|_{cb}$).

Exercise 1.5. (i) Let $\widehat{H} = \ell_2^n \otimes_2 K = \ell_2^n(K)$ and let $V\colon \ell_2^n \to \widehat{H}$ (resp. $W\colon \ell_2^n \to \widehat{H}$) be the linear map defined by $Ve_i = e_i \otimes x_i$ (resp. $We_j = e_j \otimes y_j$). Let $\pi\colon M_n \to M_n \otimes B(K)$ be the representation defined by $\pi(a) = a \otimes I$. Then (if, say, our scalar product is linear in the second variable) we have

$$V^*\pi(a)W = \sum a_{ij}V^*\pi(e_{ij})W = \sum a_{ij}e_{ij}\langle x_i, y_j\rangle.$$

Hence $u(\cdot) = V^*\pi(\cdot)W$, which implies

$$\|u\|_{cb} \leq \|V\|\|W\| \leq 1.$$

Moreover, if $x = y$, then $V = W$ and u is completely positive.

(ii) The generalization is obvious. Replacing $\ell_2(S)$ and $\ell_2(T)$ by their direct sum, we are reduced to the case $S = T$. We then define $\widehat{H} = \ell_2(T) \otimes_2 K$ and $V, W\colon \ell_2(T) \to \widehat{H}$ by $Ve_s = e_s \otimes x_s$, $We_t = e_t \otimes y_t$, and $\pi\colon B(\ell_2(T)) \to B(\widehat{H})$ by $\pi(a) = a \otimes I$. Then the same identity holds and yields $\|u\|_{cb} \leq 1$.

Exercise 2.1.1. Assume $E \subset B(H)$, $F \subset B(K)$. We can write

$$\|x\| = \sup\{|\langle xs, t\rangle|\},$$

where s, t run over the unit ball of $H \otimes_2 K$. By density, we may clearly restrict the sup to s, t in $H \otimes K$. Then, for some finite-dimensional subspace $H_n \subset H$, we have $s, t \in H_n \otimes K$. Hence, with v associated to H_n as above,

$$\langle xs, t\rangle = \left\langle \left(\sum a_i \otimes b_i\right) s, t \right\rangle = \left\langle \left(\sum v(a_i) \otimes b_i\right) s, t \right\rangle,$$

this yields

$$\|x\| \leq \sup_{n, v \in \mathcal{C}_n} \left\|\sum v(a_i) \otimes b_i\right\|,$$

and since the converse is obvious, we obtain (2.1.2).

When considering a directed net of subspaces, we note that, if $H_\alpha \subset H_\beta$, we have $\|v_\alpha(e)\| \leq \|v_\beta(e)\|$ and $\|\sum v_\alpha(a_i) \otimes b_i\| \leq \|\sum v_\beta(a_i) \otimes b_i\|$; thus the preceding argument leads to the two announced equalities.

Exercise 2.1.2. By the preceding exercise we have $\|u(x)\| = \sup_\alpha \|v_\alpha u(x)\|$ for any x, and hence $\|u\| = \sup_\alpha \|v_\alpha u\|$. Applying this, for each n, to $I_{M_n} \otimes u \colon M_n(F) \to M_n(E)$ (with $I_{M_n} \otimes v_\alpha$ in place of v_α) we obtain $\|I_{M_n} \otimes u\| = \sup_\alpha \|I_{M_n} \otimes v_\alpha u\|$, which implies $\|u\|_{cb} = \sup_\alpha \|v_\alpha u\|_{cb}$.

Exercise 2.1.3. The equality $\|T\| = \max\{\|T_1\|, \|T_2\|\}$ is immediate from the definition of T. Then, for any x in $M_n(E)$, we have

$$(I \otimes u)(x) \simeq (I \otimes u_1)(x) \oplus (I \otimes u_2)(x)$$

(using $\ell_2^n \otimes H = [\ell_2^n \otimes H_1] \oplus [\ell_2^n \otimes H_2]$). Hence $\|I \otimes u(x)\| = \max\{\|(I \otimes u_i)(x)\| \mid i = 1, 2\}$, whence $\|u\|_{cb} = \max\{\|u_1\|_{cb}, \|u_2\|_{cb}\}$.

Exercise 2.1.4. Let $b_i = x_i \otimes 1$ and $a_i = 1 \otimes y_i$, so that $b_i a_i = a_i b_i = x_i \otimes y_i$. The first inequality then follows immediately from (1.11). Exchanging the roles of a_i, b_i we get the second one. The equality cases follow from (1.10).

Exercise 2.2.1. Let $\ell_\infty(S, \mathbb{R})$ denote the space all bounded *real-valued* functions on S with its usual norm. In $\ell_\infty(S, \mathbb{R})$, the set $\mathcal{F}$ is disjoint from the set $C_- = \{\varphi \in \ell_\infty(S, \mathbb{R}) \mid \sup \varphi < 0\}$. Hence, by the Hahn-Banach Theorem (we separate the convex set $\mathcal{F}$ and the convex open set C_-), there is a nonzero $\xi \in \ell_\infty(S, \mathbb{R})^*$ such that $\xi(f) \geq 0 \ \forall f \in \mathcal{F}$ and $\xi(f) \leq 0 \ \forall f \in C_-$. Let $M \subset \ell_\infty(S, \mathbb{R})^*$ be the cone of all finitely supported (non-negative) measures on S viewed as functionals on $\ell_\infty(S, \mathbb{R})$. Since we have $\xi(f) \leq 0 \ \forall f \in C_-$, ξ must be in the bipolar of M for the duality of the pair $(\ell_\infty(S, \mathbb{R}), \ell_\infty(S, \mathbb{R})^*)$. Therefore, by the bipolar theorem, ξ is the limit for the topology $\sigma(\ell_\infty(S, \mathbb{R})^*, \ell_\infty(S, \mathbb{R}))$ of a net of finitely supported (non-negative) measures ξ_α on S. We have for any f in $\ell_\infty(S, \mathbb{R})$, $\xi_\alpha(f) \to \xi(f)$, and this holds in particular if $f = 1$; thus (since ξ is nonzero) we may assume $\xi_\alpha(1) > 0$. Hence, if we set $\lambda_\alpha(f) = \xi_\alpha(f)/\xi_\alpha(1)$, we obtain the announced result.

Exercise 2.2.2. First observe that, by the arithmetic/geometric mean inequality, we have for any $a, b \geq 0$

$$(ab)^{1/2} = \inf_{t>0}\{2^{-1}(ta + (b/t))\}.$$

In particular we have

$$\left\|\sum x_1^j x_1^{j*}\right\|^{1/2} \left\|\sum x_2^{j*} x_2^j\right\|^{1/2} \leq 2^{-1}\left(\left\|\sum x_1^j x_1^{j*}\right\| + \left\|\sum x_2^{j*} x_2^j\right\|\right).$$

Let S_i be the set of states on B_i $(i = 1, 2)$ and let $S = S_1 \times S_2$. The last inequality implies

$$\left| \sum \varphi(x_1^j, x_2^j) \right| \leq 2^{-1} \sup_{f = (f_1, f_2) \in S} \left\{ f_1 \left(\sum x_1^j x_1^{j*} \right) + f_2 \left(\sum x_2^{j*} x_2^j \right) \right\}.$$

Moreover, since the right side does not change if we replace x_j^1 by $z_j x_j^1$ with $z_j \in \mathbb{C}$ arbitrary such that $|z_j| = 1$, we may assume that the last inequality holds with $\sum |\varphi(x_1^j, x_2^j)|$ instead of $\left| \sum \varphi(x_1^j, x_2^j) \right|$. Then let $\mathcal{F} \subset \ell_\infty(S, \mathbb{R})$ be the convex cone formed of all possible functions $F \colon S \to \mathbb{R}$ of the form

$$F(f_1, f_2) = \sum_j 2^{-1} f_1(x_1^j x_1^{j*}) + 2^{-1} f_2(x_2^{j*} x_2^j) - |\varphi(x_1^j, x_2^j)|.$$

By Exercise 2.2.1, there is a net $\mathcal{U}$ of probability measures (λ_α) on S such that, for any F in $\mathcal{F}$, we have

$$\lim_{\mathcal{U}} \int F(g_1, g_2) d\lambda_\alpha(g_1, g_2) \geq 0.$$

We may as well assume that $\mathcal{U}$ is an ultrafilter. Then, if we set

$$f_i = \lim_{\mathcal{U}} \int g_i d\lambda_\alpha(g_1, g_2) \in S_i$$

(in the weak-$*$ topology $\sigma(B_i^*, B_i)$), we find that for any choice of (x_1^j) and (x_2^j) we have

$$\sum_j 2^{-1} f_1(x_1^j x_1^{j*}) + 2^{-1} f_2(x_2^{j*} x_2^j) - |\varphi(x_1^j, x_2^j)| \geq 0;$$

hence, in particular, $\forall x_1 \in F_1, \forall x_2 \in F_2$

$$2^{-1}(f_1(x_1 x_1^*) + f_2(x_2^* x_2)) \geq |\varphi(x_1, x_2)|.$$

By the homogeneity of φ, this implies

$$\inf_{t > 0} \{ 2^{-1}(t f_1(x_1 x_1^*) + f_2(x_2^* x_2)/t) \} \geq |\varphi(x_1, x_2)|,$$

and hence we obtain the desired conclusion using our initial observation on the geometric/arithmetic mean inequality.

Exercise 2.2.3. The proof of this is exactly the same as the preceding one, except that $\mathcal{F}$ should now be defined as formed of all possible functions $F \colon S \to \mathbb{R}$ of the form

$$F(f_1, f_2) = \sum_j 2^{-1} f_1(x_1^j x_1^{j*}) + 2^{-1} f_2(x_2^{j*} x_2^j) - |\psi(x_1^j, x_j, x_2^j)|,$$

where (x_j) is allowed to be an arbitrary family in the unit ball of (G, α).

Exercise 2.3.1. Let G be an arbitrary operator space (for instance, $G = M_N$). Let $x = \sum a_i \otimes b_i \in G \otimes E$, and let $u \colon E^* \to G$ be associated to x. Since (by definition) $\|x\|_{G \otimes_{\min} E^{**}} = \|u\|_{cb}$, it suffices to show that $\|x\|_{G \otimes_{\min} E} = \|u\|_{cb}$. By (2.1.6) we have

$$\|x\|_{G \otimes_{\min} E} = \sup_{n, v \in \mathcal{B}_n} \left\{ \left\| \sum v(b_i) \otimes a_i \right\|_{M_n(G)} \right\},$$

where $\mathcal{B}_n = \{v \colon E \to M_n \mid \|v\|_{cb} \leq 1\}$. By definition of E^*, $\mathcal{B}_n$ can be identified with the unit ball of $M_n(E^*)$, and if $y \in M_n(E^*)$ is the element corresponding to $v \colon E \to M_n$, we have

$$\sum v(b_i) \otimes a_i = (I \otimes u)(y).$$

Hence, we obtain as announced

$$\|x\|_{G \otimes_{\min} E} = \sup_n \{ \|I \otimes u(y)\| \mid \|y\|_{M_n(E^*)} \leq 1 \} = \|u\|_{cb}.$$

This completes the proof.

Exercise 2.3.2. We have $\|u^*\|_{cb} = \sup\{ \|(I \otimes u^*)(y)\|_{M_n(E^*)} \mid \|y\|_{M_n(F^*)} \leq 1, n \geq 1 \}$. But each y in the unit ball of $M_n(F^*)$ is in one-to-one correspondence with a map $v_y \colon F \to M_n$ with $\|v_y\|_{cb} = \|y\|_{M_n(F^*)} \leq 1$. Moreover, in the identification $M_n(E^*) \simeq CB(E, M_n)$, $(I \otimes u^*)(y)$ corresponds to the composition $v_y u$, so that $\|(I \otimes u^*)(y)\| = \|v_y u\|_{cb}$. Therefore we have

$$\|I \otimes u^* \colon M_n(F^*) \to M_n(E^*)\| = \sup\{ \|vu\|_{cb} \mid \|v \colon F \to M_n\|_{cb} \leq 1 \}.$$

By Exercise 1.4, if we take the supremum of both sides over all $n \geq 1$, we obtain $\|u^*\|_{cb} = \|u\|_{cb}$.

Exercise 2.3.3. The key to this variant is Proposition 1.12. Let us denote $\|u\|_n = \|I_{M_n} \otimes u\|_{M_n(E) \to M_n(F)}$. By Proposition 1.12, each y in the unit ball of $M_n(F^*)$ is in one-to-one correspondence with a map $v_y \colon F \to M_n$ with $\|v_y\|_n = \|y\|_{M_n(F^*)} \leq 1$. Thus, as in the preceding solution, we have

$$\|(I \otimes u^*)(y)\|_{M_n(E^*)} = \|v_y u\|_{CB(E, M_n)} = \|v_y u\|_n \leq \|v_y\|_n \|u\|_n \leq \|u\|_n,$$

and we obtain $\|u^*\|_n \leq \|u\|_n$. By iteration, we find $\|(u^*)^*\|_n \leq \|u^*\|_n$. But, by Exercise 2.3.1 we have

$$\|u\|_n = \|(u^*)^*_{|E}\|_n$$

hence we also obtain $\|u\|_n \leq \|u^*\|_n$.

Exercise 2.3.4. The "only if" part is essentially obvious using (2.3.3) for u and u^{-1}, so we turn to the "if" part. Assume that $u^* \colon F^* \to E^*$ is a complete isomorphism. Then $u^{**}: E^{**} \to F^{**}$ is a complete isomorphism by the first part of the solution. By restricting u^{**} to E, we see that $u: E \to F$ must be at least a completely isomorphic embedding; but, since u^* is injective, we know u must have a dense range, so actually u must be onto F. Then we conclude that $u = u_{|E}^{**}: E \to F$ is a complete isomorphism. The completely isometric case follows from (2.3.3).

Exercise 2.3.5. It suffices to show that $I_{\mathcal{K}} \otimes u: \mathcal{K} \otimes_{\min} R^* \to \mathcal{K} \otimes_{\min} C$ is isometric. Let (x_i) be a finitely supported sequence in $\mathcal{K}$. We claim that

$$\left\| \sum x_i \otimes \xi_i \right\|_{\mathcal{K} \otimes_{\min} R^*} = \left\| \sum x_i \otimes u(\xi_i) \right\|_{\mathcal{K} \otimes_{\min} C}.$$

Let $v: R \to \mathcal{K}$ be the map defined by $v(\cdot) = \sum x_i \xi_i(\cdot)$. By definition of R^* we have

$$\left\| \sum x_i \otimes \xi_i \right\|_{\min} = \|v\|_{cb} = \|I_{\mathcal{K}} \otimes v\|$$
$$= \sup\{\|(I_{\mathcal{K}} \otimes v)(y)\|_{\mathcal{K} \otimes_{\min} \mathcal{K}} \mid y \in \mathcal{K} \otimes R, \|y\|_{\mathcal{K} \otimes_{\min} R} \leq 1\}.$$

By density, we may restrict the preceding supremum to those y for which only finitely many y_i are nonzero. Let $y = \sum y_i \otimes e_{1i}$. Then by (1.10) we have $\|y\| = \|\sum y_i y_i^*\|^{1/2}$ and $(I_{\mathcal{K}} \otimes v)(y) = \sum y_i \otimes v(e_{1i}) = \sum y_i \otimes x_i$. Thus we find

$$\left\| \sum x_i \otimes \xi_i \right\|_{\min} = \sup\left\{ \left\| \sum y_i \otimes x_i \right\| \mid \left\| \sum y_i y_i^* \right\| \leq 1 \right\}.$$

By Exercise 2.1.4, this last supremum is $= \|\sum x_i^* x_i\|^{1/2}$. Thus, by (1.10) again, we have

$$\left\| \sum x_i \otimes \xi_i \right\|_{\min} = \left\| \sum x_i \otimes e_{i1} \right\|_{\min};$$

hence, since $e_{i1} = u(\xi_i)$, we obtain our claim, which proves that u is a complete isometry. The other identities follow by exactly the same arguments.

Exercise 2.4.1. (ii) $\Rightarrow$ (i): Assume (ii). Note that $\|I_G \otimes v: G \otimes_{\min} F_1 \to G \otimes_{\min} E\|_{cb} \leq C$ by (2.1.4), and $(I_G \otimes u)(I_G \otimes v) = I_G \otimes I_{F_1}$. To show that $I_G \otimes u$ is surjective, it suffices to show that any x in $G \otimes F$ with $\|x\|_{\min} < 1$ admits a lifting $\widehat{x}$ on $G \otimes E$ with $\|\widehat{x}\|_{\min} < C$. But this is clear since we can assume $x \in G \otimes F_1$ and set $\widehat{x} = (I_G \otimes v)(x)$.

(i) $\Rightarrow$ (ii): By a routine argument (using direct sums) one can show that there is a constant C such that, for any G, $I_G \otimes u$ is C-surjective, that is, any x in $G \otimes_{\min} F$ with $\|x\| \leq 1$ can be lifted to an $\widehat{x}$ in $G \otimes_{\min} E$ with $\|\widehat{x}\| \leq C$. Let F_1 be as in (ii). Let $x \in F_1^* \otimes F_1$ be the tensor associated to the identity on F_1. Note $\|x\|_{\min} = 1$. Let $\widehat{x} \in F_1^* \otimes_{\min} E$ with $\|\widehat{x}\| \leq C$ be a lifting of x, and let $v: F_1 \to E$ be the linear map associated to $\widehat{x}$. Clearly (by (2.3.2)) $\|\widehat{x}\|_{\min} = \|v\|_{cb}$, and since $\widehat{x}$ lifts x, uv is equal to the inclusion $F_1 \subset F$.

By the extension property of $B(H)$, one shows easily that (ii) implies (iii). Finally, (iii) $\Rightarrow$ (i) is easy by restricting first to the algebraic tensor product (for any t in $G \otimes F$ with $\|t\|_{\min} < 1$, let $x_i = (I \otimes v_i)t \in G \otimes E$; then $\|x_i\|_{\min} < C$ and $(I \otimes u)x_i \to t$, so that (i) follows by the open mapping theorem.)

Exercise 2.4.2. Let v_n be as in the hint. By our assumption on u, there is a c.c. $v \colon E \to A$ such that

$$\|qvx_i - ux_i\| < 2^{-n-2} \qquad \forall i = 1, \ldots, n+1.$$

For a suitable choice of α (to be specified below) we will let

$$v_{n+1}(x) = \sigma_\alpha^{1/2} v_n(x) \sigma_\alpha^{1/2} + (1 - \sigma_\alpha)^{1/2} v(x) (1 - \sigma_\alpha)^{1/2}.$$

Note that, since

$$v_{n+1}(x) = [\sigma_\alpha^{1/2} (1 - \sigma_\alpha)^{1/2}] \begin{bmatrix} v_n(x) & 0 \\ 0 & v(x) \end{bmatrix} \begin{bmatrix} \sigma_\alpha^{1/2} \\ (1 - \sigma_\alpha)^{1/2} \end{bmatrix},$$

we have

$$\|v_{n+1}\|_{cb} \le \max\{\|v_n\|_{cb}, \|v\|_{cb}\} \le 1.$$

By Lemma 2.4.4, for any given x and $\varepsilon > 0$, α can be chosen large enough so that

$$\|v_{n+1}(x) - [\sigma_\alpha v_n(x) + (1 - \sigma_\alpha) v(x)]\| < \varepsilon;$$

hence $\|qv_{n+1}(x) - qv(x)\| < \varepsilon$, which implies $\|qv_{n+1}(x) - u(x)\| < \varepsilon + \|qv(x) - u(x)\|$. So we can choose α large enough so that

$$\|qv_{n+1}(x_i) - u(x_i)\| < \varepsilon + 2^{-n-2} \qquad \forall i = 1, \ldots, n+1.$$

Moreover, we have

$$\|v_{n+1}(x) - v_n(x)\| < \varepsilon + \|(1 - \sigma_\alpha)[v(x) - v_n(x)]\|;$$

hence, for α large enough, by (2.4.6), we can ensure that

$$\|v_{n+1}(x) - v_n(x)\| < \varepsilon + \|q[v(x) - v_n(x)]\| < \varepsilon + \|qv(x) - u(x)\| + \|qv_n(x) - u(x)\|.$$

Thus, if we now make this last choice of α valid for any x in $\{x_1, \ldots, x_n\}$, and if we take $\varepsilon = 2^{-n-2}$, we obtain the announced estimates for v_{n+1}. It follows that $v_n(x)$ is Cauchy for any x in $\{x_1, x_2, \ldots\}$. Therefore $v(x) = \lim_n v_n(x)$ exists and satisfies $qv = u$.

Exercise 2.6.1. We can identify $M_n(\ell_1(\{E_i \mid i \in I\})^*)$ with $CB(\ell_1(\{E_i \mid i \in I\}), M_n)$. By the definition of the ℓ_1-direct sum, the latter space can be identified with $\ell_\infty(\{CB(E_i, M_n) \mid i \in I\})$ or, equivalently, with

$$\oplus_{i \in I} CB(E_i, M_n) \simeq \oplus_{i \in I} M_n(E_i^*),$$

and by (2.6.1) this last space is the same as $M_n(\oplus_{i \in I} E_i^*)$. Thus we conclude

$$M_n(\ell_1(\{E_i \mid i \in I\})^*) \simeq M_n(\oplus_{i \in I} E_i^*).$$

Exercise 2.6.2. The two sides of the equality are isometric Banach spaces. By a density argument it is easy to reduce this to the case when I is finite, so we will assume I finite. The universal property of the ℓ_1-direct sum immediately implies that

$$J \colon \ell_1(\{E_i^* \mid i \in I\}) \to c_0(\{E_i \mid i \in I\})^*$$

is a complete contraction. To show that J is a complete isometry, it suffices to show the following claim: For any complete contraction $u \colon \ell_1(\{E_i^* \mid i \in I\}) \to M_n$ we also have

$$\|u \colon c_0(\{E_i \mid i \in I\})^* \to M_n\|_{cb} \leq 1.$$

But such a u defines an element in the unit ball of $M_n(\ell_1(\{E_i^* \mid i \in I\})^*)$, which, by the preceding exercise, coincides with the unit ball of $M_n(\oplus_{i \in I} E_i^{**})$. But we have (by (2.6.1)) $M_n(\oplus_{i \in I} E_i^{**}) = \oplus_{i \in I} M_n(E_i^{**})$; hence (by (2.4.1)) $= \oplus_{i \in I} M_n(E_i)^{**}$, and hence (by elementary Banach space theory recalling I is finite) $= (\oplus_{i \in I} M_n(E_i))^{**}$ or, equivalently, by (2.6.1) again, $(M_n(\oplus_{i \in I} E_i))^{**}$, which by (2.4.1) again is the same as $M_n((\oplus_{i \in I} E_i)^{**})$. Since this last space can be identified (see (2.3.1)') with $cb((\oplus_{i \in I} E_i)^*, M_n)$, we finally obtain $\|u \colon (\oplus_{i \in I} E_i)^* \to M_n\|_{cb} \leq 1$, which is equivalent to the above claim (note that since I is finite, $\oplus_{i \in I} E_i = c_0(\{E_i \mid i \in I\}))$.

Exercise 2.6.3. This is an immediate consequence of the preceding two exercises.

Exercise 2.13.1. For any x with $\|x\| = 1$, there is an i such that $\|x - x_i\| \leq \varepsilon$; hence
$$\|ux\| \geq \|ux_i\| - \|u(x - x_i)\| \geq 1 - \varepsilon' - \varepsilon.$$

By homogeneity, this implies $\|u^{-1} \colon u(E) \to E\| \leq (1 - \varepsilon - \varepsilon')^{-1}$ and a fortiori $d(E, u(E)) \leq \|u\| \cdot (1 - \varepsilon - \varepsilon')^{-1} \leq (1 - \varepsilon - \varepsilon')^{-1}$.

Exercise 2.13.2. Choose ε so that $(1 - \varepsilon)^{-1} < 1 + \delta$. Let $\{x_i \mid i \leq N\}$ be an ε-net in the unit sphere of E. Let $\xi_i \in B_{E^*}$ be such that $\xi_i(x_i) = 1$, and let $u \colon E \to \ell_\infty^N$ be the mapping defined by $u(x) = \sum_1^N \xi_i(x) e_i$. Let $\widetilde{E} = u(E)$.

Then u satisfies the assumption of the preceding exercise with $\varepsilon' = 0$. Hence $d(E, \widetilde{E}) \leq (1 - \varepsilon)^{-1} \leq 1 + \delta$.

Exercise 2.13.3. Consider $f_1, \ldots, f_n$ in A. Let $E = \{f_1, \cdots, f_n\}$. Fix $\varepsilon > 0$. Let $\{U_j \mid j \leq N\}$ be a finite open covering of K such that the oscillation of each f_i on each U_j is $< \varepsilon$. Let $\{\varphi_j\}$ be a partition of unity subordinate to $\{U_j\}$, that is, φ_j is continuous, supp $\varphi_j \subset U_j$, $\varphi_j \geq 0$ and $\sum \varphi_j \equiv 1$. For each j, we choose a point t_j in U_j. Then we define $v \colon A \to \ell_\infty^N$ by $v(f) = \sum_1^N f(t_j)e_j$ and $w \colon \ell_\infty^N \to A$ by $w(e_j) = \varphi_j$. Clearly $\|v\| \leq 1$ and $\|w\| \leq 1$. Moreover, for any $i = 1, \ldots, n$, we have $\|u(f_i) - f_i\| = \|wv(f_i) - f_i\| = \sup_{t \in K} \left|\sum_j \varphi_j(t)(f_i(t_j) - f_i(t))\right| \leq \varepsilon$. Let $\alpha = (E, \varepsilon)$ and let u_α be the map u associated to E and ε as above. Clearly this yields the desired net.

Exercise 3.1. First consider the special case when $G = \ell_\infty^N$. Then $E \overset{\vee}{\otimes} G = \ell_\infty^N(E)$ and $(E \overset{\vee}{\otimes} G)^{**} = \ell_\infty^N(E^{**}) = E^{**} \overset{\vee}{\otimes} G$. Recall that, for any subspace $Y \subset X$ of a Banach space X, we have $Y^{**} \subset X^{**}$; more precisely, Y^{**} can be identified with the $\sigma(X^{**}, X^*)$-closure of Y in X^{**}. In particular, this shows $(E \overset{\vee}{\otimes} \widetilde{G})^{**} = E^{**} \overset{\vee}{\otimes} \widetilde{G}$ for any subspace $\widetilde{G} \subset \ell_\infty^N$. For the general case we use the fact that, for any $\varepsilon > 0$, there is an integer N and a subspace $\widetilde{G} \subset \ell_\infty^N$ that is $(1 + \varepsilon)$-isomorphic to G (i.e., such that $d(G, \widetilde{G}) \leq 1 + \varepsilon$). Using $\widetilde{G}$ instead of G and applying the first part, we get the announced result up to ε, and letting $\varepsilon \to 0$ we obtain the general case. (See e.g.[DJT, p. 178] for more details). Note that $E^{**} \overset{\vee}{\otimes} G = B(E^*, G)$ isometrically and $E \overset{\vee}{\otimes} G$ corresponds to the subspace of $B(E^*, G)$ formed of all the weak-$*$ continuous maps. In particular, if $X = E \overset{\vee}{\otimes} G$, any element in $B_{X^{**}}$ is the $\sigma(X^{**}, X^*)$-limit of a net in B_X. Therefore, for any $v \colon E^* \to G$ with $\|v\| = 1$, there is a net of weak-$*$ continuous maps $v_\alpha \colon E^* \to G$ with $\|v_\alpha\| \leq 1$ such that, for any ξ in E^*, $v_\alpha(\xi) \to v(\xi)$ in the weak $(=$ norm$)$ topology of G.

Exercise 3.2. Let F be an arbitrary operator space. Consider $x = \sum b_i \otimes a_i \in F \otimes E^*$. Let $u \colon E \to F$ be the associated map. Then

$$\|x\|_{F \otimes_{\min} \max(E)^*} = \|u\|_{CB(\max(E), F)} = \|u\|_{B(E, F)} = \|x\|_{F \overset{\vee}{\otimes} E^*}$$
$$= \|x\|_{F \otimes_{\min} \min(E^*)}.$$

This proves that $(\max(E))^* = \min(E^*)$. To prove the other equality, we will use the principle of local reflexivity (see Exercise 3.1).

For any x in $F \otimes_{\min} \max(E^*)$ with associated map $u \colon E \to F$, we have, by (2.1.6), $\|x\|_{F \otimes_{\min} \max(E^*)} = \sup \|I_F \otimes v(x)\|$, where the supremum runs over all n and $v \colon E^* \to M_n$ with $\|v\| \leq 1$. By Exercise 3.1 we may restrict to v weak-$*$ continuous. But then v represents an element $t \in M_n \otimes E$ with

$\|t\|_{M_n \otimes_{\min} \min(E)} = \|v\| \leq 1$. Thus we obtain $\|x\|_{F \otimes_{\min} \max(E^*)} = \sup \|(I_{M_n} \otimes u)(t)\|$ with the sup over all such t. Equivalently,

$$\|x\|_{F \otimes_{\min} \max(E^*)} = \|u\|_{CB(\min(E),F)},$$

which means (take $F = M_n$) that $\max(E^*) = (\min(E))^*$.

Exercise 3.3. By the preceding exercise, if E is minimal (resp. maximal), then E^{**} is also. Conversely, if E^{**} is minimal, then, by Exercise 2.3.1, E (being a subspace of E^{**}) is itself minimal. If E^{**} is maximal, then, for any F and any $u : E \to F$, we have (by (2.3.3)) $\|u\|_{cb} = \|u^{**}\|_{cb} = \|u^{**}\|$ (there is equality because E^{**} is maximal), and of course $\|u^{**}\| = \|u\|$; thus we obtain $\|u\|_{cb} = \|u\|$ and we conclude that E is maximal (indeed taking for u the identity map from E to $\max(E)$, we find $\|u\|_{cb} = 1$).

Exercise 3.4. Let E_i be a family of Banach spaces, and let $E = \bigoplus_{i \in I} E_i$ be the direct sum in the ℓ_∞-sense. It is easy to check that, for any *finite-dimensional* normed space G, we have isometrically:

$$G \overset{\vee}{\otimes} \left(\bigoplus_{i \in I} E_i \right) = \bigoplus_{i \in I} (G \overset{\vee}{\otimes} E_i).$$

Now assume that $\{E_i \mid i \in I\}$ is a family of operator spaces. Consider x in $M_n \left(\bigoplus_{i \in I} E_i \right)$ and let $\{x_i \mid i \in I\}$ be the corresponding family with $x_i \in M_n(E_i)$ for each i. By (2.6.2) we have

$$\|x\| = \sup_{i \in I} \|x_i\|_{M_n(E_i)}.$$

But now, if each E_i is assumed minimal, we have

$$\|x_i\|_{M_n(E_i)} = \|x_i\|_{M_n \overset{\vee}{\otimes} E_i};$$

hence, applying the preceding observation with $G = M_n$, we find

$$\|x\| = \sup \|x_i\|_{M_n \overset{\vee}{\otimes} E_i} = \|x\|_{M_n \overset{\vee}{\otimes} E}.$$

Thus we conclude that $E = \min(E)$.

Exercise 3.5. It suffices to show that, for any $u: \ell_1(\{E_i \mid i \in I\}) \to B(H)$, we have $\|u\|_{cb} = \|u\|$. We can assume that u corresponds to a family $\{u_i \mid i \in I\}$ with $u_i: E_i \to B(H)$. Then obviously $\|u\| = \sup \|u_i\|$, and, by definition of the o.s.s. on $\ell_1(\{E_i \mid i \in I\})$, we also have $\|u\|_{cb} = \sup \|u_i\|_{cb}$. But, since each E_i is maximal, we have $\|u_i\| = \|u_i\|_{cb}$ for each i, and hence we obtain $\|u\|_{cb} = \|u\|$.

Exercise 3.6. The operators $U_i \otimes U_i$ and $U_i \otimes \overline{U_i}$ are commuting self-adjoint unitaries. In particular, they generate a commutative C^*-algebra and therefore are minimal (recall Proposition 1.10(ii)).

Exercise 3.7. The first question is treated in §9.3, to which we refer the reader. Let $T = \sum_1^n C_k \otimes C_k$. For any subset $\alpha \subset \{1,\dots,n\}$ with $\alpha = \{i_1,\dots,i_k\}$ with $i_1 < i_2 \cdots < i_k$ we denote $e_\alpha = e_{i_1} \wedge \cdots \wedge e_{i_k}$, and e_ϕ denotes the vacuum vector. Let $\omega_d = \sum_{|\alpha|=d} e_\alpha \otimes e_\alpha$. Since $C_k e_\alpha = 0$, if $k \in \alpha$, we have $(C_i \otimes C_i)(\omega_d) = \sum_{|\alpha|=d+1, i \in \alpha} e_\alpha \otimes e_\alpha$; hence $T(\omega_d) = (d+1)\omega_{d+1}$. We have clearly $\|\omega_d\| = \binom{n}{d}^{1/2}$, and therefore we find

$$\|T\| \geq (d+1)\binom{n}{d+1}^{1/2}\binom{n}{d}^{-1/2};$$

hence

$$\geq (d+1)^{1/2}(n-d)^{1/2}.$$

Finally, we choose if n is odd: $d = (n-1)/2$, and if n is even: $d = n/2$.

Exercise 3.8. By Theorem 3.1, it is easy to show that for any separable $Y_1 \subset X$ there is a separable subspace Y_2 with $Y_1 \subset Y_2 \subset X$ such that for any $u\colon Y_2 \to B(H)$ we have $\|u_{|Y_1}\|_{cb} \leq \|u\|$. (Consider a dense set in the open unit ball of $M_n(Y_1)$ and factorize each of them as in Theorem 3.1. Then, any space Y_2 containing all the diagonal entries appearing in these factorizations for all possible n has the desired property.) Now, by induction, we can build a sequence $X = Y_1 \subset Y_2 \subset \cdots \subset Y_n \subset \cdots X$ of separable subspaces such that for any $u\colon Y_{n+1} \to B(H)$ we have $\|u_{|Y_n}\|_{cb} \leq \|u\|$; then, taking $X_2 = \overline{\cup Y_n}$, we obtain for any $u\colon X_2 \to B(H)$, $\|u\|_{cb} = \|u\|$, which means that X_2 is maximal, and of course it is separable.

Exercise 5.1. Consider ξ in the unit ball of $(E_1 \otimes_h E_2)^*$. By Corollary 5.4, there are a Hilbert space H and complete contractions $\sigma_1\colon E_1 \to B(H,\mathbb{C})$ and $\sigma_2\colon E_2 \to B(\mathbb{C},H)$ such that $\xi(x_1 \otimes x_2) = \sigma_1(x_1)\sigma_2(x_2)$. For simplicity of notation we will assume H separable. Let $P_n\colon H \to H$ be the orthogonal projection onto the span of the first n vectors of a fixed orthonormal basis in H. Then let

$$\xi_n(x_1 \otimes x_2) = \sigma_1(x_1)P_n\sigma_2(x_2).$$

Note that $\xi_n(x_1 \otimes x_2) \to \xi(x_1 \otimes x_2)$ ($x_i \in E_i, i = 1,2$) when $n \to \infty$. Moreover, we have $\xi_n \in E_1^* \otimes_h E_2^*$ and $\|\xi_n\|_h \leq \|\xi\|_h \leq 1$. In addition, for any scalar sequence (ε_j) with $|\varepsilon_j| = 1$ we have $\|\sum_1^n \varepsilon_j(P_j - P_{j-1})\| \leq 1$; hence $\sup_n \left\|\sum_{j=1}^n \varepsilon_j(\xi_j - \xi_{j-1})\right\|_{E_1^* \otimes_h E_2^*} \leq 1$.

If $E_1^* \otimes_h E_2^* \not\supset c_0$, then $\xi_n = \xi_1 + \sum_2^n \xi_j - \xi_{j-1}$ must norm-converge in $E_1^* \otimes_h E_2^*$. Indeed, otherwise we find $\delta > 0$ and an increasing integer sequence

n_j such that $\|\xi_{n_j} - \xi_{n_{j-1}}\| > \delta$ for all j. But then, setting $x_j = \xi_{n_j} - \xi_{n_{j-1}}$, we have $\|x_j\| > \delta$ and $\sup_{n,|\varepsilon_j|=1} \|\sum \varepsilon_j x_j\| < \infty$, which implies (see [LT1, p. 22]) that $E_1^* \otimes_h E_2^*$ contains c_0 isomorphically. Thus, ξ_j norm-converges to ξ in $E_1^* \otimes_h E_2^*$, which shows that ξ actually belongs to $E_1^* \otimes_h E_2^*$.

Exercise 5.2. We will show that $C_n \otimes_h C_k \simeq C_{nk}$. Let $B = B(\ell_2)$. It suffices to show that, for any y in $B \otimes C_n \otimes C_k$, we have $I = II$, where $I = \|y\|_{B \otimes_{\min}(C_n \otimes_h C_k)}$ and $II = \|y\|_{B \otimes_{\min} C_{nk}}$. Assume $y = \sum_{ij} y_{ij} \otimes e_{i1} \otimes e_{j1}$. Then $II = \left\|\sum_{ij} y_{ij}^* y_{ij}\right\|^{1/2}$. On the other hand, by Corollary 5.9, I is equal to the infimum of $\|\sum y_i^* y_i\|^{1/2} \|\sum z_j^* z_j\|^{1/2}$ over all possible ways to write $y_{ij} = y_i z_j$ with y_i, z_j in B. Clearly we have $\|\sum y_{ij}^* y_{ij}\| = \left\|\sum_j z_j^* \left(\sum_i y_i^* y_i\right) z_j\right\|$, and hence $II \leq I$. Conversely, we can write

$$e_{11} \otimes y_{ij} = (e_{1i} \otimes I)\left(\sum_k e_{k1} \otimes y_{kj}\right);$$

hence, with the obvious identifications, this gives us $y_{ij} = y_i z_j$ with

$$\left\|\sum y_i^* y_i\right\|^{1/2} = \left\|\sum e_{1i}^* e_{1i}\right\|^{1/2} = 1$$

and

$$\left\|\sum z_j^* z_j\right\|^{1/2} = \left\|\sum_{kj} y_{kj}^* y_{kj}\right\|^{1/2} = II.$$

Thus we obtain $I \leq II$. This proves that $C_n \otimes_h C_k \simeq C_{nk}$ completely isometrically. The fact that $R_n \otimes_h R_k \simeq R_{nk}$ can either be proved similarly or be deduced from the preceding by duality using (5.14). This completes the proof of (5.16), or equivalently, of (5.16)' when H, K are both finite-dimensional. But then the general case follows using finite-dimensional approximations. We skip the routine details.

Exercise 5.3. Assume $E \subset B(H)$. We first show that (5.17) is isometric. Consider x in $M_n(E)$. Then, by Theorem 5.1,

$$\left\|\sum e_{i1} \otimes x_{ij} \otimes e_{1j}\right\|_h = \sup\left\{\left\|\sum \sigma_1(e_{i1})\sigma_2(x_{ij})\sigma_3(e_{1j})\right\|\right\},$$

where the sup runs over all σ_i with $\|\sigma_i\|_{cb} \leq 1$. Let $b_i = \sigma_1(e_{i1})$, $a_j = \sigma_3(e_{1j})$. Note that (see Remark 1.13) $\|\sum b_i b_i^*\|^{1/2} \leq \|\sigma_1\|_{cb} \|\sum e_{i1} e_{i1}^*\|^{1/2} \leq 1$ and similarly $\|\sum a_j^* a_j\|^{1/2} \leq 1$. Hence, by Remark 1.13, we have

$$\left\|\sum e_{i1} \otimes x_{ij} \otimes e_{1j}\right\|_h \leq \|x\|_{M_n(E)}.$$

Conversely, we have

$$\|x\|_{M_n(E)} = \sup\left\{\left|\sum\langle x_{ij}h_j, k_i\rangle\right|\right\},$$

where the sup runs over $h_j, k_i \in H$ with $\sum\|h_j\|^2 \leq 1$, $\sum\|k_i\|^2 \leq 1$. Then, if we define $\sigma_3(e_{1j}) = (h_j)_c \in B(\mathbb{C}, H)$ and $\sigma_1(e_{i1}) = (k_i)_r \in B(H, \mathbb{C})$, we have (see Exercise 1.1) $\|\sigma_1\|_{cb} \leq 1$, $\|\sigma_3\|_{cb} \leq 1$, and $\langle x_{ij}h_j, k_i\rangle = \sigma_1(e_{i1})x_{ij}\sigma_3(e_{1j})$; hence we obtain the announced isometric equality (5.17). Now, knowing that (5.17) is isometric for any E, we will show that it must be completely isometric. Indeed, for any $k \geq 1$ we have isometrically

$$M_k(C_n \otimes_h E \otimes_h R_n) = C_k \otimes_h (C_n \otimes_h E \otimes_h R_n) \otimes_h R_k;$$

hence, by associativity and by (5.16),

$$= C_{nk} \otimes_h E \otimes_h R_{nk} = M_{nk}(E) = M_k(M_n(E)).$$

Since this holds for all k, we conclude that (5.17) is completely isometric. Then (5.18) is immediate by density. Finally, let us restrict the identification $C \otimes_h E \otimes_h R \simeq \mathcal{K} \otimes_{\min} E$ to the subspaces spanned respectively by $C \otimes E \otimes e_{11}$ and by $e_{11} \otimes E \otimes R$. By the injectivity of both tensor products involved, we obtain $C \otimes_h E \simeq C \otimes_{\min} E$ (resp. $E \otimes_h R \simeq E \otimes_{\min} R$) completely isometrically. More generally, for any Hilbert space H we have $H_c \otimes_h E \simeq H_c \otimes_{\min} E$. Thus, taking $E = K_c$ with K another Hilbert space, we find $H_c \otimes_h K_c \simeq H_c \otimes_{\min} K_c$. But now, recalling $H_c = B(\mathbb{C}, H)$ and $K_c = B(\mathbb{C}, K)$, we find

$$H_c \otimes_{\min} K_c \simeq B(\mathbb{C} \otimes_2 \mathbb{C}, H \otimes_2 K) = (H \otimes_2 K)_c.$$

Thus we obtain another proof of (5.16)′.

Exercise 5.4. Fix $\varepsilon > 0$. By definition of the completion, we have $x = \sum_1^\infty x_k$ with $x_k \in E_1 \otimes E_2$ such that $\sum_1^\infty \|x_k\|_h < (1 + \varepsilon)\|x\|_h$. Then, for each k, we can write x_k as a finite sum $x_k = \sum_{i \in A_k} a(k, i) \otimes b(k, i)$ with $\left\|\sum_{i \in A_k} a(k, i)a(k, i)^*\right\| \leq \|x_k\|_h(1 + \varepsilon)$ and $\left\|\sum_{i \in A_k} b(k, i)^*b(k, i)\right\| \leq \|x_k\|_h(1 + \varepsilon)$. Finally, we may order each set A_k in any way we wish, and the resulting series $\sum_k \sum_{i \in A_k} a(k, i) \otimes b(k, i)$ does the job.

Exercise 5.5. The first and third isomorphisms were already proved in the text respectively in (5.18) and (5.23). To prove the second one, note that by (5.15) we have a completely isometric embedding $R \otimes_h E^* \otimes_h C \to (C \otimes_h E \otimes_h R)^*$. Thus it actually suffices to show that $R \otimes E^* \otimes C$ is dense in $(C \otimes_h E \otimes_h R)^*$. This can be verified in the following manner. Consider an element x in $C \otimes_h E \otimes_h R$. Let $x(n)$ be the natural projection of x to

$C_n \otimes_h E \otimes_h R$. We will denote $P_n x = x(n)$. It is easy to check that for any increasing sequence $n_1 < n_2 \ldots$ we have

$$\|x\| \leq (\|x(n_1)\|^2 + \|x(n_2) - x(n_1)\|^2 + \cdots)^{1/2}.$$

Now let φ be an element of $(C \otimes_h E \otimes_h R)^*$ and let $\varphi(n) = P_n^* \varphi$. By duality we have

$$(\|\varphi(u_1)\|^2 + \|\varphi(n_2) - \varphi(n_1)\|^2 + \ldots)^{1/2} \leq \|\varphi\|.$$

This inequality clearly implies that $\{\varphi(n) \mid n \geq 1\}$ must be a Cauchy sequence in $(C \otimes_h C \otimes_h R)^*$. Hence $\varphi(n)$ converges in $(C \otimes_h E \otimes_h R)^*$ to a limit that must be equal to φ since it coincides with it on the dense subspace $\cup_n C_n \otimes_h E \otimes_h R$. Thus we have $\|\varphi - P_n^* \varphi\| \to 0$. But we may argue in exactly the same way for the projections Q_m from $C \otimes_h E \otimes_h R$ onto $C \otimes E \otimes_h R_m$. This also gives us $\|\varphi - Q_m^* \varphi\| \to 0$. Hence $\|\varphi - Q_m^* P_n^* \varphi\|$ is arbitrarily small when n and m are suitably large. Since $Q_m^* R_n^* \varphi \in R_n \otimes E^* \otimes C_m$, this completes the solution.

Exercise 5.6. By the preceding exercise we have $(\mathcal{K} \otimes_{\min} E)^* \simeq \mathcal{K}^* \otimes^\wedge E^*$, and by (4.4) $(\mathcal{K}^* \otimes^\wedge E^*)^* \simeq CB(\mathcal{K}^*, E^{**})$. Hence, if E is finite-dimensional, by (2.3.2) this is $\simeq \mathcal{K}^{**} \otimes_{\min} E$.

If A is a C^*-algebra, Exercise 5.5 together with Theorems 2.5.2 and 4.1 gives us $(\mathcal{K} \otimes_{\min} A)^{**} \simeq \mathcal{K}^{**} \overline{\otimes} A^{**}$.

Exercise 5.7. Assume $x = \sum a_{ij} x_i \otimes y_j$. Then, using (1.12), it is easy to check that $\|x\|_\mu \leq \|x\|_h \leq \|a\| \left(\sum \|x_i\|^2\right)^{1/2} \left(\sum \|y_j\|^2\right)^{1/2}$. Assume conversely that $\|x\|_\mu < 1$. By Theorem 5.17 we may as well assume that either $\|x\|_h < 1$ or $\|{}^t x\|_h < 1$. Since the two cases are similar, we will check only the first case, so we assume $\|x\|_h < 1$. Then, by the very definition of $\| \cdot \|_h$, we can write $e_{11} \otimes x = X_1 \odot X_2$ with $X_1 \in M_{1,n}(E_1)$, $X_2 \in M_{n,1}(E_2)$ such that $\|X_1\| < 1$, $\|X_2\| < 1$. By Theorem 3.1, we can write for some integer N

$$X_1 = \alpha_0 \cdot D \cdot \alpha_1 \quad \text{and} \quad X_2 = \beta_0 \cdot \Delta \cdot \beta_1,$$

where $\alpha_0, \alpha_1, \beta_0, \beta_1$ are scalar matrices and D, Δ are diagonal matrices with entries in E_1, E_2, respectively, such that

$$\|\alpha_0\| \|D\| \|\alpha_1\| < 1 \quad \text{and} \quad \|\beta_0\| \|\Delta\| \|\beta_1\| < 1.$$

Here the product $\alpha_0 \cdot D$ (resp. $\Delta \cdot \beta_1$) is a line (resp. column) matrix of length N with coefficients $(x_1, \ldots, x_N)$ (resp. $(y_1, \ldots, y_N)$) such that $\sum \|x_j\|^2 < 1$ (resp. $\sum \|y_j\|^2 < 1$). Thus, if we define $a_{ij} = (\alpha_1 \beta_0)_{ij}$, we obtain

$$x = X_1 \odot X_2 = \sum_{ij=1}^n a_{ij} x_i \otimes y_j,$$

and $\|a\|_{M_N}(\sum\|x_i\|^2)^{1/2}(\sum\|y_j\|^2)^{1/2} < 1$. By homogeneity, this completes the proof.

Exercise 5.8. Consider $x \in A \otimes B \otimes \cdots \otimes A \otimes B$ (k times). We will show that

$$\|x\|_{T_{2k}} \leq (2k-1)^{2k-1}\|\psi_{2k}(x)\|.$$

Assume $\|\psi_{2k}(x)\| \leq 1$. By Corollary 5.3, it suffices to show that, for any family $(\sigma^1, \sigma^2, \ldots, \sigma^{2k})$ of complete contractions into the same $B(H)$, we have

$$\|\sigma^1 \cdot \sigma^1 \cdot \ldots \cdot \sigma^{2k}(x)\| \leq 1.$$

By Theorem 1.6, we may as well assume that σ^j is extended to $A\ast B$ (we view A and B as sitting inside $A\ast B$) and that it is of the form $V_j\pi_j(\cdot)W_j$ for some representation $\pi_j\colon A\ast B \to B(H_j)$ and with $\|V_j\| = \|W_j\| \leq 1$. Moreover, replacing each π_j by the single representation $\pi = \pi_1 \oplus \cdots \oplus \pi_{2k}$ and suitably modifying V_j and W_j, we may assume that, if x is either in A or B, we have $\sigma^j(x) = V_j\pi(x)W_j$. Let us denote by $L(V_j)$ and $R(W_j)$ the operator of left and right multiplication, respectively, by V_j and W_j, so that σ^j is equal to $L(V_j)R(W_j)\pi$ restricted to either A or B, depending on the parity of j. Thus we are reduced to showing

$$\|(L(V_1)R(W_1)\pi) \cdot \ldots \cdot (L(V_{2k})R(W_{2k})\pi)\,(x)\| \leq 1.$$

After some obvious simplifications, it finally suffices to show that, for any contractions $T_1, T_2, \ldots, T_{2k-1}$, we have

$$\|\pi \cdot L(T_1)\pi \cdot L(T_2)\pi \cdot \ldots \cdot L(T_{2k-1})\pi\,(x)\| \leq 1.$$

We will now use our assumption that $\|\psi_{2k}(x)\| \leq 1$. This implies that, for any pair π_1, π_2 of representations on the same Hilbert space, we have

$$\|\pi_1 \cdot \pi_2 \ldots \pi_1 \cdot \pi_2\,(x)\| \leq 1.$$

Let U be an arbitrary unitary in $B(H_\pi)$. Taking $\pi_1 = \pi$ and $\pi_2(\cdot) = U\pi(\cdot)U^*$, we find

$$\|\pi \cdot L(U)\pi \cdot L(U^*)\pi \cdot \ldots \cdot L(U)\pi\,(x)\| \leq 1.$$

Since this holds for any π, we may apply this with π replaced by $\pi \oplus 0$ and

$$U = \begin{pmatrix} T & (1-TT^*)^{1/2} \\ -(1-T^*T)^{1/2} & T^* \end{pmatrix}$$

(as in the proof of Lemma 5.14), and we obtain that for any T with $\|T\| \leq 1$ we have

$$\|\pi \cdot L(T)\pi \cdot L(T^*)\pi \cdot \ldots \cdot L(T)\pi\,(x)\| \leq 1. \tag{$*$}$$

Finally we will use the polarization formula to replace $(T, T^*, T, T^*, \ldots)$ in the last inequality by $(T_1, T_2, T_3, \ldots)$, with $\|T_j\| \leq 1$. We define for $z = (z_j) \in \mathbb{T}^{2k-1}$

$$T(z) = \sum_{j=1}^{2k-1} (z_j T_j)^{\varepsilon_j} \cdot (2k-1)^{-1},$$

where $\varepsilon_j = 1$ if j is odd and $\varepsilon_j = *$ if j is even. Note that $\|T(z)\| \leq 1$, and if we denote by dz the normalized Haar measure on $\mathbb{T}^{2k-1}$, we have

$$\int [T(z) \otimes T(z)^* \otimes \cdots \otimes T(z)^* \otimes T(z)] \overline{z_1 z_2} \ldots \overline{z_{2k-1}} dz$$

$$= (2k-1)^{2k-1} T_1 \otimes T_2 \otimes \cdots \otimes T_{2k-1}.$$

Therefore, a simple averaging argument gives us that $(*)$ implies

$$\|\pi \cdot L(T_1)\pi \cdot L(T_2)\pi \cdot \ldots \cdot L(T_{2k-1})\pi \ (x)\| \leq (2k-1)^{2k-1},$$

and we conclude as announced that

$$\|x\|_h \leq (2k-1)^{2k-1}.$$

Hence we have shown $\|\psi_{2k|\mathcal{W}_{2k}}^{-1}\| \leq (2k-1)^{2k-1}$. Using operator coefficients instead of scalar ones, the same argument yields

$$\|\psi_{2k|\mathcal{W}_{2k}}^{-1}\|_{cb} \leq (2k-1)^{2k-1}.$$

This completes the case $d = 2k$. The case $d = 2k+1$ is similar.

Exercise 5.9. To show $E_x \simeq C_n$, it suffices to show the following claim: For any $a_1, \ldots, a_n$ in $B(H)$ we have

$$\left\| \sum a_i \otimes x_i \right\|_{\min} = \left\| \sum a_i^* a_i \right\|^{1/2}.$$

Assume $\| \sum a_i \otimes x_i \|_{\min} \leq 1$. We will show $\| \sum a_i^* a_i \| \leq 1$. Note $\sum a_i \otimes x_i = \sum_{ik} w_{ik} a_i \otimes e_i \otimes e_k$. By Corollary 5.9, $\| \sum a_i \otimes x_i \|_{\min} \leq 1$ implies (assuming $\dim(H) = \infty$) that there are α_i, β_k in the unit ball of $B(H)$ such that

$$w_{ik} a_i = \alpha_i \beta_k;$$

hence $a_i = \overline{w}_{ik} \alpha_i \beta_k$ for any k, which implies $a_i = \alpha_i n^{-1} \sum_k \overline{w}_{ik} \beta_k$, or, equivalently, if we set $\widehat{\beta}_k = n^{-1/2} \beta_k$ and $\gamma_i = n^{-1/2} \sum_k \overline{w}_{ik} \widehat{\beta}_k$, we have $a_i = \alpha_i \gamma_i$. But then this implies

$$\left\| \sum a_i^* a_i \right\|^{1/2} \leq \sup \|\alpha_i\| \left\| \sum \gamma_i^* \gamma_i \right\|^{1/2} \leq \left\| \sum \gamma_i^* \gamma_i \right\|^{1/2},$$

and since $\|n^{-1}\overline{w}\|_{M_n} \leq 1$, we find

$$\leq \left\|\sum \widehat{\beta}_k^* \widehat{\beta}_k\right\|^{1/2} \leq \left(\sum \|\widehat{\beta}_k\|^2\right)^{1/2} \leq 1,$$

and we conclude as announced $\|\sum a_i^* a_i\| \leq 1$. In the converse direction we have a natural completely contractive map $C_n \to \ell_\infty^n$ taking e_{i1} to e_i; hence, if we set $X_i = e_{i1} \otimes \sum_k w_{ik} e_k$, we can write

$$\left\|\sum a_i \otimes x_i\right\|_{B(H) \otimes_{\min} E_x} \leq \left\|\sum a_i \otimes X_i\right\|_{B(H) \otimes_{\min}(C_n \otimes_h \ell_\infty^n)},$$

and by Exercise 5.3

$$\left\|\sum a_i \otimes X_i\right\|_{B(H) \otimes_{\min}(C_n \otimes_h \ell_\infty^n)} = \left\|\sum a_i \otimes X_i\right\|_{B(H) \otimes_{\min} C_n \otimes_{\min} \ell_\infty^n}$$

$$= \sup_k \left\|\sum_i a_i \otimes w_{ik} e_{i1}\right\|_{B(H) \otimes_{\min} C_n} = \left\|\sum_i a_i^* a_i\right\|^{1/2}.$$

Hence we obtain $\|\sum a_i \otimes x_i\|_{\min} \leq \|\sum a_i^* a_i\|^{1/2}$. This completes the proof of the above claim. Thus we have proved $E_x \simeq C_n$ completely isometrically. The proof for $E_y \simeq R_n$ is essentially the same.

The reader will note (if the entries of w are no longer assumed unimodular) that, if we let $c_1 = \|n^{-1/2}[w_{ik}^{-1}]\|_{M_n}$ and $c_2 = \sup_{i,k} |w_{ik}|$, then we find a completely isomorphic embedding $u: C_n \to \ell_\infty^n \otimes_h \ell_\infty^n$ taking e_{i1} to x_i such that

$$\|u\|_{cb} \leq c_2 \quad \text{and} \quad \|u_{|E_x}^{-1}\|_{cb} \leq c_1.$$

Exercise 5.10. To show that $e_{ij} \to z_{ij}$ defines a complete isometry from M_n to $\ell_\infty^n \otimes_h \ell_\infty^n \otimes_h \ell_\infty^n$ it clearly suffices to show the following claim: For any a_{ij} in $B(H)$ (with say $H = \ell_2$) we have

$$\left\|\sum a_{ij} \otimes z_{ij}\right\|_{\min} = \|(a_{ij})\|_{M_n(B(H))}.$$

Assume $\|\sum a_{ij} \otimes z_{ij}\|_{\min} \leq 1$. We will show that $\|(a_{ij})\|_{M_n(B(H))} \leq 1$. By Corollary 5.9, since $\|\sum a_{ij} \otimes z_{ij}\| = \left\|\sum_{ijk} a_{ij} w_{ik} w_{kj}' e_i \otimes e_k \otimes e_j\right\|$, our assumption $\|\sum a_{ij} \otimes z_{ij}\| \leq 1$ implies that we can write

$$a_{ij} w_{ik} w_{jk}' = \alpha_i \beta_k \gamma_j,$$

where $\alpha_i, \beta_k \gamma_j$ are in the unit ball of $B(H)$. Hence we have

$$a_{ij} = \alpha_i \left(n^{-1} \sum_k \overline{w}_{ik} \overline{w_{kj}'} \beta_k \right) \gamma_j,$$

which implies that

$$[a_{ij}] = D_1[n^{-1/2}\overline{w}]D_2[n^{-1/2}\overline{w'}]D_3,$$

where D_1, D_2, D_3 are the diagonal matrices with entries respectively (α_i), (β_k), (γ_j). Thus we obtain

$$\|[a_{ij}]\| \le \|D_1\|\|n^{-1/2}\overline{w}\|\|D_2\|\|n^{-1/2}\overline{w'}\|\|D_3\| \le 1.$$

Conversely, since we have complete contractions from C_n to ℓ_∞^n taking e_{i1} to e_i and from R_n to ℓ_∞^n taking e_{1i} to e_i, we have a complete contraction from $C_n \otimes_h \ell_\infty^n \otimes_h R_n$ to $\ell_\infty^n \otimes_h \ell_\infty^n \otimes_h \ell_\infty^n$ taking $e_{i1} \otimes e_k \otimes e_{1j}$ to $e_i \otimes e_k \otimes e_j$. Hence we have

$$\left\|\sum a_{ij} \otimes z_{ij}\right\| \le \left\|\sum_{ijk} a_{ij}w_{ik}w'_{kj} \otimes e_{i1} \otimes e_k \otimes e_{1j}\right\|_{B(H)\otimes_{\min}(C_n\otimes_h\ell_\infty^n\otimes_h R_n)}.$$

But, by (5.17), $C_n \otimes_h \ell_\infty^n \otimes_h R_n \simeq M_n(\ell_\infty^n) \simeq M_n \otimes_{\min} \ell_\infty^n$, and hence the last norm is equal to

$$\sup_k \|[a_{ij}w_{ik}w'_{kj}]_{ij}\|_{M_n(B(H))}.$$

Now, since $|w_{ik}| = |w'_{kj}| = 1$, we have for each fixed k

$$\|[a_{ij}w_{ik}w'_{kj}]\|_{M_n(B(H))} = \|[a_{ij}]\|_{M_n(B(H))};$$

hence we conclude

$$\left\|\sum a_{ij} \otimes z_{ij}\right\| \le \|(a_{ij})\|_{M_n(B(H))},$$

which completes the proof of our claim.

Note that we have established the last assertion in the course of the proof since the matrices $\overline{W}$ and $\overline{W'}$ clearly have the same properties as W and W'.

Exercise 6.1. Consider E_i as embedded into a C^*-algebra B_i and each B_i as embedded into the free product $B = B_1 * \cdots * B_N$. Then, since $B \otimes_{\min} A$ is an operator algebra, its N-fold product map defines a complete contraction

$$P_N\colon (B \otimes_{\min} A) \otimes_h \cdots \otimes_h (B \otimes_{\min} A) \to B \otimes_{\min} A.$$

If we restrict P_N to

$$(E_1 \otimes_{\min} A) \otimes_h \cdots \otimes_h (E_N \otimes_{\min} A),$$

we obtain the first announced result, by Theorem 5.13.

By embedding A into a unital operator algebra, we may assume that A is unital. Then we may restrict the preceding map to $(E_1 \otimes_{\min} A) \otimes_h (E_2 \otimes_{\min} \mathbb{C}I) \otimes_h (E_3 \otimes_{\min} A)$, and this yields the second assertion for the case $N = 3$.

Exercise 7.1. Let $(T_i)_{i \in I}$ be any orthonormal basis of $OH(I)$. For any finitely supported family $(a_i)_{i \in I}$ in (say) $B(\ell_2)$ we have

$$\left\| \sum a_i \otimes T_i \right\|^2 = \left\| \sum a_i \otimes \overline{a_i} \right\| = \left\| \sum \overline{a_i} \otimes a_i \right\|$$
$$= \left\| \sum \overline{a_i} \otimes T_i \right\|^2,$$

and on the other hand

$$= \left\| \left(\sum a_i \otimes \overline{a_i} \right)^* \right\| = \left\| \sum a_i^* \otimes \overline{a_i^*} \right\| = \left\| \sum a_i^* \otimes T_i \right\|^2.$$

Exercise 7.2. Let $(T_i)_{i \in I}$ be orthonormal in $E = OH(I)$. Let

$$\theta_i = \begin{pmatrix} 0 & T_i \\ T_i^* & 0 \end{pmatrix} \in M_2(E).$$

Clearly, if $E \subset B(\mathcal{H})$, then $\theta_i \in M_2(B(\mathcal{H}))$ is self-adjoint. Using the preceding exercise it is easy to show that $\overline{\operatorname{span}}[\theta_i \mid i \in I] \simeq E$ completely isometrically.

Exercise 7.3. If we apply the formula

$$\left\| \sum_{i \in J} a_i \otimes T_i \right\| = \left\| \sum_{i \in J} a_i \otimes \overline{a_i} \right\|^{1/2}$$

with $a_i = \overline{T}_i$, we obtain that the norm $t = \left\| \sum_{i \in J} \overline{T}_i \otimes T_i \right\|$ satisfies $t = t^{1/2}$; hence $t = 0$ or 1, and $t = 0$ is excluded if $|J| \neq 0$.

Exercise 7.4. Let $U \colon E \to E$ be a unitary operator, and let $(T_i)_{i \in I}$ be any orthonormal basis of E. Let $\theta_i = U(T_i)$. Clearly θ_i is orthonormal; hence, by the proof of Theorem 7.1, we have

$$\left\| \sum a_i \otimes \theta_i \right\| = \left\| \sum a_i \otimes T_i \right\|$$

for any finitely supported family (a_i) in $\mathcal{K}$ or $B(\ell_2)$. This shows that $U \colon E \to E$ is a completely isometric isomorphism. In particular, $\|U\|_{cb} = \|U\|$. By the Russo-Dye Theorem, this remains true for any U in $B(E)$, whence $CB(E, E) \simeq B(\ell_2(I))$ isometrically.

However, the latter identity cannot be completely isometric because, if we fix e in the unit sphere of E and consider the subspace of $CB(E, E)$ formed of all the maps of the form $x \to \langle e, x \rangle h$ with $h \in E$, then in $CB(E, E)$ we obtain $E = OH(I)$ but in $B(\ell_2(I))$ we obtain $\ell_2(I)_c$ (column Hilbert space), and it is easy to see that these are different operator spaces.

Exercise 7.5. Let $(T_\alpha)_{\alpha \in I}$ be an orthonormal basis in $E = OH(I)$. Then

$$x = \sum e_{ij} \otimes x_{ij} = \sum_\alpha \left(\sum_{i,j} e_{ij} x_{ij}(\alpha) \right) \otimes T_\alpha$$

and hence

$$\|x\|^2 = \left\| \sum_\alpha \left(\sum_{i,j} e_{ij} x_{ij}(\alpha) \right) \otimes \overline{\left(\sum_{k\ell} e_{k\ell} x_{k\ell}(\alpha) \right)} \right\|$$

$$= \left\| \sum_{ijk\ell} \langle x_{ij}, x_{k\ell} \rangle e_{ij} \otimes \overline{e_{k\ell}} \right\|$$

$$= \left\| \sum_{ijk\ell} \langle x_{ij}, x_{k\ell} \rangle e_{ij} \otimes e_{k\ell} \right\|.$$

The last line is because $\|(\overline{a_{k\ell}})\|_{M_n(\overline{E})} = \|(a_{k\ell})\|_{M_n(E)}$ for any E.

Exercise 7.6. Let (T_k) be an orthonormal basis in OH. Let $h_i = \sum_k h_i(k) T_k$. Then

$$\left\| \sum h_i \otimes x_i \right\|^2 = \left\| \sum_k T_k \otimes \sum_i h_i(k) x_i \right\|^2$$

$$= \left\| \sum_k \left(\sum_i h_i(k) x_i \right) \otimes \overline{\left(\sum_j h_j(k) x_j \right)} \right\|$$

$$= \left\| \sum_{i,j} \left(\sum_k h_i(k) \overline{h_j(k)} \right) x_i \otimes \overline{x_j} \right\|.$$

Exercise 7.7. We have $\|\sum e_{i1} \otimes T_i\| = \|\sum e_{i1} \otimes e_{i1}\|^{1/2}$. But $\sum_1^n e_{i1} \otimes e_{i1}$ obviously has the same norm as $\sum_1^n e_{i1}$; hence $\|\sum_1^n e_{i1} \otimes T_i\| = \|\sum_1^n e_{i1}\|^{1/2} = n^{1/4}$. The other estimate is proved similarly. If we identify C_n with R_n^*, then we find that $\|\sum e_{i1} \otimes T_i\|$ is equal to the c.b. norm of the identity map j from R_n to OH_n; hence this is $= n^{1/4}$. We may then view v as the composition of a unitary operator on R_n (hence a complete isometry) followed by j. This yields $\|v\|_{cb} = n^{1/4}$. The proof for u is similar.

Exercise 7.8. Let $E_\theta = (E, E^{op})_\theta$. Note that (by Theorem 2.7.4) $(E_\theta)^* = (E^*)_\theta$. By Theorem 7.8, there is a factorization

$$T_n\colon\ R_n \xrightarrow{\ w\ } E_\theta \xrightarrow{\ v\ } C_n$$

with $\|v\|_{cb}\|w\|_{cb} = \sqrt{n}$. Since $E(0) \simeq C_n \otimes_{\min} E$ and $E(1) \simeq C_n \otimes_{\min} E^{op}$, we have $E(\theta) \simeq C_n \otimes_{\min} E_\theta \simeq CB(R_n, E_\theta)$. Hence, if (x_i) and (ξ_i) are defined by $x_i = w(e_{1i})$ and $v(\cdot) = \sum \xi_i(\cdot)e_{i1}$, we find $\|w\|_{cb} = \|(x_i)\|_{E(\theta)}$ and $\|v\|_{cb} = \|v^*\|_{cb} = \|(\xi_i)\|_{E^*(\theta)}$. This gives us $\|(x_i)\|_{E(\theta)}\|(\xi_i)\|_{E^*(\theta)} = \sqrt{n}$. In the case $\theta = 1/2$, Corollary 7.12 then yields

$$\left\|\sum x_i \otimes \bar{x}_i\right\|_{\min}^{1/2} \le \|(x_i)\|_{E(1/2)}$$

$$\left\|\sum \xi_i \otimes \bar{\xi}_i\right\|_{\min}^{1/2} \le \|(\xi_i)\|_{E^*(1/2)}.$$

Hence we recover $d_{cb}(E, OH_n) \le \sqrt{n}$ as a by-product.

Exercise 7.9. Let μ be the harmonic measure of the point $z = 1/2$ in the strip $S = \{z \in \mathbb{C} \mid 0 < \mathrm{Re}(z) < 1\}$. Recall that μ is a probability measure on ∂S such that $f(1/2) = \int f\, d\mu$ whenever f is a bounded harmonic function on S extended nontangentially to $\overline{S}$. Obviously μ can be written as $\mu = 2^{-1}(\mu_0 + \mu_1)$, where μ_0 and μ_1 are probability measures supported respectively by

$$\partial_0 = \{z \mid \mathrm{Re}(z) = 0\} \quad \text{and} \quad \partial_1 = \{z \mid \mathrm{Re}(z) = 1\}.$$

Let (A_0, A_1) be a compatible pair of Banach spaces. We first need to describe $(A_0, A_1)_{1/2}$ as a quotient of a subspace of $L_2(\mu_0; A_0) \oplus L_2(\mu_1; A_1)$. The classical argument for this is as follows. Let $\mathcal{F}(A_0, A_1)$ be as defined in §2.7. We start by showing that, for any x in $(A_0, A_1)_{1/2}$, we have

$$\|x\|_{(A_0, A_1)_{1/2}} = \inf\{\max\{\|f\|_{L_2(\mu_0; A_0)}, \|f\|_{L_2(\mu_1; A_1)}\},$$

where the infimum runs over all f in $\mathcal{F}(A_0, A_1)$ such that $f(1/2) = x$. For a proof, see, e.g., [KPS, p. 224]. Then let $E = L_2(\mu_0; A_0) \oplus_\infty L_2(\mu_1, A_1)$ and let $F \subset E$ be the closure of the subspace $\{f_{|\partial_0} \oplus f_{|\partial_1} \mid f \in \mathcal{F}(A_0, A_1)\}$. The preceding equality shows that the mapping $f \to f(1/2)$ defines a metric surjection $Q\colon F \to (A_0, A_1)_{1/2}$. We now consider the couple $(A_0, A_1) = (R, C)$, where we think of R and C as operator space stuctures on the "same" underlying vector space, identified with ℓ_2. We introduce the operator space $E = L_2(\mu_0; \ell_2)_r \oplus L_2(\mu_1; \ell_2)_c$. Let F and $Q\colon F \to \ell_2$ be the same as before, so that if we assume f analytically extended inside S, we have $Q(f) = f(1/2)$.

We first claim that

$$\|Q\colon F \to OH\|_{cb} \le 1.$$

To verify this, consider x in $M_n(F)$ with $\|x\|_{M_n(F)} \leq 1$. We claim that $\|x(1/2)\|_{M_n(OH)} \leq 1$. We may view x as a sequence (x_k) of M_n-valued functions on ∂S extended analytically inside S, so that

$$\|x\|_{M_n(F)}$$
$$= \max\left\{\left\|\left(\int \sum x_k x_k^* \, d\mu_0\right)^{1/2}\right\|_{M_n}, \ \left\|\left(\int \sum x_k^* x_k \, d\mu_1\right)^{1/2}\right\|_{M_n}\right\},$$

and by $(7.3)'$

$$\|x(1/2)\|_{M_n(OH)}^2 = \left\|\sum x_k(1/2) \otimes \overline{x_k(1/2)}\right\|_{\min}$$
$$= \sup\left\{\left|\operatorname{tr}\left(\sum x_k(1/2) a x_k(1/2)^* b\right)\right|\right\},$$

where the supremum runs over all $a, b \geq 0$ in M_n such that $\operatorname{tr}|a|^2 \leq 1$ and $\operatorname{tr}|b|^2 \leq 1$. Fix a, b satisfying these conditions. Consider then the analytic function

$$F(z) = \operatorname{tr}\left(\sum x_k(z) a^{2z} x_k(\bar{z})^* b^{2(1-z)}\right)$$

on S. Note that

$$F(1/2) = \operatorname{tr}\left(\sum x_k(1/2) a x_k(1/2)^* b\right) = 2^{-1}\left(\int_{\partial_0} F \, d\mu_0 + \int_{\partial_1} F \, d\mu_1\right).$$

But for all $z = it$ in ∂_0 we have

$$F(it) = \sum_k \operatorname{tr}(b^{1-it} x_k(it) a^{2it} x_k(-it)^* b^{1-it});$$

hence, by Cauchy–Schwarz, for any z in ∂_0

$$|F(z)| \leq \left(\sum_k \operatorname{tr}(b x_k(z) x_k(z)^* b)\right)^{1/2} \left(\sum_k \operatorname{tr}(b x_k(\bar{z}) x_k(\bar{z})^* b)\right)^{1/2}.$$

A similar verification shows that for any z in ∂_1 we have

$$|F(z)| \leq \left(\sum_k \operatorname{tr}(a x_k(z)^* x_k(z) a)\right)^{1/2} \left(\sum_k \operatorname{tr}(a x_k(\bar{z})^* x_k(\bar{z}) a)\right)^{1/2}.$$

Thus we obtain by Cauchy–Schwarz

$$|F(1/2)| = \left|\int F d\mu\right| \leq 2^{-1}\left(\int_{\partial_0} |F| \, d\mu_0 + \int_{\partial_1} |F| \, d\mu_1\right)$$
$$\leq 2^{-1}\left\{\operatorname{tr}\left(b^2 \int \sum x_k x_k^* \, d\mu_0\right) + \operatorname{tr}\left(a^2 \int \sum x_k^* x_k \, d\mu_1\right)\right\}$$
$$\leq \|x\|_{M_n(F)} \leq 1,$$

which proves our claim.

It is now easy to show that Q is actually a complete metric surjection, or, equivalently, that $I \otimes Q$: $M_n(F) \to M_n(OH)$ is a metric surjection for any $n \geq 1$. Indeed, consider $x \in M_n(OH)$ with $\|x\|_{M_n(OH)} < 1$. Since $M_n(OH) = (M_n(R), M_n(C))_{1/2}$ (isometrically) by Corollary 5.9, there is a bounded continuous analytic function f on $\overline{S}$ with values in $M_n(R) + M_n(C)$ such that $\alpha_0 = \sup\{\|f(z)\|_{M_n(R)} \mid z \in \partial_0\} < 1$, $\alpha_1 = \sup\{\|f(z)\|_{M_n(C)} \mid z \in \partial_1\} < 1$ and $f(1/2) = x$. Let us write $f(z) = (f_k(z))_k$, where f_k is an M_n-valued function on $\overline{S}$. We have trivially

$$\left\| \left(\int \sum f_k(z) f_k(z)^* d\mu_0(z) \right)^{1/2} \right\|_{M_n} \leq \alpha_0 < 1$$

and

$$\left\| \left(\int \sum f_k(z)^* f_k(z) d\mu_1(z) \right)^{1/2} \right\|_{M_n} \leq \alpha_1 < 1;$$

hence $\|f\|_{M_n(F)} < 1$. Since clearly $(I \otimes Q)(f) = x$, this shows that $I \otimes Q$: $M_n(F) \to M_n(OH)$ is a metric surjection. Thus we have completely isometrically $OH \simeq F/\ker(Q)$. Finally, since $E \simeq R \oplus C$ and $F \subset E$, this completes the solution.

Exercise 8.1. Let $G = \mathbb{F}_\infty$. If C is separable, let $\{u_i \mid i \geq 1\}$ be a dense sequence in its unitary group. Let π: $G \to C$ be the unitary representation taking the i-th (free) generator g_i to u_i. By the universal property of $C^*(G)$ (see (8.1)), π extends to a representation $\widehat{\pi}$: $C^*(G) \to C$, taking $U_G(g_i)$ to u_i. Clearly $\widehat{\pi}$ has dense range; hence $\widehat{\pi}$ is onto and $C \simeq C^*(\mathbb{F}_\infty)/I$ with $I = \ker(\widehat{\pi})$. The proof of the nonseparable case is the same.

Exercise 8.2. By Corollary 1.7, any complete contraction v: $E_2 \to B(H_v)$ is the restriction of a complete contraction $\widetilde{v}$: $E_1 \to B(H_v)$. Thus any representation $C^*\langle E_2 \rangle \to B(H)$ is the restriction of a representation $C^*\langle E_1 \rangle \to B(H)$. Similarly, any completely contractive morphism u: $OA(E_2) \to B(H)$ extends completely contractively to $OA(E_1)$. This shows that $C^*\langle E_2 \rangle \to C^*\langle E_1 \rangle$ and $OA(E_2) \to OA(E_1)$ are completely isometric embeddings. The unital case is identical.

Exercise 8.3. The representation π obviously has dense range; hence it is onto, and it is automatically a complete metric surjection (see Proposition 1.5). Since $\|u\|_{cb} \leq 1$, u defines a complete contraction

$$\widetilde{u}: OA(E_1)/\ker(u) \to OA(E_1/E_2).$$

Note that the composition $E_1 \to OA(E_1) \to OA(E_1)/\ker(u)$ is a complete contraction that vanishes on E_2; hence it defines a complete contraction

$E_1/E_2 \to OA(E_1)/\ker(u)$, and hence it extends further to a completely contractive morphism $w \colon OA(E_1/E_2) \to OA(E_1)/\ker(u)$. But then it is easy to check that $\widetilde{u}w$ is the identity, so $w = \widetilde{u}^{-1}$ and $\widetilde{u}$ is a complete isometry. The unital case is identical.

Exercise 8.4. Let $\lambda = \lambda_G$ and $U = U_G$.

(i) $\Rightarrow$ (ii): Since $U \otimes \lambda \simeq 1 \otimes \lambda$ (by Proposition 8.1) for any x in $\mathbb{C}[G]$, we have

$$\left\| \sum x(t) U(t) \otimes \lambda(t) \right\| = \left\| \sum x(t) \lambda(t) \right\|.$$

Let $T = \sum x(t) U(t) \otimes \lambda(t)$. Assuming $U(t) \in B(H)$, pick h, k in the unit ball of H. We then have

$$|\langle T(h \otimes f_\alpha), k \otimes f_\alpha \rangle| \le \|T\|;$$

hence

$$\left| \sum x(t) \langle U(t) h, k \rangle \langle \delta_t * f_\alpha, f_\alpha \rangle \right| \le \|T\|,$$

and since $\langle \delta_t * f_\alpha, f_\alpha \rangle \to 1$ for any t, we obtain $|\sum x(t) \langle U(t) h, k \rangle| \le \|T\|$ and hence $\|\sum x(t) U(t)\| \le \|T\| = \|\sum x(t) \lambda(t)\|$. This establishes (i) $\Rightarrow$ (ii) since the converse inequality is trivial.

(ii) $\Rightarrow$ (iii): Let $E \subset G$ be any finite subset. If (ii) holds, since the trivial representation (identically equal to 1 on G) is included in the universal one, we have

$$|E| = \left\| \sum_{t \in E} U(t) \right\| = \left\| \sum_{t \in E} \lambda(t) \right\|,$$

and hence (iii) holds.

(iii) $\Rightarrow$ (i): Let $E \subset S$ be a finite subset containing the unit. Assuming (iii), there is a net (y_α) in the unit sphere of $\ell_2(G)$ such that $\||E|^{-1} \sum_{t \in E} \lambda(t) \cdot y_\alpha\|_2 \to 1$. By the uniform convexity of $\ell_2(G)$ (see Exercise 20.2 and recall $e \in E$) this implies $\|\lambda(t) y_\alpha - y_\alpha\|_2 \to 0$ for any t in E. Since this holds for any finite subset E, we can rearrange the resulting nets to make sure that the last convergence holds for any t in S. Then, since S generates G, the same convergence holds for any t in G. This shows (using the proposed alternate definition of amenability) that (iii) implies (i).

Exercise 11.1. Let $\rho \colon B \otimes_{\max} A \to (B/\mathcal{I}) \otimes_{\max} A$ denote the natural representation (obtained from $B \to B/\mathcal{I}$ after tensoring with the identity of A). Obviously, ρ vanishes on $\mathcal{I} \otimes_{\max} A$. Hence, denoting by $Q \colon B \otimes_{\max} A \to (B \otimes_{\max} A)/(\mathcal{I} \otimes_{\max} A)$ the quotient map, we have a factorization of ρ of the form $\rho = \widehat{\rho} Q$

$$B \otimes_{\max} A \xrightarrow{Q} (B \otimes_{\max} A)/(\mathcal{I} \otimes_{\max} A) \xrightarrow{\widehat{\rho}} (B/\mathcal{I}) \otimes_{\max} A,$$

where $\widehat{\rho}$ is a (contractive) representation.

Consider any x in $(B/\mathcal{I}) \otimes A$ of the form $x = \sum \beta_i \otimes a_i$. Let $b_i \in B$ be any lifting of β_i. We define $\pi(x) = Q \left(\sum b_i \otimes a_i \right)$. Clearly this definition is unambiguous (it depends only on x and not on the choice of representations). Moreover, π is a $*$-homomorphism from $(B/\mathcal{I}) \otimes A$ to the quotient C^*-algebra $(B \otimes_{\max} A)/(\mathcal{I} \otimes_{\max} A)$ (which can be realized inside some $B(H)$). Hence, by the definition of the max norm on $(B/\mathcal{I}) \otimes A$, we must have $\|\pi(x)\| \leq \|x\|_{\max}$. Moreover, $\widehat{\rho}(\pi(x)) = x$ for any x in $B/\mathcal{I} \otimes A$, but, since (as we just saw) π is continuous, this remains true for all x in $B/\mathcal{I} \otimes_{\max} A$. Therefore π must be injective, and, since it obviously has dense range, it defines an isomorphism from $(B/\mathcal{I}) \otimes_{\max} A$ to $(B \otimes_{\max} A)/(\mathcal{I} \otimes_{\max} A)$. Then the identity $\widehat{\rho}(\pi(x)) = x$ implies that $\widehat{\rho}$ is the inverse of π, in particular, $\widehat{\rho}$ must be injective, and therefore the kernel of Q coincides with the kernel of $\widehat{\rho}Q = \rho$, so the sequence is exact.

To prove that last assertion, observe that the isometric isomorphism

$$\pi \colon B/\mathcal{I} \otimes_{\max} A \to (B \otimes_{\max} A)/(\mathcal{I} \otimes_{\max} A)$$

induces a linear isomorphism between $(B/\mathcal{I}) \otimes A$ and the image of $B \otimes A$ in $(B \otimes_{\max} A)/(\mathcal{I} \otimes_{\max} A)$. Thus, for any x in $(B/\mathcal{I}) \otimes A$ with $\|x\|_{\max} < 1$, we can find $\widetilde{x}$ in $B \otimes A$ with $\|\widetilde{x}\|_{\max} < 1$ such that $x - \widetilde{x} \in \mathcal{I} \otimes_{\max} A$, or, equivalently, $x - \widetilde{x} \in \mathcal{I} \otimes A$.

Exercise 11.2. This follows from the preceding exercise.

Exercise 11.3. The completion of $A_1 \otimes A_2$ for the given norm is a C^*-algebra. Clearly $a_1 \geq 0, a_2 \geq 0$ imply $a_1 \otimes a_2 \geq 0$; hence $a_1 \geq 0, b_1 - a_1 \geq 0$, and $a_2 \geq 0$ imply $0 \leq a_1 \otimes a_2 \leq b_1 \otimes a_2$ and hence $\|a_1 \otimes a_2\| \leq \|b_1 \otimes a_2\|$. Then $a_1^* x^* x a_1 \leq \|x\|^2 a_1^* a_1$ gives us $\|a_1^* x^* x a_1 \otimes a_2^* a_2\| \leq \|x\|^2 \|a_1^* a_1 \otimes a_2^* a_2\| = \|x\|^2 \|a_1 \otimes a_2\|^2$, and hence $\|x a_1 \otimes a_2\| \leq \|x\| \, \|a_1 \otimes a_2\|$. Applying this twice we obtain $\|x a_1 \otimes y a_2\| \leq \|x\| \, \|y\| \|a_1 \otimes a_2\|$; hence $\|a_1 \otimes a_2\|^2 = \|a_1^* a_1 \otimes a_2^* a_2\| \leq \|a_1^*\| \, \|a_2^*\| \|a_1 \otimes a_2\|$, which, after division, yields $\|a_1 \otimes a_2\| \leq \|a_1\| \, \|a_2\|$.

Exercise 11.4.

(i) Since $\pi(A_1 \otimes A_2)$ is a $*$-algebra, both K and $K^\perp$ are invariant subspaces and $\pi(x)_{|K^\perp} = 0$ for any x in $A_1 \otimes A_2$, whence the announced decomposition, and it clearly suffices to solve the problem for $\pi(\cdot)_{|K}$.

(ii) Since we know by the preceding exercise that $\pi(x_\alpha \otimes a_2)$ and $\pi(a_1 \otimes y_\beta)$ are uniformly bounded nets, it suffices to show that they converge on a total subset of $H = K$, in particular on $\{\pi(A_1 \otimes A_2)H\}$, on which this convergence becomes obvious. We have

$$\pi_1(a_1)[\pi(x \otimes y)h] = \pi(a_1 x \otimes y)h$$

and

$$\pi_2(a_2)[\pi(x \otimes y)h] = \pi(x \otimes a_2 y)h,$$

which shows that π_1 and π_2 actually are *independent* of the choice of $\{x_\alpha\}$ and $\{y_\beta\}$.

(iii) From the preceding two identities it is immediate that π_1 and π_2 are $*$-homomorphisms with commuting ranges.

Exercise 11.5 (i) Assume the given matrix positive. For any real t we have

$$\left\langle \begin{pmatrix} p & a \\ a^* & q \end{pmatrix} \begin{pmatrix} ty \\ x \end{pmatrix}, \begin{pmatrix} ty \\ x \end{pmatrix} \right\rangle \geq 0;$$

hence $P(t) = t^2 \langle py, y \rangle + 2t \langle ax, y \rangle + \langle qx, x \rangle \geq 0$, so the discriminant of P must be ≤ 0, and we obtain $|\langle ax, y \rangle|^2 \leq \langle py, y \rangle \langle qx, x \rangle$. Conversely, if this holds, P must keep the same sign, and it is ≥ 0 for large t; hence we have $P(1) \geq 0$ and the given matrix is positive.

(ii) is an obvious consequence of (i).

(iii) Since u is c.p., the map $u_2 = I_{M_2} \otimes u \colon M_2(A) \to M_2(B)$ is positive. If A is unital, we have

$$\|a\| \leq 1 \Rightarrow \begin{pmatrix} 1 & a \\ a^* & 1 \end{pmatrix} \geq 0 \Rightarrow \begin{pmatrix} u(1) & u(a) \\ u(a^*) & u(1) \end{pmatrix} \geq 0.$$

Hence, since $u(a^*) = u(a)^*$, (ii) shows that $|\langle u(a)x, y \rangle|^2 \leq \langle u(1)x, x \rangle \langle u(1)y, y \rangle$ and hence $\|u(a)\|^2 \leq \|u(1)\|^2$. Thus we obtain $\|u\| = \|u(1)\|$. Similarly, $\|I_{M_n} \otimes u\| = \|I_{M_n} \otimes u(1)\| \leq \|u\|$, and hence $\|u\|_{cb} = \|u\| = \|u(1)\|$. Let (a_α) be as in (ii). Let $\widetilde{A}$ be the unitization of A and let $u_\alpha \colon \widetilde{A} \to B$ be the map defined by $u_\alpha(a) = u(a_\alpha^{1/2} a a_\alpha^{1/2})$. Then clearly u_α is c.p. and $u_\alpha(a) \to u(a)$ for any a in A. Hence $\|u\|_{cb} \leq \sup_\alpha \|u_\alpha\|_{cb} = \sup_\alpha \|u_\alpha(1)\| = \sup_\alpha \|u(a_\alpha)\|$. Thus we conclude $\|u\|_{cb} = \|u\|$.

Exercise 11.6. For (i), it clearly suffices to show that, if $\sigma_1 \colon A_1 \to B(H)$ and $\sigma_2 \colon A_2 \to B(H)$ are representations with commuting ranges, there is an extension $\widehat{\sigma}_1 \colon A_1^{**} \to B(H)$ such that $\widehat{\sigma}_1$ and σ_2 still have commuting ranges. But this is clear: By the universal property of the bidual (as the universal enveloping von Neumann algebra), since $\sigma_2(A_2)'$ is a von Neumann algebra, any representation $\sigma_1 \colon A_1 \to \sigma_2(A_2)'$ extends to a (normal) representation $\widehat{\sigma}_1 \colon A_1^{**} \to \pi_2(A_2)'$.

(ii) follows from (i) by iteration.

(iii) Assume A_1, A_2 unital for simplicity. Let $\pi_i \colon A_i \to (A_1 \otimes_{\max} A_2)^{**}$, $(i = 1, 2)$ be the obvious inclusions $(a_1 \to a_1 \otimes 1, a_2 \to 1 \otimes a_2)$, and let $\widetilde{\pi}_i \colon A_i^{**} \to (A_1 \otimes_{\max} A_2)^{**}$ be their canonical extensions as normal representations. Obviously (π_1, π_2) and therefore $(\widetilde{\pi}_1, \widetilde{\pi}_2)$ have commuting ranges; hence, if we let $\pi = \widetilde{\pi}_1 \cdot \widetilde{\pi}_2$, we obtain a (contractive) representation

$$\pi \colon A_1^{**} \otimes_{\max} A_2^{**} \to (A_1 \otimes_{\max} A_2)^{**},$$

which clearly coincides with j when restricted to $A_1 \otimes A_2$.

Exercise 11.7. (i) Consider $t \in A \otimes B$, where B is an arbitrary C^*-algebra. We need to show $\|t\|_{\max} \leq \|t\|_{\min}$. By density, it clearly suffices to show this for any t in $(\bigcup_\alpha A_\alpha) \otimes B$. In other words, we may assume $t \in A_\alpha \otimes B$ for some α. But then, since A_α is nuclear, we have $\|t\|_{A \otimes_{\max} B} \leq \|t\|_{A_\alpha \otimes_{\max} B} = \|t\|_{\min}$.

(ii) Let α run over the set of all the finite σ-subalgebras of $\mathcal{A}$. Let $A_\alpha = L_\infty(\Omega, \alpha, \mu)$ and $A = L_\infty(\Omega, \mathcal{A}, \mu)$. Since A_α is finite-dimensional, it is nuclear; hence (i) shows that A is nuclear. By Gelfand's Theorem, any commutative von Neumann algebra can be identified with $L_\infty(\Omega, \mathcal{A}, \mu)$ for some $(\Omega, \mathcal{A}, \mu)$.

(iii) If A is commutative, A^{**} is a commutative von Neumann algebra; hence it is nuclear by (ii). Hence $A^{**} \otimes_{\min} B = A^{**} \otimes_{\max} B$ for any B. But since $A \otimes_{\max} B \subset A^{**} \otimes_{\max} B$ (isometrically) by the preceding exercise, we have $A \otimes_{\max} B = A \otimes_{\min} B$ and A is nuclear.

Exercise 11.8. Let $\lambda = \lambda_G$ and $\rho = \rho_G$. For any finite subset $E \subset G$, we clearly have $\lambda(t)\rho(t)\delta_e = \delta_e$ for any t; hence $\sum_{t \in E} \lambda(t)\rho(t)\delta_e = |E|\delta_e$, and therefore $|E|$ is

$$\leq \left\| \sum_{t \in E} \lambda(t)\rho(t) \right\| \leq \left\| \sum_{t \in E} \lambda(t) \otimes \rho(t) \right\|_{C_\lambda^*(G) \otimes_{\max} C_\rho^*(G)}$$

$$\leq \left\| \sum_{t \in E} \lambda(t) \otimes U(t) \right\|_{C_\lambda^*(G) \otimes_{\max} C^*(G)}.$$

Hence, if $\| \cdot \|_{\min} = \| \cdot \|_{\max}$ either on $C_\lambda^*(G) \otimes C_\rho^*(G)$ or on $C_\lambda^*(G) \otimes C^*(G)$, we obtain

$$|E| \leq \left\| \sum_{t \in E} \lambda(t) \otimes U(t) \right\|_{\min},$$

and therefore, by Proposition 8.1, $|E| = \left\| \sum_{t \in E} \lambda(t) \right\|$, so that G is amenable by Exercise 8.4. Since $C_\rho^*(G) \simeq C_\lambda^*(G)$, this completes the solution.

Exercise 12.1. Clearly u is c.p. when $a = b$. Otherwise, let $v \colon A \to M_2(A)$ be the mapping defined by

$$v(x) = \begin{pmatrix} bxb^* & bxa^* \\ axb^* & axa^* \end{pmatrix}.$$

An elementary verification shows that

$$v(x) = t \begin{pmatrix} x & 0 \\ 0 & x \end{pmatrix} t^*,$$

where

$$t = 2^{-1/2} \begin{pmatrix} b & b \\ a & a \end{pmatrix}.$$

Clearly this shows that v is c.p.; hence by definition of the dec-norm (before (11.2)) we have

$$\|u\|_{\text{dec}} \leq \max\{\|v_{11}\|, \|v_{22}\|\},$$

where $v_{11}(x) = bxb^*$ and $v_{22}(x) = axa^*$. Hence we obtain

$$\|u\|_{\text{dec}} \leq \max\{\|a\|^2, \|b\|^2\}.$$

Applying this to the mapping $x \to u(x)\|a\|^{-1}\|b\|^{-1}$ we find

$$\|u\|_{\text{dec}} \leq \|a\| \, \|b\|.$$

Exercise 12.2. Let $a = (a_1, a_2, \ldots, a_n), b = (b_1, b_2, \ldots, b_n)$ be viewed as one-row matrices with entries in A. Then, for any x in $M_n, u(x)$ can be written as a matrix product:

$$u(x) = axb^*.$$

Thus it is clear that u is c.p. if $a = b$. When a_i, b_j are arbitrary, we may (again) introduce the mapping $v \colon M_n \to M_2(A)$ defined by

$$v(x) = \begin{pmatrix} bxb^* & bxa^* \\ axb^* & axa^* \end{pmatrix}.$$

Again we note

$$v(x) = t \begin{pmatrix} x & 0 \\ 0 & x \end{pmatrix} t^*,$$

where $t = 2^{1/2} \begin{pmatrix} b & b \\ a & a \end{pmatrix} \in M_{2,2n}(A)$, which shows (see Corollary 1.8) that v is c.p., so we obtain again

$$\|u\|_{\text{dec}} \leq \max\{\|v_{11}\|, \|v_{22}\|\} \leq \max\{\|b\|^2, \|a\|^2\}$$
$$= \max\left\{ \left\|\sum b_j b_j^*\right\|, \left\|\sum a_i a_i^*\right\| \right\},$$

and, by homogeneity, this yields

$$\|u\|_{\text{dec}} \leq \left\|\sum b_j b_j^*\right\|^{1/2} \left\|\sum a_i a_i^*\right\|^{1/2}.$$

Exercise 13.1. Since we may restrict ourselves to families of finite-dimensional unitaries in (8.8), the restriction of $\tilde{\sigma}$ to the span of I and $\{U(g_i) \mid i \in I\}$ is completely isometric. Therefore, $(\tilde{\sigma})_{|E}^{-1}$ is completely contractive, which implies, by Proposition 13.6, that $\|(\tilde{\sigma})^{-1}\|_{cb} \leq 1$.

Exercise 15.1. By Theorem 15.5, $A^{**} \otimes_{\min} C = A^{**} \otimes_{\max} C$ iff A^{**} has the WEP. Since A^{**} is a von Neumann algebra, this holds iff it is injective and hence, by the Connes–Choi–Effros Theorem (Theorem 11.6), iff A is nuclear.

Exercise 15.2. It is obvious that any nuclear C^*-algebra has this property by the "injectivity" of the minimal tensor product. Conversely, assume that A has the property under consideration. We will show that A^{**} is injective. We may assume A unital (see Remark 12.13). Let $\sigma\colon A \to B(H)$ be a unital embedding such that $A^{**} \simeq \sigma(A)''$. We will apply the property under consideration to $B_1 = \sigma(A)'$ and $B_2 = B(H)$. Let $\pi\colon A \otimes_{\max} B_1 \to B(H)$ be the unital C^*-representation defined by $\pi(a \otimes b) = \sigma(a)b$. Since $A \otimes_{\max} B_1 \subset A \otimes_{\max} B(H)$, by Corollaries 1.7 and 1.8, π admits a completely positive and completely contractive extension $T\colon A \otimes_{\max} B(H) \to B(H)$. Let us define

$$\forall\, x \in B(H) \qquad P(x) = T(1 \otimes x).$$

Note that P is the identity when restricted to B_1. We claim that P (which is clearly completely contractive) is a projection from $B(H)$ onto B_1. Indeed, by Lemma 14.3, T must be bimodular with respect to $A \otimes B_1$, so that for any a in A we have

$$T(1 \otimes x)\sigma(a) = T(a \otimes x) = \sigma(a)T(1 \otimes x);$$

hence $\sigma(a)P(x) = P(x)\sigma(a)$, so that $P(x) \in \sigma(A)' = B_1$. This proves our claim.

Thus we conclude that $\sigma(A)'$ is injective and, by the "well-known fact" $\sigma(A)'' = A^{**}$, must also be injective.

Exercise 15.3. Let B be a WEP C^*-algebra with an ideal $I \subset B$ so that $B/I = \mathcal{M}$. Let $q\colon B \to B/I$ be the quotient map, and let $B_1 = q^{-1}(M) \subset B$. Since $M = B_1/I$, it clearly suffices to show that B_1 is WEP, or, equivalently (by Theorem 15.5), it suffices to show that, if $C = C^*(\mathbb{F}_\infty)$, we have $C \otimes_{\max} B_1 = C \otimes_{\min} B_1$. For that purpose, consider t in $C \otimes B_1$ with $\|t\|_{\min} < 1$. Since B is WEP (by Theorem 15.5), we have $\|t\|_{C \otimes_{\max} B} < 1$, hence $\|(I \otimes q)(t)\|_{C \otimes_{\max} \mathcal{M}} < 1$, and finally, by Proposition 11.8, $\|(I \otimes q)(t)\|_{C \otimes_{\max} M} \leq 1$. By Exercise 11.1, since $M = B_1/I$, it follows that there is θ in $C \otimes B_1$ with $\|\theta\|_{\max} < 1$ and such that $(I \otimes q)(\theta) = (I \otimes q)(t)$. Therefore we have $t = \theta + x$ with $x \in C \otimes I$. But now by the triangle inequality

$$\|x\|_{\min} \leq \|t\|_{\min} + \|\theta\|_{\min} < 2,$$

and, by Corollary 11.9, we have $C \otimes_{\min} I = C \otimes_{\max} I$; hence $\|x\|_{\max} = \|x\|_{\min} < 2$.

Finally, again by the triangle inequality, we conclude $\|t\|_{\max} \leq \|\theta\|_{\max} + \|x\|_{\max} < 3$. Thus we proved that the norms of $C \otimes_{\max} B_1$ and $C \otimes_{\min} B_1$ are equivalent on $C \otimes B_1$; since these are C^*-norms, they coincide.

Exercise 15.4. By construction, $B = \ell_\infty(\{M(n) \mid n \geq 1\})$ is WEP if the algebras $M(n)$ are all WEP. Hence $\mathcal{M} = B/I_{\mathcal{U}}$ is QWEP. Since $\mathcal{M}$ has a finite faithful trace, there is a completely positive contractive conditional expectation P from $\mathcal{M}$ onto M. Thus, by the preceding exercise M is QWEP.

Exercise 15.5. We first claim that for any finite subset $E \subset G$ we always have (actually this follows from Theorem 8.2)

$$|E| = \left\| \sum_{t \in E} \lambda(t) \otimes \overline{\lambda(t)} \right\|_{C_\lambda^*(G) \otimes_{\max} \overline{C_\lambda^*(G)}}.$$

Note that, for any x in $\mathbb{C}(G)$, we have

$$\left\| \sum x(t)\overline{\lambda(t)} \right\|_{\overline{C_\lambda^*(G)}} = \left\| \sum x(t)\lambda(t)^* \right\|_{B(\ell_2(G))} = \left\| \sum x(t)\rho(t) \right\|_{B(\ell_2(G))}.$$

Hence the correspondence $\overline{\lambda(t)} \to \rho(t)$ linearly extends to an isometric representation from $\overline{C_\lambda^*(G)}$ to $C_\rho^*(G)$. Therefore, for any finite subset $E \subset G$, we have

$$\left\| \sum_{t \in E} \lambda(t) \otimes \overline{\lambda(t)} \right\|_{\max} = \left\| \sum_{t \in E} \lambda(t) \otimes \rho(t) \right\|_{\max}.$$

But now (see the solution to Exercise 11.8)

$$\left\| \sum_{t \in E} \lambda(t) \otimes \rho(t) \right\|_{\max} \geq \left\| \sum_{t \in E} \lambda(t)\rho(t) \right\|_{B(\ell_2(G))} \geq |E|.$$

Thus we obtain $\left\| \sum_{t \in E} \lambda(t) \otimes \overline{\lambda(t)} \right\|_{\max} \geq |E|$, which proves our claim (since the inverse inequality is trivial). Similarly, if we let $M = VN(G)$, we have (by the same reasoning) $|E| = \left\| \sum_{t \in E} \lambda(t) \otimes \overline{\lambda(t)} \right\|_{M \otimes_{\max} \overline{M}}$. Therefore, by Theorem 15.6, if either $C_\lambda^*(G)$ or M is WEP, we also have

$$|E| = \left\| \sum_{t \in E} \lambda(t) \otimes \overline{\lambda(t)} \right\|_{\min},$$

and by Proposition 8.1 this implies $|E| = \left\| \sum_{t \in E} \lambda(t) \right\|$. Then, by Exercise 8.4, we conclude that G is amenable.

Exercise 17.1. Let $A \subset B(H)$ be a C^*-algebra. We already know (by Theorem 17.1 and Theorem 15.5) that, if A is nuclear, then it is exact and WEP. Conversely, by Proposition 15.3, if A has the WEP, then, for any C^*-algebra C, the natural morphism

$$A \otimes_{\max} C \longrightarrow B(H) \otimes_{\max} C$$

is isometric. On the other hand, by Theorem 17.1, if A is exact, the same mapping is a contraction from $A \otimes_{\min} C$ to $B(H) \otimes_{\max} C$. Therefore, for any t in $A \otimes C$ we must have $\|t\|_{\max} \leq \|t\|_{\min}$, so that we conclude $A \otimes_{\min} C = A \otimes_{\max} C$ for any C, and hence A is nuclear.

Exercise 18.1. Let X be any operator space and let $E \subset X$ be a subspace. We will show that any complete contraction $u \colon E \to A^{**}$ admits a completely contractive (c.c. in short) extension $\widetilde{u} \colon X \to A^{**}$. It clearly suffices to show this when E is finite-dimensional. In that case, by the local reflexivity of A, there is a net of c.c. maps $u_\alpha \colon E \to A$ tending to u with respect to $\sigma(A^{**}, A^*)$. Let $i_A \colon A \to A^{**}$ denote the canonical inclusion. Since A has the WEP, $i_A u_\alpha$ admits a c.c. extension $v_\alpha \colon X \to A^{**}$. Let $\mathcal{U}$ be any ultrafilter refining our net. We let $\widetilde{u}(x) = \lim_{\mathcal{U}} v_\alpha(x)$ (with respect to $\sigma(A^{**}, A^*)$). Then $\widetilde{u} \colon X \to A^{**}$ is a c.c. map, and for any e in E we have

$$\widetilde{u}(e) = \lim v_\alpha(e) = \lim u_\alpha(e) = u(e).$$

Thus $\widetilde{u}$ is the desired extension.

Exercise 18.2. Fix $0 < \varepsilon < (1 + \lambda)^{-1}$. Let $\{x_j \mid 1 \leq j \leq N\}$ be an ε-net in the unit sphere of E, and let $\xi_j \in E^*$ be such that $1 = \xi_j(x_j) = \|\xi_j\|$ $(1 \leq j \leq N)$. Let α be large enough so that

$$1 - \varepsilon < |\langle \xi_j, u_\alpha(x_j) \rangle| \quad (1 \leq j \leq N).$$

We claim that

$$1 - (1 + \lambda)\varepsilon < \|u_\alpha^{-1}{}_{|u_\alpha(E)}\|.$$

Indeed, for any x in the unit sphere of E, pick j such that $\|x - x_j\| < \varepsilon$. We then have

$$\|u_\alpha(x)\| \geq \|u_\alpha(x_j)\| - \lambda\|x - x_j\| \geq |\langle \xi_j, u_\alpha(x_j) \rangle| - \lambda\varepsilon > (1 - (1 + \lambda)\varepsilon);$$

hence, by homogeneity,

$$\forall x \in E \qquad \|u_\alpha(x)\| \geq (1 - (1 + \lambda)\varepsilon)\|x\|,$$

which proves our claim.

Exercise 18.3. By construction we have $\widetilde{V}(gxh) = g\widetilde{V}(x)h$ for all g, h in G, from which it is easy to see that $\widetilde{V}$ has the required form. We then have $\|u\|_n = \|I \otimes u \colon M_n(E) \to M_n(X)\| = \|\widetilde{V}\|$, and, by the convexity of the norm, denoting $V_{g,h}(x) = V(gxh)$ we have $\|\widetilde{V}\| \leq \sup\{\|V_{g,h}\| \mid g, h \in G\} = \|V\|$.

Exercise 18.4. Consider $u \colon E \to (X/Y)^{**}$ with $\dim(E) < \infty$ and $\|u\|_{cb} \leq 1$. Let $v = ru \colon E \to X^{**}$. Since X is λ-locally reflexive, there is a net $v_\alpha \colon E \to X$

suitably approximating v with $\|v_\alpha\|_{cb} \le \lambda$. If we let $u_\alpha = qv_\alpha\colon E \to X/Y$, we obtain a similar approximating net for u, showing that X/Y is λ-locally reflexive. Now, if X is a C^*-algebra and Y is a closed two-sided ideal, we have $X^{**} \simeq Y^{**} \oplus (X/Y)^{**}$, and hence our assumption is valid in that case.

Exercise 18.5. Since $C^*(\mathbb{F}_\infty)$ embeds in $C^*(\mathbb{F}_2)$, it suffices to show that the latter is not locally reflexive. But if it were, then, by Exercises 18.4 and 8.1, every separable C^*-algebra would be locally reflexive. By Proposition 18.5, every C^*-algebra would be locally reflexive, which is absurd (for instance, $B(\ell_2)$ is not locally reflexive because, by Exercise 18.1, its bidual would then be injective).

Exercise 18.6. By (18.2), for any C^*-algebra B and for any t in $X^{**} \otimes B^{**}$ we have

$$\|J(t)\|_{(X \otimes_{\min} B)^{**}} \ge \|t\|_{X^{**} \otimes_{\min} B^{**}}.$$

So we only need consider the reverse inequality. Assume $\|t\|_{X^{**} \otimes_{\min} B^{**}} \le 1$. By property C'', there is a net (t_α) in $X \otimes B^{**}$ with $\|t_\alpha\|_{\min} \le \lambda''$ that w^*-tends to t (here w^* refers to the $\sigma((X \otimes_{\min} B)^{**}, (X \otimes_{\min} B)^*)$ topology). By property C', each t_α can be w^*-approximated by a net $(t_{\alpha,\beta})$ in $X \otimes B$ with $\|t_{\alpha,\beta}\|_{\min} \le \lambda''\lambda'$. Thus we conclude $\|t\|_{(X \otimes_{\min} B)^{**}} \le \lambda'\lambda''$.

Exercise 18.7. Consider E (finite-dimensional) and $u\colon E \to A^{**}$ with $\|u\|_{cb} \le 1$. We need to approximate u (pointwise in the $\sigma(A^{**}, A^*)$-sense) by completely contractive maps from E to A. Replacing u by $v_\alpha u\colon M_{n(\alpha)} \to A^{**}$, we see that we may as well assume $E = M_n$ for some n. But then, if A^{**} is injective and $E = M_n$, by (11.5) and (12.7) we have $E^* \otimes_\delta A^{**} = E^* \otimes_{\min} A^{**}$ (isometrically). But now, by (12.5) and Exercise 11.6(iii), we find $\|E^* \otimes_\delta A^{**} \to (E^* \otimes_\delta A)^{**}\| \le 1$; hence a fortiori $\|E^* \otimes_{\min} A^{**} \to (E^* \otimes_{\min} A)^{**}\| \le 1$, which shows that A is locally reflexive.

Exercise 18.8. Recall $A^{**} \simeq (A/I)^{**} \oplus I^{**}$. Let B be any C^*-algebra. We first observe that

$$(A^{**} \otimes_{\min} B^{**})/(I^{**} \otimes_{\min} B^{**}) = (A/I)^{**} \otimes_{\min} B^{**}.$$

Since A has (C), we have

$$\|A^{**} \otimes_{\min} B^{**} \longrightarrow (A \otimes_{\min} B)^{**}\| \le 1;$$

hence a fortiori

$$\|A^{**} \otimes_{\min} B^{**} \longrightarrow ((A/I) \otimes_{\min} B)^{**}\| \le 1.$$

But this last mapping clearly vanishes on $I^{**} \otimes B^{**}$; hence, passing to the quotient, we find

$$\|(A^{**} \otimes_{\min} B^{**})/(I^{**} \otimes_{\min} B^{**}) \to ((A/I) \otimes_{\min} B)^{**}\| \le 1,$$

but then, by our initial observation, this means that A/I has property (C).

Exercise 18.9. Let $i_X\colon X \to X^{**}$ be the canonical inclusion. Then $(i_{\mathcal{K}})^{**}\colon$ $\mathcal{K}^{**} \to (\mathcal{K}^{**})^{**}$ is *not* equal to $i_{\mathcal{K}^{**}}$. In short $(i_{\mathcal{K}})^{**} \neq i_{\mathcal{K}^{**}}$! Let $B = \mathcal{K}^{**}$. To show that A is locally reflexive we need to show that $\|i_B \otimes I\colon B \otimes_{\min} A^{**} \to (B \otimes_{\min} A)^{**}\| \leq 1$, but the "argument" only proves that $\|(i_{\mathcal{K}})^{**} \otimes I\colon B \otimes_{\min} A^{**} \to (B \otimes_{\min} A)^{**}\| \leq 1$.

Exercise 19.1. Consider $s = \sum a_i \otimes x_i$ in $E \otimes B_1$ and $t = \sum b_j \otimes y_j$ in $F \otimes B_2$ with $\|s\|_{\min} < 1$ and $\|t\|_{\min} < 1$. We have by (2.3.5)

$$\left\| \sum u(a_i) \otimes x_i \right\|_{CB(F, G \otimes_{\min} B_1)} \leq \|u\|_{cb}.$$

Let $v = \sum u(a_i) \otimes x_i\colon F \to G \otimes_{\min} B_1$. We have

$$\|(v \otimes I_{B_2})(t)\|_{\min} \leq \|v\|_{cb}\|t\|_{\min} \leq \|v\|_{cb} \leq \|u\|_{cb}.$$

But note that

$$(v \otimes I_{B_2})(t) = \widehat{u}(s, t);$$

hence we obtain $\|\widehat{u}(s, t)\|_{\min} \leq \|u\|_{cb}$.

Exercise 19.2. We will apply (19.1) with $a_i = b_i = e_i$ (here $\{e_i\}$ is an orthonormal basis in ℓ_2^n) and with u equal to the identity on $\max(\ell_2^n)$. Since $d_{S\mathcal{K}}(\min(\ell_2^n)) = 1$, this yields (using (10.23)) $n \leq 4 \ d_{S\mathcal{K}}(\max(\ell_2^n))\sqrt{n}$.

Exercise 20.1. By (7.2) we have

$$\left\| \sum u_i \otimes \lambda(g_i) \right\| \leq \left\| \sum u_i \otimes \overline{u_i} \right\|^{1/2} \left\| \sum \lambda(g_i) \otimes \overline{\lambda(g_i)} \right\|^{1/2};$$

hence, by Fell's absorption principle (Prop.10.1),

$$a(n) \leq \left\| \sum u_i \otimes \overline{u_i} \right\|^{1/2} a(n)^{1/2},$$

which after squaring and dividing gives the desired inequality.

Exercise 20.2. We have $\frac{1}{n}\sum_1^n \|x_k - M(x)\|^2 = \frac{1}{n}\sum_1^n \|x_k\|^2 - \|M(x)\|^2$, whence the first inequality. Therefore $\|M(x)\|^2 > 1 - \varepsilon$ implies $\Delta(x) \leq 2\sqrt{n\varepsilon}$, and the last assertion follows.

Exercise 21.1. Let $p(x) = \lim_{m \to \infty} p_m(x)$. Let (e_i) be a basis of E. By our assumption, there is a constant C such that, for all m and all $x = \sum x_i e_i \in E$, we have $p_m(x) \leq C\sum|x_i| = C\|x\|_1$. Let $\Omega = \{x \in E \mid \|x\|_1 = \sum|x_i| \leq 1\}$. Note that $|p_m(x) - p_m(y)| \leq p_m(x - y) \leq C\|x - y\|_1$ and $p_m(0) = 0$; hence, by Ascoli's Theorem, the sequence $\{p_m\}$ is relatively compact in $C(\Omega)$. Hence p_m must converge uniformly on Ω, and, by homogeneity, p_m also converges uniformly over any bounded subset of E.

Exercise 21.2. Assume (i). Let (e_i) be any basis in E. Let $u_m \colon E \to E_m$ be an isomorphism such that $\|u_m\| = d(E_m, E)$ and $\|u_m^{-1}\| = 1$. Let $e_i(m) = u_m(e_i)$. Then (ii) is easy to check. Thus (i) $\Rightarrow$ (ii). (ii)$\Rightarrow$(ii)$'$ is trivial and (ii)$' \Rightarrow$(iii) is obvious. Assume (iii). Again let (e_i) be any basis in E. Viewing e_i as an element of $\Pi E_m / \mathcal{U}$, we may select a representative of its equivalent class in $\ell_\infty(\{E_m\})$, denoted by $(e_i(m))_{m \geq 1}$. We then obtain

$$\forall \, x \in \mathbb{C}^n \qquad \lim_{\mathcal{U}} \left\| \sum x_i e_i(m) \right\|_{E_m} = \left\| \sum x_i e_i \right\|_E.$$

Therefore we may argue as in the proof of Theorem 21.1 to show that the convergence (along $\mathcal{U}$) $\|\sum x_i e_i(m)\| \to \|\sum x_i e_i\|$ is actually uniform over the unit ball of E. Then we obtain $\lim_{\mathcal{U}} d(E_m, E) = 1$, and, since this holds for any $\mathcal{U}$, (i) follows immediately.

Exercise 21.3. We will show (ii)$' \Rightarrow$ (iii). Assume (ii)$'$. Then, for any N, let $e_i^N \in \Pi E_m / \mathcal{U}$ be associated to the sequence $(e_i^N(m))_m$. For any (a_i) in M_N, we have

$$\left\| \sum a_i \otimes e_i \right\|_{M_N(E)} = \left\| \sum a_i \otimes e_i^N \right\|_{M_N(\Pi E_m / \mathcal{U})}.$$

Let $(f_1, ..., f_n)$ be a cluster point of the sequence $\{(e_1^N, ..., e_n^N) \mid N \geq 1\}$ in $\Pi E_m / \mathcal{U}$. Then we must have, for any N and for any (a_i) in M_N,

$$\left\| \sum a_i \otimes e_i \right\|_{M_N(E)} = \left\| \sum a_i \otimes f_i \right\|_{M_N(\Pi E_m / \mathcal{U})}.$$

Hence the correspondence $e_i \to f_i$ is a completely isometric isomorphism from E to $\Pi E_m / \mathcal{U}$, so that (iii) holds. The other implications can be proved exactly as in the preceding exercise, so we omit the details.

Exercise 21.4. Let $u_m \colon E_m \to M_N$ and $u \colon E \to M_N$ be the operators associated respectively to $\sum a_i \otimes \xi_i(m)$ and $\sum a_i \otimes \xi_i$. Recall (by (2.3.2))

$$\left\| \sum a_i \otimes \xi_i(m) \right\|_{\min} = \|u_m\|_{cb} \quad \text{and} \quad \left\| \sum a_i \otimes \xi_i \right\|_{\min} = \|u\|_{cb}.$$

Thus it suffices to show that $\|u_m\|_{cb} \to \|u\|_{cb}$. For any $b = (b_i)_{i \leq n}$ in $(M_N)^n$ we define $p_m(b) = \|\sum b_i \otimes e_i(m)\|_{\min}$ and $p(b) = \|\sum b_i \otimes e_i\|_{\min}$. Then, by Proposition 1.12, we have

$$\|u_m\|_{cb} = \sup \left\{ \left\| \sum a_i \otimes b_i \right\| \; \middle| \; b \in (M_N)^n, p_m(b) \leq 1 \right\},$$

and

$$\|u\|_{cb} = \sup \left\{ \left\| \sum a_i \otimes b_i \right\| \; \middle| \; b \in (M_N)^n, p(b) \leq 1 \right\}.$$

By our assumption, $p_m(b) \to p(b)$ for any b; hence, by Exercise 21.1, $p_m(b) \to p(b)$ uniformly over the subset $\{b \in (M_N)^n, p(b) \leq 1\}$. Let $\varepsilon_m = \sup\{|p_m(b) - p(b)| \mid p(b) \leq 1\}$. We then have $\varepsilon_m \to 0$ and

$$(1 - \varepsilon_m)p(b) \leq p_m(b) \leq (1 + \varepsilon_m)p(b).$$

Therefore, the preceding formulas for $\|u_m\|_{cb}$ and $\|u\|_{cb}$ yield

$$\|u_m\|_{cb} \leq (1 - \varepsilon_m)^{-1} \|u\|_{cb}$$

and

$$\|u\|_{cb} \leq (1 + \varepsilon_m) \|u_m\|_{cb}.$$

Thus we conclude that $\|u_m\|_{cb} \to \|u\|_{cb}$.

Exercise 21.5. Let $F = $ span $[x_i]$. Let $\widetilde{F} \subset M_N$ and let $v \colon F \to \widetilde{F}$ be an isomorphism. Our assumption implies

$$\lim \left\| \sum v(x_i) \otimes e_i(m) \right\|_{\min} = \left\| \sum v(x_i) \otimes e_i \right\|_{\min};$$

hence

$$\begin{aligned}
\limsup_{m \to \infty} \left\| \sum x_i \otimes e_i(m) \right\|_{\min} &\leq \|v^{-1}\|_{cb} \lim \left\| \sum v(x_i) \otimes e_i(m) \right\|_{\min} \\
&\leq \|v^{-1}\|_{cb} \left\| \sum v(x_i) \otimes e_i \right\|_{\min} \\
&\leq \|v^{-1}\|_{cb} \|v\|_{cb} \left\| \sum x_i \otimes e_i \right\|_{\min},
\end{aligned}$$

and similarly

$$\left\| \sum x_i \otimes e_i \right\|_{\min} \leq \|v\|_{cb} \|v^{-1}\|_{cb} \liminf_{m \to \infty} \left\| \sum x_i \otimes e_i(m) \right\|_{\min}.$$

Let $C = d_{SK}(F)$. Taking the infimum ver all poss ible N and v we obtain

$$\begin{aligned}
C^{-1} \left\| \sum x_i \otimes e_i \right\| &\leq \liminf \left\| \sum x_i \otimes e_i(m) \right\| \\
&\leq \limsup \left\| \sum x_i \otimes e_i(m) \right\| \leq C \left\| \sum x_i \otimes e_i \right\|.
\end{aligned}$$

This gives the first assertion, and, in particular, if $C = 1$, we obtain the second one.

Exercise 25.1. Since U dilates T, we have for all x, y in K

$$P_H U \begin{pmatrix} x \\ y \end{pmatrix} = T \begin{pmatrix} x \\ y \end{pmatrix}.$$

In particular,

$$P_H U \begin{pmatrix} x \\ 0 \end{pmatrix} = T \begin{pmatrix} x \\ 0 \end{pmatrix} = \begin{pmatrix} 0 \\ ax \end{pmatrix}.$$

Hence, if a is isometric, we have

$$\left\| P_H U \begin{pmatrix} x \\ 0 \end{pmatrix} \right\| = \|x\| = \left\| U \begin{pmatrix} x \\ 0 \end{pmatrix} \right\|,$$

which forces $U \left(\begin{smallmatrix} x \\ 0 \end{smallmatrix} \right) = P_H U \left(\begin{smallmatrix} x \\ 0 \end{smallmatrix} \right)$. Therefore $U \left(\begin{smallmatrix} x \\ 0 \end{smallmatrix} \right) = \left(\begin{smallmatrix} 0 \\ ax \end{smallmatrix} \right)$. Clearly this implies $\left(\begin{smallmatrix} x \\ 0 \end{smallmatrix} \right) = U^{-1} \left(\begin{smallmatrix} 0 \\ ax \end{smallmatrix} \right)$; so, if a is onto, we obtain the other equality.

Exercise 25.2. This boils down to the observation that commuting pairs of complete contractions $\sigma_i \colon E_i \to B(H)$ $(i = 1, 2)$ are in one-to-one correspondence with commuting pairs of completely contractive homomorphisms $u_i \colon OA(E_i) \to B(H)$ $(i = 1, 2)$.

Exercise 27.1. Let $q \colon A \to A/I$ be the quotient map. Consider y in the open unit ball of $M_n(A/I)$. Let x be a lifting of y in the open unit ball of $M_n(A)$. If $\ell(A) \leq d$, we can factorize x as indicated in (27.10). Then, applying $I \otimes q$ to each of $D_1, \ldots, D_d$, we obtain a similar factorization for $y = (I \otimes q)(x)$. This shows that $\ell(A/I) \leq d$, and hence $\ell(A/I) \leq \ell(A)$.

Let us now assume $\ell(A/I) \leq d$ and $\ell(I) \leq d$. Consider x in the open unit ball of $M_n(A)$. Let $y = (I \otimes q)(x)$. Since we assume $\ell(A/I) \leq d$, for some constant $K(A/I)$ we can factorize y as in (27.10), so that $y = \alpha_0 D_1 \ldots D_d \alpha_d$ with $D_i \in M_N(A/I)$ such that $\prod \|\alpha_i\| \leq K(A/I)$ and $\|D_i\| < 1$. Let $\Delta_i \in M_N(A)$ be liftings of D_i such that $\|\Delta_i\| < 1$, and let

$$x' = \alpha_0 \Delta_1 \alpha_1 \ldots \Delta_d \alpha_d.$$

Note that $\|x'\| < K(A/I)$. Then $(I \otimes q)(x) = (I \otimes q)(x')$ and therefore $x - x' \in M_n(I)$. Let $x'' = x - x'$. By the triangle inequality, $\|x''\|_{M_n(I)} < 1 + K(A/I)$; hence (since $\ell(I) \leq d$ by assumption) we can factorize x'' as in (27.10) with a constant $K(I)$. This gives us $\|x''\|_{(d)} \leq K(I)(1 + K(A/I))$. Recall that, by (27.11), $\| \cdot \|_{(d)}$ is subadditive. Thus we obtain for $x = x' + x''$

$$\|x\|_{(d)} \leq \|x'\|_{(d)} + \|x''\|_{(d)} \leq K(A/I) + K(I)(1 + K(A/I)),$$

and hence we conclude that $\ell(A) \leq d$.

Now assume that A is a C^*-algebra so that I has a quasi-central approximate unit (σ_α) as in Lemma 2.4.4. Consider x in $M_n(I)$ with $\|x\| < 1$. If $\ell(A) \leq d$, then x can be factorized as $x = \alpha_0 D_1 \ldots D_1 \alpha_d$ as in (27.10) but with $D_1, \ldots, D_d$ diagonal matrices all with entries in A with $\|D_i\| < 1$, $\prod \|\alpha_i\| \leq K$. We will use the following notation: For any σ in A and any $y = [y_{ij}]$ either in $M_n(A)$ or in $M_N(A)$ we denote by $\sigma \cdot y$ the matrix $[\sigma y_{ij}]$. Then we observe that the properties of quasi-central approximate units (described before Lemma 2.4.4) guarantee that

$$\|\sigma_\alpha^d \cdot x - \alpha_0 (\sigma_\alpha \cdot D_1) \alpha_1 \ldots (\sigma_\alpha \cdot D_d) \alpha_d\| \to 0$$

and

$$\|x - \sigma_\alpha^d \cdot x\| \to 0.$$

But now $\sigma_\alpha \cdot D_1, \ldots, \sigma_\alpha \cdot D_d$ have their entries in I; hence (since $\|x - \sigma_\alpha^d \cdot x\|_{(d)} \to 0$) we obtain $\ell(I) \leq d$.

Finally, if $A \simeq I \times (A/I)$, of course we have $I \simeq A/(A/I)$, and hence $\ell(I) \leq \ell(A)$.

Exercise 28.1. Let $B = (B_1, \ldots, B_n)$ be an n-tuple in M_n. We define

$$\forall \, \alpha, y, z \in \ell_2^n \qquad F_B(\alpha, y, z) = \left\langle \sum_1^n \alpha_k B_k y, z \right\rangle.$$

Let $D_n = \left\{ \alpha \in \mathbb{C}^n \mid \sum |\alpha_k|^2 \leq 1 \right\}$. Note that

$$\left\| \sum \alpha_k B_k \right\| = \sup_{y, z \in D_n} \{ |F_B(\alpha, y, z)| \}.$$

We need to produce $(B_1, \ldots, B_n)$ in M_n satisfying (28.19) and such that

$$\sup_{\alpha, y, z \in D_n} |F_B(\alpha, y, z)| \leq \beta.$$

By a well-known fact (cf., e.g., [P8, pp. 49–50]) there is a subset $A_n \subset D_n$ with cardinal $|A_n| \leq 5^{2n}$ such that

$$\sup_{\alpha, y, z \in D_n} |F_B(\alpha, y, z)| \leq 8 \sup_{\alpha, y, z \in A_n} |F_B(\alpha, y, z)|.$$

The idea is to choose $B_1, \ldots, B_n$ "at random" in M_n using Gaussian random matrices. Let $\{ g_k(i,j) \mid 1 \leq i, j, k < \infty \}$ be a collection of independent standard complex-valued Gaussian variables, that is, with mean zero and such that $\mathbb{E}|g_k(i,j)|^2 = 1$. For any $1 \leq k \leq n$, we define g_k as the $n \times n$ matrix with entries $\{ n^{-1/2} g_k(i,j) \mid 1 \leq i, j \leq n \}$.

Then, for each α, y, z, the random variable

$$X(\alpha, y, z) = \sum_{k=1}^n \alpha_k \langle g_k y, z \rangle$$

has the same distribution as

$$Y(\alpha, y, z) = n^{-1/2} g_1(1, 1) \left(\sum |\alpha_k|^2 \right)^{1/2} \left(\sum |y_j|^2 \right)^{1/2} \left(\sum |z_i|^2 \right)^{1/2}.$$

Hence, assuming α, y, z all in D_n, for any $t > 0$, we can write

$$\mathbb{P}\{ |X(\alpha, y, z)| > t \} = \mathbb{P}\{ |Y(\alpha, y, z)| > t \} \leq P\{ |n^{-1/2} g_1(1, 1)| > t \}.$$

Therefore, there are positive numerical constants $K_1, K_2, \ldots$ so that

$$\mathbb{P}\{ |X(\alpha, y, z)| > t \} \leq K_1 \exp(-K_2 n t^2),$$

which implies (recall $|A_n| \leq 5^{2n}$)

$$\mathbb{P}\{\sup_{\alpha,y,z \in A_n} |X(\alpha,y,z)| > t\} \leq K_1 5^{6n} \exp(-K_2 n t^2) \leq K_1 \exp(-(K_2 t^2 - 12)n).$$

Thus, if we choose θ so that (say)

$$K_2 \theta^2 - 12 = 1,$$

the preceding estimate guarantees (by Borel-Cantelli) that

$$\limsup_{n \to \infty} \sup_{\alpha,y,z \in A_n} |X(\alpha,y,z)| \leq \theta.$$

Hence (by the known fact recalled above)

$$\limsup_{n \to \infty} \sup_{\alpha,y,z \in D_n} |X(\alpha,y,z)| \leq 8\theta.$$

On the other hand, we claim that

$$\liminf_{n \to \infty} n^{-1} \left\| \sum_1^n g_k \otimes \overline{g_k} \right\|_{\min} \geq 1.$$

Hence, we can certainly choose ω and $N(\omega)$ so that, if we set $B_k = g_k(\omega)$, then for all $n \geq N(\omega)$ we have (28.18) with (say) $\beta = 8\theta + 1$ and $n/2 \leq \left\| \sum_{k=1}^n B_k \otimes \overline{B_k} \right\|$. Replacing B_k by $\sqrt{2}\, B_k$ we obtain the result as stated, and it is a straightforward matter to adjust the constants to include the (finitely many) cases $n < N(\omega)$.

We will now prove the preceding claim. By definition of the min-norm, we have (see Prop. 2.9.1)

$$\left\| \sum_1^n g_k \otimes \overline{g_k} \right\|_{\min} = \sup\left\{ \left| \mathrm{tr}\left(\sum_1^n g_k x g_k^* y \right) \right| \right\},$$

where the supremum runs over x,y in the unit ball of the $n \times n$ Hilbert-Schmidt matrices. Taking x and y both equal to $n^{-1/2}$ times the identity we obtain

$$n^{-1} \left\| \sum_1^n g_k \otimes \overline{g_k} \right\| \geq n^{-2} \mathrm{tr}\left(\sum_1^n g_k g_k^* \right) = n^{-3} \sum_{1 \leq i,j,k \leq n} |g_k(i,j)|^2.$$

But, by the strong law of large numbers, we have almost surely

$$n^{-3} \sum_{1 \leq i,j,k \leq n} |g_k(i,j)|^2 \to 1,$$

and hence we obtain the announced result. ∎

REFERENCES

[AB]   R. Archbold and C. Batty. C^*-tensor norms and slice maps. *J. London Math. Soc.* 22 (1980) 127–138.

[A1]   A. Arias. Completely bounded isomorphisms of operator algebras. *Proc. Amer. Math. Soc.* 124 (1996) 1091–1101.

[A2]   _____. A Hilbertian operator space without the OAP. *Proc. Amer. Math. Soc.* 130 (2002) 2669–2677.

[AFJS] A. Arias, T. Figiel, W. Johnson, and G. Schechtman. Banach spaces which have the 2-summing property. *Trans. Amer. Math. Soc.* 347 (1995) 3835–3857.

[An1] C. Anantharaman-Delaroche. Classification des C^*-algèbres purement infinies nucléaires (d'après E. Kirchberg). *Séminaire Bourbaki,* Vol. 1995/96. Astérisque No. 241 (1997), Exp. No. 805, 3, 7–27.

[An2] C. Anantharaman-Delaroche. Amenability and exactness for dynamical systems and their C*-algebras. Preprint 2000 (http://arXiv.org/pdf/math.OA/0005014).

[AO]   C. Akemann and P. Ostrand. Computing norms in group C^*-algebras. *Amer. J. Math.* 98 (1976) 1015–1047.

[AP]   C. Anantharaman-Delaroche and C. Pop. Relative tensor products and infinite C^*-algebras. Preprint, 1999.

[APe] A. B. Aleksandrov and V. Peller. Hankel operators and similarity to a contraction. *Internat. Math. Res. Notices* no. 6 (1996) 263–275.

[Ar1] W. Arveson. Subalgebras of C^*-algebras. *Acta Math.* 123 (1969) 141–224. Part II, *Acta Math.* 128 (1972), 271–308.

[Ar2] _____. *An Invitation to C^*-Algebras.* Springer-Verlag, 1976.

[Ar3] _____. *Ten Lectures on Operator Algebras.* CBMS (Regional Conferences of the A.M.S.) 55, 1984.

[Ar4] _____. Notes on extensions on C^*-algebras. *Duke Math. J.* 44 (1977) 329–355.

[AR]   A. Arias and H. P. Rosenthal. M-complete approximate identities in operator spaces. *Studia Math.* 141 (2000) 143–200.

[B1]   D. Blecher. Tensor products of operator spaces II. *Canadian J. Math.* 44 (1992) 75–90.

[B2]   _____. The standard dual of an operator space. *Pacific J. Math.* 153 (1992) 15–30.

[B3]   _____. Generalizing Grothendieck's program. *Function spaces,* edited by K. Jarosz. Lecture Notes in Pure and Applied Math. vol. 136. Marcel Dekker, 1992.

[B4]   _____. A completely bounded characterization of operator algebras. *Math. Ann.* 303 (1995) 227–240.

[B5] _____. Multipliers and dual operator algebras. *J. Funct. Anal.* 183 (2001) 498–525.

[Ba] S. Banach. *Théorie des opérations linéaires. Varsovie, 1932,* Oeuvres de Stephan Banach, vol. 2, Acad. Polon. Sc., Editions scientifiques de Pologne, Varsovie, 1979.

[BCL] K. Ball, E. Carlen, and E. Lieb. Sharp uniform convexity and smoothness inequalities for trace norms. *Invent. Mat.* 115 (1994) 463–482.

[BD] F. Bonsall and J. Duncan. *Complete Normed Algebras.* Springer-Verlag, 1973.

[Be] J. Bergh. On the relation between the two complex methods of interpolation. *Indiana Univ. Math. J.* 28 (1979) 775–777.

[BeL] J. Bergh and J. Löfström. *Interpolation Spaces. An Introduction.* Springer-Verlag, 1976.

[BF] M. Bożejko and G. Fendler. Herz-Schur multipliers and completely bounded multipliers of the Fourier algebra of a locally compact group. *Boll. Unione Mat. Ital.* (6) 3-A (1984) 297–302.

[Bla1] B. Blackadar. Nonnuclear subalgebras of C^*-algebras. *J. Operator Theory* 14 (1985) 347–350.

[Bla2] _____. *K-Theory for Operator Algebras. Second edition.* Mathematical Sciences Research Institute Publications, 5. Cambridge University Press, 1998.

[Blo] G. Blower. The Banach space $B(\ell_2)$ is primary. *Bull. London Math. Soc.* 22 (1990) 176–182.

[BLM] D. Blecher and C. Le Merdy. On quotients of function algebras, and operator algebra structures on ℓ_p. *J. Operator Theory* 34 (1995) 315–346.

[BMP] D. Blecher, P. Muhly, and V. Paulsen. Categories of operator modules (Morita equivalence and projective modules). *Mem. Amer. Math. Soc.* 143 (2000), no. 681, viii+94 pp.

[Bo1] M. Bożejko. Positive-definite kernels, length functions on groups and a noncommutative von Neumann inequality. *Studia Math.* 95 (1989) 107–118.

[BoS1] M. Bożejko and R. Speicher. An example of a generalized Brownian motion. *Commun. Math. Phys.* 137 (1991) 519–531.

[BoS2] _____. Completely positive maps on Coxeter groups, deformed commutation relations, and operator spaces. *Math. Ann.* 300 (1994) 97–120.

[Bou1] J. Bourgain. Vector valued singular integrals and the $H^1 - BMO$ duality. *Probability Theory and Harmonic Analysis,* Chao-Woyczynski. pp. 1–19. Marcel Dekker, 1986.

[Bou2] _____. Real isomorphic complex Banach spaces need not be complex isomorphic. *Proc. Amer. Math. Soc.* 96 (1986) 221–226.

[Bou3] _____. New Banach space properties of the disc algebra and H^∞. *Acta Math.* 152 (1984) 1–48.

[Bou4] _____. On the similarity problem for polynomially bounded operators on Hilbert space. *Israel J. Math.* 54 (1986) 227–241.

[Bou5] _____. *New Classes of $\mathcal{L}^p$-Spaces.* Lecture Notes in Mathematics, 889. Springer-Verlag, 1981.

[BP1] D. Blecher and V. Paulsen. Tensor products of operator spaces. *J. Funct. Anal.* 99 (1991) 262–292.

[BP2] _____. Explicit constructions of universal operator algebras and applictions to polynomial factorization. *Proc. Amer. Math. Soc.* 112 (1991) 839–850.

[BR] O. Bratelli and D. Robinson. *Operator Algebras and Quantum Statistical Mechanics II.* Springer-Verlag, 1981.

[BRS] D. Blecher, Z. J. Ruan, and A. Sinclair. A characterization of operator algebras. *J. Funct. Anal.* 89 (1990) 188–201.

[BS] D. Blecher and R. Smith. The dual of the Haagerup tensor product. *J. London Math. Soc.* 45 (1992) 126–144.

[Buc1] A. Buchholz. Norm of convolution by operator-valued functions on free groups. *Proc. Amer. Math. Soc.* 127 (1999) 1671–1682.

[Buc2] A. Buchholz. Operator Khintchine inequality in non-commutative probability. *Math. Ann.* 319 (2001) 1–16.

[Bur] D. Burkholder. Distribution function inequalities for martingales. *Ann. Prob.* 1 (1973) 19–42.

[BV] H. Bercovici and D. Voiculescu. Free convolution of measures with unbounded support. *Indiana Univ. Math. J.* 42 (1993) 733–773.

[C] A. Calderón. Intermediate spaces and interpolation, the complex method. *Studia Math.* 24 (1964) 113–190.

[Ca] T. K. Carne. Not all H'-algebras are operator algebras. *Proc. Camb. Phil. Soc.* 86 (1979) 243–249.

[CC1] J. F. Carlson and D. N. Clark. Projectivity and extensions of Hilbert modules over $A(D^N)$. *Michigan Math. J.* 44 (1997) 365–373.

[CC2] _____. Cohomology and extensions of Hilbert modules. *J. Funct. Anal.* 128 (1995) 278–306.

[CCFW] J. Carlson, D. Clark, C. Foias, and J. Williams. Projective Hilbert $A(D)$-modules. *New York J. Math.* 1 (1994) 26–38, electronic.

[CE1] M. D. Choi and E. Effros. Nuclear C^*-algebras and the approximation property. *Amer. J. Math.* 100 (1978) 61–79.

[CE2] _____. Nuclear C*-algebras and injectivity: The general case. *Indiana Univ. Math. J.* 26 (1977) 443–446.

[CE3] _____. Injectivity and operator spaces. *J. Funct. Anal.* 24 (1977) 156–209.

[CE4] _____. Separable nuclear C^*-algebras and injectivity. *Duke Math. J.* 43 (1976) 309–322.

[CES] E. Christensen, E. Effros, and A. Sinclair. Completely bounded multilinear maps and C^*-algebraic cohomology. *Invent. Math.* 90 (1987) 279–296.

[Ch1] M. D. Choi. A Schwarz inequality for positive linear maps on C^*-algebras. *Illinois J. Math.* 18 (1974) 565–574.

[Ch2] M. D. Choi. A simple C^*-algebra generated by two finite order unitaries. *Canadian J. Math.* 31 (1979) 887–890.

[Che] L. Chen. An inequality for the multivariate normal distribution. *J. Mult. Anal.* 12 (1982) 306–315.

[Chr1] E. Christensen. Extensions of derivations. *J. Funct. Anal.* 27 (1978) 234–247.

[Chr2] _____. Extensions of derivations II. *Math. Scand.* 50 (1982) 111–122.

[Chr3] _____. On non self adjoint representations of operator algebras *Amer. J. Math.* 103 (1981) 817–834.

[Chr4] _____. Similarities of II_1 factors with property Γ. *J. Operator Theory* 15 (1986) 281–288.

[Chr5] _____. Perturbation of operator algebras II. Indiana *Math. J.* 26 (1977) 891–904.

[Chr6] _____. Finite von Neumann algebra factors with property Γ. *J. Funct. Anal.* 186 (2001) 366–380.

[ChS] A. Chaterjee and R. Smith. The central Haagerup tensor product and maps between von Neumann algebras. *J. Operator Theory* 112 (1993) 97–120.

[CK] E. Carlen and P. Krée. On martingale inequalities in noncommutative stochastic analysis. *J. Funct. Anal.* 158 (1998) 475–508.

[CL] E. Carlen and E. Lieb. Optimal hypercontractivity for Fermi fields and related non-commutative integration inequalities. *Comm. Math. Phys.* 155 (1993) 27–46.

[Co1] A. Connes. Classification of injective factors, Cases $II_1, II_\infty, III_\lambda, \lambda \neq 1$. *Ann. Math.* 104 (1976) 73–116.

[Co2] _____. *Non-Commutative Geometry.* Academic Press, 1995.

[Co3] _____. A factor not anti-isomorphic to itself. *Bull. London Math. Soc.* 7 (1975), 171–174.

[CoH] M. Cowling and U. Haagerup. Completely bounded multipliers of the Fourier algebra of a simple Lie group of real rank one. *Invent. Math.* 96 (1989) 507–549.

[CS1] E. Christensen and A. Sinclair. Representations of completely bounded multilinear operators. *J. Funct. Anal.* 72 (1987) 151–181.

[CS2] _____. A survey of completely bounded operators. *Bull. London Math. Soc.* 21 (1989) 417–448.

[CS3] _____. On von Neumann algebras which are complemented subspaces of $B(H)$. *J. Funct. Anal.* 122 (1994) 91–102.

[CS4] _____. Module mappings into von Neumann algebras and injectivity. *Proc. London Math. Soc.* (3) 71 (1995) 618–640.

[CS5] _____. Completely bounded isomorphisms of injective von Neumann algebras. *Proc. Edinburgh Math. Soc.* 32 (1989) 317–327.

[Cu] J. Cuntz. Simple C^*-algebras generated by isometries. *Comm. Math. Phys.* 57 (1977) 173–185.

[Da1] K. Davidson. *Nest Algebras*. Wiley, 1988.

[Da2] _____. *C^*-Algebras by Example*. Fields Institute, Toronto, AMS publication, 1996.

[DCH] J. de Cannière and U. Haagerup. Multipliers of the Fourier algebras of some simple Lie groups and their discrete subgroups. *Amer. J. Math.* 107 (1985) 455–500.

[DHV] P. de la Harpe and A. Valette. *La Propriété T de Kazhdan pour les Groupes Localement Compacts*. Astérisque, Soc. Math. France 175 (1989).

[Di1] J. Dixmier. *Les Algèbres d'Opérateurs dans l'Espace Hilbertien (Algèbres de von Neumann)*. Gauthier-Villars, 1969. (In translation: *von Neumann Algebras*. North-Holland, 1981).

[Di2] _____. *Les C^*-algèbres et leurs représentations*. Gauthier-Villars, 1969.

[Di3] _____. Formes linéaires sur un anneau d'opérateurs. *Bull. Soc. Math. France* 81 (1953) 9–39.

[Dix1] P. Dixon. *Q*-algebras. Unpublished Lecture Notes. Sheffield University, 1975.

[Dix2] _____. Varieties of Banach algebras. *Quart. J. Math. Oxford* 27 (1976) 481–487.

[DJT] J. Diestel, H. Jarchow, and A. Tonge. *Absolutely Summing Operators*. Cambridge University Press, 1995.

[DP] K. Davidson and V. Paulsen. On polynomially bounded operators. *J. für die reine und angewandte Math.* 487 (1997) 153–170.

[E1] E. Effros. Aspects of non-commutative order. Notes for a lecture given at the Second US–Japan seminar on C^*-algebras and applications to physics (April 1977) .

[E2] _____. Advances in quantized functional analysis. *Proceedings International Congress of Mathematicians*, Berkeley, 1986, pp. 906–916.

[EE]	E. Effros and R. Exel. *On Multilinear Double Commutant Theorems. Operator Algebras and Applications, Vol. 1*, edited by D. Evans and M. Takesaki. London Math. Soc. Lecture Notes Series 135, pp. 81–94.

[EH]	E. Effros and U. Haagerup. Lifting problems and local reflexivity for C^*-algebras. *Duke Math. J.* 52 (1985) 103–128.

[EJR]	E. Effros, M. Junge, and Z. J. Ruan. Integral mappings and the principle of local reflexivity for noncommutative $L271$-spaces. *Ann. of Math.* 151 (2000) 59–92.

[EK]	E. Effros and A. Kishimoto. Module maps and Hochschild-Johnson cohomology. *Indiana Univ. Math. J.* 36 (1987) 257–276.

[EKR]	E. Effros, J. Kraus, and Z. J. Ruan. On two quantized tensor products. *Operator Algebras, Mathematical Physics and Low Dimensional Topology (Istanbul, 1991)*. A. K. Peters, 1993, pp. 125–145.

[EL]	E. Effros and C. Lance. Tensor products of operator algebras. *Adv. Math.* 25 (1977) 1–34.

[El]	G. Elliott. On approximate finite dimensional von Neumann algebras I and II. *Math. Scand.* 39 (1976) 91–101, and *Canad. Math. Bull.* 21 (1978) 415–418.

[En]	P. Enflo. A counterexample to the approximation problem in Banach spaces. *Acta Math.* 130 (1973) 309–317.

[EOR]	E. Effros, N. Ozawa, and Z. J. Ruan. On injectivity and nuclearity for operator spaces. *Duke Math. J.* 110 (2001) 489–521.

[ER1]	E. Effros and Z.J. Ruan. On matricially normed spaces. *Pacific J. Math.* 132 (1988) 243–264.

[ER2]	———. A new approach to operator spaces. *Canadian Math. Bull.* 34 (1991) 329–337.

[ER3]	———. On the abstract characterization of operator spaces. *Proc. Amer. Math. Soc.* 119 (1993) 579–584.

[ER4]	———. Self duality for the Haagerup tensor product and Hilbert space factorization. *J. Funct. Anal.* 100 (1991) 257–284.

[ER5]	———. Recent development in operator spaces. *Current Topics in Operator Algebras*. Proceedings of the ICM-90 Satellite Conference Held in Nara (August 1990). World Sci. Publishing, 1991, pp. 146–164.

[ER6]	———. Mapping spaces and liftings for operator spaces. *Proc. London Math. Soc.* 69 (1994) 171–197.

[ER7]	———. The Grothendieck-Pietsch and Dvoretzky-Rogers Theorems for operator spaces. *J. Funct. Anal.* 122 (1994) 428–450.

[ER8]	———. On approximation properties for operator spaces. *Int. J. Math.* 1 (1990) 163–187.

[ER9] _____. Representations of operator bimodules and their applications. *J. Operator Theory* 19 (1988) 137–157.

[ER10] _____. Operator space tensor products and Hopf convolution algebras, *J. Operator Theory*. To appear.

[ER11] _____. *Operator Spaces*. Oxford Univ. Press, 2000.

[ER12] _____. $\mathcal{OL}_p$-spaces, *Contemporary Math.* 228 (1998) 51–77.

[EvK] D. E. Evans and Y. Kawahigashi. *Quantum Symmetries on Operator Algebras*. Oxford University Press, 1998, xvi+829 pp.

[EvL] D. Evans and J. Lewis. Dilations of irreversible evolutions in algebraic quantum theory. *Commun. Dublin Inst. for Advanced Studies*, Series A (Theoretical Physics) 24 (1977).

[EW] E. Effros and S. Winkler. Matrix convexity: operator analogues of the bipolar and Hahn-Banach theorems. *J. Funct. Anal.* 144 (1997) 117–152.

[Fi] P. Fillmore. *A User's Guide to Operator Algebras*. CMS Series of Monographs and Advanced Texts. Wiley, 1996.

[Fol] G.B. Folland. *A Course in Abstract Harmonic Analysis*. Studies in Advanced Mathematics. CRC Press, 1995, x+276 pp.

[FLM] T. Figiel, J. Lindenstrauss, and V. Milman. The dimensions of the spherical sections of convex bodies. *Acta Math.* 139 (1977) 53–94.

[Fo] S. Foguel. A counterexample to a problem of Sz. Nagy. *Proc. Amer. Math. Soc.* 15 (1964) 788–790.

[FTP] A. Figa-Talamanca and M. Picardello. *Harmonic Analysis on Free Groups*. Marcel Dekker, 1983.

[GaR] D. Gaşpar and A. Rácz. An extension of a theorem of T. Ando. *Michigan Math. J.* 16 (1969) 377–380.

[GH] L. Ge and D. Hadwin. Ultraproducts of C^*-algebras. Preprint, 1999.

[GHJ] F. Goodman, P. de la Harpe and V. F. R. Jones. *Coxeter Graphs and Towers of Algebras*. MSRI Publications, 14. Springer-Verlag, 1989.

[Gl] E. Gluskin. The diameter of the Minkowski compactum is roughly equal to n. *Funct. Anal. Appl.* 15 (1981) 72–73.

[Gr] A. Grothendieck. Résumé de la théorie métrique des produits tensoriels topologiques. *Boll. Soc. Mat. São-Paulo* 8 (1956) 1–79.

[Gro] M. Gromov. Random walk in random groups. Preprint, IHES, January 2002.

[Gu1] A. Guichardet. Tensor products of C^*-algebras. *Dokl. Akad. Nauk. SSSR* 160 (1965) 986–989.

[Gu2] _____. *Algèbres d'observables associées aux relations de commutation*. Armand Colin, 1968.

[Gu3] _____. *Symmetric Hilbert Spaces and Related Topics.* Springer Lecture Notes 261, Springer-Verlag, 1972.

[H1] U. Haagerup. Injectivity and decomposition of completely bounded maps. *Operator Algebras and Their Connection with Topology and Ergodic Theory.* Springer Lecture Notes in Math. 1132, Springer-Verlag, 1985, pp. 170–222.

[H2] _____. An example of a non-nuclear C^*-algebra which has the metric approximation property. *Inventiones Math.* 50 (1979) 279–293.

[H3] _____. Decomposition of completely bounded maps on operator algebras. Unpublished manuscript, Sept. 1980.

[H4] _____. Self-polar forms, conditional expectations and the weak expectation property for C^*-algebras. Unpublished manuscript, 1995.

[H5] _____. A new proof of the equivalence of injectivity and hyperfiniteness for factors on a separable Hilbert space. *J. Funct. Anal.* 62 (1985) 160–201.

[H6] _____. Solution of the similarity problem for cyclic representations of C^*-algebras. *Ann. Math.* 118 (1983) 215–240.

[H7] _____. The standard form of von Neumann algebras. *Math. Scand.* 37 (1975) 271–283.

[H8] _____. Group C^*-algebras without the completely bounded approximation property. Unpublished manuscript, May 1986.

[Ha] L. Harris. *Bounded Symmetric Domains in Infinite Dimensional Spaces.* Springer Lecture Notes 364. Springer-Verlag, 1974 pp. 13–40.

[Hal] P. Halmos. Ten problems in Hilbert space. *Bull. Amer. Math. Soc.* 75 (1970) 887–933.

[Ham] M. Hamana. Injective envelopes of C^*-algebras. *J. Math. Soc. Japan* 31 (1979) 181–197.

[Har1] A. Harcharras. On some stability properties of the full C^*-algebra associated to the free group F_∞. *Proc. Edinburgh Math. Soc.* (2) 41 (1998) 93–116.

[Har2] _____. Fourier analysis, Schur multipliers on S^p and non-commuta tive $\Lambda(p)$-sets. *Studia Math.* 137 (1999) 203–260.

[HiP] F. Hiai and D. Petz. *The Semicircle Law, Free Random Variables and Entropy.* Math. Surveys 77. Amer. Math. Soc., 2000.

[HK] U. Haagerup and J. Kraus. Approximation properties for group C^*-algebras and group von Neumann algebras. *Trans. Amer. Math. Soc.* 344 (1994) 667–699.

[HP1] U. Haagerup and G. Pisier. Factorization of analytic functions with values in non-commutative L_1-spaces. *Canadian J. Math.* 41 (1989) 882–906.

[HP2] _____. Bounded linear operators between C^*-algebras. *Duke Math. J.* 71 (1993) 889–925.

[HT1] U. Haagerup and S. Thorbjørnsen. Random matrices and K-theory for exact C^*-algebras. *Documenta Math.* 4 (1999) 341–450.

[HT2] _____. A new application of random matrices: $Ext(C_r^*(F_2))$ is not a group. (Preliminary version, August 30, 2002).

[HWW] P. Harmand, D. Werner, and W. Werner. *M-Ideals in Banach Spaces and Banach Algebras.* Lecture Notes in Mathematics 1547. Springer-Verlag, 1993, viii+387 pp.

[J1] M. Junge. Factorization Theory for Spaces of Operators. Habilitation thesis. Kiel University, 1996.

[J2] _____. The projection constant of OH_n and the little Grothendieck inequality. Preprint (2002).

[JLM] M. Junge and C. Le Merdy. Factorization through matrix spaces for finite rank operators between C^*-algebras. *Duke Math. J.* 100 (1999) 299–319.

[JNRX] M. Junge, N. Nielsen, Z. J. Ruan, and Q. Xu. $\mathcal{COL}_p$-spaces-The local structure of non-commutative L_p-spaces. *Adv. Math.* To appear.

[Jo] W.B. Johnson. A complementary universal conjugate Banach space and its relation to the approximation problem. *Israel J. Math.* 13 (1972) 301–310 (1973).

[JO] W. Johnson and T. Oikhberg. Separable lifting property and extensions of local reflexivity. *Illinois J. Math.* 45 (2001) 123–137.

[Jol] P. Jolissaint. A characterization of completely bounded multipliers of Fourier algebras. *Colloquium Math.* 63 (1992) 311–313.

[Jon] V. F. R. Jones. *Subfactors and Knots.* CBMS Regional Conference Series in Mathematics, 80. American Math. Soc., 1991.

[JOR] M. Junge, N. Ozawa, and Z. J. Ruan. $\mathcal{OL}_\infty$ spaces. *Math. Ann.* To appear.

[JP] M. Junge and G. Pisier. Bilinear forms on exact operator spaces and $B(H) \otimes B(H)$. *Geometric and Functional Analysis (GAFA J.)* 5 (1995) 329–363.

[JR] M. Junge and Z. J. Ruan. Approximation properties for non-commutative L_p-spaces associated with discrete groups. Preprint (2001). To appear in *Duke Math. J.*

[JS] V. Jones and V. S. Sunder. *Introduction to Subfactors.* Cambridge Univ. Press, 1997.

[K] H. Kesten. Symmetric random walks on groups. *Trans. Amer. Math. Soc.* 92 (1959) 336–354.

[Ka] Y. Katznelson. *An Introduction to Harmonic Analysis.* Wiley, 1968 (republished by Dover in 1976).

[Kad] R. Kadison. A generalized Schwarz inequality and algebraic invariants for operator algebras. *Ann. of Math.* 56 (1952) 494–503.

[KaR] R. Kadison and J. Ringrose. *Fundamentals of the Theory of Operator Algebras, Vol. II, Advanced Theory.* Academic Press, 1986.

[Ki1] E. Kirchberg. On subalgebras of the CAR-algebra. *J. Funct. Anal.* 129 (1995) 35–63.

[Ki2] _____. On non-semisplit extensions, tensor products and exactness of group C^*-algebras. *Invent. Math.* 112 (1993) 449–489.

[Ki3] _____. C^*-nuclearity implies CPAP. *Math. Nachr.* 76 (1977) 203–212.

[Ki4] _____. Positive maps and C^*-nuclear algebras. *Proc. Inter. Conference on Operator Algebras, Ideals and Their Applications in Theoretical Physics (Leipzig 1977).* Teubner Texte, 1978.

[Ki5] _____. Commutants of unitaries in UHF algebras and functorial properties of exactness. *J. Reine Angew. Math.* 452 (1994) 39–77.

[Ki6] _____. Personal communication.

[Ki7] _____. On the matricial approximation property. Preprint.

[Ki8] _____. Exact C^*-algebras, Tensor products, and Classification of purely infinite algebras. *Proceedings ICM 94, Zürich.* Vol. 2, Birkhäuser, 1995, pp. 943–954.

[Ki9] _____. The derivation and the similarity problem are equivalent. *J. Operator Theory* 36 (1996) 59–62.

[KiP] E. Kirchberg and N.C. Phillips. Embedding of exact C^*-algebras in the Cuntz algebra $\mathcal{O}^2$. *J. Reine Angew. Math.* 525 (2000) 17–53.

[KiW] E. Kirchberg and S. Wassermann. Exact groups and continuous bundles of C^*-algebras. *Math. Ann.* 315 (1999) 169–203.

[Kis1] S. Kislyakov. Operators that are (dis)similar to a contraction: Pisier's counterexample in terms of singular integrals. Zap. Nauchn. Semin. S.- Peterburg. Otdel. Mat. Inst. Steklov (POMI).

[Kis2] _____. Similarity problem for certain martingale uniform algebras. Preprint.

[KiV] E. Kirchberg and G. Vaillant. On C^*-algebras having linear, polynomial, and subexponential growth. *Invent. Math.* 108 (1992) 635–652.

[KiW] E. Kirchberg and S. Wassermann. C^*-algebras generated by operator systems. *J. Funct. Anal.* 155 (1998) 324–351.

[Ko] O. Kouba. Interpolation of injective or projective tensor products of Banach spaces. *J. Funct. Anal.* 96 (1991) 38–61.

[Kö] H. König. On the complex Grothendieck constant in the n-dimensional case. *Proc. of the Strobl Conf. Austria 1989*, edited by P. Mueller and W. Schachermayer. London Math. Soc. Lect. Notes 158, 1990, pp. 181–198.

[KPS] S. G. Krein, Yu. Petunin, and E. M. Semenov, *Interpolation of Linear Operators*. Translations of Mathematical Monographs 54. American Mathematical Society, 1982.

[Kr] J. Kraus. The slice map problem and approximation properties. *J. Funct. Anal.* 102 (1991) 116–155.

[KTJ1] H. König and N. Tomczak-Jaegermann. Bounds for projection constants and 1-summing norms. *Trans. Amer. Math. Soc.* 320 (1990) 799–823.

[KTJ2] _____. Norms of minimal projections. *J. Funct. Anal.* 119 (1994) 253–280.

[Ku] W. Kuratowski. *Topology Vol. 1*. Academic Press, 1966 (new edition translated from the French).

[Ky] S-H. Kye. *Notes on Operator Algebras*. Lecture Notes Series 7, Seoul National Univ., 1993.

[KyR] S-H. Kye and Z-J. Ruan. On the local lifting property for operator spaces. *J. Funct. Anal.* 168 (1999) 355–379.

[La1] C. Lance. On nuclear C^*-algebras. *J. Funct. Anal.* 12 (1973) 157–176.

[La2] _____. Tensor products and nuclear C^*-algebras. *Operator Algebras and Applications*. Amer. Math. Soc. Proc. Symposia Pure Math., 1982, Vol. 38, part 1, pp. 379–399.

[Lac] H. E. Lacey. *The Isometric Theory of Classical Banach Spaces*. Springer-Verlag, 1974.

[Leh1] F. Lehner. A characterization of the Leinert property. *Proc. Amer. Math. Soc.* 125 (1997) 3423–3431.

[Leh2] _____. Free operators with operator coefficients. *Colloq. Math.* 74 (1997) 321–328.

[Le] M. Leinert. Faltungsoperatoren auf gewissen diskreten Gruppen. *Studia Math.* 52 (1974) 149–158.

[Leb] A. Lebow. A power bounded operator which is not polynomially bounded. *Michigan Math. J.* 15 (1968) 397–399.

[LeM1] C. Le Merdy. Factorization of p-completely bounded multilinear maps. *Pacific J. Math.* 172 (1996) 187–213.

[LeM2] _____. Representations of a quotient of a subalgebra of $B(X)$. *Math. Proc. Cambridge Phil. Soc.* 119 (1996) 83–90.

[LeM3] _____. On the duality of operator spaces. *Canadian Math. Bull.* 38 (1995) 334–346.

[LeM4] _____. Self-adjointness criteria for operator algebras. *Arch. Math.* 74 (2000) 212–220.

[LeM5] _____. A strong similarity property of nuclear C^*-algebras. *Rocky Mountain J. Math.* 30 (2000) 279–292.

[LeM6] _____. An operator space characterization of dual operator algebras. *Amer. J. Math.* 121 (1999) 55–63.

[LeM7] _____. The weak* similarity property on dual operator algebras. *Integr. Equ. Oper. Theory* 37 (2000) 72–94.

[Li] J. Lindenstrauss. *Extension of Compact Operators.* Mem. Amer. Math. Soc. No. 48, 1964, pp. 1–112.

[LiR] J. Lindenstrauss and H. P. Rosenthal. The $\mathcal{L}_p$ spaces. *Israel J. Math.* 7 (1969) 325–349.

[LPP] F. Lust-Piquard and G. Pisier. Non-commutative Khintchine and Paley inequalities. *Ark. för Mat.* 29 (1991) 241–260.

[LPS] A. Lubotzky, R. Phillips, and P. Sarnak. Hecke operators and distributing points on S^2, I. *Comm. Pure and Applied Math.* 39 (1986) 149–186.

[LR] J. Lopez and K. Ross. *Sidon Sets.* Marcel Dekker, 1975.

[LT1] J. Lindenstrauss and L. Tzafriri. *Classical Banach Spaces, Vol. I. Sequence Spaces.* Springer-Verlag, 1976.

[LT2] _____. *Classical Banach Spaces, Vol. II. Function Spaces.* Springer-Verlag, 1979.

[LT3] _____. *Classical Banach Spaces.* Springer Lecture Notes 338, Springer-Verlag, 1973.

[Lu] A. Lubotzky. *Discrete Groups, Expanding Graphs and Invariant Measures.* Progress in Math. 125. Birkhäuser, 1994.

[LuP] F. Lust-Piquard. Inégalités de Khintchine dans C_p $(1 < p < \infty)$. *C.R. Acad. Sci. Paris* 303 (1986), 289–292.

[M] P. A. Meyer. *Quantum Probability for Probabilists.* Springer Lecture Notes 1538, Springer-Verlag, 1993.

[Ma1] B. Magajna. Strong operator modules and the Haagerup tensor product. *Proc. London Math. Soc.* 74 (1997) 201–240.

[Ma2] _____. The Haagerup norm on the tensor product of operator modules. *J. Funct. Anal.* 129 (1995) 325–348.

[MaP] B. Maurey and G. Pisier. Séries de variables aléatoires vectorielles indépendantes et propriétés géométriques des espaces de Banach. *Studia Math.* 58 (1976) 45–90.

[Mat1] B. Mathes. A completely bounded view of Hilbert-Schmidt operators. *Houston J. Math.* 17 (1991) 404–418.

[Mat2] _____. Characterizations of row and column Hilbert space. *J. London Math. Soc.* 50 (1994), no. 2, 199–208.

[Mi] K. Miyazaki. Some remarks on intermediate spaces. *Bull. Kyushu Inst. Tech.* 15 (1968) 1–23.

[MN] P. Meyer-Nieberg. *Banach Lattices.* Universitext. Springer-Verlag, 1991.

[Mor] M. Morgenstern. Existence and explicit constructions of $q+1$ regular Ramanujan graphs for every prime power q. *J. Combin. Theory Ser. B* 62 (1994) 44–62.

[MP] B. Mathes and V. Paulsen. Operator ideals and operator spaces. *Proc. Amer. Math. Soc.* 123 (1995) 1763–1772.

[MS] V. Milman and G. Schechtman. *Asymptotic Theory of Finite Dimensional Normed Spaces.* Springer Lecture Notes 1200, Springer-Verlag, 1986.

[Ne] E. Nelson. Notes on non-commutative integration. *J. Funct. Anal.* 15 (1974) 103–116.

[Ni] N. Nikolskii. *Treatise on the Shift Operator.* Springer-Verlag, 1986.

[NR1] M. Neal and B. Russo. Contractive projections and operator spaces. *C. R. Acad. Sci. Paris* 331 (2000) 873–878.

[NR2] _____. A holomorphic characterization of ternary rings of operators. Preprint, 2001. To appear.

[O1] T. Oikhberg. Direct sums of homogeneous operator spaces. *J. London Math. Soc.* 64 (2001) 144–160.

[O2] _____. Geometry of operator spaces and products of orthogonal projections. Ph.D. Thesis, Texas A&M Univ., 1997.

[O3] _____. Subspaces of maximal operator spaces. *Integral Equations Operator Theory.* To appear.

[O4] _____. Completely complemented subspace problem. *J. Operator Theory* 43 (2000) 375–387.

[OP] T. Oikhberg and G. Pisier. The "maximal" tensor product of two operator spaces. *Proc. Edinburgh Math. Soc.* 42 (1999) 267–284.

[OR] T. Oikhberg and H. P. Rosenthal. On certain extension properties for the space of compact operators. *J. Funct. Anal.* 179 (2001) 251–308.

[ORi] T. Oikhberg and E. Ricard. Operator spaces with few completely bounded maps. Preprint (2002).

[Oz1] N. Ozawa. On the set of finite-dimensional subspaces of preduals of von Neumann algebras. *C. R. Acad. Sci. Paris Sr. I Math.* 331 (2000) 309–312.

[Oz2] _____. Amenable actions and exactness for discrete groups. *C. R. Acad. Sci. Paris Sr. I Math.* 330 (2000) 691–695.

[Oz3] _____. On the lifting property for universal C^*-algebras of operator spaces. *J. Operator Theory* 46 (2001) 579–591.

[Oz4] _____. A non-extendable bounded linear map between C^*-algebras. *Proc. Edinburgh Math. Soc.* 44 (2001) 241–248.

[Oz5] _____. An application of expanders to $B(\ell_2) \otimes B(\ell_2)$. *J. Funct. Anal.* To appear.

[Oz6] _____. Local Theory and Local Reflexivity for Operator Spaces. Ph.D. Thesis, Texas A&M Univ., May 2000.

[P1] G. Pisier. The operator Hilbert space OH, complex interpolation and tensor norms. *Memoirs Amer. Math. Soc.* 122, 585 (1996) 1–103.

[P2] _____. Non-commutative vector valued L_p-spaces and completely p-summing maps. *Soc. Math. France. Astérisque* 247 (1998) 1–131.

[P3] _____. Espace de Hilbert d'opérateurs et interpolation complexe. *C. R. Acad. Sci. Paris S. I* 316 (1993) 47–52.

[P4] _____. *Factorization of Linear Operators and the Geometry of Banach Spaces.* CBMS (Regional Conferences of the AMS) 60, 1986; reprinted with corrections 1987.

[P5] _____. Completely bounded maps between sets of Banach space operators. *Indiana Univ. Math. J.* 39 (1990) 251–277.

[P6] _____. Exact operator spaces. Colloque sur les algèbres d'opérateurs. *Recent Advances in Operator Algebras (Orléans 1992) Astérisque (Soc. Math. France)* 232 (1995) 159–186.

[P7] _____. Projections from a von Neumann algebra onto a subalgebra. *Bull. Soc. Math. France* 123 (1995) 139–153.

[P8] _____. *The Volume of Convex Bodies and Banach Space Geometry.* Cambridge Univ. Press, 1989.

[P9] _____. Dvoretzky's theorem for operator spaces and applications. *Houston J. Math.* 22 (1996) 399–416.

[P10] _____. *Similarity Problems and Completely Bounded Maps, Second, Expanded Edition.* Springer Lecture Notes 1618, Springer-Verlag, 2001.

[P11] _____. A simple proof of a theorem of Kirchberg and related results on C^*-norms. *J. Operator Theory* 35 (1996) 317–335.

[P13] _____. Remarks on complemented subspaces of von-Neumann algebras. *Proc. Royal Soc. Edinburgh* 121A (1992) 1–4.

[P14] _____. Quadratic forms in unitary operators. *Linear Algebra Appl.* 267 (1997) 125–137.

[P15] _____. *Riesz Transforms: A Simpler Analytic Proof of P.-A. Meyer's Inequality. Séminaire de Probabilités XXII.* Springer Lecture Notes 1321, Springer-Verlag, 1988.

[P16] _____. Holomorphic semi-groups and the geometry of Banach spaces. *Ann. of Math.* 115 (1982) 375–392.

[P17] _____. The similarity degree of an operator algebra. *St. Petersburg Math. J.* 10 (1999) 103–146.

[P18] _____. The similarity degree of an operator algebra II. *Math. Zeit.* 234 (2000) 53–81.

[P19] _____. Remarks on the similarity degree of an operator algebra. *Intern. J. Math.* 12 (2001) 403–414.

[P20] _____. A polynomially bounded operator on Hilbert space which is not similar to a contraction. *J. Amer. Math. Soc.* 10 (1997) 351–369.

[P21] _____. A simple proof of a theorem of Jean Bourgain. *Michigan Math. J.* 39 (1992) 475–484.

[P22] _____. The Operator Hilbert Space *OH* and TYPE III von Neumann Algebras. Preprint (2002).

[Pa1] V. Paulsen. *Completely Bounded Maps and Dilations.* Pitman Research Notes 146. Pitman Longman (Wiley) 1986.

[Pa2] _____. Representation of function algebras, Abstract operator spaces and Banach space geometry. *J. Funct. Anal.* 109 (1992) 113–129.

[Pa3] _____. Completely bounded maps on C^*-algebras and invariant operator ranges. *Proc. Amer. Math. Soc.* 86 (1982) 91–96.

[Pa4] _____. Completely bounded homomorphisms of operator algebras. *Proc. Amer. Math. Soc.* 92 (1984) 225–228.

[Pa5] _____. The maximal operator space of a normed space. *Proc. Edinburgh Math. Soc.* 39 (1996) 309–323.

[PaP] V. Paulsen and S. Power. Tensor products of non-self-adjoint operator algebras. *Rocky Mountain J. Math.* 20 (1990) 331–350.

[Par1] S. Parrott. On a quotient norm and the Sz.-Nagy Foias lifting theorem. *J. Funct. Anal.* 30 (1978) 311–328.

[Par2] _____. Unitary dilations for commuting contractions. *Pacific J. Math.* 34 (1970) 481–490.

[PaS] V. Paulsen and R. Smith. Multilinear maps and tensor norms on operator systems. *J. Funct. Anal.* 73 (1987) 258–276.

[PaSu] V. Paulsen and C-Y. Suen. Commutant representations of completely bounded maps. *J. Operator Theory* 13 (1985) 87–101.

[Pat] A. Paterson. *Amenability.* Math. Surveys 29. Amer. Math. Soc., 1988.

[Pe1] V. Peller. Estimates of functions of power bounded operators on Hilbert space *J. Operator Theory* 7 (1982) 341–372.

[Pe2] V. Peller. Estimates of functions of Hilbert space operators, similarity to a contraction and related function algebras. *Linear and Complex Analysis Problem Book*, edited by Havin, Hruscev, and Nikolskii. Springer Lecture Notes 1043, Springer-Verlag, 1984, pp. 199–204.

[Ped] G. Pedersen. *C*-Algebras and Their Automorphism Groups.* Academic Press, 1979.

[Pes] V. Pestov. Operator spaces and residually finite-dimensional C^*-Algebras. *J. Funct. Anal.* 123 (1994) 308–317.

[PiS]   G. Pisier and D. Shlyakhtenko. Grothendieck's theorem for operator spaces. *Inventiones Math.* 150 (2002) 185–217.

[Po1]   S. Popa. A short proof of injectivity implies hyperfiniteness for finite von Neumann algebras. *J. Operator Theory* 16 (1986) 261–272.

[Po2]   _____. On amenability in type II_1 factors. *Operator Algebras and Applications*, edited by D. Evans and M. Takesaki, volume 2, LMS Lecture Notes Series 136. Cambridge Univ. Press, 1988.

[Po3]   S. Popa. *Classification of Subfactors and Their Endomorphisms.* CBMS Regional Conference Series in Mathematics, 86. American Math. Soc., 1995.

[Pop]   C. Pop. Bimodules normés représentables sur des espaces hilbertiens. Ph.D. Thesis, University of Orléans, 1999.

[Pow]   S. C. Power. *Hankel Operators on Hilbert Space.* Research Notes in Mathematics, 64. Pitman, 1982.

[Pu1]   G. Popescu. Von Neumann inequality for $(B(\mathcal{H})^n)_1$. *Math. Scand.* 68 (1991) 292–304.

[Pu2]   _____. Universal operator algebras associated to contractive sequences of non-commuting operators. *J. London Math. Soc.* 58 (1998) 469–479.

[Pu3]   _____. Poisson transforms on some C^*-algebras generated by isometries. *J. Funct. Anal.* 161 (1999) 27–61.

[PV1]   M. Pimsner and D. Voiculescu. Exact sequences for K-groups and Ext-groups of certain cross-products of C^*-algebras. *J. Operator Theory* 4 (1980) 93–118.

[PV2]   _____. K-groups of reduced crossed products by free groups. *J. Operator Theory* 8 (1982) 131–156.

[PX]   G. Pisier and Q. Xu. Inégalités de martingales non commutatives. *C. R. Acad. Sci. Paris Sr. I Math.* 323 (1996) 817–822.

[R]   H. P. Rosenthal. The complete separable extension property. *J. Operator Theory* 43 (2000) 329–374.

[Ri1]   E. Ricard. Décomposition de H^1, multiplicateurs de Schur et espaces d'opérateurs. Thèse, Université Paris VI, 2001.

[Ri2]   _____. A tensor norm on Q-spaces. *J. Operator Theory* 48 (2002) 431–445.

[Rie]   M. Rieffel. Induced representations of C^*-algebras. *Adv. Math.* 13 (1974) 176–257.

[RLL]   M. Rørdam, F. Larsen, and N. Laustsen. *An Introduction to K-Theory for C27*-Algebras.* Cambridge Univ. Press, 2000.

[Ro]   A. R. Robertson. Injective matricial Hilbert spaces. *Math. Proc. Cambridge Philos. Soc.* 110 (1991) 183–190.

[RS] M. Rørdam and E. Størmer. *Classification of Nuclear C*-Algebras. Entropy in Operator Algebras*. Springer-Verlag, 2002.

[RW] A. R. Robertson and S. Wassermann. Completely bounded isomorphisms of injective operator systems. *Bull. London Math. Soc.* 21 (1989) 285–290.

[Ru1] Z. J. Ruan. Subspaces of C*-algebras. *J. Funct. Anal.* 76 (1988) 217–230.

[Ru2] _____. Injectivity of operator spaces. *Trans. Amer. Math. Soc.* 315 (1989) 89–104.

[Sa] S. Sakai. *C*-Algebras and W*-Algebras*. Springer-Verlag, 1974.

[Sar] P. Sarnak. *Some Applications of Modular Forms*. Cambridge Univ. Press, 1990.

[Se] I. Segal. A non-commutative extension of abstract integration. *Ann. Math.* 57 (1953) 401–457.

[Sh] Y. Shalom. Bounded generation and Kazhdan's property (T). *Inst. Hautes Études Sci. Publ. Math.* No. 90 (1999), 145–168.

[Sk] G. Skandalis. *Algèbres de von Neumann de groupes libres et probabilités non commutatives*. Séminaire Bourbaki, Vol. 1992/93. Astérisque No. 216 (1993) 87–102.

[Sm1] R. R. Smith. Completely contractive factorization of C*-algebras. *J. Funct. Anal.* 64 (1985) 330–337.

[Sm2] _____. Completely bounded maps between C*-algebras. *J. London Math. Soc.* 27 (1983) 157–166.

[SN] B. Sz.-Nagy. Completely continuous operators with uniformly bounded iterates. *Publ. Math. Inst. Hungarian Acad. Sci.* 4 (1959) 89–92.

[SNF] B. Sz.-Nagy and C. Foias. *Harmonic Analysis of Operators on Hilbert Space*. Akademiai Kiadó, 1970.

[SS1] A. M. Sinclair and R.R. Smith. The Haagerup invariant for operator algebras. *Amer. J. Math.* 117 (1995) 441–456.

[SS2] _____. *Hochschild Cohomology of von Neumann Algebras*. London Math. Soc. Lecture Notes Series. Cambridge Univ. Press, 1995.

[SS3] _____. Factorization of completely bounded bilinear operators and injectivity. *J. Funct. Anal.* 157 (1998) 62–87.

[St] W. Stinespring. Positive functions on C*-algebras. *Proc. Amer. Math. Soc.* 6 (1955), 211–216.

[Sta] J. D. Stafney. The spectrum of an operator on an interpolation space. *Trans. Amer. Math. Soc.* 144 (1969) 333–349.

[StZ] S. Stratila and L. Zsidó. *Lectures on von Neumann Algebras*. Editura Academiei, Bucharest; Abacus Press, Tunbridge Wells, 1979.

[Su] C-Y. Suen, Completely bounded maps on C*-algebras. *Proc. Amer. Math. Soc.* 93 (1985) 81–87.

[Sun] V. S. Sunder. *An Invitation to von Neumann Algebras.* Universitext. Springer-Verlag, 1987.

[Sz] A. Szankowski. $B(H)$ does not have the approximation property. *Acta Math.* 147 (1981) 89–108.

[T] S. Thorbjørnsen. *Mixed Moments of Voiculescu's Gaussian Random Matrices.* Preprint, Odense Univ., 1999.

[Ta1] M. Takesaki. A note on the cross-norm of the direct product of C^*-algebras. *Kodai Math. Sem. Rep.* 10 (1958) 137–140.

[Ta2] _____. On the cross-norm of the direct product of C^*-algebras. *Tôhoku Math. J.* 16 (1964) 111–122.

[Ta3] _____. *Theory of Operator Algebras I.* Springer-Verlag, 1979.

[TJ1] N. Tomczak-Jaegermann. *Banach-Mazur Distances and Finite-Dimensional Operator Ideals.* Pitman-Longman (Wiley), 1989.

[TJ2] N. Tomczak-Jaegermann. The moduli of convexity and smoothness and the Rademacher averages of trace class S_p. *Studia Math.* 50 (1974) 163–182.

[To1] J. Tomiyama. On the projection of norm one in W^*-algebras. *Proc. Japan Acad.* 33 (1957) 608–612.

[To2] _____. *Tensor Products and Projections of Norm One in von Neumann Algebras.* Lecture Notes, Univ. of Copenhagen, 1970.

[To3] _____. Applications of Fubini type theorem to the tensor product of C^*-algebras. *Tôhoku Math. J.* 19 (1967) 213–226.

[Tor] A. M. Torpe. *Nuclear C^*-Algebras and Injective von Neumann.* Odense Univ. Preprint, 1981.

[Tr] S. Trott. A pair of generators for the unimodular group. *Canadian Math. Bull.* 3 (1962) 245–252.

[Va] A. Valette. An application of Ramanujan graphs to C^*-algebra tensor products. *Discrete Math.* 167/168 (1997) 597–603.

[V1] N. Varopoulos. Some remarks on Q-algebras. *Ann. Inst. Fourier* 22 (1972) 1–11.

[V2] _____. A theorem on operator algebras. *Math. Scand.* 37 (1975) 173–182.

[VDN] D. Voiculescu, K. Dykema, and A. Nica. *Free Random Variables.* CRM Monograph Series, Vol. 1, Amer. Math. Soc., 1992.

[Vo1] D. Voiculescu. Property T and approximation of operators. *Bull. London Math. Soc.* 22 (1990) 25–30.

[Vo2] _____. Limit laws for random matrices and free products. *Invent. Math.* 104 (1991) 201–220.

[Wa1] S. Wassermann. On tensor products of certain group C^*-algebras. *J. Funct. Anal.* 23 (1976) 239–254.

[Wa2] _____. *Exact C^*-Algebras and Related Topics.* Lecture Notes Series, 19. Seoul National Univ., 1994.

[Wa3] _____. On subquotients of UHF algebras. *Math. Proc. Cambridge Phil. Soc.* 115 (1994) 489–500.

[Wa4] _____. The slice map problem for C^*-algebras. *Proc. London Math. Soc.* 32 (1976) 537–559.

[Wa5] _____. Injective W^*-algebras. *Math. Proc. Cambridge Phil. Soc.* 82 (1977) 39–47.

[Wat1] F. Watbled. Interpolation complexe d'un espace de Banach et de son antidual. *C. R. Acad. Sci. Paris Sr. I Math.* 321 (1995) 1437–1440.

[Wat2] F. Watbled. Complex interpolation of a Banach space with its dual. *Math. Scand.* 87 (2000) 200–210.

[We] J. Wermer. Quotient algebras of uniform algebras. *Symp. on function algebras and rational approximation.* Univ. of Michigan, 1969.

[WeO] N. Wegge-Olsen. *K-Theory and C^*-Algebras.* Oxford Univ. Press, 1993.

[Win] S. Winkler. Matrix convexity. Ph.D. Thesis, UCLA. 1996.

[Wit1] G. Wittstock. Ein operatorwertigen Hahn-Banach Satz. *J. Funct. Anal.* 40 (1981), 127–150.

[Wit2] _____. Extensions of completely bounded module morphisms. *Proc. Conference on Operator Algebras and Group Representations.* Neptum, Pitman, 1983.

[WW] C. Webster and S. Winkler. The Krein-Milman theorem in operator convexity. *Trans. Amer. Math. Soc.* 351 (1999) 307–322.

[Xu] Q. Xu. Interpolation of operator spaces. *J. Funct. Anal.* 139 (1996) 500–539.

[Y] K. Ylinen. Representing completely bounded multilinear operators. *Acta Math. Hungar.* 56 (1990) 295–297.

[Z1] C. Zhang. Representations of operator spaces. Preprint, Univ. of Houston, 1993.

[Z2] _____. Completely bounded Banach-Mazur distance. *Proc. Edinburgh Math. Soc.* (2) 40 (1997) 247–260.

[Z3] _____. Representation and geometry of operator spaces. Ph.D. Thesis, Univ. of Houston (1995).

SUBJECT INDEX

NOTATION INDEX